DATE DUE

Advances in
Insect Physiology

Volume 11

Advances in Insect Physiology

edited by

J. E. TREHERNE
M. J. BERRIDGE
and V. B. WIGGLESWORTH

Department of Zoology, The University
Cambridge, England

Volume 11

1975

ACADEMIC PRESS
LONDON NEW YORK SAN FRANCISCO
A Subsidiary of Harcourt Brace Jovanovich, Publishers

ACADEMIC PRESS INC. (LONDON) LTD
24-28 Oval Road
London NW1

US edition published by
ACADEMIC PRESS INC.
111 Fifth Avenue,
New York, New York 10003

Library of Congress Catalog Card Number: 63-14039
ISBN: 0-12-024211-7

PRINTED IN GREAT BRITAIN BY
WILLIAM CLOWES & SONS LIMITED
LONDON, COLCHESTER AND BECCLES

Contributors

L. Barton Browne

Division of Entomology, CSIRO, Canberra City, Australia

A. Clive Crossley

School of Biological Sciences, University of Sydney, Australia

William H. Telfer

Department of Biology, University of Pennsylvania, Philadelphia, Pennsylvania

John A. Thomson

Department of Genetics, University of Melbourne, Parkville, Victoria, Australia

Contents

Regulatory Mechanisms in Insect Feeding

L. Barton Browne

Division of Entomology, CSIRO, Canberra City, Australia

1 Introduction

There is abundant evidence that insects possess mechanisms which enable them to regulate their intake of food and water with a considerable degree of precision (Dethier, 1969; Gelperin, 1971a). The literature relevant to the understanding of the regulation of feeding by insects is too large to be dealt with fully in one review and I have therefore selected only two aspects for detailed discussion.

The total feeding behaviour of most insects is made up of a number of components, and the first topic I will discuss is how the performance of these may vary according to the insect's state of deprivation. The second part of the review consists of a discussion of the long-term regulation of feeding, especially in relation to some of the previously discussed behavioural variations in the components of feeding. Since the emphasis is on the role in the regulation of feeding of behavioural changes resulting from feeding and deprivation, much of the review is concerned with the regulation of intake of materials taken repeatedly by the insect and which are usually required for the maintenance of life, rather than with special foods required for particular purposes such as reproduction. These two topics were selected for emphasis because it seems clear that the basis for regulation of the intake of food and water over a period by most, and perhaps all insects, is that food-deprived individuals behave differently from recently fed ones. Moreover, considerable information is available about the physiological bases of some of these behavioural differences.

A number of factors other than feeding and deprivation are known to influence the feeding behaviour of insects and, although I have placed the detailed discussion of these beyond the scope of this review, I have included a brief section on some of these influences. In this, I have included enough of the more important references to allow entry into the literature relating to these aspects.

2 Regulatory changes in components of feeding behaviour

The number of behavioural components involved in the total feeding behaviour of an insect depends upon its temporal and spatial relationships with its food. An insect which feeds intermittently and ranges widely from its food between bouts of feeding would probably have, as components of feeding behaviour, "random" locomotor activity, orientated movements towards food or food sources in response to visual or olfactory stimuli, a variety of responses concerned with the initiation of feeding when the food is reached, responses which are responsible for the maintenance and then the cessation of feeding and, finally, locomotor activity again which takes it away from the food source. An insect which feeds intermittently but remains in contact with its food shares the components concerned with initiation, maintenance and termination of feeding, but not those related to locomotor behaviour in the period between feeding episodes. An insect which feeds more or less continuously lacks all components except those concerned with the maintenance of feeding. The bases for the regulation of intake by an insect might be differences, according to its state of deprivation, in any or all of the components of its total feeding behaviour, with insects showing a greater number of components having greater possibilities for exhibiting regulatory behaviour than ones with behaviour patterns with fewer components.

In this section, I shall discuss examples of behavioural regulation in the components of feeding. The discussion is divided into four parts: the first consisting of a general discussion of several kinds of experiments commonly used in the investigation of these regulatory processes; the second deals with the regulation of locomotor pre-ingestion behaviour; the third with that of nonlocomotor pre-ingestion behaviour; and the fourth with the regulation of ingestion. The division of pre-ingestion behaviour into locomotor and nonlocomotor is somewhat arbitrary, since even when an insect is in contact with its food it usually makes some movement before beginning to ingest. In the first category, I shall discuss the regulation of movement whether clearly orientated or not, which involves considerable displacement of the whole insect. Any kinetic component of the behaviour assigned to the second category usually involves the movement of only part of the body or the displacement of the whole body only over a short distance.

2.1 GENERAL COMMENTS ON THE DESIGN AND INTERPRETATION OF EXPERIMENTS

Investigations into the regulation of the components of feeding behaviour and of its physiological bases have involved the use of a relatively small number of general types of experiments. Several of those which have commonly been used have a number of inherent problems relating to the interpretation of the results.

Investigations of the effect of feeding and deprivation on components of feeding behaviour have involved either a comparison of the behaviour of deprived insects with that of insects immediately after feeding, or the monitoring of behaviour during a period of deprivation, or both. Difficulties in interpretation occur when the only evidence for behavioural changes has been obtained from experiments in which behaviour has been monitored throughout a period of deprivation. The problems arise because insects increase in age during the period of the study and may, therefore, change their physiological characteristics in ways which are unrelated to deprivation. It is important, therefore, that experimental designs should be such that the effects of deprivation are clearly distinguishable from those of ageing. This is most easily achieved by having available for comparison recently fed insects which are otherwise strictly comparable to the ones undergoing deprivation. Another satisfactory but somewhat more complex method is to subject cohorts of insects of different ages to deprivation. If then the behaviour of the cohorts is similar, it can fairly be stated that the behavioural changes are due to effects of deprivation.

Investigations into the physiological mechanisms underlying behavioural changes with feeding and deprivation are usually concerned with determining which of the many internal factors, that vary according to the state of deprivation, might play a role in bringing about the behavioural differences. The experiments performed fall into two basic categories. The first is that in which the aim is to obtain, and to determine the behavioural characteristics of, insects which are in a satiated condition with respect to one parameter, but deprived with respect to all others, or vice versa. This may be achieved either by altering artificially one parameter while keeping others constant (e.g. by injection), or by preventing one parameter from changing during a period of deprivation, or following ingestion. Two kinds of clear-cut results have been obtained from experiments in this general category, it having been found that the insect displays behavioural characteristics which accord either with the state of the one parameter being manipulated or of the remaining parameters. Results of the first type are usually taken as evidence that the factor under investigation is involved in the particular facet of behavioural regulation being studied, and those of the second as evidence for its noninvolvement. The first of these conclusions is soundly based and it remains only to caution against believing that the factor under examination is necessarily the only one involved. Conclusions of the second kind concerning noninvolvement require rather more comment in that the strict interpretation of the negative result is only that this factor in question is not *alone* responsible for the behavioural regulation. It is possible, at least in theory, that the regulatory system might be such that no single factor has any detectable effect on behaviour, when caused to vary independently of other factors with which it normally changes in concert. If this were so, the successive manipulation of single parameters would not reveal the controlling mechanism.

In the second type of experiment commonly used, nerves suspected of carrying input from receptors monitoring various parameters, which change according to the state of deprivation of the insect, are sectioned. Implicit in the design of these experiments is the often unstated belief that the inputs are maximal when the insect is fully fed, and that these are inhibitory to the performance of the component of feeding behaviour being investigated. Again, two types of fairly clear-cut results have been obtained. In some instances it has been found that the particular operation has no detectable effect on behaviour, whereas in others the operation results in the insect behaving, in some respects at least, as if deprived even though it is fully fed, with the result that its ability to regulate its feeding is diminished. The lack of effect of an operation is usually interpreted as meaning that input normally travelling via the nerve which was sectioned plays no important part in the regulation of the behaviour being investigated. There is, of course, the additional possibility that input via this nerve is only one of several sources of inhibition and that no significant loss of control occurs when the central nervous system (CNS) is deprived of any one of them. The finding that nerve section does cause a fed insect to behave as if deprived certainly indicates that input via that nerve is involved in, and indeed is essential for, the regulation of the particular component of behaviour under investigation. The possibility cannot be excluded, however, that other inputs might play a part but that they are able to express their effect only if the input normally carried via the sectioned nerve is reaching the CNS. In these circumstances loss of control might result from the sectioning of any one of a number of nerves.

Recognition of the limitations of the kinds of experiments frequently performed necessitates re-examination, and in some instances re-interpretation, of some of the results which have been obtained.

2.2 REGULATION OF LOCOMOTOR PRE-INGESTION BEHAVIOUR

It is well known that the locomotor behaviour of a number of insects changes according to their state of deprivation in ways which enhance the deprived insect's chances of making contact with food. Changes have been demonstrated in the general level of "spontaneous" apparently randomly directed locomotor activity, in behaviour involving usually orientated movement in response to stimuli provided by the food itself, and in orientated behavioural responses to physical factors of the environment. Data relating to these three behavioural categories are discussed separately.

2.2.1 *Level of locomotor activity*

The effects of feeding and deprivation on apparently random locomotor activity have now been examined in a number of species, and it has generally been found that deprived insects are more active than fed ones. It should be realized, however, that, under almost any set of conditions, the level of locomotor activity displayed by an insect has two components, spontaneous activity and reactivity (or responsiveness) to features of its environment, and that the importance of each will vary according to the type of experimental situation. Findings discussed in sections 2.2.2 and 2.2.3 show that the readiness of insects to make orientated movements in response to various kinds of stimulation changes according to the state of deprivation, and it seems certain therefore that the reactivity of insects to stimulation which results in their engaging in nonorientated movement would also change. It is probable, therefore, that changes in observed activity with feeding and deprivation would usually be reflecting changes in both spontaneous activity and reactivity. It would, however, seem unwise to assume *a priori* that the physiological mechanism controlling each would be identical, particularly in view of the finding by Connolly (1967) that there was no correlation between the two parameters in three strains of *Drosophila melanogaster* selected for differences in spontaneous activity and in reactivity to inanimate features of the environment. For this reason, reference is made whenever possible to the probable roles played by the two components of the activity in the particular experimental situations in which the measurements were made.

The effects of feeding and deprivation on the locomotor activity of the blowfly *Phormia regina* were extensively studied by Barton Browne and Evans (1960), and by Green (1964a, 1964b) and attempts were made to elucidate the underlying physiological mechanisms. The results of Green are the more readily interpretable in terms of spontaneous activity in that he scored the activity of single flies in a rocking actograph. Under these conditions, there was no stimulation from other individuals and it is likely that the level of effective stimulation from the inanimate environment would have been fairly low because the insects remained in the actograph chambers throughout the period of deprivation, and therefore would probably have become, to some extent, habituated to their surroundings. In addition, according to Green, the flies did not perceive the tilting action of the actograph. The experiments of Barton Browne and Evans (1961) are, superficially at least, less readily interpretable since these workers determined the rates at which groups of flies dispersed along a line of boxes, connected by funnels, in response to light stimulus. It was shown, however, that the relationship between the rate of dispersal of fed flies and that of deprived flies obtained with the light stimulus was similar to that in darkness, Barton Browne and Evans having chosen to conduct their experiments using the light stimulus rather than in darkness only because of the higher rate of dispersal obtained and the consequently lower variability. It seems therefore that the relative rates under the directed stimulus can be taken as a measure of locomotor activity. It is not certain, however, to what extent reactivity to the presence of other individuals played a role, but since both sexes were present interactions between males and females may have played some part in determining the rate of dispersal. The relationships between the amount of locomotor activity and the state of deprivation obtained by Barton Browne and Evans are generally similar to those obtained by Green and it seems valid, therefore, to discuss the two sets of results together, largely in terms of effects of feeding and deprivation on spontaneous activity.

Both Barton Browne and Evans (1960) and Green (1964a) showed that the activity of flies which had recently been fed to repletion on any of a variety of sugar solutions was very low compared with that of flies which had been deprived of food for 24 h. Barton Browne and Evans (1960) found that the ingestion of glucose, mannose or fucose reduced the activity of flies and Green (1964b) showed that the rate at which their activity increased again after feeding was inversely related to the concentration of sucrose solution consumed. Green showed, further, that feeding and deprivation affected the proportion of the time the flies engaged in locomotor activity rather than the speed of walking, which was the main method of progression in his actograph chambers.

The attempts by Barton Browne and Evans (1960) and by Green (1964b) to elucidate the underlying mechanisms in *P. regina* were not completely successful, but their experiments eliminated, more or less satisfactorily, a number of possibilities and gave some indications as to what the underlying mechanisms might be. Barton Browne and Evans concluded that no significant regulatory role was played by an inability of the flies to move because of increased weight after feeding, by the metabolic state of the fly, by the total concentration of sugar in the haemolymph, or by the haemolymph potassium level. Green concluded, further, that no part was played by input concerning the state of distension of the abdomen, crop, or posterior portion of the crop duct, by input from the receptors of the labellar lobes which would have been stimulated during regurgitation, or by possible limitation in the amount of oxygen reaching the thoracic musculature because of the collapsed state of the abdominal air sacs after feeding.

The elimination of two of the above factors depended, however, upon evidence from experiments in which one parameter was held essentially constant at a level more or less typical of satiated flies, a type of experiment about which some general remarks were made earlier. The conclusion that the concentration of carbohydrates in the haemolymph was not involved was based on the finding by Barton Browne and Evans (1960) that flies are active despite the presence in the haemolymph of high concentrations of the non-metabolizable sugar fucose, and that of Green (1964b) concerning the noninvolvement of the state of distension of the crop or of the abdomen, was drawn from his finding that fed flies with subsequently ligated crop ducts became active within a short time. These results should be reinterpreted as showing, strictly, only that neither a high fucose concentration in the haemolymph nor the possession of a full crop *alone* causes a reduction of locomotor activity. The elimination of the other factors appears acceptable without such qualification.

Two positive results, in the sense that treatments other than actual ingestion reduced the activity of starved flies, were obtained. Barton Browne and Evans (1960) found that injection of less than 3 μl of water or 2.0 M glucose into the haemolymph markedly reduced the activity of the flies as measured one hour later, the injection of glucose being rather more effective. On the basis of this result they suggested that changes in the composition of haemolymph due to the absorption of material from the mid-gut was an important factor in bringing about the post-feeding reduction in activity. Green (1964b) joined flies parabiotically and found that the activity of the starved "motile" fly was reduced when the fly riding inverted on its back was fed. Green interpreted this result as indicating that hormonal material released into the haemolymph of the fed fly was responsible for the reduced activity in its parabiotic partner. The result would, however, seem to be explicable equally well in terms of changes of composition of the shared haemolymph supply.

Finally, I will make brief reference to the conclusion reached by Barton Browne and Evans (1960) that the mechanism controlling locomotor activity after a fly has ingested sugar solution is different from that controlling tarsal taste threshold to sugar (see section 2.3) after a sugar meal. They based their argument on the lack of correspondence between the curves relating threshold to crop volume and activity to crop volume in flies previously fed 2.0 M mannose or 2.0 M glucose. This comparison, although valid, is somewhat circumstantial in that the two sets of data were obtained at different times and for different purposes and perhaps more convincing evidence can be drawn from the observation that activity but not tarsal threshold is influenced by alteration of the composition of the haemolymph. The result obtained on a single fly by Green (1964b) that recurrent nerve section did not influence the activity pattern may be further evidence that the mechanisms are distinct but, as pointed out below (section 2.3), the exact effects of recurrent nerve section on tarsal taste threshold are somewhat uncertain.

It is apparent that the investigations so far have gone only part way towards elucidating the mechanism by which feeding inhibits subsequent locomotor activity in *P. regina*. The available evidence, however, is consistent with the view that the level of locomotor activity is related to changes in the composition of the haemolymph. Not only is there evidence for this from injection and parabiosis experiment, but also from the data of Green (1964b) which show that locomotor activity remains depressed only so long as sugar solution is being released from the crop, and hence is passing from the mid-gut to the haemolymph. The inverse relationship between crop emptying rate and locomotor activity in flies fed 0.5 M sucrose is quite striking. A comparison of Green's (1964a) results relating locomotor activity to the concentration of imbibed sucrose solution with those of Gelperin (1966a), who established that a dilute solution emptied from the crop more rapidly than a concentrated one, lends further support to this view. On the basis of the available evidence, one can say no more than that this hypothesis that activity is related to the composition of the haemolymph seems the most likely one, if it is assumed that one factor dominates the causal mechanism. It may yet be shown, however, that such is not the case and that the control mechanism is more complex than previously believed.

The effect of feeding and deprivation on the levels of various kinds of locomotor activity exhibited by the adults or larvae of several species of locusts have been investigated. One study, concerning the orientated movements of larvae in response to grass odour, will be discussed in detail in section 2.2.2, but will be referred to briefly in this. The remainder, which were investigations of the effects of deprivation on several somewhat different kinds of nondirected locomotor activity, are discussed here.

Blaney and Chapman (1970) allowed single 5th instar larvae of *Locusta migratoria ad lib.* access to palatable food and found that the insects took their food in the form of discrete meals separated by considerable periods, during which no feeding occurred. Observations during this inter-meal period showed that the proportion of the time for which the insects were moving declined progressively after the completion of one meal until just before the beginning of the next. Bernays and Chapman (1974a) have provided evidence that hormonal material released from the storage lobes of the corpus cardiacum (CC) as a result of distension of the fore-gut is at least partially responsible for the reduction in activity following feeding. They demonstrated that the injection into the haemolymph of homogenates of the CC storage lobes caused a reduction in the proportion of the time for which larvae were active and, further, that a comparable reduction in locomotor activity occurred when the fore-gut was artificially distended by filling it with agar, the determination of activity in these experiments being carried out under similar conditions to those used by Blaney and Chapman (Bernays and Chapman, personal communication). It is well known that feeding in locusts causes the release of neurosecretory material from the storage lobes of the CC. Mordue (1969) demonstrated the release of material with diuretic activity and Bernays and Chapman (1972a) and Bernays and Mordue (1973) showed that material released from the CC was responsible for the closure of the apical pore of chemoreceptors on the palps. It is not certain whether the hormonal material responsible for the reduction in activity is identical with that responsible for either or both of the other known effects. It can be said, however, that the time course of changes after feeding in locomotor activity and in the proportion of chemoreceptors with closed apical pores are rather different. The lowest level of locomotor activity is reached some time after feeding (Blaney and Chapman, 1970), whereas the proportion of closed pores is at a maximum within a short time after feeding ceases. This difference cannot, however, be taken as proof that different hormones are responsible, since one effect concerns the CNS, whereas the other is probably entirely peripheral. It is reasonable to expect that central nervous responses to hormones might be less immediate than the responses of receptors.

More recent experiments by Bernays and Chapman (personal communication) indicate that yet another factor might play a part in causing the reduction in activity which follows feeding. They found that the activity levels of larvae of *L. migratoria* when measured $1\frac{1}{2}$-2 h after the injection of a variety of solutions, which caused an increase in the osmotic pressure of the haemolymph, was generally less than that of either water-injected insects or stabbed controls.

The activity measurements made by Blaney and Chapman (1970) and Bernays and Chapman (1974a) were carried out under conditions purposely designed to reduce to a minimum the reactivity component, and therefore probably give a good indication of the levels of spontaneous activity. The only complication would appear to be the possibility that the observed activity might include orientated movement to visual and olfactory stimuli provided by the food present in the cages.

Ellis (1951) made a detailed study of the effects of food deprivation on the marching behaviour of 5th instar larvae of the African migratory locust, *Locusta migratoria*. Marching is a social form of locomotor activity engaged in by groups of locust larvae, which depends for its maintenance on interactions between individuals (Ellis and Hoyle, 1954). Any measurement of marching activity would, therefore, have in it strong elements of reactivity, even though it would undoubtedly contain some elements of spontaneous activity.

Ellis (1951) performed a number of experiments in which she determined the effects of feeding and deprivation on the percentage of the time spent marching and on the speed of marching. Because marching is a social activity requiring a build up of excitation by mutual stimulation (Ellis and Hoyle, 1954), neither fed nor deprived larvae began marching immediately after the beginning of a trial. The time taken for the insects to reach a given level of marching activity was found, however, to decrease markedly with increasing period of previous deprivation up to about 4 h. The rate at which the larvae marched increased rather more rapidly with deprivation reaching a maximum after 3 h.

Hoyle (1954) demonstrated that the potassium ion concentration in the haemolymph of fed locusts was significantly higher than in the haemolymph of ones which had been deprived of food, and showed that this reduced the resting membrane potential of the leg muscle fibres. Further, he demonstrated that fed insects jumped somewhat shorter distances than deprived ones. On the basis of these results, Ellis and Hoyle (1954) sought to explain Ellis' (1951) previous results in terms of a direct effect of the post-feeding increase in potassium concentration in the haemolymph on the response of muscles to motor output from the CNS. In an experiment designed to test this hypothesis, they showed that the proportion of time spent marching by larvae which had fed on filter paper soaked with sucrose solution and either 2.0 M or 3.0 M potassium chloride was at least as low as that by grass-fed locusts. Others which had consumed filter paper soaked with only sugar solution began marching just as rapidly as insects which had been deprived of all food. Unfortunately, the potassium ion concentration of the haemolymph of the variously fed groups was not determined. It is not known, therefore, whether the concentrations reached after the locust had fed on these high concentrations of potassium chloride were in any way comparable with those normally reached by insects feeding on plant materials, which contain much lower concentrations of potassium. Although the guts of the larvae fed on potassium-laden filter paper were less full than that of those fed grass, their intake was apparently quite substantial and would have resulted in their consuming relatively large amounts of potassium.

Other workers have attempted to determine whether the concentration of potassium in the haemolymph plays an important role in explaining the relationships they observed between the state of deprivation of larvae of locusts and some kind of locomotor activity other than marching. Their findings, as well as those of Bernays and Chapman (1974a) which implicate hormones and of Bernays and Chapman (personal communication) which implicate haemolymph osmotic pressure in the regulation of post-feeding locomotor activity, suggest that the potassium concentration of the haemolymph may not be solely, if at all, responsible for the deprivation-dependent changes in the tendency of larvae of *L. migratoria* to march.

As argued above, the results of Blaney and Chapman (1970) and Bernays and Chapman (1974a) would relate largely to spontaneous activity and it seems reasonable to assume that spontaneous activity also plays some role in determining the percentage of time for which locust larvae engage in marching. The findings of Bernays and Chapman (1974a) that the filling of the fore-gut with agar, a procedure which could not have directly influenced the haemolymph potassium ion concentration, and the injection of CC homogenate both caused reductions in activity comparable with those following feeding on grass (Bernays and Chapman, 1974a), show that the potassium ion concentration of the haemolymph is certainly not the sole explanation of the post-feeding reduction in spontaneous locomotor activity and indicate that its role is at most a minor one. It is unlikely, therefore, that the haemolymph potassium concentration plays an important part in determining the level of any spontaneous activity component of marching.

Moorhouse (1969), in an investigation which will be discussed more fully later, showed that haemolymph potassium concentration was certainly not alone, if at all, responsible for deprivation-dependent changes in the responsiveness of larvae of *Schistocerca gregaria* to the odour of grass. The relationship of these results to the hypothesis of Ellis and Holye (1954) that the potassium concentration of the haemolymph determines marching activity is not certain, but two features suggest that some common mechanisms might be involved in these two aspects of locomotor behaviour. The first is that both involve the reactivity or responsiveness of larvae to stimuli, in marching to stimulation from other locusts, and in upwind movement to an odour-bearing stream of moving air. The second is that the insects, during their period of unresponsiveness to either kind of stimulation, assume a characteristic post-prandial resting position. It seems reasonable, therefore, to take the results of Moorhouse as an indication that the potassium concentration of the haemolymph may not be playing a central role in determining the level of the reactivity component of marching.

On the basis of these results, it would seem that the relationship between the state of deprivation of locust larvae and the proportion of their time that they spend marching cannot be explained adequately by direct effects of the potassium ion concentration in the haemolymph on either the spontaneous activity or the reactivity component of marching behaviour. In addition, I believe that more general considerations also tend to argue against the possibility that changes in the potassium level of the haemolymph are solely responsible for the deprivation-dependent behavioural changes observed by Ellis (1951), since it seems very unlikely that an effect so nonspecific and lacking in integrative possibilities as that of potassium concentration on the response of muscle could control anything so complex as the temporal patterning of locomotor behaviour. It is likely, however, that the changes in the haemolymph potassium concentration would influence the speed of marching through its effect on jumping distance.

Chapman (1958, 1959) found that adults of the red locust, *Nomadacris septemfasciata*, which, on the evidence of their gut contents, had fed recently, were less likely to take flight than those which had not done so. In field and laboratory experiments, he examined the possibility that this relationship might be explicable in terms of changes in the haemolymph potassium concentration, and found that the potassium concentration was remarkably independent of feeding and deprivation. He concluded, therefore, that the relationship observed in the field between gut contents and the tendency to fly in *N. septemfasciata* was not explicable in terms of changes in the potassium concentration in the haemolymph.

Data are available about the relationships between food deprivation and locomotor activities in several insects other than the blowfly and locusts but generally no information is available about the physiological bases for the effects.

Bursell (1957) and Brady (1972a) have shown that the tsetse fly, *Glossina morsitans*, becomes progressively more active over a number of days of deprivation. Brady found that, in mature males, the number of bursts of activity increases with increasing deprivation but that the duration of the individual bursts remains extremely constant. He found that the increase in activity was greatest in pregnant females and showed that similar results were obtained with females of three different ages. This last finding indicates, that the observed increase was, in fact, at least in pregnant females, the result of deprivation and was not merely an effect of increasing age. Both Bursell and Brady used single flies and therefore probably obtained reasonable measures of spontaneous activity.

Hans and Thorsteinson (1961) showed that the tendency of the weevil *Sitona cylindricollis* to take flight was greater in deprived than in fed individuals, the insects being tested in groups of 5. These authors suggested that the reduced activity of the fed individuals might have been due to the cooling effects of the presumed high transpiration rate of the more fully hydrated fed beetles. Their data, which show activity of deprived and fed beetles at different temperatures, do not really seem to bear this out, as deprivation appears to have little effect on the relative levels of activity at different temperatures. It appears that, in this instance, as was the case with attempts to explain the activity of locust larvae in terms of the potassium concentration of the haemolymph, the modification of behaviour cannot be adequately explained in terms of direct gross physiological effects.

Connolly (1966) studied the effect of deprivation on the normal diurnal rhythm of activity in *Drosophila melanogaster* kept in groups containing 5 males and 5 females. He showed that the activity of fed individuals fell progressively over the period from 1 to 7 h after the onset of light in a light-dark cycle, whereas the activity of starved individuals remained constant at the initial high level during this period. The difference in activity of the two groups diverged as the period of deprivation increased. The interpretation of these results in terms of effects of deprivation is, however, complicated by the continuous presence of food with the fed controls and the consequent possibility that sensory input provided by the food might have had a depressing effect upon activity. In this study with *D. melanogaster* as in the previously discussed one with *S. cylindricollis*, it is not possible to assess the relative parts played by effects of deprivation on spontaneous activity and reactivity, because of the use, in the activity trials, of groups of insects.

Dingle (1968), as part of a programme to investigate the migratory strategies of the milkweed bug, *Oncopeltus fasciatus*, examined the effects of food deprivation on its flight activity in the laboratory. The flight testing procedure (Dingle, 1965) employed consisted of using various techniques to induce the insects to begin flying and then to record the total duration of the first 5 flights. If an insect became refractory after fewer than 5 flights, the total durations of the 4 or fewer flights were recorded. Dingle (1968) found that deprivation for several days increased the proportion of bugs flying for a total of 30 min or more, but that there was no increase in the number of long cruising flights of 1 h or more. He interpreted this as an indication that deprivation did not switch the insect from nonmigratory to migratory flight behaviour, but that it increased the tendency of the bugs to make relatively short flights. The method of assay used by Dingle is such that both spontaneous activity and reactivity would probably be involved. The reactivity of the insects would determine the readiness with which they would respond to the various procedures used to initiate flight and would therefore determine the proportion of insects which took fewer than 5 flights. The duration of individual flights would probably be a measure of spontaneous activity.

In all the examples dealt with so far, deprivation has been accompanied by an increase in locomotor activity. Recently, Reynierse *et al.* (1972) reported that the activity of the cockroach *Nauphaeta cinerea* was reduced by 16-18 days of complete deprivation or of access to only food or water. They showed also that activity declined gradually but erratically throughout a 15-day period during which cockroaches were completely deprived of food and water. Adults of *Phormia regina* (Green, 1964a), and teneral males of *Glossina morsitans* (Brady, 1972a) also show a reduction in activity after prolonged deprivation but, in contrast to *N. cinerea*, both these insects show increased activity after shorter periods of deprivation.

It is well known that insects such as lepidopterous larvae, which usually remain on their food between meals, will wander widely if the food supply is exhausted or becomes unpalatable. A recently documented example is that of *Porthetria dispar* larvae (Leonard, 1970). The physiological basis of this wandering has never been seriously investigated and the question arises as to whether the increased locomotor activity shown is the result of deprivation or whether it occurs because the insect is no longer receiving stimulation from the food, or both. One study which does have bearing on this question is that of Wensler (1971) on the locomotor and feeding activity of the subterranean larva of the scarabaeid *Sericesthis geminata*. Wensler showed that larvae moved further each day in soil containing no root material than in soil containing roots of growing plants, and that there was an immediate increase in locomotion if the roots were removed. She found, however, that there was no further increase in activity over a further 9 days in the absence of roots. It seems clear that, in this instance, the lower rate of movement of larvae in the presence of roots was due to the effects of immediate sensory stimulation and that the increased activity after removal of the roots was due to the removal of stimuli inhibitory to locomotion. There is no evidence of any effect of deprivation upon activity. The importance of these results is that they provide substance for the reservations expressed earlier in the discussion of the result of Connolly (1966) with *Drosophila melanogaster* and emphasize the necessity for determining the activity of fed and deprived insects under identical conditions.

2.2.2 *Responsiveness to olfactory and visual stimuli provided by food*

There have now been a number of demonstrations that food-deprived insects are more responsive than fed ones to the odour of their food. The most detailed information available concerns the variation in responsiveness of the late instar larvae of the desert locust, *Schistocerca gregaria*, to the odour of grass. Haskell *et al.* (1962) studied the responses of 4th instar larvae of this species in a flat-bed wind tunnel and showed that the proportion of insects making upwind movements towards grass, situated in the intake of the tunnel, increased progressively over a period of 8 h of deprivation. Moorhouse (1969, 1971) made a more intensive study of the relationship, this time using 5th instar larvae. He showed that larvae which had been allowed to feed to repletion after having been deprived of food for the previous 17 h were almost completely unresponsive to the wind-borne grass odour but that the proportion which moved upwind began to increase after 1 h, until by 4 h after feeding all insects responded. Larvae which were not given an opportunity to feed at the end of the 17 h period of standardization remained fully responsive throughout the 4-h period. After being starved for 17h, larvae of *S. gregaria* take about 25-35 min to feed to repletion. By allowing them access to food for periods of from 0 to 25 min, Moorhouse showed that the amount by which the responsiveness to wind-borne grass odour was reduced increased with increasing meal size.

Using, as a starting point, the previously discussed potassium hypothesis which Ellis and Hoyle (1954) proposed to explain the deprivation-dependent changes in the marching activity of larvae of *L. migratoria*, Moorhouse (1969) investigated the possibility that the changes in responsiveness of larvae of *S. gregaria* to grass odour with feeding and deprivation might be explicable in terms of the effect of feeding on the potassium concentration in the haemolymph. He injected into the haemocoel of deprived larvae an amount of potassium chloride which increased the potassium ion concentration in the haemolymph to approximately that of grass-fed individuals, and found that there was only a small depressing effect on responsiveness to grass odour. He found, furthermore, that similar reductions resulted when larvae were injected with sodium chloride or water. In each case, the reduction achieved was much less than that which resulted when the locusts fed on grass. It was shown also that the responsiveness of larvae which had ingested filter paper laden with both potassium chloride and sucrose was only slightly depressed and that the effect of the ingestion of sodium chloride treated filter paper was similar. Again the reduction in the level of responsiveness was small compared with that which occurred after larvae had consumed grass.

The injection of distilled water into the haemocoel of replete locusts did not hasten the return of responsiveness to olfactory stimuli even though the injection would almost certainly have produced a reduction in the potassium ion concentration of the haemolymph. On the contrary, it was found that the post-prandial resting position was maintained longer by water injected larvae than by stabbed controls.

These results show that the responsiveness of larvae of *S. gregaria* to grass odour was not affected in any specific way by the manipulation of the potassium concentration of the haemolymph and that changes in the potassium ion concentration were not alone responsible for the deprivation-dependent changes in responsiveness. The results do not, however, strictly rule out the possibility that the potassium ion concentration of the haemolymph might play some role when acting in concert with other factors which vary according to the state of deprivation.

The results of experiments performed by Kennedy and Moorhouse (1969) provide information of a somewhat different kind about the mechanisms which might be involved in the regulation of the responsiveness of larvae of *S. gregaria* to food odours. They compared the responsiveness of larvae which had been tumbled in a jar before being tossed into the wind tunnel with that of larvae which had been gently herded together before the trial. This latter group, as could be expected from the results of Haskell *et al.* (1962) and Moorhouse (1968, 1971), were essentially indifferent to the odour-free wind and responded to wind bearing a grass odour by upwind movement, with an intensity depending on their state of deprivation. The tumbled locusts, on the other hand, moved upwind in response to both odour-free and odoriferous wind, the response in the presence of grass odour being somewhat stronger. Furthermore, there was no clear cut difference in the responsiveness of starved and fed individuals. These results indicate that the upwind movement is the result of an elevation in the level of excitation either throughout the CNS or in a specific part of it. When larvae have not been tumbled, the relevant part of the nervous system is "aroused" by the odour of grass more readily in starved than in fed individuals. Tumbling, however, appears to have the effect of increasing the excitation in all larvae to a level which approaches that achieved by grass odour in starved individuals.

The effects of tumbling are present for some time after treatment whereas in previously undisturbed larvae the upwind response ceases immediately the wind becomes odour free (Haskell *et al.*, 1962). This suggests that the upwind movement is the result of an increase in the state of excitation in a specific part of the nervous system, and that this declines rapidly when the excitatory input ceases. Tumbling, it appears, causes more general excitation which persists for a time. It seems likely that the continued execution of upwind movement by tumbled larvae may be the result of the continued excitation of the part of the nervous system concerned specifically with this behaviour by input from other parts of the CNS in which a high level of excitation has been generated by the tumbling.

Locust larvae display a higher level of locomotor activity, both as regards speed and continuity of locomotion, when making upwind movement than when drifting downwind in the absence of grass odour (Haskell *et al.*, 1962). This relationship caused Kennedy and Moorhouse (1969) to consider the question of whether the level of locomotor excitation, as evidenced by overt characteristics of the locomotor activity, might not be what governed the sense of orientation to the wind. They concluded that this was not so, both on the basis of their own finding that antennectomized larvae, which although always exhibiting a low level of locomotor activity, would, in some circumstances, show a fairly strong upwind response, and that of Haskell *et al.* (1962) who showed that valeric acid in the air stream caused larvae to move briskly downwind.

Using similar techniques to those used by Moorhouse (1971), de Wilde *et al.* (1969) studied the upwind movements of the Colorado potato beetle, *Leptinotarsa decemlineata*, in response to clean air and to air containing the odour of various plants. They did not compare the responses of fed and deprived insects to food odour but they did show that deprived insects of either sex made stronger upwind responses to moving clean air than fed ones.

Wensler (1972) investigated factors limiting the responsiveness of adults of the mosquito *Aedes aegypti* to an ethyl ether extract of white honey. She showed that the proportion of insects of either sex congregating under a beaker containing the extract increased with increasing deprivation, the insects first becoming responsive after 5 to 6 days and reaching full responsiveness by 7 to 10 days. The finding that females allowed to feed after 7 or 8 days of deprivation immediately lost their responsiveness for 6 days, showed that the effects observed in the original experiment were not merely those of ageing. Wensler found further that the honey odour stimulated the mosquitoes to general flight activity and that this effect also increased with increasing deprivation.

Nadel and Pelag (1965) showed that the proportion of females of the Mediterranean fruit fly, *Ceritītis capitata*, caught in traps baited with "trimedlure", a powerful attractant for males of the species, increased with increasing deprivation until, after 24 h, the percentage of females in the catch rose to about 38 per cent. After sucrose or sucrose and protein hydrolysate had been supplied to the mixed cage population, the percentage of females caught fell to less than 10 per cent but rose again about 20 h after removal of the food. It seems that females of this species become responsive to odour of "trimedlure" only after a period of deprivation. It was shown (Hardee *et al.*, 1966) that adults of the boll weevil, *Anthonomus grandis*, fed only sugar were more responsive to the odour of an attractive extract from cotton than were weevils which had had access to only cotton bolls continuously before the trial or had had bolls and sugar solution on alternate days. Pienkowski and Golik (1969) showed that 48 h starved adult alfalfa weevils made more frequent turns in an atmosphere containing the odour of alfalfa than did ones deprived of food for 24 h.

It is well known that the responses of insects to water vapour vary according to their state of water or food deprivation and, in two instances, the behaviour can be directly related to the insect's subsequent ingestion of water. Mellanby and French (1958) describe the behaviour of water-deprived and water-satiated larvae of the meal worm, *Tenebrio molitor*, and the tomato moth, *Diataraxia oleracea*, in the vicinity of a drop of water. In each species, water-deprived larvae turned towards a drop of water when they passed within 1-2 cm of it and ingested it vigorously when they established contact with it. Water satiated larvae, on the other hand, made no orientated movements towards water in their immediate vicinity. Mellanby and French state that the responsiveness of these larvae to water vapour is related to their total water content, but supply no formal supporting evidence.

It has been shown also, in quite a large number of species, that individuals which are fully supplied with food or food and water show a preference for relatively dry air when given a choice between two or more relative humidities, but that deprived individuals show a preference for moist air (e.g. Bentley, 1944; Roth and Willis, 1951; Dodd and Ewer, 1952; Riegert, 1958; Syrjamaki, 1962; Youdeowei, 1967; Arbogath and Carthon, 1972). It is often implied that this response is concerned only with the reduction of water loss by transpiration and, in those insects which are capable of doing so, with absorption of water from the atmosphere. However, it can be said also that orientated movement towards areas of high humidity might lead an insect to water or an aqueous solution which could then be ingested. It is relevant therefore to discuss very briefly what little is known about the mechanisms responsible for the reversal.

In all instances where the effects of deprivation on humidity responses have been studied in any detail, the insects have been deprived of all kinds of ingestible material. Under these conditions, the stored products beetles, *Tribolium castaneum* and *T. confusum* (Roth and Willis, 1951) and *Oryzaephilus surinamensis* (Arbogast and Carthon, 1972) maintained an almost constant ratio of water to dry matter over the period of deprivation during which the reversal of humidity response occurred. This finding indicates that the reversal might depend on a recognition of the total loss of material from the body, rather than of any state specifically related to desiccation. Riegert (1958) similarly studied the effect of complete deprivation on the humidity response, total weight loss and water loss in the grasshopper *Melonoplus bivittatus* and found that water loss occurred at a somewhat faster rate than did loss of solids, and therefore that the total water expressed as a percentage of live weight gradually declined during prolonged deprivation. The data, however, do not provide convincing evidence that the changes in humidity reaction are related to gross changes in water balance, as especially in females, the initial dry response declined to zero over a period during which loss in live weight and water were essentially parallel.

The results of experiments in which humidity responses of fed and food deprived insects were examined before and after the removal of all or parts of the antennae and palps have provided evidence that the wet and dry responses are mediated by at least partially different sets of receptors, which have different spatial distributions upon the body of the insect (Bentley, 1944; Dodd and Ewer, 1952; Roth and Willis, 1951). It seems probable that the change from the wet to the dry response or vice versa involves a change of the "attention" of the CNS from the input of one set of receptors to the input of another set. The possibility that changing internal balances might directly determine which kinds of receptor are operative at different stages of desiccation cannot, however, be completely excluded, but it can be considered unlikely in view of the finding that reversal of the humidity response occurs, at least in some insects, in the absence of gross changes in water balance, as indicated by the percentage of the live weight which is accounted for by the water content of the insect.

Tsetse flies are known to respond to visual cues in their search for food and Brady (1972b) has examined the changes with deprivation in the responsiveness of *Glossina morsitans* to a slowly moving visual stimulus, as measured by the number of insects taking off in a given time. He found that the responsiveness of teneral males and virgin females, mature males, pregnant females and virgin mature females all increased progressively over 4 days of deprivation. Although there was no fed control, the fact that the effect was detectable in various categories of fly rules out the possibility that the increase was primarily an effect of ageing or development.

2.2.3 *Responsiveness to light and gravity*

In addition to the variations they display in their locomotor responses to stimuli provided by the food itself, insects are known to show deprivation-dependent changes in their orientated responses to light and gravity, the changes being such that they affect the insects' chances of encountering further food.

Wellington (1948) studied the responses of starved and fed larvae of the spruce budworm, *Choristoneura fumiferana*, to a discrete light source. He demonstrated that late instar larvae showed a photopositive reaction when fed and a photonegative one after being deprived of food for some hours, but that the response of the first two instars was always photopositive and that of the third changed during a period of deprivation from a photopositive response to indifference to light. In an investigation of the physiological basis of this reversal, Wellington demonstrated that photonegative, deprived late instar larvae became photopositive after injection of ringer solution into the gut and concluded that the gut distension was responsible for the reversal. Further, he showed the ligation across the 5th segment caused photonegative larvae to become photopositive, whereas ligation across any other segment did not result in behavioural changes. He showed also that when a local anaesthetic was injected into segment 5 of fed photopositive larvae, most of the larvae became photonegative. No behavioural changes resulted from injection into any other segment, except for some rather equivocal effects produced by injections in the region of the sub-oesophageal ganglion. These results seem to provide clear evidence that distension of the gut is responsible for the reversal of light response and that the sensory system monitoring the distension is situated in segment 5. It is not known whether the change is the result of direct pressure on the ventral ganglion, as Wellington suggested, or whether specific receptors are involved. Wellington argues convincingly that changes in responsiveness to discrete light, coupled with an invariable photopositive response to diffuse light and a light compassing reaction when partially starved, enables the larva to escape from the part of the host tree which has been eaten out.

Green (1954) made a similar study of the reactions of the sawflies *Neodiprion americanus banksianae* and *N. lecontei*. He found that, when fed, most larvae in the first three instars were indifferent to light but became strongly photopositive to discrete light source after a period of deprivation. In contrast, larvae in later instars were strongly photopositive when fed but tended to become less so with starvation, especially in the case of *N. lecontei*. Green again shows clearly how the changing reactions play an important part in the insect's feeding behaviour in the field.

Madge (1964) examined the light response of starved and fed larvae of the cutworm *Tryphaena pronuba*. He found that a period of deprivation markedly weakened the photonegative response of 7th instar larvae and suggested that this weakening of response would allow these normally noctural larvae, if starved, to feed above ground in daylight.

Dixon (1959) studied the searching behaviour of the coccinellid *Adalia decempunctata*, which preys on aphids, and showed that the rapidity with which a larva was able to abandon a stem on which it found no aphids varied according to its state of deprivation. A larva, when placed at the base of a thin cylinder, would climb to the top and, upon finding the apex devoid of prey, would begin to descend. It would not, however, descend all the way to the bottom, but would turn upwards again to go to the apex. It would repeat this behaviour a number of times, gradually coming further down the cylinder with each progressive downward excursion until it finally reached the bottom. It was demonstrated by Dixon that the number of turns executed by the larvae before reaching the bottom decreased as the period for which the larvae had previously been deprived of food increased. This behavioural change means that a starved larva would be able to escape from a stem apex which bore no prey more rapidly than a fed one.

2.3 REGULATION OF NONLOCOMOTOR PRE-INGESTION BEHAVIOUR

When an insect is in contact with or in the immediate vicinity of food, it must, in order to commence feeding, make movements which will cause its mouthparts to contact the food. Commonly, and perhaps universally, these movements are in response to stimuli provided by the food and it is known that the responsiveness of insects to these may vary according to their state of deprivation. These changes in responsiveness have been determined in two related ways. In the first, thresholds of response are obtained for insects in different states of deprivation by determining the strength of stimulus required to elicit a positive response for stated percentages (usually 50 per cent) of insects. The second approach has been to determine the proportion of insects responding to a stated stimulus.

Many insects have, on their tarsi, contact chemoreceptors, the suitable stimulation of which will cause insects, which are in a responsive state, to exhibit behaviour which brings the mouthparts in contact with the stimulating material. Stimulation of tarsal receptors may elicit several related responses. First, it has been demonstrated that when the tarsi on one side of the blowfly *Phormia terraenovae* (Pflumm, 1970) receive stimulation, the insect turns towards that side, and it seems probable that this response is common among insects whose tarsi bear contact chemoreceptors sensitive to food stimuli. The second response, which also has been demonstrated in blowflies (Dethier, 1961; Pflumm, 1970), but which is also certainly shown by other insects, is for the insect to stop walking, presumably when the two fore-tarsi are receiving approximately equal stimulation. The third response is for the insect to extend or otherwise move the mouthparts so as to bring them in contact with the food. This latter act will either allow the insect to begin ingesting the food immediately or, probably more commonly, will bring into contact with the food the chemoreceptors situated on the mouthparts, which are involved in eliciting the next phase of behaviour leading towards ingestion.

Dethier (1961) showed that 24 h deprived adults of *P. regina* stopped when they made tarsal contact with water and drank to repletion but that when they again encountered water they would either ignore it or would turn away from it depending on whether the surface to which the water had been applied was water absorbent or not. It is clear that feeding to repletion on water reduced the responsiveness of the flies to this material. Dethier suggested that the "turning away" response observed on the nonabsorbent material was mediated by input from tarsal mechanoreceptors stimulated by water which sits upraised above the general surface, and that this input was overridden in flies responsive to water by input from the tarsal water receptors. On the absorbent material, on which the water is continuous with the general surface, there would be no stimulation of the mechanoreceptors and a water satiated fly would receive no input to which it was responsive. Avoidance of a drop of water by water satiated individuals was also reported by Mellanby and French (1958) for larvae of *Tenebrio molitor* and *Diataraxia oleracea*.

A number of workers, using proboscis extension as the criterion for a positive response, have examined the effects of feeding and deprivation on tarsal threshold of various species of blowflies to sugars. Minnich (1929) examined threshold changes with deprivation in adults of *Calliphora vomitoria* which had fed to repletion on 1.0 M sucrose after having previously been starved for the 24 h following emergence, and found that flies deprived of food for 144 h would respond to sucrose at about $\frac{1}{500}$ of the concentration required to elicit a response from flies deprived for 24 h. Haslinger (1935), who examined the decline in threshold in *C. erythrocephala (= C. vicina)* which had fed *ad lib.* on sucrose for the 5 days following emergence, found that the tarsal threshold for sucrose fell by a factor of about 128 between the first and tenth day of deprivation. He found also that the sensitivity of flies to other sugars increased during the period of deprivation, but to somewhat smaller extents than for sucrose. In neither of these investigations were thresholds determined for flies which had been deprived of food for less than 24 h.

Evans and Dethier (1957) made a detailed study of the changes in tarsal threshold with deprivation in *Phormia regina*. In their experiments, the first threshold determinations were made 1 h after flies, which had previously been deprived for 24 h, had fed to repletion on one of a number of sugar solutions. It was found that threshold was elevated in flies which had fed to repletion on 2.0 M glucose or mannose, or 1.0 M fucose or lactose, and that it did not return to a level typical of starved flies for 20-100 h, depending on which sugar solution the flies had ingested. When flies were fed fucose or mannose solutions, threshold did not reach a maximum until about 2 h after feeding. Evans and Dethier (1957) found that the threshold recorded 1 h after flies had fed on a 2.0 M glucose solution was not subsequently exceeded, but Evans and Barton Browne (1960) showed that the threshold of glucose fed flies was higher 45 min after feeding than it was 25 min after. Evans and Barton Browne established also that the amount by which threshold increased depended on both the concentration and volume of the sugar solution ingested. They showed that the ingestion of 10 μl of 2.0 M glucose induced a greater rise in tarsal threshold than the ingestion of 3 μl of the same solution, and that 3 μl of 0.01 M glucose caused no detectable increase.

The physiological mechanisms underlying the observed increases after feeding in the tarsal thresholds of *P. regina* to sugars have been extensively studied, mainly by Dethier and his co-workers, and most of these investigations have been discussed in some detail by Dethier (1969). The present discussion, even though it examines critically some of Dethier's conclusions, should be looked upon as being complementary to his more extensive treatment of the topic.

Some consideration has been given to the possibility that events during feeding might directly cause, in some part of the nervous system, long-term changes which render *P. regina* more or less refractory to subsequent tarsal stimulation with sugar solutions. Evans and Dethier (1957) pointed out and argued fairly convincingly against the possibility that a long-lasting state of refractoriness might be induced in the external chemoreceptors of the tarsi. Similarly Dethier and Bodenstein (1958) considered and regarded as unlikely the possibility that the CNS may have been rendered unresponsive by changes induced by the motor activity on sensory feedback which accompanies the actual act of ingestion. A third possible kind of neural effect, which has not previously been mentioned, is that the sensory input from the chemoreceptors, which receive stimulation during feeding, might directly induce changes in the CNS. In addition, the possibility exists that hormonal material, which renders the fly less responsive to tarsal stimulation with sugars, might be released as a result of neural events during feeding.

The results of several investigations are relevant to the question of whether perseverating effects of events during feeding might be responsible for post-feeding increases in threshold and, if so, to what extent. In several experiments, studies were made of the effect on the time course of changes in threshold of giving flies different treatments which altered the time course after feeding of changes in certain internal parameters related to the nutritional state of the fly. Hudson (1958) examined the effects of flight on the rate of decline of the glucose threshold in flies which had ingested 10 μl of 1.0 M glucose. After 110 min of flight, threshold was only 10 per cent of the initial value whereas that of control flies mounted on waxed sticks did not fall appreciably during this period. It is clear, therefore, that, when the insects were forced to fly, any perseverating effects resulting from events during feeding had ceased to be important by less than 2 h afterwards. While this result might reasonably be interpreted as indicating that any effects of events during feeding would cease to be apparent within 2 h, under any conditions, the possibility must be admitted that flight may have shortened the period for which any effects of this kind might persist. Evans and Dethier (1957) found that the tarsal threshold to glucose of flies which had the crop duct ligated just after feeding on 2.0 M glucose began to fall after about 4-6 h of relative inactivity while attached to waxed sticks, whereas that of unoperated controls remained elevated. This result shows that tarsal threshold is influenced by the distribution of solution within the fly, and therefore that events during feeding are certainly not solely responsible for post-feeding threshold elevation. In addition, they indicate that any effect of this kind cannot be a dominant cause of threshold elevation for any more than 4-6 h in inactive flies.

Also relevant to the assessment of the possible role of direct events during feeding is the statement by Dethier and Bodenstein (1958) that flies which had undergone recurrent nerve section failed to show the normal post-feeding rise in tarsal threshold. If this is so, it would follow that direct effects of events during ingestion play no significant role in bringing about the post-feeding elevation in threshold. There is, however, conflicting information, which will be discussed later, about the specific effect of recurrent nerve section on the tarsal thresholds of *P. regina* to sugars, even though it is clear that the operation impairs the ability of flies to regulate their intake of sugar solution.

Gelperin (1966b) injected 5 μl of either 2.0 M fructose or Bodenstein's saline into the haemocoel of recently fed flies and determined the sizes of meals of 1.0 M fructose they would consume 12 h later. He found that the meals taken by the sugar-injected flies were smaller than those taken by the saline-injected controls and concluded that the presence of a high concentration of sugar in the haemolymph delayed the normal post-ingestion decline in the threshold. This he related to his finding (Gelperin, 1966a) that the elevation of the haemolymph sugar concentration delayed crop emptying. These conclusions should, however, be treated with some caution since, while it is apparent that the taste threshold and meal size are normally well correlated, it seems unlikely, on the basis of what is known about the physiological regulation of meal size in *P. regina* (see section 2.4), that threshold directly determines meal size, or even that both are under the control of the same physiological mechanisms. It is possible therefore that procedures such as the injection of material into the haemocoel might cause the two measures of responsiveness to sugars to become less well correlated than under normal circumstances. It can fairly be stated, however, that Gelperin's results, like those relating to the effect of recurrent nerve section, provide evidence that factors other than effects of events during feeding play a role in the general regulation of subsequent feeding behaviour, even it it is somewhat uncertain as to whether the results should be interpreted as relating strictly to the regulation of tarsal threshold.

The general finding that the time course of changes in threshold can be substantially altered by giving flies various treatments after feeding shows clearly that perseverating effects of events during feeding are certainly not solely responsible for threshold elevation. Furthermore, it can be said, on the basis of the results discussed, that any long-term effects of events during ingestion cannot exert a significant influence upon threshold beyond, at the most, the first few hours of the prolonged period of threshold elevation, which follows the ingestion of concentrated sugar solutions.

The finding that maximum threshold is often not achieved for some hours after feeding (Evans and Dethier, 1957; Evans and Barton Browne, 1960) constitutes evidence, of a different kind, that direct neural effects of events during feeding cannot play a dominant role even in the relatively short term, since it would seem reasonable to expect that any such effects would be maximal immediately after the cessation of feeding. Of particular relevance is the finding by Evans and Barton Browne (1960) that the tarsal threshold of *P. regina* declined by a factor of two between 15 and 25 min after feeding on 2.0 M glucose but had increased again by 45 min after. It seems likely that the relatively high threshold 15 min after feeding would probably be due, in parts at least, to effects of events occurring during feeding, but that these were declining rapidly at this time. The increase between 25 and 45 min clearly cannot be explained in terms of the direct neural effects of events during feeding.

It is less certain that a similar argument can be used validly in relation to possible effects of any hormones released during ingestion. It would be reasonable to expect that the titre of any released hormone would be maximal at the time of the cessation of feeding, but it seems unsafe to assume that the time course of changes in the neural consequences of hormone action would necessarily be simply related to the time course of changes in the hormone titre. It seems reasonable therefore to draw the general conclusion that direct neural effects of events during feeding do not play a dominant role in bringing about threshold elevation, but that hormonal material released during feeding might possibly play a part, bearing in mind the previously mentioned temporal limits on any such effects.

In summary, it can be stated that there is no positive evidence favouring the view that any perseverating effects of events during feeding play a role in the elevation of threshold which occurs after a fly ingests sugar solution and that there is strong evidence indicating that effects of this kind are not solely responsible. The case against their partial involvement would, however, be more complete if it were shown rigorously that various operations involving nerve section do, in fact, prevent threshold elevation.

A number of the investigations into the physiological basis for the regulation of tarsal threshold in *P. regina* have examined the possibility that feedback concerning the amount of sugar in the various parts of the gut might be involved. It was established by Evans and Dethier (1957), who used flies restrained on waxed sticks, that threshold was generally related to the amount of sugar in the crop and that it returned to its lowest level only when this amount was negligible. In addition, Hudson (1958) showed that flight, which greatly increased the rate of crop emptying, caused a corresponding increase in the rate at which threshold declined. These two pieces of information provided circumstantial evidence that the volume of the crop might directly affect threshold. Evans and Dethier (1957) directly investigated this possibility. They ligated the crop duct either just before or just after flies fed to repletion on 2.0 M glucose and found that, in both instances, threshold rose to a high level following feeding, albeit for a shorter period than in normal flies. The result with flies ligated before feeding indicates that the threshold can rise even if the crop is empty, whereas that obtained with flies ligated after feeding shows that threshold can fall even when the crop remains full, and therefore that the possession of a full crop does not alone result in threshold elevation.

The possibility that threshold elevation might be brought about by the presence of sugar solution in some part of the gut other than the crop has been investigated in a number of ways. Dethier and Bodenstein (1958) ligated the guts of flies just behind the proventriculus either just before or just after feeding them with 1.0 M sucrose. They found that the threshold of both groups of ligated flies remained high for about 3 h and concluded that the presence of sugar in the mid-gut is not necessary for the maintenance of an elevated threshold. In a complementary experiment, these workers examined the thresholds of flies which had been given enemas of 1.0 M glucose. They found that there was no elevation of threshold even though the glucose solution, which had been introduced via the anus, filled the gut from the rectum forward to the proventriculus. Similar results were obtained when 2.0 M glucose was injected through the wall of the mid-gut. The presence of concentrated sugar solution in the mid-gut and hind-gut therefore did not alone cause an elevation in the tarsal threshold of *P. regina* to sugar.

In experiments discussed above, it was shown that threshold elevation occurred when ligation before feeding allowed sugar solution either into only the crop and fore-gut (Dethier and Bodenstein, 1958) or only the fore-gut, mid-gut and hind-gut (Evans and Dethier, 1957). These results, taken together with the findings that the presence of sugar solution in either the crop (Evans and Dethier, 1957) or mid-gut and hind-gut (Dethier and Bodenstein, 1958) produced no increase in threshold, show that elevation of threshold occurs only when the fore-gut contains sugar solution and suggests strongly that its presence in only the fore-gut is sufficient to cause threshold elevation. Dethier and Bodenstein (1958) attempted to demonstrate this directly by injecting sugar solution into the fore-gut but succeeded in doing this only in one insect which did, however, show a substantial increase in threshold.

Gelperin (1967) showed that peristalsis or controlled enlargement of the fore-gut produced spike activity in the recurrent nerve and that the origin of these spikes was two bipolar neurones situated in a nerve branch connecting the recurrent nerve and the fore-gut. More recently he showed (Gelperin, 1972) that there was more spike activity in the recurrent nerve of flies which had been fed 1.0 M sucrose than that of flies fed 0.1 M sucrose, even though the 0.1 M solution is emptied from the crop more rapidly than the 1.0 M solution (Gelperin, 1966a). He suggested that flies fed 1.0 M sucrose shuttled slugs of solution back and forth between the fore-gut and the crop duct but that material was only occasionally passed to the mid-gut. This situation could be expected to stimulate the mechanoreceptors more intensely and frequently than when material released from the crop passes directly to the mid-gut, as probably happens with the less concentrated solution. These findings by Gelperin show, therefore, that the level of sensory input via the recurrent nerve varies with concentration in the same way as threshold (Evans and Barton Browne, 1960), and consequently provide strong circumstantial evidence that input via the recurrent nerve, from stretch receptors associated with the fore-gut, plays an important role in determining tarsal thresholds to sugars.

If, as is suggested by the results so far discussed, the elevation of tarsal threshold in *P. regina*, which follows feeding on sugar solutions, is solely or even largely the result of inhibitory input reaching the CNS via the recurrent nerve, it could reasonably be expected that normal threshold elevation would not occur in flies which had undergone recurrent nerve section before feeding. The effects of this operation on tarsal threshold elevation were investigated by Dethier and Bodenstein (1958) and by Evans and Barton Browne (1960). Their results, which relate specifically to tarsal threshold, are somewhat conflicting even though both groups of authors, as well as Dethier and Gelperin (1967) and Gelperin (1972), agree that flies operated in this way become markedly hyperphagic when allowed either continuous access to sugar solution for a period, or when sugar solution is repeatedly applied to their labellar lobes by means of a brush. It is also agreed that repeated proboscis extension occurs in operated flies which are allowed to feed *ad lib.* and that this is at least in part responsible for their hyperphagia.

Dethier and Bodenstein (1958) state that operated flies responded to tarsal contact with 0.1 M sucrose after having been allowed to become hyperphagic on 1.0 M sucrose. Evans and Barton Browne (1960), on the other hand, reported that operated flies displayed high thresholds ($>$ 1.0 M) to glucose 1 h after the consumption of 1.0 M glucose but that these flies showed abnormally persistent response to tarsal contact with water. The discrepancies in the results are discussed in some detail by Dethier (1969) and, at this point, I can only agree with his conclusion that additional experimental information is required about the precise effect of recurrent nerve section in *P. regina*. However, even in the absence of certain knowledge about the effects of recurrent nerve section, it seems safe to assume that input signalling the presence of fluid in the fore-gut plays an important part in bringing about the elevation of tarsal threshold to sugar. Evidence that the presence of solution in the fore-gut might alone be sufficient to cause elevation of tarsal threshold in *P. regina* is, however, very slender since in only one fly did Dethier and Bodenstein (1958) succeed in injecting sugar solution into only the fore-gut. Therefore, on the basis of the evidence discussed so far, the possibility cannot be completely eliminated that input concerning fore-gut distension is effective in causing threshold elevation only if the CNS is receiving input from other monitoring systems indicating that the fly is in a fed state.

There have been a number of direct investigations into the possibility that changes with feeding and deprivation other than those occurring in the gut might also play a part in the regulation of threshold. The possibility that the concentration of sugar in the haemolymph might play a role in determining the tarsal thresholds of *P. regina* to sugars was investigated by Evans and Dethier (1957) and Hudson (1958). Evans and Dethier (1957), using flies restrained on waxed sticks, compared the time courses of changes in taste threshold and in the concentration in the haemolymph of trehalose, which is the main blood sugar of adult *P. regina*, and of the sugar on which the flies had fed. They demonstrated that there was no predictable relationship between threshold and the concentration of either sugar. When flies were fed 2.0 M glucose, for instance, threshold declined steadily over about 70 h whereas the trehalose and total haemolymph carbohydrate levels rose sharply for 10 h after feeding and did not begin to decline until 50 h after. Haemolymph glucose was always at a low level and had not declined after 30 h. In flies fed 1.0 M fucose, a sugar which is not utilized by *P. regina*, threshold had fallen to a low level by 30 h after feeding, at which time the fucose and total carbohydrate concentration of the haemolymph remained at a high level. Hudson (1958) also compared the time courses of changes in threshold level and total carbohydrate in the haemolymph, but her experiments differed from those of Evans and Dethier (1957) in that she examined the relationships in flies which had been forced to fly. Again no consistent relationship was found between threshold and carbohydrate concentration in the haemolymph. It was demonstrated, for instance, that threshold fell sharply between 15 and 30 min after the commencement of flight but that there was no corresponding fall in the total concentration of haemolymph carbohydrate. Evans and Dethier (1957) performed a series of experiments in which they injected glucose, trehalose or fucose solution directly into the haemocoel of 24 h starved flies. None of the injections caused an increase in threshold even though the sugar concentrations produced in the haemolymph were at least comparable with the highest attained during the post-feeding period of elevated threshold. All these findings show that a high level of sugar in the haemolymph does not alone cause the elevation of the taste threshold of flies previously deprived for 24 h. They suggest, further, that this factor may play no part in determining threshold, since it seems somewhat improbable that the fly would use information concerning any parameter which does not change progressively throughout the period of progressive reduction in threshold.

The relationship between threshold and the glycogen content of the fly was examined by Hudson (1958) who demonstrated that glycogen was still increasing at a time when threshold had fallen substantially. By a similar argument to that used above regarding haemolymph sugar concentration, it seems likely that the glycogen content of the fly is not a controlling factor for sugar threshold. The finding by Evans and Dethier (1957), that feeding on the non-nutritional sugars lactose and fucose caused prolonged threshold elevation generally similar to those obtained when flies were fed on the nutritional sugars glucose and mannose, indicates that the threshold regulating mechanism is independent of the nutritional state of the tissues.

The results discussed above suggest that it is very unlikely that post-absorptive consequences of sugar ingestion play any significant part in the determination of the tarsal threshold of *P. regina* to sugar.

Several characteristics of the time course of changes in threshold after feeding in normal flies, and in flies which received various treatments involving ligation of some part of the gut, suggest that the tarsal threshold at any given time might not be dependent simply upon the current level of sensory input reaching the CNS at that time. The first of these is that, as previously mentioned, peak threshold is not reached for some time after feeding (Evans and Dethier, 1957; Evans and Barton Browne, 1960), and the second that the thresholds of insects whose crop ducts were ligated either before or immediately after feeding remain elevated for 4 or more hours, even though the insects, respectively, have no capacity to store solution or are unable to release stored material into the fore-gut after feeding. Finally, there is the point made by Dethier (1969) that even though the fore-gut contains sugar during ingestion, threshold does not rise sufficiently to cause feeding to be immediately terminated. The possible involvement of long-lasting effects of events during feeding have already been discussed. There is, however, the additional possibility that input from sensory systems monitoring nutritional state might generate perseverating effects. I discuss here, therefore, more generally, the question of whether perseverating effects might be involved in the regulation of tarsal threshold in *P. regina*.

It must first be pointed out that at least some of the findings regarding the time course of threshold changes are not necessarily incompatible with the view that tarsal threshold is immediately and directly related to current sensory input from fore-gut receptors. The level of input from these receptors is known to depend on the degree of distension of the fore-gut which, as pointed out by Gelperin (1972), depends not only on the rate at which material enters the mid-gut from the crop, but also on the rate at which it passes through the cardiac valve to the mid-gut. Since we lack precise information about the degree of distension at various times after feeding, we are in no position to exclude the possibility that the time course of threshold elevation after feeding might be following, faithfully, changes in the degree of fore-gut distension.

It seems also that the relatively long period of threshold elevation in flies which had undergone ligation of the crop duct immediately before feeding might also be explicable in terms of instantaneous effect of continuing sensory input. Evans and Dethier state that, under these conditions, considerable amounts of solution enter the mid-gut and fore-gut and, since the insects were fed 2.0 M sucrose, it is probable that osmotic changes in the haemolymph consequent upon absorption from the mid-gut would be sufficient to ensure that the cardiac valve remained closed for a considerable time (Gelperin, 1966a). This might well result in the retention of solution in the fore-gut for sufficient time to account for the period of threshold elevation. Prolonged threshold elevation in flies whose crop ducts were ligated after feeding is, however, less easily reconciled with the view that only current input is responsible for threshold elevation. The results of Knight (1962) show that it is usual for all of ingested sugar solution to enter the crop immediately and that release does not begin for 2-12 min. If it is assumed that ligation was carried out in the experiments of Evans and Dethier (1957) less than 2 min after feeding, it seems unlikely that the prolonged elevation of threshold can be explained in terms of the continuing presence of sugar solution in the fore-gut. The possible influence of sugar solution in the fore-gut during ingestion will be discussed in section 2.4 when the factors regulating meal size are discussed.

There would seem to be two possible alternatives to the hypothesis that tarsal threshold in *P. regina* is a reflection of the current level of sensory input reaching the CNS via the recurrent nerve. The first is that events during feeding and/or changes consequent upon feeding generate not only neural events but also cause the release of hormonal material which plays a part in causing threshold elevation. The second possibility is that the regulatory mechanism is entirely neural but that threshold elevation depends on the build-up of a state of inhibition in the CNS, the level of which depends on the amount of inhibitory input reaching the CNS during some defined preceding period.

Dethier and Bodenstein (1958) directly investigated the possibility that a hormonal mechanism might be involved. They transferred haemolymph diluted with saline from flies fed 2 h previously to flies which had been deprived of food for 24 h and found that the threshold of the vast majority of the starved flies was not changed by the transfusion of haemolymph from the fed insects. On the basis of this result they concluded, tentatively, that threshold elevation after feeding was not hormonally mediated but, as they suggested, the negative result might be attributable to the relatively small amount of haemolymph transferred. There is now substantial evidence that feeding in a number of insects results in hormonal release, especially from the CC (e.g. Dadd, 1961; Davey, 1962a, 1962b; Clark and Langley, 1963; Berridge 1966; Highnam *et al.*, 1966; Mordue, 1969; Bernays and Chapman, 1972b). In view of this general evidence and especially that of Bernays and Chapman who demonstrated that the material released from the CC in *L. migratoria* is involved in the post-feeding decline in sensitivity of the gustatory receptors of the maxillary palps, it would seem inadvisable to reject completely the possibility that hormones, released in response to feeding, may play a part in the determination of threshold to sugar, or for that matter in the regulation of other processes involved in the control of feeding in *P. regina*. It seems that parabiosis experiments of the kind performed by Green (1964b) in his investigation into the effect of feeding on locomotor activity in *P. regina* might usefully be repeated, this time to determine whether the feeding of one fly influences the tarsal threshold of its parabiotic partner.

The possibility that perseverating effects generated in the CNS by inputs from internal sources might play a role in the regulation of tarsal threshold has not been directly investigated in *P. regina*. Evidence has been obtained, however, that both excitatory and inhibitory input from external chemoreceptors generate perseverating effects of short duration (Dethier, 1957; Dethier *et al.*, 1965, 1968) in parts of the nervous system of *P. regina* concerned with feeding behaviour; and this suggests the possibility that input from internal sources might also be capable of producing effects of this kind.

So far, in this discussion, it has been implied that any inhibitory feedback involved in the regulation of tarsal threshold influences the responsiveness of the CNS to excitatory input from the tarsal chemoreceptors and that the receptor sensitivity itself is not markedly affected by the state of deprivation. There is no specific information available concerning the effects of feeding and deprivation on the responsiveness of tarsal receptors to chemical stimulation, but information, to be discussed later, concerning labellar receptors in *P. regina* casts some doubts as to whether constancy of sensory input can be assumed.

In summary, it can be said that feedback concerning the presence of fluid in the fore-gut plays a major role in determining the tarsal threshold of *P. regina* to sugars and may even be solely responsible. It is not possible to say at this stage whether the chain of events is exclusively neural or whether hormones play a part in threshold determination. Nor is it possible to say for certain whether neural changes responsible for tarsal threshold elevation occur only in the CNS or whether changes in receptor sensitivity and hence in sensory input play a role.

The tarsal thresholds to sugars of several insects other than blowflies are known to decline with increasing sugar deprivation but in none of these has the physiological bases for the changes been studied. Minnich (1922) demonstrated that tarsal thresholds of adults of the red admiral butterfly, *Pyrameis atalanta*, to sugars declined during deprivation and it has been shown that similar changes occur in the tabanid *Hybomitra lasiophthalma* (Lall, 1969; Davis and Lall, 1970) and the mosquito *Aedes aegypti* (Salama, 1969).

Evans (1961) and Dethier and Evans (1961) studied the responses of *P. regina* to tarsal stimulation with water. They found that flies which had been deprived of food and water in a low humidity for 24 h gave a positive proboscis response to tarsal contact with water, whereas those kept at a high humidity during the period of deprivation were unresponsive. They showed that the injection of water, 2.0 M glucose either alone or in a physiological saline, a 4 times concentrated physiological saline, or mineral oil into the haemocoel of flies which had initially responded positively to tarsal contact with water, caused a proportion of the flies to become unresponsive, the proportion generally increasing with increasing volume of injected material. Flies which were initially unresponsive did not become so after the injection of 2.0 M glucose or 4 times concentrated saline. Evans (1961) also bled flies by squeezing the thorax after the removal of the fore-legs. He found that a high percentage of flies, which were the unresponsive portion of a cohort of flies which had been completely deprived under desiccating conditions, were rendered responsive by bleeding, but that the treatment had this effect in only a small proportion of well hydrated flies. On the basis of these results, Evans (1961) and Dethier and Evans (1961) concluded that responsiveness to tarsal stimulation with water is dependent on the volume of the haemolymph and that the rapidity with which altering the volume caused changes in responsiveness suggested that the effect is neurally mediated. No evidence was found that changes in the responsiveness of *P. regina* to water are related to the changes in the composition of the haemolymph.

Dethier and Evans found that flies which had suffered recurrent nerve section became bloated through taking abnormally large amounts of water when allowed access to water-soaked filter paper for 2 h, and that this polydypsia resulted from the repeated ingestion of water throughout this period. They found also that water intake, under these conditions of continuous access, was normal in flies which had had the corpora allata or the median neurosecretory cells of the brain removed. Dethier and Evans did not formally determine the responsiveness of previously water-fed flies with sectioned recurrent nerves to tarsal contact with water, but Evans and Barton Browne (1960) found that operated flies which had previously taken large amounts of 2.0 M glucose would respond to water repeatedly. These results are consistent with the view that recurrent nerve section prevents the normal post water-ingestion reduction in responsiveness to tarsal contact with water. Dethier and Evans suggest that the recurrent nerve carries input from neurones which respond to haemolymph volume or pressure.

Barton Browne (1964), Barton Browne and Dudziński (1968) and Barton Browne (1968) studied the regulation of water intake by the

Australian sheep blowfly, *Lucilia cuprina*. This insect is markedly more responsive to water after 24 h of deprivation than is *P. regina*, all individuals showing a strong response to tarsal stimulation, irrespective of the humidity conditions in which they were kept during the period of deprivation. Because of this, the criteria used for water responsiveness in their studies were the volume of water ingested and the duration of continuous ingestion. It is not strictly appropriate, therefore, to discuss the results obtained with *L. cuprina* in terms of tarsal threshold and the findings will be detailed later when the regulation of meal size is discussed. At this point, however, it can be said that this fly differs from *P. regina* in that the injection of concentrated sodium chloride solutions increases the responsiveness to water of flies which have fed to repletion on water 1 h earlier. Although the data were not presented in terms of the proportion of the flies responding, it is clear that, whereas many 1-h deprived noninjected flies were unresponsive to water, the majority and perhaps all of those which received injections of 0.3 or 0.5 M sodium chloride (Barton Browne, 1968) were responsive. Also, it was shown that, whereas the injection of water into flies deprived of both food and water rendered many of them unresponsive, the injection of equal volumes of sodium chloride solutions did not have this effect. In *L. cuprina*, therefore, in contrast to the situation in *P. regina*, the composition of the haemolymph seems to be more important in controlling water responsiveness than is its volume.

So far, in the discussion of tarsal thresholds, only the responsiveness of insects to substances which stimulate them to proceed to the next phase of their feeding behaviour have been considered. The question arises as to whether insects might not become less sensitive to materials which deter them from doing so, as the period of deprivation increases. It is known from *ad lib.* feeding experiments that a number of insect species will, when deprived, ingest materials which they had rejected when in a less deprived state. Normally, rejection thresholds have been obtained by examining the responses of insects to a series of concentrations of the deterrent substance, each presented as part of a mixture containing a fixed concentration of a feeding stimulant. Under these conditions, an apparent loss of sensitivity to a deterrent might be due either to an increase in the rejection threshold to the deterrent or to a decrease in the acceptance threshold to the feeding stimulant, or both. Haslinger (1935) using *Calliphora erythrocephala* (= *C. vicina*) attempted to overcome this problem by presenting a series of concentrations of deterrent made up in concentrations of fructose which were 3 times those known to be threshold concentrations for flies in each of the test states of deprivation. His results with hydrochloric acid as the deterrent substance show that the flies did not, in fact, become less sensitive to it, the concentration of acid required to obtain rejection remaining essentially constant throughout prolonged deprivation. It is not known whether this is true of all deterrents for all insects, but it seems that further experiments along the lines of those of Haslinger would yield valuable information.

Proboscis extension and related responses are elicited in blowflies and some other insects by the stimulation of the labellar chemoreceptors. It is possible therefore to obtain labellar thresholds, and to examine their changes in relation to feeding and deprivation.

Minnich (1931) determined the labellar thresholds to sugars of *Calliphora vomitoria* between 24 h and 6-7 days after feeding to repletion on 1.0 M sucrose. During this period, the sensitivity of the flies to labellar stimulation with sucrose solutions changed little, a number of flies being no more sensitive at the end of the period of deprivation than they were 24 h after feeding. In absolutely parallel experiments referred to earlier, Minnich (1929) found that tarsal thresholds declined markedly during the same period. Haslinger (1935) conducted experiments in which he compared the changes with increasing deprivation in labellar and tarsal thresholds in *Calliphora erythrocephala* (= *C. vicina*). Like Minnich, he made his earliest determination after the flies had been deprived for 24 h. He found that, during the following 9 days, the labellar threshold to sucrose fell somewhat more slowly than the tarsal threshold, but that this relationship was not true for thresholds to all sugars tested. For some sugars, in fact, he found that the labellar threshold fell somewhat more rapidly than the tarsal threshold.

The effects of deprivation on labellar thresholds to sugars have been studied also in *P. regina*. Arab (1957) reported that the labellar sugar thresholds varied with feeding and deprivation in a similar manner to tarsal thresholds, but Getting and Steinhardt (1972) obtained results which would appear to conflict with this finding. These latter workers made no formal determinations of threshold, but found that a 0.1 M sucrose solution applied to a single labellar chemoreceptor hair elicited proboscis extension from flies which had consumed up to 20 μl of 1.0 M sucrose from 0 to 60 min previously. Getting (1971) had previously shown that 0.1 M sucrose when applied to a single hair was just above the minimum concentration required to elicit proboscis extension in a starved fly, and Getting and Steinhardt (1972) concluded, therefore, that the threshold to the input from the labellar chemoreceptors is not greatly elevated by feeding. This conclusion is supported by the results of experiments in which Getting and Steinhardt showed that the relationship between sensory input from a single labellar hair and the activity of the extensor muscle of the haustellum was unaffected by previous feeding. This result shows that there is no threshold regulatory system interposing inhibition into the pathway linking the chemoreceptors and the motor neurones concerned with labellar extension. These electrophysiological results leave only the possibility that the receptors themselves have been rendered less sensitive by feeding, but this seems unlikely in view of the behavioural results and because it could be expected that Getting and Steinhardt would have commented had they found any electrophysiological evidence for markedly reduced receptor sensitivity.

Reference must, however, be made at this point to the findings of Omand (1971) who showed that the sensitivity, as measured electrophysiologically, of the largest type of labellar hair in *P. regina* can, in fact, be influenced by the previous feeding history of the flies. She demonstrated that the sensitivity of these hairs to sucrose, water and sodium chloride increased markedly during 72 h of sugar deprivation following emergence, but that there was little increase if the flies had continuous access to 0.5 M sucrose during the 72 h period. She demonstrated, further, by comparing the responses of the receptors of flies which had been given 0.5 M sucrose for 42 h after 32 h of sugar deprivation with those of the receptors of 26 h deprived flies, that access to sucrose solution not only prevented the receptors from increasing in sensitivity, but that it actually caused a reduction. She showed also that the responsiveness of chemoreceptors increased with increasing period of deprivation in flies which previously had prolonged access to sugar and that access to D-arabinose, a sugar which is nonnutritive for *P. regina*, caused at least as great a reduction as did access to sucrose, glucose or fructose. In all her experiments, Omand deprived the flies of sucrose for several hours before testing and was therefore able to eliminate the possibility that the effect was due to adaptation of the receptors.

It is difficult to assess the relevance of these results to the kinds of threshold changes with feeding and deprivation which have been dealt with so far in this section, since Omand did not determine whether the ingestion of a single meal had any effect on peripheral sensitivity. If, in fact, a single meal were found to be effective in causing changes in the responsiveness of these receptors, or of those on the tarsi, it would be necessary to make considerable modification to the generally accepted scheme put forward to explain the regulation of taste threshold in *P. regina*, since implicit to this scheme is that sensory input, even if not constant (Dethier, 1974), does not vary systematically according to the state of deprivation of the insect. If, however, the kind of effects observed by Omand occur only as a result of prolonged contact, the findings indicate the possible existence of a new dimension to the control mechanisms which, because of the design of previous behavioural experiments, had not been suspected. The interpretation of these results is further complicated by the fact that access to sucrose solution not only rendered the sugar- and water-sensitive cells less responsive but also had a depressing effect on salt-sensitive cells. In this respect, the results of Omand differ from the possibly generally analogous results of Schoonhoven (1969) who showed with *Manduca sexta* that feeding on artificial media containing feeding deterrents reduced the sensitivity of cells sensitive to these compounds but did not affect the sensitivity of another type of cell to sucrose. The finding by Stürckow *et al.* (1967) that the tips of the labellar hairs of *P. regina* can open and close suggests a possible mechanism for variation in receptor sensitivity, but it has not been demonstrated that the state of the hair is related in any way to the state of deprivation.

Also relevant to the question of whether labellar thresholds in *P. regina* alter according to the state of deprivation is the finding of Dethier *et al.* (1965) that the changes in the level of central excitation, as measured by subsequent responsiveness to water, generated by the brief stimulation of one labellar hair with sucrose was greater in deprived flies than in recently sugar fed ones, and that the effect increased up to about 60 h of deprivation. This result appears, at first sight at least, to be incompatible with that of Getting and Steinhardt (1972), but it seems that the two might be reconciled by assuming that the sensory input from the labellar receptors has access to more than one part of the CNS, and that the part directly involved in proboscis extension is not that in which the persevering effect is being generated.

It is difficult to draw any useful conclusion from the conflicting results that have been obtained concerning the effects of feeding and deprivation on the labellar thresholds of blowflies, other than that the whole question needs detailed re-examination. It may well be that there are real differences between the species used, but this explanation cannot be invoked to reconcile the results of Arab (1957) with those of Getting and Steinhardt (1972). It seems possible that differences in the method of application of the stimulus to the labellum might have been responsible for some of the differences between the results. The method used by Getting and Steinhardt is very strictly defined and their results should therefore be the more readily interpretable. The possibility must be recognized, however, that the CNS may not react typically to the input from only one hair since it seems very unlikely that a single hair would commonly be stimulated under natural conditions. The other workers applied the materials in ways which would have stimulated many of the chemoreceptor hairs representative of different classes (Wilczek, 1967) and perhaps the interpseudotracheal papillae as well.

Manjra (1971) examined the effects of deprivation and the consumption of sucrose solution or a mixture of blood and sucrose on the labellar threshold to sugar of the mosquito *Culiseta inornata*. Thresholds were tested by applying sucrose solution to 3 or 4 labellar hairs and the spreading of the labellum was taken as a positive response. It was found that threshold to sucrose decreased with increasing deprivation and that a meal of 0.5 M sucrose, which is directed to the diverticulum of the gut, and blood plus 0.03 M sucrose, which goes immediately to the mid-gut, both caused substantial increases in threshold, there being about a 16-fold increase after sugar feeding and a 7-fold increase after the blood meal. In experiments designed to elucidate the mechanisms involved, Manjra showed that threshold did not rise after sugar feeding if the ingested material was immediately withdrawn from the diverticulum with a hypodermic needle. The findings are consistent with the view that distension of the abdomen is important in bringing about elevation of labellar thresholds but there is a possibility that the situation might be more complex, with different mechanisms operating after sugar and blood meals.

Blaney and Chapman (1970), as part of an investigation into the role of the maxillary palps in the feeding behaviour of late instar larvae of *Locusta migratoria*, showed that, as the post-feeding period of deprivation increased from less than 1 h to between 2 and 3 h, the proportion of insects which palpated upon the normally rejected plant *Bellis perennis*, but which did not proceed to the next stage in the feeding sequence, biting, fell from about 68 to 10 per cent. It can be said, therefore, that the threshold of response to stimulation of sensilla of the palps changes according to the state of deprivation. The experiment provided no evidence, however, as to whether there was a decrease in threshold to feeding stimulants or an increase in threshold to deterrents or both. Experiments with extracts showed clearly that the relevant stimuli were of a chemical nature.

Blaney and Chapman (1969) examined the fine structure of the terminal sensilla of the maxillary palps of the desert locust, *Schistocerca gregaria*, and observed that the pores at their tips were sometimes open and sometimes closed. Their suggestion that this might be related to feeding and deprivation was followed up by Bernays *et al.* (1972), Bernays and Chapman (1972a), and Bernays and Mordue (1973). These workers, working with larvae of *L. migratoria*, found that the total electrical resistance across the palps and the proportion of sensilla which would not respond electrophysiologically to sodium chloride decreased sharply during the first 2 h of deprivation, but that those sensilla which did respond had normal sensitivity. They found, further, that the increase in resistance could be induced by causing distension of the fore-gut by the introduction of agar via the oesophagus through a cannula, but that there was an increase in resistance only if the crop was completely filled. Experiments involving the sectioning of various nerves of the stomatogastric nervous system and the injection of haemolymph from fed insects, or CC homogenate into the haemocoel of deprived locusts, showed that information about crop distension is relayed via the posterior pharyngeal nerves and the frontal connectives to the brain and that material released by the storage lobes of the CC is responsible for causing closure of the sensilla.

The precise role of the closure of the sensilla following feeding is unknown. It is not clear, for instance, whether the CNS is sensitive to differences in the total sensory input from a number of receptors, or whether its response is more closely related to the level of sensory input from individual receptors. Furthermore, even if the total amount of input is important, it is not known whether there might not be in addition a central mechanism involved in the control of the threshold of response to stimulation of the receptors of the palps. It can be said, however, that the timing of the opening and closing of the sensilla and the frequency of feeding by larvae of *L. migratoria* seem to be concordant, in that the interval between meals is usually of the order of about 1 h and almost always less than 2 h (Blaney and Chapman, 1970; Bernays *et al.*, 1972) by which time the number of sensilla in the open state has again reached a maximum. It is known that larvae of *L. migratoria* spend a small but significant percentage of their time palpating (Blaney and Chapman, 1970) during the inter-meal period, and it could well be that the closure of the palp sensilla plays a role in excluding inappropriate input from reaching the CNS.

Changes with feeding and deprivation in the probing responses of some blood-sucking insects to volatile stimuli have been studied. Hopkins (1964) examined the responses of the stable fly, *Stomoxys calcitrans*, to ammonia vapour and found that the proportion of flies which responded by probing their substrate was greatly reduced immediately after feeding to repletion on either blood or 0.5 M sucrose, and that the proportion responding increased progressively over 30 h of deprivation following a blood meal. Little is known about the regulatory mechanisms involved. It was shown that the ingestion of 2.0 M sucrose solution reduced responsiveness more effectively than did 0.18 M or 0.25 M sucrose, even though the volumes consumed were comparable, being of the order of 2 μl in all instances. This finding shows that any input concerning the volume of ingested material is certainly not the only factor involved in bringing about the reduction in responsiveness, and suggests that it is unimportant relative to input concerning the composition of the ingested material. The finding that most flies were still responsive immediately after being partially satiated by the ingestion of 5 μl of blood is consistent with this view, but it possibly must be admitted that the continued probing by these flies might be an indication that the partial feeding had produced in them a state of central excitation which rendered them responsive to odours typical of blood sources. Finally, it should be noted that Hopkins believed that the observed responses were to ammonia alone, but Gatehouse (1967) has shown that the response is to a synergistic effect produced by the presence of ammonia and a sudden rise in the moisture content of the air.

Khan and Maibach (1970) studied the effects of the ingestion of sugar solution and water on the probing response of females of the mosquito *Aedes aegypti* in response to either the human forearm or a warm moist surface. All classes of mosquito tested showed a high level of responsiveness to the human arm irrespective of whether they had been kept supplied with the sugar solution or water or whether they had been deprived of one or both of these materials. The lowest response was obtained from insects which had been allowed to ingest sugar solution at the start of the experimental period after having been deprived of both sugar and water for 24 h. The effects of feeding and deprivation on the responses of mosquitoes to the warm moist surface were much more marked. Insects which had previously been deprived of food and water for 24 h remained very responsive throughout a following 6-h test period. The responsiveness, after 1 h of the test period, was markedly lower in insects which had had access to sugar solution up to the beginning of the experiment, lower still in ones which had been allowed to ingest water at the beginning of the experimental period after having been deprived of both sugar and water, and even lower in similarly deprived mosquitoes which had been given access to sugar solution rather than water at the beginning of the experimental period. The responsiveness of the water-fed mosquitoes increased somewhat more rapidly during the 6-h experimental period than did that of either group fed sugar solution.

It is clear from these results that the responsiveness of the mosquitoes to water vapour was affected by the ingestion of water or sugar solution, but that they remained responsive to other volatiles emanating from the human arm irrespective of their state of sugar or water deprivation. Nothing definite can be said about the physiological mechanisms underlying the regulation of responsiveness to water vapour, but the finding that it increased more rapidly after the ingestion of water than after sugar solution suggests that, if the crop emptying rate of mosquitoes is dependent upon the osmotic pressure of the ingested material as it is in *P. regina* (Gelperin, 1966a) and *Periplaneta americana* (Treherne, 1957), the level of responsiveness to water vapour might well be related to the volume of material in the diverticulum.

In further experiments, Khan and Maibach (1971) compared the effects of the ingestion of blood with those of the ingestion of sugar solution on the probing responses of *Aedes aegypti* to the human palm and a moist warm surface. They found that mosquitoes which were first allowed to feed on sugar solution, thus filling the diverticulum, and then on blood, thus filling the mid-gut, became unresponsive to both the palm and the artificial target and remained so until the time of oviposition, if continuously supplied with sucrose solution. They found, however, that when no food or water was available, the insects began to become responsive to both stimuli after about the same period of deprivation, the required period being shorter at a low than at a high ambient humidity. When these deprived mosquitoes were allowed to consume water, they again became immediately unresponsive to both kinds of target. Insects kept at the low humidity were equally responsive to the two types of target but those kept at the higher humidity were more responsive to the human palm than to the warm moist artificial target. These results indicate that the probing by mosquitoes which had had a period of deprivation following the ingestion of both sugar solution and blood was primarily a response to water vapour but that under some circumstances the palm was a more powerful stimulus than the artificial target. Mosquitoes given only a blood meal were initially unresponsive to both targets, but became progressively more responsive during the following 24 h, the response throughout the whole period of deprivation being somewhat greater to the palm of the hand than to the artificial target.

On the basis of these results, Khan and Maibach (1971) suggest that the feedback controlling the responsiveness of mosquitoes to stimuli emanating from the skin is not controlled by the volume of material in the diverticulum or by distension of the abdominal wall, because probing occurs even when the diverticulum is full of sugar solution. The finding that completely deprived mosquitoes responded more readily to the palm or forearm than did sugar-fed ones suggests, however, that one or both of these factors may play some role, even though it is probable that, as suggested by Khan and Maibach, mid-gut distension is a more important factor. It is possible that the reduced responsiveness to the human arm of mosquitoes fed sugar solution might be attributable to their lack of responsiveness to the water vapour component of the total volatiles emanating from the target, but the low level of response of these mosquitoes immediately after feeding on sugar solution suggests that this is not, in fact, the complete explanation.

Brady (1973) studied changes with deprivation in the proportion of males of the tsetse fly, *Glossina morsitans*, making a probing response when a warm ball of plastic foam was brought into contact with their tarsi. He found that the percentage of mature flies responding increased linearly with time, from a low initial value, over a period of 4 days of deprivation, and that that of teneral flies responding increased over a 3-day period from a higher initial value of 30-40 per cent. He found also that the time taken for the flies to respond decreased with increasing deprivation, thus demonstrating that the responsiveness of individuals increased progressively. He pointed out that these changes in responsiveness were well correlated with those he demonstrated in the locomotor activity (Brady, 1972a) and the responsiveness to visual stimuli (Brady, 1972b) of *G. morsitans*.

Holling (1966) obtained a great deal of information on the relationship between deprivation and the responsiveness of mantids, notably *Hierodula crassa*, to visual stimuli provided by the potential prey. By bringing a housefly gradually closer to the front of the mantid's head, he showed that a mantid would neither stalk nor strike at the prey within the first 8 h after satiation, but that from 8 h onwards the distance over which the mantid would do so increased steadily, until by 48 h it would stalk and strike at prey 6-7 cm away. In a related experiment, he found that the maximum distance at which a mantid showed interest in prey by making head movements increased, over 48 h, from about 4 to 18 cm. In experiments with *Mantis religiosa*, Holling examined the effect of deprivation on the number of strikes that a mantid would make towards a fly dangling behind a glass screen before it gave up, and found that the number of strikes increased with increasing deprivation. All these changes reported by Holling seem very largely analogous with the changes in responsiveness to chemical stimuli discussed previously and it seems fair to state that visual thresholds in mantids change with feeding and deprivation.

2.4 REGULATION OF INGESTION

In this section, the regulation of meal size and rate of ingestion are considered with special reference to the effects of feeding and deprivation and the kind of food being consumed.

2.4.1 *Regulation of meal size*

The term "meal" as used here denotes a discrete intake of either food or water by an insect of the kind that spends only a relatively small proportion of its time feeding. The concept of a meal implies that the insect shall, in response to physiological regulatory mechanisms, cease to feed after ingesting a certain amount of food or after feeding for a particular period, even though food still remains. It implies also that there is a period following the meal during which the insect will not attempt to feed if left to behave normally with free access to food. The ingestion of food during a meal may be continuous or it may consist of a bout of not quite continuous feeding in which, usually, an initial continuous period of ingestion is followed, within a short time, by a number of shorter periods.

It has been shown for a number of species that meal size varies according to the state of deprivation of the insect and the type of material being ingested. The mechanisms underlying these relationships have been extensively studied in blowflies and locusts, and the information obtained indicates that the control mechanisms are fairly complex. For this reason I shall begin by discussing a conceptual framework against which the findings relating to particular insects might usefully be viewed.

Central to the problem of understanding the regulation of meal size is the identification of factors that cause an insect to terminate a meal, and, in the scheme outlined here, factors involved are categorized according to the precise role they might play in bringing about the cessation of feeding. This scheme, like others proposed for particular insects (Dethier, 1966; Dethier and Gelperin, 1967; Dethier, 1969; Bernays and Chapman, 1972b, 1973, 1974), is based on the premise that a meal is terminated when inhibitory influences in the CNS are sufficiently powerful to oppose input-generated or endogenous influences which are excitatory to feeding. Any inhibitory influences generated would be the result of inhibitory inputs from sensory systems monitoring various indicators of the insect's nutritional state, from chemoreceptors sensitive to any deterrents in the food and perhaps from mechanoreceptors stimulated during ingestion. Any inhibitory input might produce only instantaneous effects in the CNS, in which case the magnitude of the inhibitory influence generated by it would reflect, at any particular time, only the current level of inhibitory input. It might, on the other hand, produce perseverating effects in the CNS, in which case the magnitude of the resulting influence would reflect a time-weighted integral of past and current levels of input. Excitatory influences would similarly be generated by immediate or perseverating effects of excitatory input, usually from chemoreceptors responding to feeding stimulants. In addition, the possibility exists that ingestion might result also from neural activity which is endogenous to the CNS and which is released by suitable input. In these circumstances, ingestion would, once started, be expected to continue, even in the absence of continuing excitatory input, until switched off by inhibitory influences.

I will first outline the scheme proposed for the regulation of meal size in the situation in which endogenous influences are absent and in which therefore only excitatory and inhibitory influences generated by the immediate or perseverating effects of sensory input play a part. According to this, factors involved in bringing about the cessation of feeding may be divided into two basic categories. The first contains those which produce in the CNS influences whose levels are changing at the time of termination of a meal. Factors of this type are defined here as playing an *active* role in bringing about the cessation of feeding. The second category contains factors which produce influences whose levels are not changing at the end of the meal. These may be factors that produce influences in the CNS which remain constant throughout a meal, at a level depending on the initial state of deprivation of the insect, or ones that produce levels of excitatory or inhibitory influences which change early in the meal, but which have ceased to do so by the time the insect stops feeding. Such factors are defined as playing a *passive* role since they influence the amount by which factors playing an active role must change in order to bring about the cessation of feeding.

Figure 1 shows a hypothetical example illustrating one way in which active and passive influences in the CNS might interact to determine the

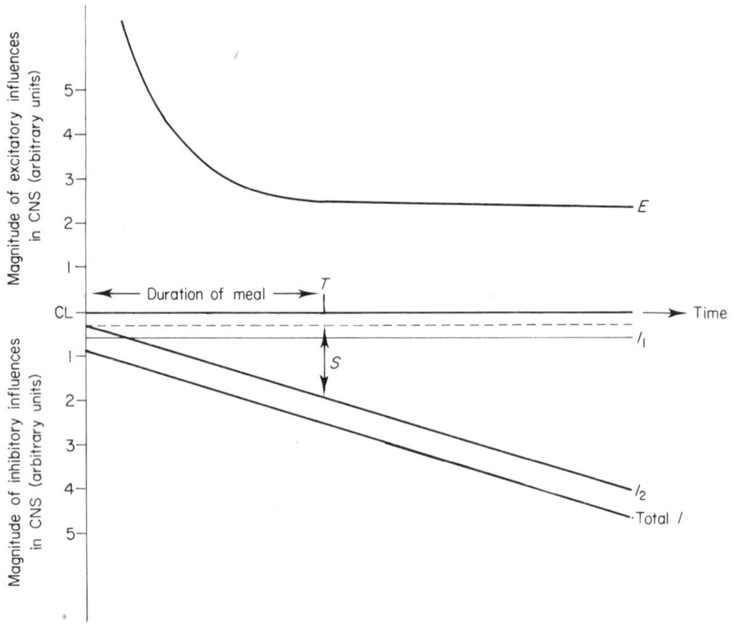

Fig. 1. Hypothetical example illustrating a way in which excitatory and inhibitory influences generated in the CNS by excitatory and inhibitory inputs might interact to determine meal duration and size. E represents the time course of change in the magnitude of an excitatory influence and I_1 and I_2 those of two inhibitory influences with differing characteristics. Total I represents the changes in time of the total magnitude of inhibitory influences. The meal is terminated at time T when the algebraic sum of E and total I reaches a critical level (CL). Under certain circumstances, S may be an indication of meal size (see text). The lines representing the magnitudes of the various influences are, for the sake of clarity, extrapolated beyond the point at which the meal is terminated.

duration of a meal. In this, there is one factor responsible for the generation of an excitatory influence in the CNS and two responsible for generating inhibitory influences. The excitatory influence (E) initially declines in magnitude but ceases to do so at some time before the termination of the meal, and therefore plays a passive role in bringing about the cessation of feeding. Of the two inhibitory influences, one (I_1) remains constant throughout the meal and hence also plays a passive role. The second inhibitory influence (I_2) increases in magnitude in a linear manner throughout the period of ingestion, and therefore plays an active role. The meal is terminated at time T when the algebraic sum of the three influences reaches the critical level (CL).

An absolute measure of meal size can be obtained from the duration of the meal and the mean rate of ingestion, but a less precise, but still useful indication of meal size can often be obtained directly from diagrams such as Fig. 1. This is possible, however, only when at least one of the inhibitory influences is generated by a sensory system whose input increases with the amount of material ingested. It seems reasonable, however, to assume that an inhibitory input such as I_2, which increases progressively while an insect is feeding would usually, or perhaps always, be generated by input from such a monitoring system. If this assumption is made for I_2, it follows that the amount (S) by which its magnitude increases during a meal is an indication of the size of the meal, except possibly when the input generates very long-lasting perseverating effects.

The situation, in which the motor output involved in ingestion is the result only of endogenous activity released by excitatory input, differs from that discussed above only in that the timing of the cessation would be completely under the control of inhibitory influences of feeding. The concept of the release of endogenous activity implies that motor output is evoked by any level of excitatory input above a threshold value, and that the amount and duration of the resulting activity is not influenced by the amount by which threshold is exceeded. It follows that, in these circumstances, meal size would not vary according to the level of excitatory input elicited by the food, provided this is above threshold. Stimulation elicited by the food could only influence meal size by means of inhibitory inputs elicited by feeding deterrents and possibly by mechanoreceptor input related to the food's physical properties. Endogenous activity may be represented, in a diagram such as Fig. 1, as being equivalent to an excitatory influence which has a constant magnitude equal to that of the inhibitory influence needed to switch off the activity.

The regulation of sizes of meals of sugar solution taken by the blowfly, *P. regina*, has been extensively studied and the dependence of meal size, as measured by the volume or duration of the initial uninterrupted intake, on both the period of previous deprivation and the concentration and type of sugar being ingested is well documented. Gelperin (1966b) showed that the amount of 1.0 M fructose consumed at a single meal increased with increasing periods of sugar deprivation and that the time course of this increase varied according to the concentration and volume of the sugar solution consumed prior to deprivation. Dethier *et al.* (1956) showed that the sizes of meals taken by flies which had been deprived of sugar for 24 h varied according to the concentration and type of sugar offered, the more stimulating the solution the greater being the duration of the meal and, generally, its volume. They demonstrated that the rate of ingestion of solution of any one sugar increased with increasing concentration up to the point where viscosity began to interfere. Because of its higher viscosity, the rate of intake of 1.0 M sucrose was, for example, reduced to such an extent that, even though flies fed for substantially longer periods than on 0.5 M sucrose, they ingested only about the same volume.

Adaptation of sugar-sensitive receptors and similar central phenomena, which result in a reduction in the level of excitatory sensory input reaching, and hence the magnitude of excitatory influences in, that part of the CNS concerned with ingestion, has long been believed to play an important role in determining the point at which *P. regina* terminates the initial period of continuous intake of sugar solution, and at one time was considered to be the only factor involved. The most relevant data in support of this belief were obtained by Dethier *et al.* (1956) who determined the duration and volume of meals taken by 24-h deprived flies of 1.0 M sucrose solution containing various concentrations of glycerol, which is believed to be nonstimulating to the taste receptors of this fly. They found that the duration of meals was not influenced by the addition of glycerol, even though the greater viscosity of the solutions containing higher concentrations of this material caused a reduction in the rate of intake and hence the volume of fluid ingested. Taken at face value, this result would seem to indicate that feedback concerning the volume of fluid imbibed played no part in determining the duration of feeding, and that the period for which a meal continued in previously starved flies was determined solely by the nature of the sensory input from the chemoreceptors. More recently, however, evidence has been obtained that feedback from internal sources does, in fact, play a role in determining meal size (Dethier and Gelperin, 1967; Gelperin, 1971b, 1972), but details of these findings have not been integrated with the previous evidence for the importance of the role of adaptation. In attempting this integration, I will begin by discussing the evidence for the involvement of feedback from internal sources and will then relate this to the previously mentioned evidence for the role of adaptation.

A number of investigations have been made of the effects of the sectioning of various nerves on the subsequent feeding behaviour of *P. regina* and it has been found that the severance of the recurrent nerve, the median abdominal nerve, or both causes flies to become markedly hyperphagic when left with access to a sugar solution for a period of from 15 to 60 min (Dethier and Bodenstein, 1958; Dethier and Gelperin, 1967; Gelperin, 1971b, 1972; Belzer, 1970). The precise basis for the increased intake of the operated flies is known to vary according to the particular type of operation performed, hyperphagia having been shown to result from an increase in the duration of the initial period of continuous intake, in the duration and frequency of periods of supplementary intake or in both (Dethier and Gelperin, 1967; Gelperin, 1972). The data concerning meal size in *P. regina* that have been discussed up to this point deal specifically with the duration of the initial period of continuous intake following a substantial period of deprivation. For this reason, it is strictly relevant to the present discussion to consider only the effects of the various operations on the duration of this intake. The effects of the operations on the supplementary periods of ingestion will be discussed later.

It must be pointed out first that the number of individual insects contributing to the published data concerning the effects of nerve section, which refers strictly to the duration or size of this initial period of ingestion, is disconcertingly small. The results, however, merit detailed consideration because of the magnitude of some of the effects obtained, and because it must be presumed that the authors believed their results to be typical. Dethier and Gelperin (1967) obtained information about the duration of the first continuous period of ingestion in 24 h starved flies which had undergone various operations. They presented two figures each depicting the feeding patterns of single operated and control flies. These showed that a fly sham-operated for recurrent nerve section imbibed 2.0 M sucrose continuously for 99 s and that one which had had its recurrent nerve sectioned imbibed for 116 s. These findings, which suggest that recurrent nerve section does not alter the period of the initial continuous intake are supported by the statement of Dethier and Bodenstein (1958) that flies which had suffered recurrent nerve section, when feeding *ad lib.*, "began to feed in a normal fashion and ceased, as may be expected, when their receptors became adapted". Dethier and Gelperin also examined the effects of the sectioning of all the abdominal nerves and stated that duration of the initial period of ingestion of an undisclosed number of operated flies was about 4 times longer than that of controls. Gelperin (1972) published the detailed feeding record of one fly with its median abdominal nerve sectioned which shows that the fly's initial period of intake lasted for 130 s as opposed to 60 s for two controls. He found, however, that a fly with both the recurrent nerve and the median abdominal nerve sectioned fed continuously for more than 20 min, having burst at the 13 min mark. Since a number of similarly operated flies burst during *ad lib.* access to sugar solution, it seems clear that this double operation causes flies to lose all control over meal size. These results suggest that the severance of the two nerves deprives the CNS of all inhibitory input which might play a role in bringing about the cessation of feeding. Although there is some possibility that the CNS might also receive other relevant input, which when acting alone has no controlling effect, the seemingly justified assumption is made in the rest of this discussion that the sectioning of the median abdominal nerve allows an estimate to be made of the regulator effects of input via the recurrent nerve and vice versa.

Electrophysiological recordings have shown that the levels of sensory input reaching the brain via the recurrent nerve (Gelperin, 1967, 1972) and the median abdominal nerve (Gelperin, 1971b) increase with increasing distension of the fore-gut and abdomen, so it seems reasonable to believe that a lack of input via these nerves is an indication to the CNS that the fly has, respectively, no food in the fore-gut or crop. It seems, however, that the level of excitatory input via either one of these nerves generates an inhibitory influence in the CNS of sufficient magnitude to oppose the excitatory influence generated by input from external chemoreceptors. The results suggest that the control mechanisms which bring about the cessation of the initial intake of food operate normally when the CNS is receiving input via the median abdominal nerve from the abdominal stretch receptors (Gelperin, 1971b) but that there is some loss of control when it receives inhibitory input only via the recurrent nerve. These findings would seem to indicate that input via the median abdominal nerve normally plays a role in bringing about the termination of ingestion, but that the input via the recurrent nerve becomes important only when the ingestion of a larger than normal meal produces a degree of distension in the fore-gut which might never occur during normal ingestion. This degree of distension might be more typical of that during crop emptying following the ingestion of a concentrated sugar solution. It seems possible that the inhibitory effect of input via the recurrent nerve which is revealed by the sectioning of the median abdominal nerve might not be part of the regulatory mechanism which normally determines the duration of and the amount of material taken in the first period of continuous ingestion, but rather that it is part of the mechanism responsible for making fed flies unresponsive to further contact with food during the normal inter-meal period.

The results just discussed show that inhibitory inputs play an important part in limiting meal size in *P. regina*, and in particular that an active role is played by influences generated by input which varies according to the degree of distension of the abdomen. This conflicts with the conclusion that receptor adaptation is the sole factor responsible for the cessation of feeding in *P. regina*, which was based on the previously mentioned results of Dethier *et al.* (1956). It is necessary therefore to seek other possible explanations for the finding that increased viscosity due to the addition of glycerol did not result in the prolongation of the duration of feeding on a particular sucrose solution, even though it caused a reduction in the rate of intake. First, it can be suggested that the input via the median abdominal nerve may not increase indefinitely with increasing volume. Gelperin (1971b) monitored input via the lateral branch of this nerve throughout the intake of only 6 μl of solution and it is possible, therefore, that input might not increase much beyond that seen at the end of this atypically small meal. This explanation seems unlikely, however, since the amount of sensory input was rising rapidly at the time feeding ceased. Another possible explanation is that the difference in viscosity could well be affecting the characteristics of inhibitory input from the fore-gut receptors by affecting the degree of fore-gut distension during ingestion. If this is the explanation, the similarities in duration observed by Dethier *et al.* (1956) would have been merely fortuitous. In addition, it is known that the cibarial pump of blowflies possesses mechanoreceptors which might allow the fly to monitor the viscosity of the fluid being ingested (Rice, 1970, 1973). It is possible, therefore, that feedback concerning the greater amount of work required to be done by the pump during the ingestion of a more viscous solution might itself be inhibitory to further ingestion. It would be interesting, in this regard, to know whether a fly with sectioned recurrent and median abdominal nerves would feed indefinitely on a very viscous solution. Another possibility is that interactions in the solution, or even at the receptor sites, between the molecules of sucrose and glycerol might cause the apparent concentration of sucrose, as seen by the receptors, to decrease with increasing glycerol concentration. The effect of any happening of this kind would be that, because of the reduction in the level of excitatory input from the external chemoreceptors, the amount of inhibitory input from internal sources required to bring about the termination of a meal would decrease with increasing glycerol concentration. In this event the cancelling out of the two effects of the presence of glycerol in the sucrose solution could well, wholly or partly, explain the similarity in the duration of ingestion.

Even though the evidence suggests that the effects of sensory adaptation might not play the dominant role previously assigned to them in bringing about the cessation of feeding, it seems likely that they must play some part, since all sugar-sensitive receptors of *P. regina* which have been studied electrophysiologically are known to adapt. The level of excitatory input reaching the CNS from these receptors would, therefore, be less at the end of a meal than at the beginning. The question which needs to be asked is whether the role is an active or a passive one. For the role to be classified as being merely passive, it is necessary, not only that the total amount of excitatory input from chemoreceptors would have ceased declining before the end of the meal, but that any perseverating effects in the CNS of previous changes in input would also have ceased.

Before making an attempt to assess the precise role of adaptation in bringing about the termination of feeding, it is necessary to consider which receptors would be providing input during ingestion. There is evidence for the involvement of four broad categories of contact chemoreceptors in the pre-ingestion and ingestion behaviour of *P. regina*. These are the tarsal chemoreceptor hairs, the labellar chemoreceptor hairs, the interpseudotracheal papillae (Dethier *et al.*, 1956) and receptors in the pharyngeal region for which behavioural evidence was obtained by Arab (1957). It is likely that these pharyngeal receptors are those described by Rice (1973) as being in the wall of the cibarial pump. Of these sets of receptors, the interpseudotracheal papillae and the pharyngeal receptors would certainly be receiving stimulation throughout a meal as would probably the receptors on one or more of the tarsi. Most of the labellar receptors no longer contact the solution once the labellar lobes are spread (Dethier, 1967) and their contribution, whatever their adaptation characteristics, would be largely a pulse of input just before ingestion begins.

A limited amount of electrophysiological data is available about the adaptation characteristics of the sugar-sensitive cells of some tarsal chemoreceptor hairs (McCutchan, 1969) and the interpseudotracheal papillae (Dethier and Hanson, 1965) of *P. regina*. McCutchan illustrated the adaptation patterns of 5 tarsal hairs representative of class D of Grabowski and Dethier (1954). The most rapidly adapting of these reached a very low rate of firing after less than 1 s of stimulation, whereas the most slowly adapting reached half its initial rate by about 300 ms. In addition, McCutchan states that the response of the class B hairs adapted to zero within half a second or less. Although a number of classes of hairs, including some of the most numerous ones, remain uncharacterized electrophysiologically, it seems safe to deduce that the greatest rate of decrease in input from the tarsal receptors occurs within the first second of a meal.

The rate of adaptation of the interpseudotracheal papillae is generally slower than for the tarsal receptors. Dethier and Hanson (1965) found that the rate of firing of some cells did not decline to half the original rate until after about 3 s of stimulation, but that most had done so by 300-400 ms. It seems, therefore, that input from these relatively slowly adapting receptors also has its maximum rate of decline within the early part of a meal.

Dethier (1952) obtained behavioural data relevant to an assessment of the possible contribution of input from tarsal receptors. He immersed the tarsi of *P. regina* in various concentrations of sucrose solution and recorded time between extension and retraction of the proboscis, which at no time was allowed to touch the solution. The period of extension increased with increasing concentration, but even with 1.0 M sucrose it was only about 12 s which is short compared with the period of about 50 s (Dethier *et al.*, 1956) taken by 24 h starved *P. regina* to complete a meal of this solution. Dethier's finding that a fly, which was no longer extending its proboscis in response to the stimulation of one foreleg, would not respond when the other foreleg was stimulated immediately afterwards indicates that refractoriness of central units was involved, at least to some extent, in the termination of the original period of proboscis extension. Taken at face value, this result would suggest that the input from tarsal receptors reaching the feeding centres of the CNS falls to a low level considerably before the termination of a normal meal. It might be argued, however, that the retraction of the proboscis in these circumstances is not because of receptor adaptation, or a similar central phenomenon, but may rather be a positive response to the situation in which a fly receives adequate tarsal stimulation but does not contact a stimulating solution with its labellar hairs. If this is so, it could well be that the period of extension would in fact be related to the strength of the stimulus applied to the tarsal hairs, and at this stage it is not possible, therefore, to exclude this second possibility. If, however, the first explanation is correct, the amount of effective sensory input reaching the relevant part of the CNS from the tarsal receptors must have fallen to a very low level, since it is known that the stimulation of a single labellar hair with 0.1 M sucrose elicits proboscis extension (Getting and Steinhardt, 1972). Furthermore, Getting (1971) demonstrated that proboscis extension, as measured by activity in the extensor aductor muscle complex, can be elicited by as few as two spikes in 20 ms.

Although the available evidence is very fragmentary, it suggests that the major reduction in total input from the external sugar-sensitive receptors occurs relatively early in the course of a meal and that the level is probably changing only slowly, if at all, when the meal is terminated. If, therefore, receptor input were having only an instantaneous effect on the part of the nervous system concerned with ingestion, its effect could well be playing a largely passive role in determining the point of termination of a meal. Evidence has been obtained, however, that chemoreceptor input generates perseverating effects in the CNS of *P. regina*.

Dethier *et al.* (1965, 1968) demonstrated that the stimulation of labellar and tarsal chemoreceptors with the feeding stimulant, sucrose, generated a state of excitation in the CNS which could still be detected after 45 s and that a central inhibitory state was generated by the feeding deterrent sodium chloride. The existence of these central states was demonstrated in an ingenious series of experiments in which an investigation was made of the effects of previous stimulation with these compounds on the subsequent responsiveness of flies to water. It was shown that flies which had previously failed to respond by proboscis extension to the application of water to a single labellar hair would do so when tested again shortly after another labellar or tarsal hair had been stimulated with a sugar solution. A less marked excitatory effect was obtained when stimulation with sodium chloride was intercalated between the initial stimulation with water and the stimulation with sucrose.

Evidence of a different kind which shows that contact with food or the act of feeding produces perseverating effects was obtained by Dethier (1957). He showed that a walking fly would make frequent clockwise and anticlockwise turning for some time after ingesting an amount of sugar solution which had been insufficient to allow it to feed to repletion. The intensity and duration of this turning behaviour was stated to vary according to the concentration of the sugar solution and the initial state of deprivation of the fly, but no data on the actual period for which the flies exhibited this characteristic pattern of locomotor activity were given. It was clear, however, that it must have continued for a considerable time relative to the duration of a normal meal.

It is clear from the results of Dethier *et al.* (1965, 1968) that sensory input from external sugar-sensitive chemoreceptors generates perseverating excitatory effects in the CNS which last for a considerable time. This finding means that changes occurring in the level of actual input from the chemoreceptors would be being reflected by the occurrence of changes in the central excitatory state, which would lag behind those in input by amounts which would depend on the decay time of the perseverating effects. If the decay time were such that the level of central excitation was still declining when feeding ceased, it could be said that sensory adaptation, through the perseverating effects generated in the CNS, was playing an active role in bringing about the cessation of feeding.

In summary, it can be said that there is a distinct possibility that changes in the magnitude of excitatory influences resulting from receptor adaptation, and in the case of tarsal receptors, similar central phenomena, may play an active role in determining the point at which a meal is terminated even though the actual changes in input may well have been more or less complete some time before the cessation of feeding. Finally, it should be noted that the finding of Dethier *et al.* (1965, 1968), that brief stimulation of labellar hairs generated a state of central excitation, suggests that even the short period of stimulation of these hairs which occurs before the opening of the labellar lobes may have contributed to any perseverating effects. Input from even these hairs may therefore play some role in determining the duration of feeding, even though they are not actually contributing input during ingestion.

A result obtained by Dethier and Gelperin (1967), which has not been mentioned so far, gives further insight of perhaps a different kind into the physiological mechanisms responsible for bringing about the cessation of feeding. They found that flies which had had the ventral nerve cord cut between the brain and the thoracic ganglionic mass fed continuously until they burst. This contrasts with the finding, which was discussed earlier, that flies with sectioned abdominal nerves ceased feeding after a period of about 2 to 4 times that of controls. In terms of sensory input from the abdomen, the two types of operation are equivalent, with the regulatory mechanism for meal size probably being left, as argued earlier, under the sole control of input via the recurrent nerve. Apparently the severance of the ventral nerve cord renders this mechanism inoperative. Dethier and Gelperin interpreted this as meaning that input from the thoracic ganglionic mass, deriving either from the thoracic locomotor centre or from tarsal receptors, plays a part in bringing about the cessation of feeding. This result, together with the general observation that flies do not feed and walk at the same time, suggests that the cessation of feeding might, in the final analysis, be brought about by the commencement of the antagonistic activity, walking, and that it is walking, rather than proboscis retraction, which is the direct result of the previously discussed interplay between excitatory and inhibitory influences. Alternatively, it might be that inhibitory input from mechanoreceptors on the tarsi plays an important role in terminating a meal.

Using as basis the general scheme outlined earlier in this section, I will now discuss the effects of deprivation and type of sugar solution on meal size and duration of *P. regina* in relation to the available information about the factors responsible for bringing about the cessation of feeding. The finding that meal duration was positively correlated with the stimulating power of the sugar solution being ingested is clearly generally explicable in terms of the different levels of excitatory input elicited by the solution from the total population of sugar-sensitive receptors which receive stimulation during feeding. It is typical of phasic-tonic receptors that more powerful stimuli not only elicit a greater initial response than less powerful ones, but that the level of the tonic response is similarly related to the strength of the stimulus. This is known to be true for the salt-sensitive cell of the chemoreceptors of *P. regina* (Gillary, 1966) and is almost certainly so for the sugar-sensitive cells. At any given time after the commencement of a meal, therefore, the total level of excitatory input elicited from all sugar-sensitive receptors, and hence the magnitude of the resulting excitatory influences in the CNS, would be greater for a highly stimulating solution than for a less stimulating one, irrespective of whether the excitatory influences were the result of only current input or whether perseverating effects were being generated. The magnitude of the inhibitory influences required to bring about the termination of feeding would therefore have to be higher in order to bring about the termination of meals of more highly stimulating foods. A major component of the input responsible for the generation of the inhibitory influences in the CNS, which bring about the cessation of feeding in *P. regina*, is known to originate from a sensory system in the abdomen which monitors the amount consumed. Differences in rate of ingestion due to differences in stimulating power or viscosity of the solution would therefore influence the rate of increase of the magnitude of the inhibitory influence, and hence of the duration of meals. Meals of a material consumed more rapidly than another would therefore be of shorter duration than would be expected from a consideration of the relative amounts of excitation they elicit from the chemoreceptors.

Figure 2 depicts three situations which illustrate the likely bases of the observed relationships between type of solution and meal duration and size. The first is that of a fly feeding on a solution of low stimulating power and low viscosity, the second of one feeding on a more stimulating solution, still with relatively low viscosity, and the third of a fly feeding on a highly stimulating solution, the rate of intake of which is limited by viscosity. It is assumed, for simplicity, that the inhibitory influences in the CNS increase linearly throughout the meal and that input from a sensory system monitoring the amount consumed is responsible for the increase. A meal is terminated when the algebraic sum of the excitatory (E) and inhibitory (I) influences is equal to the critical level (CL). Time T is the duration of the

Fig. 2. Hypothetical illustration of the likely basis for the relationship between the concentration of sugar solution and meal duration and size in *P. regina*. E represents the magnitude of excitatory influences in the CNS generated by the different levels of excitatory input elicited by the solutions from the external chemoreceptors. I represents the magnitude of inhibitory influences generated by input from a sensory system monitoring the amount of solution consumed. (a) depicts the situation in which the solution has low stimulating power and low viscosity, (b) that in which the solution is more stimulating but still with low viscosity, the rate of ingestion therefore being higher than in (a). (c) depicts the situation in which the solution is highly stimulating but in which the rate of ingestion is limited by the high viscosity of the solution. It can be seen that meal duration (T) increases progressively with increasing stimulating power of the solution, but that the relationship between meal size (as indicated by the magnitude of S) and duration is complicated by the effects of different rates of intake.

meal and, because of the nature of the input generating the inhibitory influences, *S* gives a relative measure of meal size. It can be seen that the relationship between meal duration and size, and characteristics of the stimulating solution predicted by the hypothetical scheme are similar to those which were found experimentally by Dethier *et al.* (1956). According to the hypothetical scheme given here, and in practice, both duration and meal size increase with increasing stimulus strength up to the point where the effect of increasing viscosity intervenes, after which duration continues to increase but the size of the meal does so less rapidly or not at all.

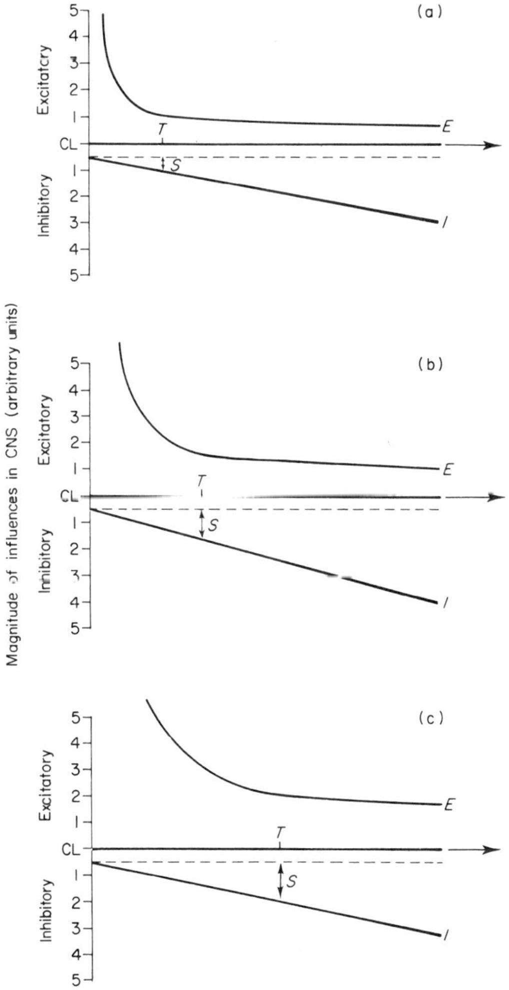

Fig. 2.

The basis for the relationship between meal size and the degree of deprivation would seem to be explicable in terms of the magnitude of inhibitory influences in the CNS at the beginning of the meal and hence at any given time after its commencement. The results of Gelperin (1966b) concerning the relationship between meal and deprivation, when taken together with those he obtained concerning the relationship between crop volume and deprivation (Gelperin, 1966a) make it clear that meal size increases over a period during which crop volume decreases. A certain basis, therefore, for deprivation-dependent differences in the magnitude of inhibitory influences, initially and during the course of a meal, is the dependence on the volume of the crop of input from stretch receptors of the abdomen. The possibility exists also that input via the recurrent nerve concerning fore-gut distension might also contribute to the build-up of inhibitory influences in the CNS. Figure 3 illustrates the likely effects, according to the proposed general scheme, of different degrees of deprivation. The situations illustrated are one in which the fly initially has an empty crop and one in which the insect has some material in the crop, having suffered less deprivation. Again, the predictions of the proposed scheme agree with experimental results obtained (Gelperin, 1966b) concerning the relationship between the state of deprivation and meal size.

Before leaving the subject of the regulation of meal size in *P. regina*, I will consider aspects of hyperphagia that were not dealt with in the earlier discussions of the effects of recurrent and median abdominal nerve section on duration of, and the size of meal taken during, only the initial period of continuous intake by previously deprived flies. Dethier and Bodenstein (1958), Dethier and Gelperin (1967), and Gelperin (1971b, 1972) performed experiments in which the intake of a concentrated sugar solution by nerve-sectioned flies was compared with that of sham operated controls over a period of from 15 to 60 min of *ad lib.* feeding, the flies having previously been deprived of food for 24 h. As was indicated earlier, recurrent nerve section and median abdominal nerve section both caused flies to become markedly hyperphagic even though the operations had no effect, or only a moderate effect, on the duration of the initial continuous period of intake. The basis for the hyperphagia exhibited by these operated flies is apparent from the feeding records of individual operated and control flies given by Dethier and Gelperin (1967), and Gelperin (1972). Whereas unoperated flies had only brief and infrequent periods of ingestion during the 30 to 60 min following the termination of the initial period of ingestion, operated flies fed quite frequently and often for considerable periods. It is clear that this increase in the duration and frequency of the supplementary periods of ingestion is wholly responsible for the hyperphagia which results from recurrent nerve section, and is partly responsible for that resulting from the sectioning of the median abdominal nerve.

It is almost certain that the increased frequency of these supplementary periods of ingestion after recurrent nerve section can be attributed to the continued post-feeding responsiveness of the operated flies to tarsal stimulation with sugar solution. As indicated previously, the only doubt which exists is whether the response is to sugar or water, or both. The effects of abdominal nerve section on the responsiveness of *P. regina* to

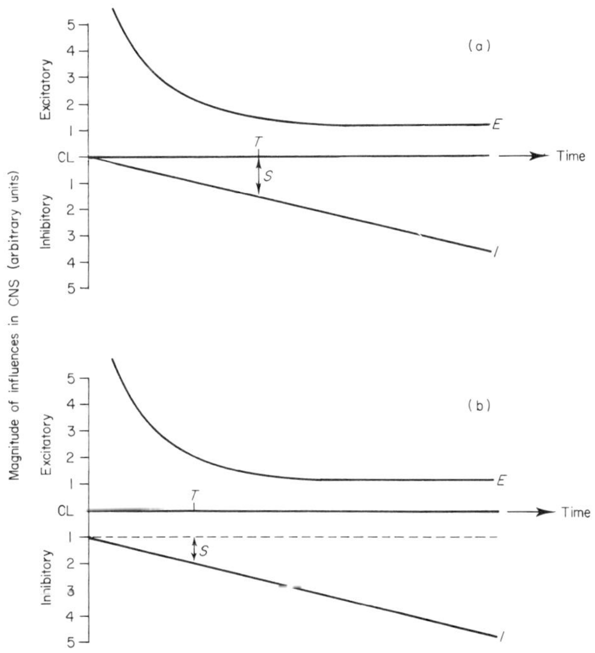

Fig. 3. Hypothetical illustration of the likely basis for the relationship between meal size and duration and the state of deprivation in *P. regina*. *E*, *I*, *T*, *S* and CL have the same meanings as in Figs 1 and 2. (a) depicts the situation in which the fly is severely deprived and in which the initial level of inhibitory influences is very low, and (b) that in which the fly is less deprived and in which, therefore, the initial magnitude of the inhibitory influences is higher. It is assumed that the rate of intake, and hence the rate of increase of *I*, is similar in the two cases. The flies are feeding on identical solutions. Meal size and duration are greater in the more deprived insect.

tarsal stimulation have not been directly investigated, but the general similarity between the effects of recurrent nerve section and abdominal nerve section on the patterning of ingestion, which follows the initial continuous periods, suggests that abdominal nerve section also impairs the regulatory mechanism responsible for bringing about the normal post-ingestion period of refractoriness to tarsal stimulation with sugar solution. The finding by Evans and Dethier (1957) that the sugar thresholds to tarsal stimulation were higher in flies with empty crops but full fore-guts, suggests, however, that median abdominal nerve section does not prevent the elevation of the tarsal sugar threshold of *P. regina*, but rather that it causes flies to remain responsive to the water component of the solution.

When a fly, whether operated or not, extends its proboscis and touches a substrate-bearing sugar solution with its labellum, its labellar and pharyngeal receptors will be in a more or less unadapted state depending on the interval which has elapsed since the last period of feeding. This means that the level of excitatory input is higher than at the time of termination of the previous period of ingestion, with the result that both operated and control flies would begin to feed each time the labellum touched the substrate. The duration of these supplementary feeds would, according to the scheme proposed earlier in this section, depend on the total magnitude of the inhibitory influences in the CNS, the increased durations of feeding by operated flies being a reflection of the smaller amount of inhibitory input reaching the CNS. Since the total intakes of recurrent nerve-sectioned and abdominal nerve-sectioned flies were approximately equal, at about rather more than twice that of controls, in the experiments of Dethier and Gelperin (1967) and Gelperin (1972), it appears that the inhibitory influences generated by input via the two nerves were approximately equal, under the conditions of their experiments.

Belzer (1970) also examined the effects of recurrent nerve and abdominal nerve section on the amount ingested by *P. regina* during access to sugar solution for 1 h. His experimental procedure differed from that of the other workers in that flies were, before being operated upon, deprived of food for only a short time, or not at all, after feeding *ad lib.* on 0.1 M sucrose, and in that the solution offered during the feeding trial was also 0.1 M sucrose. He found that the intake of flies which received either of the surgical treatments was greater than that of controls but that recurrent nerve section produced a much greater increase than the sectioning of the abdominal nerves, a result which differs from that obtained for previously starved flies feeding on a concentrated solution. Belzer gives no records of the temporal feeding patterns of operated and control flies under the conditions of his experiments and it is therefore not possible to offer an informed opinion as to why the apparent relative effects of the two operations differed according to the procedure adopted. All that can usefully be said at this stage is that the results of further experiments to examine the intake and temporal feeding patterns of flies in different nutritional states, feeding on a variety of sugar solutions, would almost certainly provide valuable insights into the mechanisms regulating ingestion in *P. regina*.

Another approach which has been used to study the problem of nerve-section induced hyperphagia in *P. regina* has been to determine the intake of flies which were force fed, in the sense that sugar solution was repeatedly applied by the experimenter to the labellum. Under these conditions, the different tendencies for operated and control flies to exhibit proboscis extension would play no part in determining the amount consumed. Evans and Barton Browne (1960) compared the intakes of recurrent nerve-sectioned flies with those of control operated ones, when 2.0 M glucose was repeatedly applied to the labellum, the proboscis being retracted at the time of the application. They found that, whereas the intake of none of the controls exceeded 40 μl, that of 15 out of 48 of the recurrent nerve section individuals exceeded 50 μl. This result shows directly that the increased frequency of proboscis extension by recurrent nerve-sectioned flies in an *ad lib.* feeding situation is not the only reason for their hyperphagia, and confirms, therefore, that the operation also affects a mechanism concerned with regulation of the volume consumed.

Dethier and Gelperin (1967) measured the intake of 2.0 M glucose of unoperated flies fed by the same method as that used by Evans and Barton Browne (1960), as well as that of legless but otherwise intact flies which were placed on sugar-soaked filter paper in such a way that their labellar lobes were usually in contact with the substrate. Dethier and Gelperin (1967) state that the intakes of both sets of flies satisfied their criterion for hyperphagia, namely that the flies consumed twice the amount taken by *ad lib.*-feeding, unoperated controls. Judging from the results of Evans and Barton Browne, however, it seems certain that recurrent nerve-sectioned flies subjected to the feeding treatments given by Dethier and Gelperin would have taken even greater amounts.

The effects of deprivation (Sinoir, 1968; Bernays and Chapman, 1972b) and type of food (Bernays and Chapman, 1972b) on the sizes of meals taken by larvae of the African migratory locust, *Locusta migratoria*, have been extensively investigated. Sinoir (1968) found that the number of minutes out of 20 for which 5th instar larvae fed on acceptable plant material increased with increasing deprivation up to a period of 3 h. Bernays and Chapman (1972b) measured the actual amounts of food consumed by larvae deprived for different periods and showed that meal size increased sharply with increasing period of deprivation up to 3 h and thereafter increased more slowly up to 6 h. Bernays and Chapman also demonstrated that the duration and size of meals taken by larvae of *L. migratoria* varied according to the type of food plant offered (Table 1), and that there was a positive correlation between meal size and the rate of ingestion.

Some of the physiological mechanisms underlying the regulation of meal size in larvae of *L. migratoria* have been investigated. Sinoir (1968) made measurements, which paralleled his investigation of the effects of deprivation on meal size, of the time course of emptying of the crop and the whole alimentary canal during deprivation, and found that the duration of feeding appeared to be more closely related to the amount of material in the crop than to that in the gut as a whole. Bernays and Chapman (1972b) also found that meal size was generally well correlated with the amount of

TABLE 1

The relationship between type of plant material and meal size and duration and the rate of ingestion by larvae of *Locusta migratoria* (Data from Bernays and Chapman, 1972b).

Plant	Meal size mg	Meal duration min	Rate of ingestion mg min^{-1}
Agropyron	116	17	7
Dactylus	113	16	7
Poa	95	15	6.5
Triticum (seedling)	42	8	5
Lolium (seedling)	37	8	4.5
Trifolium	15	5	3

food from the previous meal which had left the crop during the period of deprivation, but that meals taken after 30 min, 1 h, 2 h or 3 h of deprivation were not quite of sufficient size to completely replace the amount of material lost, the deficit being somewhat greater for insects fed on *Poa* than for those fed the somewhat more acceptable *Agropyron*.

The relationships obtained by Sinoir, and Bernays and Chapman provide strong circumstantial evidence that the effects of inhibitory sensory input concerning the degree of distension of the crop plays an important and active role in bringing about the termination of feeding. Direct evidence for this was obtained by Bernays and Chapman (1973) who showed that an operation in which both the posterior pharyngeal nerves and the recurrent nerve were sectioned (see Fig. 4) causing larvae, which at the beginning of the trial had empty crops, to take larger than normal meals of the palatable grass *Poa*. It was noted that the operated larvae continued to make attempts to feed for an hour or more whereas normal insects ceased doing so after feeding for 15-20 min. None of the other kinds of surgery performed on the stomatogastric nervous system, including the sectioning of only the recurrent nerve, caused significant increases in meal size but a number caused a reduction. On the basis of these results, it was concluded that inhibitory influences generated by input carried via the posterior pharyngeal nerves from stretch receptors in the anterior part of the fore-gut were responsible for bringing about the termination of meals of highly palatable plant materials. Rowell (1963) stated that larvae of the desert locust, *Schistocerca gregaria*, took larger than normal meals after the sectioning of the recurrent nerve but Bernays and Chapman (1973) are of the opinion that the kind of operation performed by Rowell would almost certainly have resulted in the cutting of the posterior pharyngeal nerves as well.

The results of the experiments involving the sectioning of nerves and ganglia of the stomatogastric nervous system seem to provide good evidence that input via the posterior pharyngeal nerves plays an important role in bringing about the termination of feeding but, since the sectioning of only the frontal connectives or of only the recurrent nerve did not affect meal size, they would appear to leave unanswered some questions about the route by which the input reaches the CNS. There would seem to be two possible explanations for results obtained, if it is assumed that the pathway by which information passes from the receptors of the fore-gut to the CNS is entirely neural. Firstly, it might be that input which travels via the posterior pharyngeal nerve is effective in bringing about the cessation of feeding, irrespective of which of the two possible pathways it then traverses in order to reach the CNS. Secondly, it might be that the severance of one or both nerves might produce nonspecific effects which cancel out specific ones which otherwise would result in hyperphagia. Alternatively, it is possible that the cessation of feeding might be mediated by a hormone produced by the frontal ganglion in response to posterior pharyngeal nerve input. As mentioned earlier (section 2.3), it was found that fore-gut receptor input, which results in the release of the hormonal material from the CC responsible for the closure of the terminal pores of the receptors on the maxillary palps is carried via the posterior pharyngeal nerves and the frontal connectives. It seems most likely therefore, that the input which brings about the cessation of feeding also reaches the CNS by this route, but that nonspecific effects on meal size of frontal connective section mask the potentially hyperphagia-producing effects of posterior pharyngeal nerve section.

Bernays and Chapman (1973) made a detailed examination of the changes during the course of a meal in the degree of distension in various regions of the fore-gut. They found that the crop filled in a regular manner

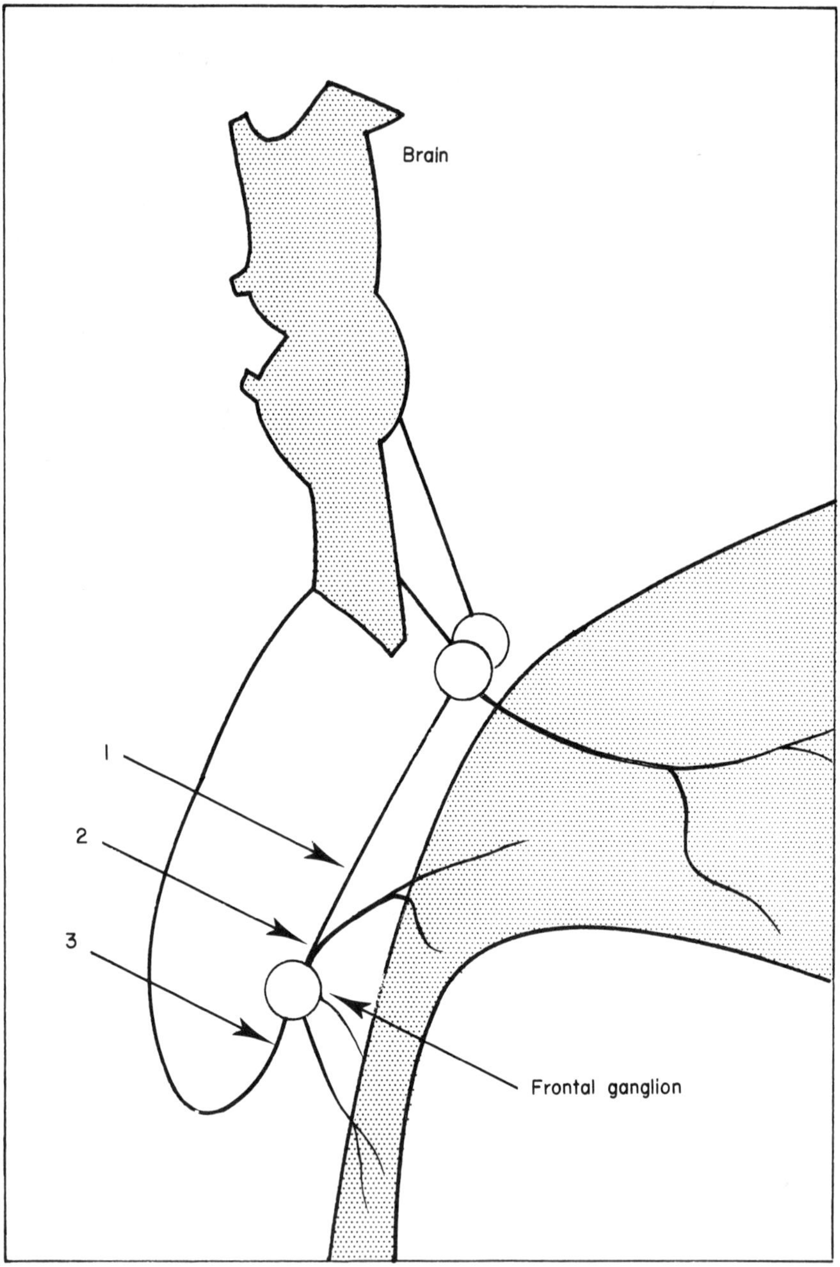

Fig. 4.

with the anterior part being the last to do so completely. The result of this filling pattern is that only the degree of distension of the anterior region, which is that innervated by the posterior pharyngeal nerves, is still increasing up to the end of a meal. Additional strong circumstantial evidence, in support of the view that input concerning the degree of distension of the anterior region of the fore-gut plays an important role in bringing about cessation of feeding was provided by a number of experiments in which the degree of distension of this region was determined at the time when larvae, which had received various treatments that altered the filling pattern of the crop, stopped feeding. Locusts with the inner oesophageal nerve cut do not transport ingested food to the posterior part of the fore-gut in the normal manner and it was found that they took smaller meals of *Agropyron* than control insects. It was found, however, that the degree of distension in the part of the fore-gut innervated by the posterior pharyngeal nerve in these operated insects was, at the end of their meal, identical with that in unoperated insects at the end of their larger meals. Similar results were obtained with larvae whose fore-gut had been ligated about two-thirds of the way back.

Bernays and Chapman (1973) found that the posterior pharyngeal nerve section did not significantly increase the size of meal taken by larvae feeding on *Trifolium* or seeding *Triticum*, and that the sizes of meals of these plants were smaller than those taken by similar insects feeding on *Poa*. They examined the degree of distension of various regions of the crop immediately after previously deprived larvae had taken a complete meal of *Trifolium* and found that, in the anterior region, it was less than at the end of larger meals of *Poa* or *Agropyron*. They concluded, therefore, that the stretch receptor input travelling via the posterior pharyngeal nerves is evoked only when the degree of distension of the anterior fore-gut is greater than that which occurs when larvae feed on *Triticum*.

Bernays and Chapman (1973) also examined the effects on meal size of the sectioning of the connectives of the ventral nerve cord posterior of the prothoracic or metathoracic ganglia and found that this operation did not result in increased intake. It would appear unlikely, therefore, that input from any source posterior to the prothoracic ganglion has a significant role in determining meal size in larvae of *L. migratoria*. The possibility must be admitted however that, as with any kind of treatment involving trauma, there is a possibility that nonspecific effects of the operation may have

Fig. 4. Diagram of part of the stomatogastric nervous system of *L. migratoria* showing the points at which Bernays and Chapman (1973) sectioned the recurrent nerve only (1), both the recurrent and posterior pharyngeal nerves (2), and the frontal connectives (3). Only operation 2 resulted in hyperphagia.

caused a reduction in meal size which balances specific effects potentially capable of allowing an increase. Of relevance, however, to this negative result is the finding by Bernays and Chapman (1973) that feeding does not cause an increase in the overall volume of *L. migratoria* larvae, the volume being accommodated by the collapsing of air sacs.

In more recent investigations, Bernays and Chapman (1974b) demonstrated that the osmotic pressure of the haemolymph of larvae of *L. migratoria* increased progressively during the course of a meal, being 24 per cent higher at the end of a meal lasting 15 min than at the beginning. It was shown that this was due to water loss from the haemolymph as a result of salivation during feeding. These authors showed further (Bernays and Chapman, 1974c) that meal size was reduced after the osmotic pressure of the haemolymph had been increased by the previous injection of various hypertonic solutions, with the degree of reduction increasing with increasing osmotic pressure. The degree of reduction was, however, independent of the species of compound used to achieve the increase. Injections of water or locust saline into the haemocoel of larvae had no effect on meal size. Increased osmotic pressure resulted in a reduction also in the rate of ingestion, but this was less marked than the reduction in meal size. It appears, therefore, that the increased osmotic pressure reduces the duration of meals but to a lesser extent than is indicated by the data for meal size (Bernays and Chapman, personal communication). Reduction in meal size due to increased osmotic pressure was found to be greater in locusts allowed to feed 20 min after injection than in ones which fed 10 min after. An increase in osmotic pressure of the same order as that which occurs during the course of a meal caused a reduction of 10 per cent in the size of a meal taken 10 min after injection and about 25 per cent in that of meals taken after 20 min. Bernays and Chapman (1974c) found that meal size was closely correlated with the integral, over the 20 min before feeding, of the amount by which the osmotic pressure exceeded the base level. They concluded, therefore, that meal size is affected not so much by the osmotic pressure prevailing at the end of the meal, but rather by the total osmotic effect over 20 min before feeding. This view was supported by the finding that injections, causing an increase in osmotic pressure, given 3 min after the beginning of a meal, had no effect on the ultimate size of the meal.

Bernays and Chapman (1974a) also investigated the role, in the determination of meal size, of sensory input elicited by the food being ingested. They found that the sizes of meals of seedling grasses taken by previously deprived larvae were less than half of those taken by similar larvae feeding on mature grasses. They demonstrated that the removal of material from seedling grass by washing it with chloroform resulted in an increase in meal size, and that the application of the extract obtained to ordinarily acceptable food resulted in a reduction. This result is reasonably interpreted by Bernays and Chapman as indicating that one or more compounds in seedling grasses elicit inhibitory input from the oral chemoreceptors, i.e. is a feeding deterrent, and that this generates inhibitory influences in the CNS. In another series of experiments, Bernays and Chapman (1974a) investigated the possibility that sensory input from the receptors of the mouthparts might generate perseverating effects in the CNS of larvae of *L. migratoria*. They describe experiments in which dried seedling grass extract, dispersed in water, was dripped on to the mouthparts of larvae before they were offered either seedling grass, which would elicit the same inhibitory input as the dripped fluid, or plant material lacking feeding deterrents. They found that larvae which had been pretreated with the extract would not feed at all on the deterrent-containing seedling grass, even though it is known that locusts which have not received this prior treatment take quite substantial meals of this material (Bernays and Chapman, 1972b). The intake of the entirely acceptable food was, on the other hand, not affected by the pretreatment with the extract. It seems, therefore, that deterrent(s) present in the seedling grass generated a perseverating inhibitory influence in the nervous system of the larvae which prevented ingestion when inhibitory input was elicited by the food, but that it did not do so when the input was largely or entirely excitatory. The finding that the sizes of meals of acceptable materials were unaffected indicated that the excitatory input was able to erase the perseverating inhibitory effect during the course of the meal so that, by the end of the meal, it was no longer a factor in determining a point at which feeding would cease. In a similar experiment, the effects of dripping whole grass extracts over the mouthparts prior to feeding was investigated and it was found that the size of the subsequent meal was increased by 17 per cent by this treatment. It seems therefore that the pretreatment with material eliciting wholly or largely excitatory input generated a perseverating excitatory influence in the nervous system. The experiments give no indication of the likely decay times of these perseverating effects.

I will now attempt to categorize the factors discussed above according to the kind of role they might play in bringing about the cessation of feeding in *L. migratoria*. It seems clear that inhibitory influence resulting from input via the posterior pharyngeal nerves concerning crop volume plays an active role, as defined earlier, in bringing about the termination of meals of highly palatable materials by insects deprived of food for more than 3 h. The evidence suggests, however, that input from this source plays no significant role in determining the point of termination of the smaller meals of less palatable materials.

The question as to whether input via these nerves is elicited so abruptly that it may be looked upon as being an all or none phenomenon or whether the input is graded over a range of degree of distension cannot be answered with certainty. It can be said, however, that the degree of distension caused in the relevant region of the fore-gut by the amount of material which is present at the end of meals of *Triticum* taken by previously deprived locusts is not sufficient to stimulate the stretch receptors involved. Two pieces of information suggest, however, that input from these receptors might vary according to the degree of distension over a limited range. The first of these is that, as has previously been mentioned, meals of the palatable grasses *Agropyron* and *Poa* taken within 2 h of a previous meal are not of sufficient size to replace the material which has been lost from the crop (Bernays and Chapman, 1972b). There are several possible explanations for this finding, but one which appears likely, and which would require that the input from fore-gut receptors should be graded over a limited range, is that inhibitory influences generated by input via the posterior pharyngeal nerves is actively responsible for the termination of feeding but that this is integrated with other inhibitory influences such as, for example, that resulting from input concerning the osmotic pressure of the haemolymph (Bernays and Chapman, 1974b, 1974c) or any effects on the CNS of hormones which are known to be released from the CC as a consequence of feeding (Bernays and Chapman, 1972a, 1974a). According to this hypothesis, feeding would cease when the input via the posterior pharyngeal nerve, and therefore the inhibitory influences generated by it in the CNS, was at a lower level than if none of the other influences existed. There are, however, at least three other possible explanations. The first is that the input is, in fact, abruptly evoked but that the resulting inhibitory influences build up progressively over a period during which the CNS is receiving this input. If this were the case, the period required for a build-up of inhibition insufficient to bring about the cessation of feeding would be shorter if other inhibitory influences were present, and consequently the meal would be smaller. Secondly, it is possible that the pattern of filling of an already partly filled crop differs from that of an initially empty one in such a way that a smaller total crop volume is required to cause a given degree of distension in the critical anterior region. A third possibility is that input via the posterior pharyngeal nerves plays, in fact, no role in the termination of meals taken by locusts which had been satiated less than 2 h earlier. Further experimentation is needed to differentiate between these possibilities.

A second piece of evidence which indicates the possibility that input from the fore-gut receptors is to some extent graded is that meal size is influenced by the previous feeding history of the larvae. Bernays and

Chapman (1972b) found that locusts, which had previously been caused to take smaller than usual meals over a period of two days, took smaller meals after 5 h deprivation than did larvae which had taken larger meals during the previous two days. This result would seem to indicate that the difference in the previous feeding either altered the response threshold of the receptors to distension of the fore-gut, or changed the responsiveness of the CNS to their inhibitory input. If this latter explanation is correct, as seems rather more likely, it would follow that the input must, to some extent, be graded according to the degree of distension.

It would seem that influences generated by input concerning the haemolymph osmotic pressure (Bernays and Chapman, 1974c) must play essentially a passive role, since the elevation of osmotic pressure by injection during the course of a meal causes no reduction in meal size. The available evidence suggests that the magnitude of the inhibitory influence generated by input concerning osmotic pressure of the haemolymph is set by the level of the osmotic pressure during the 20 min before feeding but that any changes during feeding play no significant role.

On the basis of the available evidence, it is difficult to assess the role, if any, played by receptor adaptation in bringing about the termination of meals of plant material by larvae of *L. migratoria.* The finding by Bernays and Chapman (1974a) that stimulation of the receptors of the mouthparts with extract of grass for 20 min did not reduce the sizes of meals of the same grass taken immediately afterwards suggests that receptor adaptation does not play an active role, but it should be realized that the extract probably did not contact the sensilla of the hypopharynx which are known to play a role in feeding behaviour (Haskell and Mordue, 1969; Haskell and Schoonhoven, 1969).

There are no experimental data relevant to the question of whether adaptation might play a passive role. It can be said, however, that there are only two circumstances in which adaptation can fail to do so. The first is that in which the motor output, which results in ingestion, is independent of the continuing existence of excitatory influences generated by excitatory input from the receptors of the mouthparts, but rather is driven by endogenous activity released probably by the initial stimulation. It is assumed, in these circumstances, that the period for which the endogenous activity continues is independent of the level of the initial stimulation and that it continues until turned off by inhibitory influences. It has been stated by Bernays and Chapman (1974a) that endogenous activity is important in causing ingestion to continue in *L. migratoria*, but the context of their statement suggests that they believe rather in the importance of perseverating effects which would reflect the level of previous input and the time course of changes in this level.

The second circumstance in which sensory adaptation would play no role is that in which inhibitory influences of sufficient magnitude to oppose excitatory influences produced by input from the chemoreceptors, irrespective of their state of adaptation, are abruptly evoked in the CNS. The only certainly known source of input inhibitory to feeding in *L. migratoria* is that via the posterior pharyngeal nerves and it is not known whether it generates inhibitory influences in the CNS of sufficient magnitude to counter the excitatory influences resulting from input from nonadapted receptors. Also, there is uncertainty as to the abruptness of its evocation. It is clear, however, that the effect is graded only over a limited range of degrees of distension of the crop, if at all, and that, under conditions in which the crop is filled to capacity, the passive role of adaptation might be relatively minor. It seems, on the other hand, reasonable to assume that receptor adaptation would play at least a passive role in those circumstances in which feeding is terminated before the crop is completely filled.

Bernays and Chapman (1974a) suggest that meals of seedling *Triticum* and probably of other relatively unpalatable plants are terminated primarily by inhibitory influences generated in the CNS by input elicited from the contact chemoreceptors by feeding deterrents. While it seems certain that such inhibitory influences must play at least some part in bringing about the cessation of feeding, the extent to which they are responsible is far from clear. Chemoreceptor input would be the only factor involved only if there were no inhibitory input from any other source at the time of termination of meals which do not fully distend the crop. If, however, other sources of inhibitory input were to exist, the inhibitory influences resulting from chemoreceptor input would be only partly responsible. To date, there is no direct evidence of the existence of any input of internal origin which is inhibitory to feeding other than that via the posterior pharyngeal nerve, which is known not to be elicited by the degree of distension which results from meals of plants containing deterrents. The available direct evidence is, therefore, consistent with the view that inhibitory input from chemoreceptors plays a major and active role in bringing about the termination of meals of plants containing feeding deterrents. There is, however, circumstantial evidence which possibly argues against this view. Bernays and Chapman (1972b) have shown that the sizes of meals taken by female larvae of *L. migratoria* of any plant material, irrespective of whether or not it contains deterrents, exceeds the sizes of those taken by male larvae by approximately the same percentage as that by which the body weight of the female exceeds that of the male. This finding suggests that input from a sensory system monitoring the volume of some part of the insect is likely to be playing some role in determining the sizes even of meals of plants containing deterrents.

Finally, on this topic, it is relevant to point out that for inhibitory influences generated by input from chemoreceptors to be the sole cause of the cessation of feeding, it would be necessary for the magnitude of the resulting inhibitory influences to increase relative to that of excitatory ones during the meal. If the only part played by excitatory input is to start feeding, which then continues as a result of endogenous activity until switched off by influences resulting from inhibitory input, it would be necessary only that inhibitory input produces perseverating effects in the CNS with a decay time which is long enough to allow the progressive build-up of inhibition. If, however, feeding is to any extent driven by excitatory influences generated by excitatory input, it is necessary to postulate either that deterrent-sensitive receptors are slower adapting than the stimulant-sensitive ones, or that the inhibitory influences generated in the CNS have a longer decay time than the excitatory influences.

Barton Browne *et al.* (in press, a) performed a series of experiments designed to determine the role of the consequences of sensory adaptation in bringing about the termination of meals in adults of the Australian plague locust, *Chortoicetes terminifera.* In these, water or one of a variety of concentrations of sucrose solution were applied directly to the mouthparts of locusts which had been deprived of food for 24 h following *ad lib.* feeding on plants. Under these conditions, locusts readily imbibe the fluids and mark the termination of their meal with refusal to ingest material which still remains on the mouthparts, and by characteristic leg movements. It was found that the insects took fairly substantial meals of water but that both meal size and duration increased with increasing sucrose concentration up to 0.5 M. Meals of 1.0 M sucrose were smaller and of shorter duration than those of 0.5 M. These results indicated that size and duration of meals were almost certainly related to the characteristics of the excitatory sensory input from chemoreceptors being stimulated by the food material. To investigate directly any effect of adaptation, meal sizes were compared when locusts had fluid applied to their mouthparts continuously and when fluid was applied as standard sized discrete drops with a 30-s interval between the disappearance of one drop and the application of the next, the experiment being based on the assumption that the discontinuous method of application would allow some degree of recovery from adaptation between drops. It was found that meals of 0.5 M or 1.0 M sucrose were substantially larger when application was discontinuous than when the insects were fed continuously but that the sizes of meals of water were unaffected. It was shown, by means of radiographs of locusts fed continuously or discontinuously with sugar solution containing suspended barium sulphate, that the amount of crop emptying which took place during the relatively protracted discontinuous meals was no greater than during continuous ones. There is no possibility, therefore, that the increased intakes of discontinuously fed locusts can be explained in terms of feedback concerning distension of the fore-gut and it seems certain that the increased intake can be taken as evidence that the effects of sensory adaptation on the magnitude of excitatory influences in the CNS plays at least a passive role in determining, in *C. terminifera*, the point of termination of meals of sugar solution.

Several possible explanations can be offered for the lack of increase in meal size when locusts are fed discontinuously with water. It is possible, for instance, that water-sensitive receptors do not adapt or, if they do, that the 30 s between drops is insufficient time to allow any significant degree of disadaptation in these receptors. Alternatively, there may be some abruptly invoked inhibitory input from some internal source which is able to counter the amount of excitatory input from either àdapted or partly adapted water-sensitive receptors, but which cannot counter the total input elicited by sugar solutions from sugar-sensitive and probably also water-sensitive receptors. Only if a detailed knowledge of the relevant sensory inputs were available could the relative likelihoods of these possible explanations be assessed. The possibility must be recognized, however, that receptor adaptation might play no role, either active or passive, in regulation of the size of water meals taken by *C. terminifera* and that input from one or more internal sources might be entirely responsible.

Their previously discussed results having indicated that meal size was related to the level of excitatory input, Barton Browne *et al.* (in press, b) investigated the possibility that input from oral chemoreceptors might generate perseverating excitatory influences in the nervous system of *C. terminifera*. Evidence of perseverating effects of stimulation with sucrose solutions was obtained from experiments in which locusts were fed various sequences of drop of 1.0 M sucrose and water. It was found that locusts fed single drops of sucrose solution and water alternately, or single drops of sugar solution alternated with groups of four consecutive drops of water, consumed larger meals than locusts fed only water, or locusts offered water after having been partially or wholly satiated with 1.0 M sucrose. Almost invariably, the locusts fed either of the two kinds of alternating sequences ceased feeding when a drop of water was on the mouthparts, and it was therefore the insects' lack of responsiveness to water which limited the size of these kinds of meals. The finding that alternation resulted in the locusts taking larger meals means, therefore, that some characteristics of the alternating sequences caused the insects to continue to respond to water when they had already consumed a total volume of fluid which, under other circumstances, would have rendered them unresponsive. It was found that the locusts fed groups of four water drops between consecutive sugar drops never ceased feeding on the first drop of a group and, in females particularly, rarely on the second and that the first and second water drops of a group were often consumed more rapidly than the last one of the previous group. These two pieces of information, taken together with the finding by Moorhouse *et al.* (in preparation) that, in the terminal stages of a meal, the rate of ingestion of drops was related to stimulus strength, show that the first one or two drops of a group of four water drops were recognized by the insect as being more stimulating than the remaining water drops of the group. This was interpreted as being due to the existence of a perseverating state of excitation generated by the input elicited by the preceding drop of sucrose.

The experiments of Barton Browne *et al.* (in press, b) provide no information as to the site of the perseverating effects, but in view of the demonstration by Dethier *et al.* (1965, 1968) that stimulation with sugar generated a perseverating effect in the CNS of *P. regina* and the finding by Bernays and Chapman (1974a) that contact with food by *L. migratoria*, as in *P. regina* (Dethier, 1957), caused changes in the insects behaviour which lasted for some time, it seems likely that the perseverating effects in *C. terminifera* are generated in the CNS.

The perseverating effects produced by sugar stimulation in *C. terminifera* appear to be relatively brief. In the experiments just discussed, the effects of one sugar drop were apparent for only a few seconds. In another experiment, Barton Browne, Moorhouse and van Gerwen (unpublished) compared the total intake by locusts, which were allowed to consume a volume of 0.5 M sucrose equal to approximately half the volume of normal meals of water and were then immediately fed drops of water, with intake by insects fed only water. It was found that the means of total amounts consumed by individuals in the two groups did not differ significantly (Table 2). Meal size under these circumstances was, therefore, determined entirely by the kind of material being offered at the time of cessation of feeding and was uninfluenced by a previous period of stimulation with material, which when offered to locusts throughout a meal, resulted in their taking larger meals than of water (Barton Browne *et al.*, in press, a). This result shows that any perseverating effect generated during the initial period of stimulation with 0.5 M sucrose was not apparent at the end of the water portion which took usually of the order of 15 s. This result is significant also in that it provides an answer, in *C. terminifera* at least, to the question raised by Bernays and Chapman (1974a) as to whether continued ingestion depends on the presence of continuing excitatory input from the chemoreceptors or whether, once started, feeding will continue until switched off by inhibitory inputs. The results obtained with *C. terminifera* show clearly that, even though perseverating effects of short duration can be demonstrated,

TABLE 2

The effects of partial satiation with either water or 0.5 M sucrose on the subsequent water intake of adult *Chortoicetes terminifera*. The volumes (μl) of water subsequently consumed by males which received 10 μl and of females which received 16 μl of either water or a sucrose solution in the first part of the meal (Barton Browne *et al.*, unpublished)

	Material on which locusts were partially satiated before being given water			
	Water		0.5 M sucrose	
	Mean	S.E.	Mean	S.E.
Males	11.6	±1.6	12.0	±2.4
Females	15.4	±2.4	20.6	±4.0

the size of a meal can be influenced during its course by changing the stimulus. Barton Browne *et al.* (in press, a) have shown that water is a feeding stimulant for *C. terminifera*, so it is unlikely that the drops of water were eliciting inhibitory input. It seems, therefore, that a change in the level of excitatory input during the course of a meal does influence its size, and that continued ingestion depends on the arrival at the CNS of continuing excitatory input. The perseverating effects would, however, probably result in meals continuing for a little longer than in their absence.

The results obtained with *C. terminifera* must also be discussed in relation to the findings by Bernays and Chapman (1973) that input concerning crop distension was important in bringing about the cessation of feeding in *L. migratoria*, even though there is no direct evidence for such a mechanism in *C. terminifera*. The question arises as to whether the ingestion of fluids causes crop distension comparable with that produced by feeding on plant material. Radiographs, made of adults of *C. terminifera* immediately after the consumption of different volumes of 0.5 M sucrose containing suspended barium sulphate, showed that except for a small amount of fluid which was apparently passed into the mid-gut caeca early in the meal, the whole of meals which weighed up to about 8 per cent of the insects live weight were retained in the crop. Some part of larger meals, however, passed to the mid-gut and, in extreme cases, even to the hind-gut (Barton Browne *et al.*, in press, a). Bernays and Chapman (1972b) found that female 5th instar larvae of *L. migratoria* took about 150 mg and male larvae about 118 mg of the acceptable grass *Agropyron*. These intakes would seem, on the basis of the data of Davey (1954), to represent for both sexes about 10 per cent of body weight. Allowing for the probability that a given weight of plant material would almost certainly occupy a greater volume than an equal weight of fluid, it seems reasonable to believe that the maximum degree of crop distension achieved by adults of *C. terminifera* feeding on fluids may have been rather less than that in the grass fed *L. migratoria* of Bernays and Chapman. If *C. terminifera* does have fore-gut receptors which monitor crop distension, and if these are set to respond only when the crop distension is at a maximum, it is possible that the ingestion of fluids might not elicit input from them.

Barton Browne *et al.* (in preparation) demonstrated that there is a general relationship, which differs somewhat according to sex, between the weight lost by locusts during the 24-h period of deprivation and the sizes of meals of water or sugar solution at the end of this period. Since all 24-h deprived locusts, irrespective of their weight loss, had empty fore-guts and mid-guts, it seems impossible that input concerning gut distension was responsible for the relationship. The findings must be taken, therefore, as evidence for the involvement of input concerning one or more internal parameters other than the volume of material in the gut.

In males, the relationship between weight loss and intake is most marked when locusts are fed in ways which result in their taking relatively large meals. It was found, for instance, that the coefficient for the regression relating the two parameters was significantly greater for locusts fed continuously on 0.5 M sucrose than for those fed similarly with water or 1.0 M sucrose. Related to this is the finding that the sizes of meals taken by males which lost less than 10 per cent of their body weight during the 24-h period of deprivation did not vary according to the type of fluid offered or the method of its presentation, whereas the intake of individuals which lost 14 per cent or more was markedly dependent on the type of fluid and on the feeding regime. It was found also that the intake of more than half of the males fed in ways which resulted in their taking large meals was approximately equal to their weight loss and that the intake of less than 10 per cent of the insects substantially exceeded their loss. On the basis of these results, Barton Browne *et al.* (in preparation) postulated that feeding in males was terminated, in those individuals which took sufficient fluid to replace the weight lost during the period of deprivation, by the effects of inhibitory input evoked suddenly when the volume of solid and fluid within the insect reached that maintained during *ad lib.* feeding on acceptable plant material. Inhibitory influences resulting from such input would then be considered to play an active role in bringing about the cessation of feeding.

In females, the situation is somewhat different from that in males in that the relationship between weight loss and intake was found generally to be unaffected by the type of fluid consumed and the manner of its presentation, the only exception being that the relationship was less pronounced for locusts fed 1.0 M sucrose in a continuous manner than for locusts given any other feeding treatment. Arguments will be advanced below in favour of the view that the point at which meals of continuously applied 1.0 M sucrose are terminated depends largely on the characteristics of the input elicited from the chemoreceptors of the mouthparts, and therefore that the result is not relevant to the present discussion on the probable basis of the weight-loss dependent relationship in females. Of particular significance is the finding that the regression coefficient for water-fed females is very similar to those obtained for individuals fed in ways which resulted in their taking much larger meals, a result which contrasts sharply with that obtained for males. Other respects in which the results for females differ from those for males are that the intakes of females which lost less than 10 per cent of body weight are just as dependent upon the feeding regime as those of females which lost more than 14 per cent, and that the intake by a substantial proportion of females fed in ways which resulted in their taking large meals exceeded their loss by a considerable amount.

Barton Browne *et al.* (in preparation) consider that the relationships for females might be explicable in terms of influences generated either by input from a sensory system monitoring the progressive increase during feeding in the total amount of fluid or solid material within the body or by input concerning the state of hydration of the insect. In the event of the involvement of input concerning the total amount of material within the locust, the basis for weight loss-dependence of meal size would be that the level of input after the ingestion of any particular amount of material would depend upon the volume of material within the body at the beginning of the meal. The possibility that the state of hydration of the body might play a role is indicated by the finding that there is a significant relationship between the weight loss during the 24-h period of deprivation by an individual insect and its percentage water content. Since relatively little of the ingested fluid passes to the mid-gut during a meal, it is unlikely that there would be great changes in the state of tissue hydration during feeding. It seems therefore that any role played by input concerning the state of tissue hydration would be a passive one, with the input remaining essentially constant throughout a meal, at a level determined by the initial state of hydration of the insect.

It must be pointed out also that, as in *L. migratori,* feeding on plant material, the meal sizes of females exceeded those of males by approximately the same proportion as that by which the body weight of females exceeded that of males, irrespective of the particular type of fluid or fluids offered and the method of application. This finding is circumstantial evidence, as it is in *L. migratoria,* that some input concerning the volume of some part of the insect probably plays an active role in bringing about the cessation of feeding.

Finally, some comment needs to be made about the finding that the meals of continuously applied 1.0 M sucrose taken by *C. terminifera* were smaller than those of 0.5 M sucrose. It was shown by Barton Browne *et al.* (in press, a), in experiments in which locusts were fed to repletion on one of the solutions before being offered the other, that the 1.0 M solution is at least as stimulating as the 0.5 M. It seems, therefore, that the reduced intake of locusts fed 1.0 M sucrose as compared with 0.5 M cannot be explained in terms of the relative abilities of the two solutions to stimulate oral chemoreceptors which are in a similar state of adaptation. Nor can the results be explained in terms of any reduction in rate of intake due to the greater viscosity of the 1.0 M solution, since the overall rates of ingestion of the two solutions are identical (Moorhouse *et al.,* in preparation). One possible explanation is that 1.0 M sucrose is outside the physiological range of the receptors with the result that they virtually cease functioning after a period of exposure to the solution. If this is the explanation, the finding that there is little relationship between weight loss during the 24-h period of deprivation and in the size of meals of 1.0 M sucrose taken by females means that the decline in sensory input must be abrupt, and to a level such that it could be counteracted by inhibitory influences generated by the time the insects had ingested as little as the smallest amount of 1.0 M solution taken. If this were not so, it could be expected that, for females, the regression relating meal size to weight loss would be parallel to that for insects fed water or 0.5 M sucrose.

The findings relating to regulatory mechanisms for meal size in the two species of locusts can be summarized briefly as follows. Bernays and Chapman (1973) demonstrated, in *L. migratoria*, that an active role is played by inhibitory influences generated by input from a sensory system monitoring distension of the fore-gut and by input from chemoreceptors responding to any feeding deterrents in the food. They showed also (Bernays and Chapman, 1974b, 1974c) that a passive role is played by inhibitory influences whose magnitude is determined by the osmotic pressure of the haemolymph during a period before feeding, the magnitude of this influence apparently remaining constant throughout the meal. Barton Browne *et al.* (in press, a) demonstrated that in *C. terminifera* the effects of adaptation of the chemoreceptors of the mouthparts play at least a passive role. Additionally, they obtained evidence that, in males, influences generated by abruptly evoked input concerning the volume of the whole insect play an active role (Barton Browne *et al.*, in press, b).

In neither species is there, as yet, sufficient information to justify the presentation of diagrams of the kind drawn previously for *P. regina* and it seems unwise, at this stage, to ascribe, to a composite locust, all the mechanisms for which evidence has been obtained. It seems that the techniques applied to *L. migratoria* could usefully be applied to *C. terminifera*, and vice versa, to determine whether all mechanisms are detectable in both species.

Kusano and Adachi (1969a, 1969b) investigated factors affecting the duration of meals of sugar solution taken by adults of the cabbage butterfly. *Pieris rapae crucivora.* They showed that the duration of meals of water increased during four days of deprivation following emergence. A similar trend was apparent up to 7 days in the duration of meals of 1.0 M sucrose but, because of very high variances, the increase was possibly not significant. The duration of feeding on the 1.0 M sucrose was consistently greater than on water. The insects fed only briefly on 1.0 M sodium chloride, and the duration of meals of this solution did not increase throughout 5 days of deprivation. The duration of feeding and the volume ingested was greater when 2-day deprived insects were fed 1.0 M sucrose or fructose than when they were fed 1.0 M glucose. Increases in both parameters were found to occur with increasing sucrose concentrations from 0.06 M to 1.0 M.

The findings in relation to the effects of deprivation are of doubtful value, since the design of the experiments did not allow the effect of ageing and the effect of deprivation to be separated. Probably of greatest interest, in this regard, is the finding that there was no increase with increasing deprivation in the duration of meals of 1.0 M sodium chloride. The other results establish, however, that the level of sensory input is important in determining meal size and duration. Further information on this aspect was provided by the finding that butterflies which had had the tip of the proboscis removed fed more briefly on 0.25 M sucrose than did intact individuals. The removal of all the legs of insects which had already commenced to feed was stated, on the other hand, to have no effect on the duration of feeding. It seems therefore that sensory input from receptors from the tip of the proboscis is important in determining meal size but that that from the tarsal receptors is not.

Kusano and Adachi (1969b) also investigated some of the internal mechanisms concerned with the regulation of meal size. They showed that the duration of proboscis extension in response to 0.25 M sucrose was many times greater in butterflies in which fluid was unable to enter the crop because of the presence of a ligature between the thorax and abdomen, than in controls. It was shown also that perforation of the crop in such a way that the fluid entering it leaked into the haemocoel did not affect the duration of feeding but that complete extirpation of the crop resulted in an increase in the duration of sucking. Kusano and Adachi interpreted their results as indicating that meal size and duration in *P. rapae crucivora* is determined by a balance between the effects of excitatory input from chemoreceptors on the mouthparts and presumably inhibitory input from chemoreceptors in the crop, the normal duration of meals taken by insects with perforated crops being taken as an indication that input concerning crop volume is unimportant. It would seem improbable, however, that input from chemoreceptors in the crop, acting in the way postulated by Kusano and Adachi, would play an active role in bringing about cessation of feeding, since it seems that they could provide information only about the presence or absence of the particular food material in the crop. Even if it were to be accepted that chemoreceptors are involved, it would be necessary to postulate that influences generated by one or more other inputs were playing an active role. Possible candidates would seem to be a decline of excitatory influences as a result of sensory adaptation or perhaps an increase of inhibitory influences generated by input concerning the total volume of material within the insect. However, it could well be that any actively changing input is effective in bringing about the cessation of feeding only if input indicating the presence of at least some material in the crop is also reaching the CNS.

Ma (1972) made a detailed study of the effects of deprivation and stimulus strength on the duration of meals taken by later instar larvae of *Pieris brassicae* feeding on a meridic diet to which different quantities of sugars or other compounds were added. He found that, for insects feeding on diet containing 0.001, 0.01, 0.1 or 1.0 M sucrose, the duration of the initial continuous period of ingestion increased with increasing periods of deprivation up to 78 h. A similar relationship, except that there was little difference between the feeding performances of 54-h and 78-h deprived larvae, was found between deprivation period and the total duration of the initial continuous period of ingestion and the shorter supplementary periods. The number of these supplementary periods increased with increasing deprivation up to 54 h at the lowest sucrose concentration. At the other three concentrations, however, fewer supplementary periods were taken by 4-h deprived insects than by insects deprived for 30 h or longer but no increase was apparent between 30 and 78 h.

The duration of meals of diet was found to increase with concentration of added sucrose up to 0.1 M, irrespective of whether assessed in terms of the duration of the initial period of ingestion or of the total duration. With insects deprived for 54 or 78 h, the durations of meals of diet containing

1.0 M sucrose were substantially less than of 0.1 M and a somewhat less pronounced but similar relationship was apparent in insects deprived for shorter periods. There was, however, a tendency for the number of supplementary periods of ingestion by insects deprived for 30 h or more to rise progressively with sucrose concentration up to 1.0 M. In other experiments with sugars, Ma demonstrated that at any given concentration, meals of diet containing sucrose were of longer duration than of those containing glucose, which were in turn longer than of those containing fructose. Meal duration increased with concentration of all three sugars up to 0.1 M or 0.3 M. Beyond this, there was no further increase in meal size with glucose and fructose and, as with sucrose, there was a marked reduction in duration both in total and in the initial period of ingestion of meals of 1.0 M. Ma investigated also the effects of sinigrin and sucrose, separately and together, on meal duration and demonstrated a positive interaction between the effects of the two compounds. He showed also that the addition of salts to diet containing 0.1 M sucrose reduced meal duration: with calcium chloride, there was a progressive decrease from 0.05 M and sodium chloride from 0.5 M upwards.

Ma (1972) did not investigate directly the physiological mechanisms responsible for the regulation of meal size in larvae of *P. brassicae*. The results do, however, indicate that influences generated by both excitatory and inhibitory inputs from external chemoreceptors play a part in determining the duration of meals and that the levels of influences generated by inputs of internal origin vary according to the degree of deprivation. The reduced meal times of larvae feeding on diet with 1.0 M sucrose as compared with that of larvae feeding on diet with 0.5 or 0.1 M sucrose is interesting in view of the similar result obtained with *Chortoicetes terminifera* by Barton Browne *et al.* (in press, a). It must be pointed out, however, that Ma measured only the duration of intake and not the amount of food eaten. The possibility must be borne in mind, therefore, that the rate of ingestion might vary according to sucrose concentration and that the duration of feeding might not give a good indication of the weight of food eaten.

Holling (1966) examined the relationship between the period of deprivation and the weight of houseflies eaten by the mantid *Hierodula crassa*. Crushed houseflies were repeatedly applied to the mouthparts of the mantids until they indicated satiation by refusing three flies in succession. It was found that the weight of flies consumed increased progressively with increasing deprivation up to about 32 h, recently satiated mantids refusing to accept any flies and those which had been deprived for 32 h or more taking about 1 g of flies before indicating satiation. It is interesting to note that the mantids ate significant quantities of flies presented in this way after periods of deprivation of less than 8 h, up to which time the insects were completely unresponsive to visual stimuli provided by living flies, and therefore would not voluntarily capture prey.

Barton Browne and Dudziński (1968) investigated the effects of water deprivation on the sizes of water meals taken by adult females of the Australian sheep blowfly, *Lucilia cuprina*. In their experiments, flies were deprived of all food and water for 24 h and were then allowed to satiate themselves with water. The durations and volumes of water meals taken by the flies at various times up to 10 h after the initial meal of water were then determined. The sizes of meals taken were plotted, not against the period of deprivation, but against the weight lost by the flies, since their initial meal of water, expressed as a percentage of the amount of water consumed at that meal. Barton Browne and Dudziński (1968) showed that the flies lost negligible amounts of solids during this period of deprivation and, therefore, that total weight loss gave a good estimate of water loss.

It was found that the responsiveness of *L. cuprina* to water remained low until they had lost at least 70 per cent of their initial intake of water. Up to this point many flies failed to respond to tarsal contact with water and only a few took more than 1 mg. As the weight of material lost increased above 70 per cent of their initial water intake, the mean sizes of water meals taken increased sharply. This pattern of change in meal size with increasing deprivation differs from that in all insects for which detailed information is available, in that there is a prolonged period after satiation during which the flies fail to consume significant amounts and during which, therefore, they do not replace their loss. In all other cases studied, meal size increases progressively throughout a period of deprivation and generally reflects the amount of material lost from that part of the alimentary canal in which food is stored, or from the whole body.

Barton Browne and Dudziński (1968) and Barton Browne (1968) made an investigation of the physiological bases for the deprivation-dependent changes in the amount of water taken by *L. cuprina*. They examined crop volume in relation to the percentage weight loss and found that meal size began to increase only some time after the crops of most flies had completely emptied, a result which indicates that changes taking place after water has ceased to pass from the mid-gut to the haemolymph are important in determining the sizes of water meals.

Determinations were made by Barton Browne and Dudziński (1968) of the volume and osmotic pressure of the haemolymph and the concentrations of sodium and chloride ions therein, after different periods of deprivation. These showed that the increase in meal size was well correlated with increasing concentration of the ions, but was less well correlated with changes in volume and osmotic pressure. Barton Browne (1968) examined the effects, on sizes of meals of water taken by *L. cuprina,* of altering the ionic concentrations in, and the osmotic pressure and volume of the haemolymph by injecting water or solutions of various salts or glycerol into the haemocoel, or, in the case of effecting a reduction in volume, by bleeding the fly. The results of the experiments in which various materials were injected into 24 h and 1 h water-deprived flies provide evidence that the chloride concentration and, probably to a lesser extent, osmotic pressure of the haemolymph play a part in determining the size of water meals, the higher the ionic concentration and the osmotic pressure, the higher being the amount of water taken. The rapidity with which the injections altered the responsiveness of the flies to water indicated that the effects were probably neurally mediated. Evidence was obtained also that the volume of the haemolymph or of the total amount of material in the fly has a small effect on the sizes of water meals. The apparent differences between the relative importance of the volume and composition of the haemolymph in *L. cuprina* and *P. regina* (Evans, 1961; Dethier and Evans, 1961) as factors regulating the responsiveness of flies to tarsal stimulation with water were mentioned earlier (section 2.3). In *P. regina*, the effect of volume was found to be more important than those of haemolymph composition whereas in *L. cuprina* the reverse appears to be true.

The results of experiments in which Barton Browne (1968) injected various materials into the haemocoel of water-satiated or deprived flies not only demonstrated the involvement of input concerning certain haemolymph parameters, but also provided direct evidence that input concerning the amount of water in the gut is not a significant factor in determining the sizes of meals taken by *L. cuprina*. Satiated flies, which therefore had substantial amounts of water in the crop, consumed fairly large amounts of water after receiving injections which increased the chloride concentration and osmotic pressure of the haemolymph, and deprived flies, with empty crops, were rendered unresponsive to water by injections which decreased the chloride ion concentration and osmotic pressure. Thus, flies can be caused, by suitable injections, to act as if deprived when they have a substantial amount of water in the crop, and to act as if satiated even when the crop is completely empty.

The rapidity with which dilution of the haemolymph by injection with water caused flies to become unresponsive to water, taken together with the finding by Barton Browne and Dudziński (1968) that the crops of flies dissected within 30 s of the cessation of feeding on water contained less than 80 per cent of the ingested material, suggests that influences generated by input from monitoring systems sensitive to variations in blood dilution might play an active role in bringing about the termination of feeding. If, however, there is no significant dilution of the haemolymph during a meal of water, the effects of input concerning the chloride concentration and osmotic pressure of the haemolymph would play a passive role by determining the level which must be reached by actively changing influences in order to bring about the cessation of feeding. With regard to these haemolymph parameters, it is interesting to note that flies are least responsive to water when the chloride concentration and osmotic pressure of the haemolymph are minimal. The question arises, therefore, as to whether the responsiveness of water satiated flies is reduced by inhibitory influences generated by input concerning these parameters, or whether the large sizes of water meals taken by deprived flies is explicable in terms of the generation of excitatory influences by input elicited by high levels of haemolymph chloride ion concentration and osmotic pressure which would integrate with influences generated by chemoreceptor input. The finding that the volume of the haemolymph or of the whole fly is one of the factors influencing meal size indicates that effects of input from sensory systems monitoring some aspect of volume would play an active role in bringing about the cessation of feeding.

Dethier and Evans (1961) and Belzer (1970) have shown that individuals of *P. regina* with sectioned recurrent nerves become polydypsic after access to water for 1 or 2 h. Because of lack of information on the patterning of the feeding which resulted in polydypsia, it is not possible to interpret the occurrence meaningfully in terms of mechanisms which bring about the cessation of feeding, nor is it possible to relate this finding usefully to those discussed earlier concerning the hyperphagia which results when recurrent nerve section flies are given access to sugar solutions. There is no information about the effect of recurrent nerve section on the water intake of *L. cuprina*.

Some information is available about factors affecting the duration of continuous periods of ingestion in the milkweed bug, *Oncopeltus fasciatus*. Feir and Beck (1963) demonstrated that the average duration of individual bouts of ingestion depended on the stimulation provided by the food. They showed, for instance, that the average duration of individual periods of continuous ingestion on an artificial diet covered with a coat of milkweed seed were greater than when the diet was not so covered. It is interesting to note that this occurred despite the fact that the frequency of feeding was less, and hence the mean inter-meal period greater, for bugs feeding on the uncovered diet. It was also shown that the duration of the periods of ingestion was greater when a mixture of cellulose powder and glucose was presented beneath the milkweed seed coat than when only cellulose powder was present. There is no formal evidence on the relationship between the period of deprivation and the duration of the following meal in *O. fasciatus*, but the finding that antennectomized bugs, or bugs with coated antennae, which had greater inter-meal periods, tended to have longer individual feeds than intact bugs, suggests that meal size might be affected by the duration of the preceding period of deprivation.

Feir and Beck favour the view that cessation of feeding in *O. fasciatus* is brought about by the effects of adaptation of gustatory receptors and sought to explain differences in inter-meal periods in terms of the time required for disadaptation. They provided no direct supporting evidence and the long duration of individual meals (*c.* 30 min) would seem to make it unlikely that effects of receptor adaptation play an active role. These authors suggest, by implication, a seemingly more likely explanation for their results, namely that the insects became dehydrated during feeding because of the use of water for the production of saliva, and that, as a consequence, their responsiveness to the competing presence of a water source within the arena increased until they responded to the water vapour rather than continued feeding. The longer the duration of individual periods of ingestion when there was a high level of stimulation would presumably be, then, a reflection of the greater ability of the food to compete with the stimulus of water vapour. The observation of Bongers (1969) that *O. fasciatus* alternated between feeding on seeds and water and the finding by Saxena (1967) that the seed feeding bug *Dysdercus koengii* became increasingly unresponsive of food stimuli with increasing water deprivation, even in the absence of a competing water source, supports this second explanation.

In conclusion, it must be stated also that the application of the concept of meal size might be inappropriate when applied to seed feeding bugs such as *O. fasciatus* when feeding on natural food, since it has been shown by both Beck *et al.* (1958) and Feir and Beck (1963) that the bugs spend a considerable proportion of their time feeding with the inter-meal period often being shorter than the individual bouts of ingestion. For this reason, it seems that these bouts of feeding may be more reasonably looked upon as segments in an interrupted but basically continuous feeding regime, the interruptions being caused by the insect's need to obtain water or by the exhaustion of the food supply at the point of penetration.

Moloo and Kutuza (1970) examined the effect of deprivation on the sizes of meals of blood taken by males of the tsetse fly *Glossina brevipalpis* and found that there was an increase from about 80 mg after one day of deprivation to about 110 mg after 4 days, the greatest increases occurring between the first and second days of deprivation. Since, in most males of *G. brevipalpis,* crop emptying is completed within one hour, it is clear that actively changing feedback concerning crop volume would have played no part in the observed increase in meal size with deprivation. The results obtained do, however, provide circumstantial evidence that influences generated by input relating to the total volume or weight of material within the flies may play an active role, since the amount of weight loss between the first and fourth days of deprivation was of the same order as the increase in meal size, and since both the greatest weight loss and greatest increase in meal size occurred between the first and second days. Further circumstantial evidence in support of this view is provided by the finding of Tobe and Davey (1972) that the sizes of blood meals taken by another tsetse fly, *G. austeni*, varies according to the state of the reproductive system and that the basis for this variation is that flies terminate their meals when they reach a particular total weight.

Factors responsible for the regulation of the sizes of blood meals in two kinds of blood feeding insects have been investigated in experiments involving nerve section. Maddrell (1963) investigated the effect of the sectioning of the ventral nerve cord between the prothoracic and mesothoracic ganglionic masses on the sizes of blood meals taken by *Rhodnius prolixus*, and found that the operation markedly increased both the volume of blood consumed and the duration of feeding. He determined also the effect of cutting a hole in the abdominal wall and mid-gut such that the ingested blood could leak out, and again found that meal size and duration were substantially increased. From these results and those of Anwyl (1972), who characterized abdominal stretch receptors in *R. prolixus*, it is clear that inhibitory influence generated by feedback concerning the degree of distension of the abdomen plays an active role in bringing about the cessation of feeding. The finding of Maddrell would seem to make untenable the hypothesis of Bennet-Clark (1963) that feeding ceases in *R. prolixus* because back pressure causes the pump to stop operating.

The effect of ventral nerve cord section on the sizes of blood meals taken by several species of mosquito was investigated by Gwadz (1969). He found, in all species examined, that the sectioning of the ventral nerve cord anterior to the second abdominal ganglion resulted in about a four-fold increase in the volume consumed. In experiments with *Aedes aegypti*, he showed that meal size depended on the point at which the cord was sectioned, the more anterior the cut the greater being the size and duration of the meal taken. From these results, it is apparent that inhibitory influences, generated by input concerning the abdominal distension caused by the filling of the mid-gut, play an active role in bringing about the cessation of feeding and that a number of sense organs feeding input to different ganglia are involved.

A number of investigations have been made into the chemical identity of the blood components which induce blood-feeding insects to become engorged. The concept of a gorging stimulus has implicit in it that engorgement is an all-or-none response and that partial engorgement rarely, if ever, occurs, even in circumstances in which the stimulus is one which causes only some insects to feed. There is evidence that this is so in *Rhodnius prolixus* and tsetse fly, *Glossina austeni*. Friend and Smith (1971) state, for instance, with reference to feeding trials to determine whether certain nucleoside phosphates are gorging stimulants for *R. prolixus* that "under the experimental conditions, more than 99 per cent of insects had either engorged, ingesting from 5 to 7 times their body weight, or had ingested very little by the end of the test period. Thus, the response can be considered as either positive or negative". Galun and Margalit (1969), working with *G. austeni*, found that, although different solutions of adenine nucleotides induced different percentages of flies to feed, the mean meal size of those which did feed was similar irrespective of the type of solution offered, a finding which again strongly suggests that gorging is an all-or-none response.

It seems that, in these insects, a solution, which must be very little above the threshold concentration for an individual insect, is just as effective a feeding stimulant as a more concentrated one. Several possible explanations can be offered for this. First, there is the perhaps unlikely possibility that the relevant receptors are such that they give an all-or-none response to stimulation depending on whether or not a threshold is exceeded. The second possibility is that an adequate level of excitatory input triggers endogenous activity in the CNS whose level is independent of the amount of input and which continues until switched off by inhibitory influences. The remaining possibility is that ingestion is, in fact, driven by excitatory influences resulting from the immediate or perseverating effects of excitatory input but that the magnitude of the influence is maximal at any level of input at threshold or above. It is not possible to eliminate any of these possibilities on the basis of available data.

The all-or-none relationship found in *R. prolixus* and *G. austeni* is not, however, found in all blood-feeding insects. Hosoi (1959) found that a relatively high percentage of adults of the mosquito *Culex pipiens*, feeding on saline containing 5′-adenylic acid at concentrations which induced only some insects to feed, were only partially satiated. It seems, therefore, that in this mosquito, differences in level of excitatory input influence meal size.

It can be said in summary, concerning the general question of the regulation of meal size, that the scheme proposed at the beginning of this section would seem to provide an adequate general basis for explaining the effects of deprivation and type of food on meal size. Although considerable information is available about the regulation of meal size in a few species, in no case is sufficient information available, as yet, to allow the construction of anything like a complete and detailed model. In no species is it yet possible, for instance, to assign all the factors involved to the various categories according to the precise roles the influences they generate play in bringing about the termination of meals, to assess the relative importances of each, and to differentiate between roles of influences generated in the CNS as a result of immediate and perseverating effects of sensory input and those of neural activity which is endogenous to the CNS itself. In order to resolve these uncertainties, considerable work involving behavioural and electrophysiological techniques would be required.

2.4.2 *Regulation of the rate of ingestion*

Reference was made during the previous discussion of the regulation of meal size to the relationship in *P. regina* and *L. migratoria* between the rate of ingestion and the type of food being consumed. I discuss here more general aspects of the regulation of the rate of ingestion by insects during a single continuous period of ingestion.

It seems that the relationship between deprivation and rate of ingestion has been studied in only two insects. Holling (1966) investigated the effect of deprivation on the time which the mantid *Hierodula crassa* took to consume one housefly applied to its mouthparts, and found that this remained constant throughout more than 72 h of deprivation. The rate was just as rapid during the period following feeding to satiation during which mantids were refractory to visual stimulation provided by prey, and hence would not normally feed, as when they were fully responsive. The second species for which there is information about the relationship between the period of deprivation and rate of ingestion is also a predator, the bug *Podisus maculiventris*. Gallopín and Kitching (1972) found that the rates of ingestion for this insect feeding, on larvae of *Galleria mellonella*, were similar after 24 and 48 h of deprivation, but that the rate was somewhat less after 72 h, presumably because of a reduction in general vigour as a result of the prolonged period of deprivation. It should be pointed out that, in contrast to the situation in Holling's experiments, the relationship in this bug was investigated only during the period when it would attack prey voluntarily. In neither predator, however, was there any indication that deprivation influences the rate of ingestion.

Somewhat indirect evidence for an effect of deprivation on the rate of ingestion is provided by some results of Bernays and Chapman (personal communication) which have already been referred to briefly. They found that increases in the osmotic pressure of the haemolymph of larvae of *L. migratoria* not only caused a reduction in meal size but also a reduction in the rate of ingestion. Thus, the alteration of one parameter, such that it approximates the condition in recently fed insects, caused a reduction in the rate of intake. An increase in osmotic pressure to a level typical of that in insects which had just fed to repletion caused a reduction in the rate of ingestion of only about 8 per cent, but the degree of reduction was greater at higher osmotic pressures.

Rather more information is available on the relationship between the rate of ingestion of discrete meals taken by severely deprived insects and the type of food on which they are feeding. Dethier *et al.* (1956), as mentioned earlier, found that the rate of ingestion of various sugar solutions by *Phormia regina* was, except when high viscosity interfered, generally well correlated with the stimulating power of the solution, as indicated by the duration of the meal. It is clear, therefore, that the rate of ingestion by *P. regina* is related to the level of excitatory input reaching the CNS.

Also mentioned earlier (Table 1) was the finding by Bernays and Chapman (1972b) that the rate of intake by larvae of *L. migratoria* varied according to the type of plant material being ingested, the rate being positively correlated with meal size and duration. Since it is known that feeding deterrents are responsible for the relatively small sizes and short durations of meals of seedling grasses (Bernays and Chapman, 1974a), it seems reasonable to believe that inhibitory input from chemoreceptors stimulated by the food material plays a role in determining the rate of ingestion by these larvae.

Moorhouse *et al.* (in preparation) studied the rate of intake of water and various sucrose solutions by adults of the Australian plague locust, *C. terminifera*. They found that the overall rates of intake throughout a meal were similar for insects fed water, 0.125, 0.5 or 1.0 M sucrose, as were the rates measured over a specific period early in the meal, despite the facts that the sucrose solutions were more powerfully stimulating than water, and that their stimulating power increased with concentration at least up to 0.5 M (Barton Browne *et al.*, in press, a). Further information on the relationship between stimulus and rate of ingestion in *C. terminifera* was, however, obtained in an experiment in which locusts were fed alternated drops of water and the more powerful feeding stimulant 1.0 M sucrose. It was found that drops of the two solutions were consumed at equal rates for most of the meal but that, as the end of the meal was approached, drops of sucrose solution were consumed more rapidly than drops of water, until the insects finally ceased feeding with water remaining unconsumed on the mouthparts. It seems, therefore, that under the conditions which exist towards the end of a meal of alternated drops of sucrose solution and water, the rate of ingestion is related to stimulus strength. It should be noted, however, that the differential between the rates of ingestion did not become apparent until the insect had consumed a total volume which is greater than that taken by locusts fed meals consisting of only water.

Two possible explanations can be suggested for the results obtained by Moorhouse *et al.* The first is that there might be, in fact, a relationship throughout the whole meal between the level of motor output responsible for ingestion and that of the excitatory input from the chemoreceptors, but that the maximum rate of intake is limited by some physical parameter such as the diameter of oesophagus but that, as back pressure builds up, differences in the vigour of pumping manifest themselves as differences in the rate of ingestion. This possibility can be dismissed as unlikely on the grounds that no significant relationship were found between the degree of the decline in the rate of intake over the course of a meal and the size of the meal. A second possibility, namely that the motor output is, in fact, independent, over a wide range, of the amount by which excitatory influences exceed inhibitory ones but that a relationship becomes apparent when this margin becomes very small, seems to be more likely.

Saxena (1963) examined the rate of ingestion of water and sugar solutions by the bug *Dysdercus koenigii* and found that the rates of ingestion were almost identical irrespective of the type of fluid being consumed. If it is assumed that sugars are feeding stimulants for this bug and that any effects of increased stimulation are not being offset by effects of increased viscosity, the results can be taken as an indication that the level of motor output involved is independent of the level of excitatory input from chemoreceptors. This conclusion is supported by Saxena's finding that the rate of ingestion did not decrease even just before the insects voluntarily ceased to feed on the fluid.

The information available on factors controlling rate of intake, although very fragmentary, is sufficient to indicate that the dependence of rate of intake on the characteristics of the input reaching the CNS differs between species. Casual observations on feeding insects indicate that the motor output concerned with ingestion, such as the action of the mandibles in insects feeding on solid materials and of the cibarial pump of fluid feeders, is often rhythmical. This suggests strongly that it might represent the output of a central oscillator. If this is so, it seems that the oscillator in some insects might always operate at constant frequency whereas, in others, the frequency of the oscillator might be influenced by the magnitude of excitatory influences in the CNS. In any future studies of factors affecting the rate of ingestion, it might be profitable to consider that one is possibly dealing with the output of an oscillator and to consider the problem of the control of ingestion rate in the light of the body of information which now exists about oscillators in general (see Miller, 1974).

3 Long-term regulation of intake

In the previous sections, much has been said about the variations in the behaviour of insects according to their state of deprivation. In these discussions, the implicit assumption has been made that behavioural changes of the kinds discussed allow insects to regulate, at appropriate levels, their intake of materials on which they feed throughout their lives, or throughout one stage of their life cycle, and which are required for maintaining life. In other words, it has been implied that insects use behavioural means to achieve a measure of metabolic homeostasis. I discuss here several aspects of the long-term regulation of intake in order to assess the degree to which precise regulation of intake over a period is achieved.

3.1 CONSTANCY OF INTAKE

It is well known that physiological changes related to events in an insect's life-cycle often cause changes in feeding behaviour, an aspect of the regulation of feeding whose detailed discussion I have placed beyond the scope of this review. There are, however, periods in the life-cycle when insects remain for considerable periods in a fairly constant overall physiological state. If effective regulation of intake is occurring, it could be expected that, during periods such as these, the intake of individual insects would show some constancy. Whether such constancy would be apparent in terms of the amount of food consumed per hour, per day or per week or per any other unit of time would depend on the way in which the insect patterns its feeding. It seems that on only one occasion has the intake of one material by individual insects been determined over a considerable period. Gelperin and Dethier (1967) measured the daily intake of 0.1 M sucrose by an individual male of *Phormia regina* over its full life span of 60 days. They found that there was a general decline in intake throughout the period and that considerable day-to-day variation was apparent, much of which the authors state was correlated with changes in ambient temperature. Over some periods within the 60-day period, notably from 4-15, days, however, there is evidence that intake might have been regulated at a fairly constant general level, though day-to-day fluctuations were quite large.

A number of workers have reported that the combined intake of groups of insects remains constant over a period. Although this constancy may be an indication of constancy at the level of individual insects, it need not necessarily be so, since constancy of intake by groups could equally well be explained by the occurrence of out-of-phase systematic inconstancy of intake by individuals. Constancy by groups indicates, strictly, only that there is no overall trend during the period under consideration and is evidence, therefore, for only a very general kind of regulation of intake. For this reason no detailed discussion is given of this type of finding. Amongst insects, various sized groups of which have been shown to maintain a constant level of intake of one or more materials over a considerable period, are males or nonprotein fed females of the blowflies, *Calliphora erythrocephala* (= *C. vicina*) (Strangways-Dixon, 1961a, 1961b) and *Lucilia cuprina* (Roberts and Kitching, 1974), males of the cockroach *Blatella germanica* (Gordon, 1968), mixed sexes of the mantid *Hierodula crassa* (Holling, 1966) and the predacious coccinellid *Chitoconus bipustulatis* (Yinon, 1969).

3.2 EFFECT OF DEPRIVATION ON SUBSEQUENT *AD LIB.* FEEDING

Several workers have examined the effects of periods of deprivation on either the amount of food consumed by insects over a period, or on their pattern of feeding, when again given *ad lib.* access to food.

McLean and Kinsey (1969) showed that the feeding activity of the pea aphid *Acyrthosiphon pisum* was influenced by a previous period of food deprivation. They used an electronic method to compare the feeding behaviour of unstarved aphids with that of individuals which had been deprived of food for various periods, and found that deprived aphids began feeding more rapidly than fed individuals, and that their total period of ingestion, over the 24 h of the trial, was greater. It was shown that the feeding pattern of the fed and 24 h deprived aphids did not become similar until about 16 h after the end of their period of deprivation. It is clear therefore that this insect which, when given continuous access to food, imbibes continuously for protracted periods, feeding under some circumstances for up to 21 out of 24 h, is capable of responding to a period of deprivation by increasing the amount of food it subsequently ingests.

Gordon (1968) examined the intake of sugars by males of the German cockroach, *Blattella germanica*, for several days after they had been deprived of food altogether, or had had access to another sugar or some other material. He found that previously deprived cockroaches, given access to stimulating sugars such as sucrose or glucose, consumed larger amounts on the first day of access than on subsequent days, but that this effect was not apparent when the insects were given access to a number of other less stimulating sugars after deprivation. Also, he found that cockroaches consumed larger amounts on their first day of access to sucrose after having previously had available only one of the less stimulating carbohydrates. It seems, therefore, that this cockroach can compensate for deprivation, but is able to do so only if the food is sufficiently stimulating.

Somewhat similar, but rather more readily interpretable, experiments were performed with the same insect by van Herrewege (1971). He selected for a study a 14-day period over which the daily consumption by males was constant when given continuous access to food. He deprived his experimental group of insects of food for 3 days and found that their intake on the first day during which they again had access to food was more than twice the normal amount and that the intake was somewhat greater than usual during the following 2 days. When he plotted the cumulative intake during the 14-day period, he found that the total amount taken by the controls which had continuous access to food was identical with that of the experimental group which suffered 3 days of deprivation. Clearly, the feeding pattern over the days following the period of deprivation was such that the insects almost exactly made up for their lack of intake during the period of deprivation.

Sinoir (1968) examined the effects of periods of deprivation, greater than the 8 h required for complete emptying of the alimentary canal, on the total food intake by larvae of *Locusta migratoria* during the 24-h period following deprivation. When acceptable plant material was made available, the total intakes of 8 h and 30 h deprived insects were similar. It was found, however, that the more deprived insects consumed significantly more when the food provided was wet or dry ground paper soaked in sucrose at a concentration of about 0.002 M, the insects fed on this material ingesting much less in total than the similarly treated ones which had access to the plant material. The result would seem to indicate that lack of excitatory input from the chemoreceptors was limiting the intake of the sugar-bearing paper but that changes in the locust, which take place after the emptying of the alimentary canal is complete, cause the level of input from some internal origin to alter so that the insect is rendered more responsive to the small amount of excitatory input provided. The lack of difference shown between the two groups which fed on the highly palatable plant material indicates that this difference in the input due to the different states of deprivation is not important when the level of excitatory input from the chemoreceptors is higher.

The patterning of protein feeding by the blowfly, *Phormia regina*, is known to be closely related to the insect's ovarian cycle (Dethier, 1961; Belzer, 1970). It has been shown, however, by Belzer (1970) that the amount of protein consumed by these insects is influenced also by the period of protein deprivation and that similar effects can be demonstrated in males of this species. He observed there was a small peak in protein feeding from about 10-40 h after emergence in both males and females, which had continuous access to protein, the peak in females being distinct from a later one associated with the onset of ovarian development. When access to protein was delayed, it was found that flies immediately fed on protein as soon as it was provided and that their intake was greater than that observed during the normal peak shown by flies which had continuous access. In males which had been deprived of protein for the 4 days following emergence, the amount of protein consumed during the first 5 h after they were again given access to it was of the same order as the total amount taken during the four days following emergence by flies which had continuous access. In females, the situation is more complicated because delayed access causes the early peak of protein intake to coalesce with the one associated with vitellogenesis, the intake rate of the previously protein-deprived flies being greater for some considerable time after provision of the protein than that during either peak when flies have continuous access.

In other experiments, Belzer deprived flies of protein for a considerable period later in adult life and found that there was in females, whose protein intake was generally at a low level because they were still carrying the first fully developed batch of eggs, an increase in protein intake after periods of protein deprivation of greater than 100 h. No convincing increase was apparent in males. In females, the increase in the amount taken would again seem to be of the same order as the amount that would have been consumed during the period of deprivation.

The small number of examples discussed in this section show clearly that some insects, at least, have regulatory mechanisms which allow them to compensate by increased intake for food not consumed during a period of deprivation. The finding that such a compensatory mechanism exists even in an apterous aphid, which it seems would rarely suffer periods of deprivation and which normally feeds more or less continuously, indicates that the ability to compensate for periods of deprivation may be universal among insects, or nearly so.

3.3 EFFECT OF DILUTION OF THE FOOD ON INTAKE

Investigations have been made of the effect, on intake over a period, of the dilution of a particular food usually with material lacking in powerful feeding stimulants or deterrents and in nutritional value. In some experiments, cellulose was used as the diluent and in others water.

Dadd (1960) found that the total consumption by larvae of the locusts *S. gregaria* and *L. migratoria*, over a 3-day period, of an artificial diet containing 26, 40 or 53 per cent cellulose powder increased with increasing quantity of diluting cellulose such that the weight of utilizable material consumed was essentially constant. McGinnis and Kasting (1967) examined the effect of the dilution of sprout meal with cellulose on the total intake of the grasshopper *Melanoplus sanguinipes* over 5 days, and found that the increase in intake in response to dilution was such that the actual amount of sprout meal ingested remained unchanged so long as the degree of dilution was 7-fold or less.

Gordon (1968) compared the total intake, over a period of some days, by males of the cockroach *Blattella germanica* feeding on a dry sugar with that of males feeding on the same sugar diluted with an equal amount or greater of cellulose powder. For each of the sugars tested, it was found that the addition of the cellulose increased the total amount of material ingested, but that the increase was insufficient to maintain the actual intake of carbohydrate at a constant level.

The precise physiological basis for the compensation of these insects for the dilution of their food with cellulose is unknown. It can be stated, however, that, since cellulose alone is consumed in only small quantities by *M. sanguinipes* and *B. germanica*, it is improbable that the increased intake can be explained in terms of increased palatability of the cellulose diluted material. It seems, rather, that it is related in some way to the decreased digestibility or utilizability of the cellulose-containing food. No specific information is available, however, about the feedback mechanisms involved. A possible general mechanism for the compensation by insects for the dilution of the food will be discussed later in relation to results obtained with the blowfly *P. regina* feeding on sugar solutions of different concentrations.

House (1965) examined the effect of the dilution of the nutrient component of an artificial diet on the total intake by larvae of the sphingid *Celerio euphorbiae* over a period of 4 days. He found that the diets with lower nutrient contents were consumed in greater quantities, but that the increase in intake was insufficient to offset the degree of dilution, the intakes of diets diluted by factors of 1.2, 1.4 and 2.0 being increased by factors of 1.1, 1.2 and 1.3 respectively. It is interesting, however, that the growth rates of the larvae were similar on the 4 diets.

David and Gardiner (1961) determined the amounts of different concentrations of honey solution consumed by adults of the cabbage white butterfly, *Pieris brassicae*. They found that the total volume consumed by the butterflies over a period of 5 or 6 days was larger when a 1 per cent solution was available than when the concentration was 10 per cent and that this, in turn, was taken in larger quantities than a 20 per cent solution. For the first one or two days, the intake of the 10 per cent solution exceeded that of the 1 per cent solution but thereafter the intake of the 1 per cent solution was greater by up to about 50 per cent. Thus, even though the insects increased their intake in response to dilution, at no stage did they come close to compensating for the 10-fold dilution. However, a comparison between the intake of 10 and 20 per cent indicates that, although the intake of the two solutions was similar for the first 2 days, the intake of 10 per cent solution on days 3-5 was approximately twice that of the 20 per cent solution. Thus, during this period, the insects compensated adequately for this 2-fold dilution.

Generally similar results have been obtained with the blowfly *P. regina*. Dethier and Rhoades (1954) showed that flies with access to 1.0 M sucrose consumed a greater volume on the first day than those with access to 0.1 M sucrose but consumed less thereafter. The difference in the rate of consumption after the first day was insufficient, however, to compensate for the reduction in concentration, the 10-fold reduction resulting in only about a 2-fold increase in intake. Additional experiments were performed in which flies were allowed to choose between a particular concentration of a given sugar and water. In these, the insects took relatively little water, the amount varying only little according to the kind of sugar solution provided. It was found, for several different sugars, that the volume of solution consumed decreased with increasing concentration above 0.01 M. In no instance, however, was the compensation adequate to prevent the actual weight of sugar consumed from increasing with increasing sugar concentration.

More recently, Gelperin and Dethier (1967) carried out a further series of experiments to investigate these effects in *P. regina*, and examined their findings in the light of information concerning the regulation of feeding in this fly. In one experiment, they determined the intake of flies which received 0.1 and 1.0 M sucrose alternately, access to each solution being for 2 or more days. They found that the mean rate of consumption when the flies had access to the 0.1 M solution was about 1.4 times that on days when the 1.0 M solution was provided. In another kind of experiment, they examined the effects of sugar concentration on the sizes of meals taken by flies given access to a solution only once per day. They found that flies which were given 0.5 M fructose each day consumed about 18 μl each time the solution was offered whereas those offered a 2.0 M solution took meals averaging only 11 μl. As was found when flies were allowed to feed *ad lib.*, the intake of the lower concentration was greater, but again the increase was insufficient to compensate for the degree of dilution.

Superficially, these findings would appear hard to reconcile with previously discussed findings that the sizes of meals of solutions of any particular sugar taken by 24 h starved *P. regina* generally increased with increasing concentration. Gelperin and Dethier (1967) have, however, provided a scheme, since formalized by Gelperin (1972), which adequately reconciles the two sets of data. Central to this is the finding by Gelperin (1966a) that the rate of crop emptying in *P. regina* is inversely related to the osmotic pressure of the haemolymph, which is, in turn, influenced by the osmotic pressure of the solution absorbed from the mid-gut. The result of this relationship is that ingested solutions of high osmotic pressure empty from the crop more slowly than ones of lower osmotic pressure. This information, coupled with an assumption that whenever a fly feeds on a certain kind of solution it will cease to do so when its crop volume reaches a level characteristic for that particular solution, irrespective of the crop volume at the beginning of the meal, provides an adequate general explanation for the observed relationship between total intake over a period and the concentration of sugar solution provided. Although there is no direct evidence that feeding always ceases when the crop volume reaches a level which is typical for the solution being ingested, the assumption appears justified in view of the finding by Dethier and Gelperin (1967) that flies feeding *ad lib.* on a single solution have a steady-state crop volume which varies according to the stimulating power of the solution. Further justification is provided by the knowledge that excitatory influences generated by input from the chemoreceptors and inhibitory influences generated by input from a sensory system monitoring abdominal distension are important in the regulation of meal size.

In order to illustrate the scheme in specific behavioural terms, I will consider the feeding behaviour that could be expected when a fly, which had previously been maintained on a dilute sugar solution, is given access to a concentrated and highly stimulating solution, and then that of a fly given access to a less concentrated solution of the same sugar. The fly, having been maintained on a dilute solution, initially has a low crop volume, so when it first contacts a concentrated sugar it takes a large meal, ceasing to feed when the crop volume reaches a specific high level. Initially, crop emptying rate is relatively high, because of the fly's initial low haemolymph osmotic pressure, so it is likely that the fly will ingest more solution after a relatively short time, again ceasing to feed when the crop volume is approximately the same as that at the end of the initial meal of the concentrated solution. After this initial phase, crop emptying rate settles down to a slower steady-state characteristic of the osmotic conditions induced in the haemolymph by the concentrated solution being absorbed. When this phase is reached, the fly will feed recurrently at intervals determined by the time course of changes in locomotor activity and tarsal thresholds to the particular sugar, each meal being terminated at approximately the same crop volume on each occasion. According to this scheme, flies feeding on a concentrated and highly stimulating solution would be expected to have a high rate of ingestion on the first day of access to the solution and would, thereafter, consume food at a slower rate reflecting almost exactly the rate of loss of material from the crop. This expectation is in accord with the actual findings of Dethier and Rhoades (1954) for flies feeding on 1.0 M sucrose.

When a fly is given access to a more dilute and less stimulating solution its initial intake is smaller but the steady-state rate of crop emptying is, because of the low osmotic pressure of haemolymph, greater than for a more concentrated solution. It could be expected, therefore, that the peak of intake during the first day of access would be smaller than with the concentrated solution, or even nonexistent, but that subsequent intake, being an accurate reflection of crop emptying rate, would be higher. This again accords with the actual results obtained by Dethier and his co-workers.

Complete as the scheme provided by Gelperin (1972) appears to be, it should be pointed out that it could be made more so if direct experimental information were available about the crop volume at the end of meals taken by flies with different initial crop volumes, and on the details of the temporal patterning of feeding by flies feeding on different solutions. It seems certain, however, that the greater intake by flies of dilute solutions results from their having a greater rate of loss of ingested material from the crop and that, after the first day of access to the particular sugar solution, the rate of crop emptying becomes the sole factor controlling the rate of ingestion.

It is not known whether the increased intake of diluted foods by other insects can be similarly explained. For this to be so, the regulatory mechanism of the insect concerned would have to have two characteristics in common with *P. regina*. It would first be necessary that the rate of loss of food from the part of the gut (fore-gut or crop) in which ingested material is held prior to its entering the mid-gut should be more rapid for food containing smaller proportions of digestible and/or utilizable material and secondly that the greater rate of loss should result in increased ingestion. This second requirement implies that, whenever the insect feeds, it consumes an amount which is approximately equal to the amount lost from the fore-gut or crop.

It appears that information about the relationship between the rate of emptying of the crop or fore-gut and the composition of its contents is available for only one insect other than *P. regina*. The rate of crop emptying in the cockroach, *Periplaneta americana*, was shown by Treherne (1957) to increase with decreasing concentration of glucose solution. It is uncertain whether the osmotic pressure of the crop contents themselves is the direct determinant of crop emptying rate in this insect or whether the osmotic pressure of the haemolymph is the important factor, as has been demonstrated in *P. regina*, but results of nerve sections performed by Davey and Treherne (1963) on the stomatogastric nervous system suggest that receptors in the pharynx, the input from which is carried via nerve 5, might directly monitor the osmotic pressure of the ingested fluid.

The amount of information concerning the relationship between the amount of fore-gut emptying and meal size is equally small. The late instar larva of *L. migratoria* is the only insect on which a detailed study of this relationship has been made. Sinoir (1968) and Bernays and Chapman (1972b) have shown that this insect takes meals which do, in fact, approximately replace the material lost from the fore-gut during the preceding inter-meal period. As pointed out above, Dadd (1960) demonstrated that this insect was able to compensate very effectively for the dilution of its diet with cellulose. In view of the findings of Sinoir (1968) and Bernays and Chapman (1972b), the conclusion seems inescapable that crop emptying rate must be greater when the diet is diluted. Bernays and Chapman (1974a) express the opinion that the basic mechanisms for the regulation of feeding are probably basically similar for all acridids. If this is so, it seems likely that the increased intake of *Melanoplus sanguinipes* in response to dilution, reported by McGinnis and Kasting (1967), is also a reflection of increased crop emptying rate.

On the basis of the small amount of evidence available, it seems to be reasonable to conclude that mechanisms underlying the ability of insects to compensate for the dilution of their food with a nondigestible material by increasing their total intake may prove to be generally similar to those which are believed to be responsible for this ability in *P. regina*. Much more evidence about these aspects in insects with diverse patterns of feeding is required, however, before this statement can be made with any certainty.

The finding by Latheef and Harcourt (1972) that larvae of *Leptinotarsa decemlinata* consume more tomato than potato foliage per day is perhaps an illustration under less artificial conditions of a phenomenon similar to the effects of dilution discussed above. For tomato, the approximate digestibility (AD), and the efficiency of conversion of ingested food (ECI) and of digested food to body substance (ECD) is much lower than for potato, all indices being based on dry weight values. It seems that the markedly larger quantity of tomato ingested (1.5 to 3 times depending on instar) as compared with potato is related either to its higher digestibility or its lower efficiency of utilization. This view is supported by the finding that adults of this species, for which tomato is somewhat more nutritious but slightly less digestible than potato, consumed slightly more potato than tomato over the first 5 days after emergence.

In a somewhat similar study, Mehta and Saxena (1973) examined the relationship between the consumption of various plants by 3rd instar larvae of the bollworm *Earias fabia* and various parameters relating to the nutritive value of the food and the degree to which it was utilized. They showed that the intake over 24 h was generally negatively correlated with ECI and ECD, based on wet weights, except that the intake of *Zea mays*, which is the most efficiently utilized plant, appears to be higher than would be expected if intake and nutritive value were causally related. A more convincing relationship is the correlation between AD, again on a wet weight basis, and intake. The correlation suggests that the causal relationship might therefore be between digestibility and intake. If this were so, it would seem to indicate that the rate of gut emptying might be important in determining the rate of ingestion.

Before leaving the topic of the effects of dilution, it is necessary to point out that, in two of the insects which are known to compensate for dilution, the total amount of food ingested over a period does not increase indefinitely with decreasing concentration. Dethier and Rhoades (1954) demonstrated, for instance, that the volume of sucrose solution consumed over 4 days by *P. regina* decreased with decreasing concentration below 0.01 M, intake being maximal at this concentration. Similar maxima were demonstrated for flies feeding on glucose (Dethier and Rhoades, 1954), fucose, sorbose and mannose (Dethier *et al.*, 1956). Similarly, McGinnis and Kasting (1967) showed that the total intake of sprout meal diluted with 15 times its weight of cellulose powder by *Melanoplus sanguinipes* was less than that of meal diluted with 7 times its weight of cellulose, even though total intake was inversely related to the degree of dilution up to 7-fold. In both these examples the diluent is a much less powerful feeding stimulant than the material being diluted, with the result that the amount of excitatory sensory input elicited would decrease progressively with decreasing concentration of the stimulating material. At moderate levels of dilution it seems that the internal regulatory mechanisms are able, in terms of total volume intake, to more than compensate for the decline in excitatory input but that, at greater dilutions, the lack of excitatory input becomes limiting.

A number of other examples are known in which the intake by insects over a period is apparently limited by the inadequate nature of the sensory input provided by the food, the quantities of inert substances eaten by several insects having been shown to increase with the addition of increasing amounts of feeding stimulants. Meisner *et al.* (1972) have shown, for example, that, over 72 h, larvae of *Spodoptera littoralis* generally consumed more of an inert plastic carrier soaked with 0.25 M of a particular sugar than of material soaked with a 0.0625 M solution of the same sugar. Similarly, Dadd (1960) has demonstrated that the amount of filter paper ingested by larvae of *Schistocerca gregaria* over a period of 3 days increased with increasing concentration of added sucrose solution up to 0.1 M, and Norris (1970) showed that the feeding by *Scolytus multistriatus* on pith discs over 2 days was enhanced by the addition of *p*-hydroquinone.

In a number of insects, also, it has been shown that the intake, over a period, of artificial diets of high nutritive value may be limited by their low level of feeding stimulant activity. This has been shown for several species of aphids feeding on artificial diets from which one or more amino acids were omitted or in which they were included at lower than usual levels (Srivastava and Auclair, 1971; Mittler, 1967, 1970; Harrewijn and Noordink, 1971; Arn and Cleere, 1971). The addition of feeding stimulants to diets is known to increase the food intake of larvae of a number of lepidopterous and coleopterous insects. In many of the investigations, however, the period of feeding trials was 24 h or less and there must be some doubt, therefore, as to whether their intake during the trials is necessarily an accurate reflection of their intake over a longer period. In a few instances, however, intake has been estimated over some days or even weeks of larval life. David and Gardiner (1966) showed for example that the intake of diet by larvae of *Pieris brassicae* over 3 days was increased by the addition of dried cabbage or sinigrin and Norris and Baker (1969) have shown that ethanol increased the total amount of tunnelling in an artificial diet over a period of 3 weeks by larvae of *Xyleborus ferrugineus*.

It is known also that the intake of insects over a period may be limited by the presence of feeding deterrents. Norris (1970) showed, for example, that feeding by *Scolytus multistriatus* over 10 days on pith soaked with the feeding stimulant *p*-hydroquinone was reduced by the addition of the feeding deterrent *p*-benzoquinone.

3.4 TEMPORAL PATTERNING OF INGESTION

Another aspect of long-term regulation about which there is some information is the relationship between the temporal patterning of feeding and what is known about mechanisms regulating ingestion. Blaney *et al.* (1973) have made a detailed study of the patterning of feeding by larvae of *L. migratoria* and it is possible to relate, more or less successfully, their information to the previously discussed data concerning the changes with increasing deprivation of the components of feeding behaviour in this species. These authors examined the patterning of feeding behaviour at different times in the 5th instar, but since detailed discussion of the effects of life cycle on feeding behaviour is beyond the scope of this review, I will discuss only the results they obtained from larvae mid-instar.

The most clear-cut patterning of feeding occurred under stable environmental conditions rather than under ones which would have tended to stimulate the insect to locomotor activity. Under conditions in which the general level of locomotor activity was relatively low, the insects usually aggregated their feeding activity into discrete meals separated by fairly long inter-meal periods. When larvae were under conditions which promoted locomotor activity, feeding was more frequent and was less clearly aggregated into discrete meals. Since it is likely that the role of endogenous factors is most clearly revealed under stable conditions not stimulatory to locomotion, it is the performance of the insects under these conditions which will form the basis for the present discussion. The patterning of feeding by larvae under a constant temperature of 30°C and in the absence of a radiant heat source is typical of that under stable conditions. In these circumstances, the average duration of meals, whether continuous or made up of several periods of ingestion, was of the order of 9-14 min and on 88 per cent of occasions the inter-meal period exceeded 20 min and on more than half of the occasions exceeded 50 min.

The behavioural changes shown by larvae of *L. migratoria* as the period of deprivation increases have been discussed in some detail earlier in this review. Briefly, it is known that, as the period of deprivation increases, larvae show increasing locomotor activity, become more responsive to food odours and palpate more frequently. There is also a progressive increase in the number of functional sensilla on the maxillary palps, in the number of insects which proceed from palpating to biting and in the size of meal taken. The time courses of a number of these changes have been established either in absolute terms or in relation to the inter-meal period.

In a series of observations made under conditions very similar to those under which the patterning of behaviour was investigated, Blaney and Chapman (1970) demonstrated that the larvae, although engaging in some locomotor activity and palpation of the substrate throughout the inter-meal period, show a very marked and sudden increase in both activities just prior to the beginning of feeding. It is clear from these findings that rarely, if ever, do larvae begin a meal without first exhibiting increased locomotor activity and an increase in palpation. An important result of this increased activity and palpation is that it increases the chances that the sensilla on the tip of the maxillary palps will contact food, a happening which is known to play an important role in the initiation of feeding (Bernays and Chapman, 1974a). That some locomotion and palpation occur earlier in the inter-meal period, especially soon after the cessation of the previous meal, indicates, however, the probability that these sensilla might contact food a number of times before feeding actually occurs. Presumably during this period the reduction in the number of operative sensilla and probably also central refractoriness to the normally excitatory input from these and other sensilla prevent the insect from feeding.

Blaney *et al.* (1973) found that the durations of meals taken by larvae feeding *ad lib.* at 30° C under stable conditions were remarkable independent of the previous inter-meal period so long as this exceeded about 30 min, a result which appears to conflict with the finding by Sinoir (1968) and Bernays and Chapman (1972b) that meal size in larvae of *L. migratoria* increases progressively with increasing period of deprivation up to 3 or 4 h after their having fed to repletion. It would seem that the basis for this apparent discrepancy lies in differences between *ad lib.* feeding and a feeding regime in which the insects feed at times decided upon by the experimenter. The data on the relationship between meal size and period of deprivation were obtained from experiments in which larvae were deprived of food for various periods at the end of which they were transferred to containers with food. It is certain that any larva, which at the time of its transfer had not yet reached the critical state of deprivation which would have caused it to feed spontaneously if food were available, would have been put into a locomotorily active state by the disturbance during transfer. On the evidence of the *ad lib.* feeding patterns displayed by larvae under conditions stimulatory to locomotion, it is certain that these larvae, which had been rendered prematurely active, would have been put into a state in which they would feed immediately if they encountered food. In the experiments of Sinoir (1968) and Bernays and Chapman (1972b), therefore, after shorter periods of deprivation, a proportion of larvae would have been induced to feed prematurely and therefore would almost certainly have taken smaller meals than those normally taken by larvae feeding *ad lib.* under conditions nonstimulatory to locomotion. Conversely, with longer periods of deprivation, a number of larvae would have been artificially prevented from feeding until after they would normally have done so, with the result that they would possibly have taken larger than normal meals. It seems reasonable to suppose, therefore, that the basis for the relationship between meal size and period of deprivation demonstrated by Sinoir (1968) and Bernays and Chapman (1972b) is that, as the period of deprivation increased, the proportion of insects which had been induced to feed prematurely progressively decreased and the proportion of insects whose feeding was artificially postponed increased. When feeding *ad lib.*, under conditions nonstimulatory to locomotion, it seems, on the evidence of Blaney *et al.* (1973), that insects feed when a particular state of deprivation is reached and that differences in inter-meal periods indicate the variability in time taken for the individual insect to reach this state. This variability can probably be ascribed largely to physiological differences between individuals, but may be due in part to variation between individuals in the amount of material in the crop at the end of the previous meal, or in the level of locomotor activity during the inter-meal period.

In summary, it seems that larvae of *L. migratoria* feeding *ad lib.* under stable conditions nonstimulatory to locomotion, remain quiescent and, therefore, have a reduced chance of contacting food until they reach a critical state of deprivation. When this is reached, the insects become active, palpate frequently, bite readily and then feed for a standard time. If, however, conditions are such that the insect is prematurely stimulated to locomotion, its chances of contacting food with the receptors of the maxillary palps before it reaches the critical stage of deprivation are increased. When this occurs the insect may feed for a period which is less than the standard time for which spontaneously feeding insects continue to ingest.

Another insect, the feeding patterns of which have been studied in some detail, is the larva of the cabbage butterfly, *Pieris brassicae* (Ma, 1972). This insect, like the larva of *L. migratoria*, takes its food in the form of discrete meals, but its feeding behaviour differs from that of *L. migratoria* in the important respect that there is a significant relationship between meal duration and the length of the preceding inter-meal period. Because of this difference, I will briefly discuss the results of Ma in relation to those obtained with *L. migratoria*, even though Ma did not investigate the effects of feeding and deprivation on any components of feeding behaviour to which the feeding patterns in *P. brassicae* can be related.

The longest inter-meal period shown for *P. brassicae* in Ma's data is about 50 min. On 8 occasions the inter-meal period in *P. brassicae* exceeded 30 min and on 7 of these the duration of feeding exceeded 100 s, whereas the meal duration exceeded this figure on only 4 out of 22 of the occasions on which the inter-meal period was less than 30 min. An examination of Ma's data for the relationship between individual inter-meal periods and duration of the following meals shows that 6 out of 9 of the insects with inter-feed periods of between 20 and 30 min subsequently fed for less than 100 s. Thus, on most occasions on which the inter-meal period corresponded to about half the maximum recorded, feeding was for less than the maximum observed duration. This contrasts with the situation in *L. migratoria* in which, judging from the data of Blaney *et al.*, shorter meals were commonly seen only if the duration of the inter-feed period was less than 30 min, which represents only a small proportion of the maximum observed inter-meal period of more than 150 min. If it is assumed that meal size in *P. brassicae* is related to its physiological state of deprivation, as it appears to be in *L. migratoria*, it follows that *P. brassicae* takes meals when in a less standardized state of deprivation than *L. migratoria*. Because of the lack of information about the effects of feeding and deprivation on the components of feeding behaviour in *P. brassicae*, it is not possible to be at all certain about the basis for this difference. It seems possible, however, that it might be related to differences in behaviour between the two insects upon the completion of a meal. *P. brassicae*, like most lepidopterous larvae, remains on its food during inter-meal periods. It seems, however, that *L. migratoria*, on the evidence of the statement by Blaney *et al.* (1973) about the importance of perforated panels for "roosting" and the observation by Barton Browne *et al.* (in press, a) that the locust *C. terminifera* marks the end of a meal by making vigorous leg movements, commonly leaves the food upon the termination of a meal. This means that, at the end of an inter-meal phase, *L. migratoria* must become active in order again to achieve contact with the food, whereas *P. brassicae* is already in contact. Thus, commonly, *L. migratoria* does not receive gustatory stimulation from the food during the inter-meal period, whereas *P. brassicae* can do so at any time. It is likely that the result of this would be that endogenous mechanisms would be relatively more important in determining the feeding pattern of *L. migratoria* than of *P. brassicae*.

4 Some factors other than feeding and deprivation which affect feeding behaviour

The bulk of this review has been concerned with the effects of feeding and deprivation on subsequent feeding behaviour. It is known, however, that a number of factors other than the state of deprivation affect the feeding behaviour of insects. This section consists of a brief survey of the effects of some of these other factors.

The feeding behaviour of insects is known to be influenced by events in their life cycle. It has been shown that the intake of a variety of insects is lower for a time before and after moulting than in the middle of larval instars (e.g. Beck *et al.*, 1958; Hill and Goldsworthy, 1968; Saxena, 1967; Blaney and Chapman, 1970; Srihari, 1970). Another kind of life-cycle related effect is the cyclical intake of protein by anautogenous females of Diptera, the cycles of protein ingestion being related to events in the reproductive system (e.g. Lavoipierre, 1958; Dethier, 1961; Strangways-Dixon, 1961a, 1961b; Judson, 1967; Wilkens, 1968; Belzer, 1970; Tobe and Davey, 1971; Roberts and Kitching, 1974). Furthermore, ovarian development and protein ingestion have been shown to affect the intake of carbohydrates (e.g. Dethier, 1961; Strangways-Dixon, 1961; Belzer, 1970; Roberts and Kitching, 1974). It has also been found that female locusts, which have no such specific requirements of protein for purposes of reproduction, make quantitatively different choices between foods according to whether they are in a phase of somatic or of reproductive growth (Hill *et al.*, 1968; Walker *et al.*, 1970). Of these life-cycle related effects, only the cyclical intake of protein by the blowfly *Phormia regina* is, in any substantial way, understood in physiological terms (Belzer, 1970). The data of Blaney *et al.* (1973) concerning changes in the patterns of feeding within larval instars in *Locusta migratoria* does, however, give some insight into the bases of the changes in rates of ingestion in this species. A somewhat different, but related effect, is that of adult diapause on feeding. In a recent review, Stoffolano (1974) gave examples of the lack of responsiveness to food of diapausing insects and discussed available evidence about its physiological basis in *Phormia regina*. In addition, he discussed the possibility that a period of intense feeding might precede diapause.

Another factor known to affect the rate of ingestion in some species is the presence of other individuals of the same species (Norris, 1961; Bongers and Eggerman, 1971; Hori, 1971) or of a different species (Banks and Nixon, 1958). The effect of temperature on the feeding rate of insects is well known, but this is largely by implication from data concerning the relationship between temperature and rate of development. There have been a surprisingly few occasions on which direct measurements of food intake have been made at different temperatures (e.g. Husain *et al.*, 1946; David and Gardiner, 1961; Beenakkers *et al.*, 1971; Delvi and Pandian, 1972). Similarly, the almost certain effect of locomotor activity on feeding, especially of carbohydrates, is implied by the knowledge that active insects use reserves more rapidly than quiescent ones, but the only direct information concerning the effect of locomotor activity on feeding behaviour appears to be that obtained for *P. regina* by Hudson (1958) and for *Pieris brassicae* by David and Gardiner (1961).

It has been shown that the range of food materials acceptable to a number of oligophagous phytophagous insects is influenced by the kind of food on which they have fed previously (e.g. David and Gardiner, 1966; Schoonhoven, 1967; Jermy *et al.*, 1968; Ma, 1972). It seems that previous feeding may influence the sensitivity of chemoreceptors (Schoonhoven, 1969) or may produce long-lasting changes in the CNS, or both.

Another aspect of the regulation of feeding is that of the relationship between feeding and other behaviours exhibited by an insect. Only one systematic study has been made of the relationship between feeding and another kind of behaviour in an insect, namely, the very extensive study by Kennedy (Kennedy and Booth, 1963, 1964; Kennedy, 1965, 1966, 1974) of the relationship between flight and settling in the bean aphid, *Aphis fabae*. More general aspects of the integration of different behaviours, including feeding, are illustrated by findings such as those of Caldwell and Rankin (1974) that certain kinds of behavioural activities of the milkweed bug, *Oncopeltus fasciatus*, tend to occur at different times of day. This aspect, in turn, leads to a consideration of the general question of the physiological bases of the implied shifts in behavioural priority which must be the basis of the temporal separation of different behaviours.

5 Concluding remarks

Much of the work discussed in this review has been concerned with the quantification of, and investigation into, the mechanisms underlying changes, according to feeding and deprivation, in components of feeding behaviour, and with mechanisms responsible for the regulation of meal size. There is now a considerable body of information available, particularly about the physiological mechanisms operating in adults of blowflies and in larvae and adults of locusts.

Investigations into the physiological mechanisms involved in the regulation of feeding in blowflies have spanned more than 20 years, with most of the work being performed on *P. regina*. I have discussed this work in considerable detail and my general conclusion is that, even though a very large amount of information exists, we know rather less than has been generally believed. The reasons for this seem to be that there has been insufficient integration of old data with new, and that there has been a tendency for some doubtfully based conclusions to have become "fact" through repetition. A good example of this would seem to be the general belief that adaptation plays a dominant role in bringing about the termination of meals in *P. regina*, when it now seems clear that it is only one of several factors involved, and that it may be playing only a passive role as defined in this review. Even though the work on locusts had its beginnings relatively recently, the amount of available information is now substantial. Generally, the results have been presented in relatively short papers in which little attempt has been made to create a synthesis of the findings. This deficiency has, however, been overcome to some extent by the recent review by Bernays and Chapman (1974a).

The investigations into the physiological mechanisms underlying changes with feeding and deprivation in the components of feeding behaviour have, however, been fruitful and it seems unlikely that any radical change in approach is required, at least in the study of regulatory mechanisms in insects which take their food in the form of discrete meals. The prospects of elucidating underlying mechanisms by employing approaches and techniques similar to those already used appear bright. Perhaps one of the most interesting fields for possible future study would be to determine how an insect, which requires to feed on more than one type of food, behaves when deprived of one requisite and not of another, and to determine the physiological bases for its behaviour directed towards the obtaining of each of the requisites.

I have discussed at some length the relationship, in both *P. regina* and *L. migratoria*, between the available information about the effects of feeding and deprivation on the components of feeding behaviour and what is known about the insect's ability to regulate its intake over a long period. It is apparent that, although the attempts to establish these relationships have been limited to some extent by paucity of information concerning the changes in components of feeding behaviour, greater limitations are imposed by lack of precise information about the *ad lib.* feeding behaviour of the insects. In *P. regina*, for instance, information on the patterning of feeding is almost completely lacking and in *L. migratoria* the information about the patterning of feeding relates only to a limited number of experimental situations. Information on the *ad lib.* feeding behaviour by insects under a variety of experimental situations would undoubtedly enable their long-term regulation of feeding to be related in a more meaningful way to available information about changes, according to the state of deprivation, in the components of feeding behaviour.

Finally, it seems appropriate to make mention of the relationship between the state of knowledge about the regulation of feeding and the more general question of the integration of an insect's total behaviour. It is clear that behavioural inconstancy must be the basis for integration of different behaviours with one another, since integration requires that an insect should exhibit different behavioural priorities at different times. It can fairly be said that, of all the known causes of behavioural inconstancy in insects, those related to feeding and deprivation are almost certainly the best understood. It seems, therefore, that feeding behaviour could profitably be included in any studies, in the foreseeable future, designed to elucidate mechanisms underlying the shifting priorities which must be the raw material of behavioural integration.

Acknowledgements

I wish to thank Dr J. E. Moorhouse for stimulating and useful discussions during the early planning stages of this review, and Dr R. M. M. Traynier for his detailed and constructive criticism of the first draft and for his useful suggestions in subsequent discussions. Thanks are also due to Dr K. R. Norris and Dr A. Ioannides for their comments on later drafts. I am indebted also to Drs E. A. Bernays and R. F. Chapman for allowing me to read, and make reference to, several papers in preparation.

References

Anwyl, R. (1972). The structure and properties of an abdominal stretch receptor in *Rhodnius prolixus. J. Insect Physiol.* **18**, 2143-2153.

Arab, Y. M. (1957). A study of some aspects of contact chemoreception in the blowfly. Ph.D. Thesis, Johns Hopkins University, Baltimore, Maryland.

Arbogast, R. T. and Carthon, M. (1972). Effect of starvation and desiccation on the water balance and humidity response of adult *Oryzaephilus surinamensis* (Coleoptera: Cucujidae). *Entomologia exp. appl.* **15**, 488-498.

Arn, H. and Cleere, J. S. (1971). A double-label choice-test for the simultaneous determination of diet preference and ingestion by the aphid *Amphorophora agathonica. Entomologia exp. appl.* **14**, 377-387.

Banks, C. J. and Nixon, H. L. (1958). Effects of the ant, *Lasius niger* L., on the feeding and excretion of the bean aphid, *Aphis fabae* Scop. *J. exp. Biol.* **35**, 703-711.

Barton Browne, L. (1964). Thirst in the blowfly, *Lucilia cuprina. Nature, Lond.* **202**, 1137-1138.

Barton Browne, L. (1968). Effects of altering the composition and volume of the haemolymph on water ingestion of the blowfly, *Lucilia cuprina. J. Insect Physiol.* **14**, 1603-1620.

Barton Browne, L. and Dudziński, A. (1968). Some changes resulting from water deprivation in the blowfly, *Lucilia cuprina. J. Insect Physiol.* **14**, 1423-1434.

Barton Browne, L. and Evans, D. R. (1960). Locomotor activity of the blowfly as a function of feeding and starvation. *J. Insect Physiol.* **4**, 27-37.

Barton Browne, L., Moorhouse, J. E. and van Gerwen, A. C. M. (In press, a). Sensory adaptation and the regulation of meal size in the Australian plague locust, *Chortoicetes terminifera. J. Insect Physiol.*

Barton Browne, L., Moorhouse, J. E. and van Gerwen, A. C. M. (In press, b). An excitatory state generated during feeding in the locust, *Chortoicetes terminifera. J. Insect Physiol.*

Barton Browne, L., Moorhouse, J. E. and van Gerwen, A. C. M. (In preparation). A relationship between weight loss during a period of food deprivation and the sizes of subsequent meals taken by the Australian plague locust, *Chortoicetes terminifera.*

Beck, S. D., Edwards, C. A. and Medler, J. T. (1958). Feeding and nutrition of the milkweed bug, *Oncopeltus fasciatus* (Dallas). *Ann. ent. Soc. Am.* **51**, 283-288.

Beenakkers, A. M. T., Meisen, M. A. H. Q. and Scheres, J. M. J. C. (1971). Influence of temperature and food on growth and digestion in fifth instar larvae and adults of *Locusta. J. Insect Physiol.* **17**, 871-880.

Belzer, W. R. (1970). The control of protein ingestion in the black blowfly, *Phormia regina* (Meigen). Ph.D. Thesis, University of Pennsylvania, Philadelphia, Pennsylvania.

Bennet-Clark, H. C. (1963). The control of meal size in the blood sucking bug, *Rhodnius prolixus. J. exp. Biol.* **40**, 741-750.

Bentley, E. W. (1944). The biology and behaviour of *Ptinus tectus* Boie (Coleoptera, Ptinidae). V. Humidity reactions. *J. exp. Biol.* **20**, 152-158.

Bernays, E. A. and Chapman, R. F. (1972a). The control of changes in peripheral sensilla associated with feeding in *Locusta migratoria* (L.). *J. exp. Biol.* **57**, 755-763.

Bernays, E. A. and Chapman, R. F. (1972b). Meal size in nymphs of *Locusta migratoria. Entomologia exp. appl.* **15**, 399-410.

Bernays, E. A. and Chapman, R. F. (1973). The regulation of feeding in *Locusta migratoria*: internal inhibitory mechanisms. *Entomologia exp. appl.* **16**, 329-342.

Bernays, E. A. and Chapman, R. F. (1974a). The regulation of food intake by acridids. *In* "Experimental Analysis of Insect Behaviour" (Ed. L. Barton Browne), pp. 48-59. Springer-Verlag, Berlin, Heidelberg, New York.

Bernays, E. A. and Chapman, R. F. (1974b). Changes in haemolymph osmotic pressure in *Locusta migratoria* L. larvae in relation to feeding. *J. Ent.* [A], 149-155.

Bernays, E. A. and Chapman, R. F. (1974c). The effects of haemolymph osmotic pressure on the usual size of nymphs of *Locusta migratoria* L. *J. exp. Biol.* **61**, 473-480.

Bernays, E. A. and Mordue (Luntz), A. J. (1973). Changes in the palp tip sensilla of *Locusta migratoria* in relation to feeding: the effects of different levels of hormone. *Comp. Biochem. Physiol.* **45** A, 451-454.

Bernays, E. A., Blaney, W. M. and Chapman, R. F. (1972). Changes in chemoreceptor sensilla on the maxillary palps of *Locusta migratoria* in relation to feeding. *J. exp. Biol.* **57**, 745-753.

Berridge, M. J. (1966). The physiology of excretion in the cotton stainer, *Dysdercus fasciatus* Signoret, IV. Hormonal control of excretion. *J. exp. Biol.* **44**, 553-566.

Blaney, W. M. and Chapman, R. F. (1969). The fine structure of the terminal sensilla on the maxillary palps of *Schistocerca gregaria* (Forskål) (Orthoptera, Acrididae). *Z. Zellforsch. mikrosk. Anat.* **99**, 74-97.

Blaney, W. M. and Chapman, R. F. (1970). The functions of the maxillary palps of Acrididae (Orthoptera). *Entomologia exp. appl.* **13**, 363-376.

Blaney, W. M., Chapman, R. F. and Wilson, A. (1973). The pattern of feeding of *Locusta migratoria* (L.) (Orthoptera, Acrididae). *Acrida*, **2**, 119-137.

Bongers, J. (1969). Saugverhalten und Nahrungsaufnahme von *Oncopeltus fasciatus* Dallas (Heteroptera, Lygaeidae). *Oecologia*, **3**, 374-389.

Bongers, J. and Eggermann, W. (1971). Der Einfluss des Subsozialverhaltens der spezialisierten Samensauger *Oncopeltus fasciatus* Dall. und *Dysdercus fasciatus* Sign. auf ihre Ernährung. *Oecologia*, **6**, 293-302.

Brady, J. (1972a). Spontaneous circadian components of tsetse fly activity. *J. Insect Physiol.* **18**, 471-484.

Brady, J. (1972b). The visual responsiveness of the tsetse fly *Glossina morsitans* Westw. (Glossinidae) to moving objects: the effects of hunger, sex, host odour and stimulus characteristics. *Bull. ent. Res.* **62**, 257-279.

Brady, J. (1973). Changes in the probing responsiveness of starving tsetse flies (*Glossina morsitans* Westw.) (Diptera, Glossinidae). *Bull. ent. Res.* **63**, 247-255.

Bursell, E. (1957). The effect of humidity on the activity of tsetse flies. *J. exp. Biol.* **34**, 42-51.

Caldwell, R. L. and Rankin, M. A. (1974). Separation of migratory from feeding and reproductive behaviour in *Oncopeltus fasciatus*. *J. comp. Physiol.* **88**, 383-394.

Chapman, R. F. (1958). A field study of the potassium concentration in the blood of the red locust *Nomadacris septemfasciata* (Serv.), in relation to its activity. *Anim. Behav.* **6**, 60-67.

Chapman, R. F. (1959). Observations on the flight of the red locust, *Nomadacris septemfasciata* (Serville). *Behaviour*, **14**, 300-334.

Clarke, K. U. and Langley, P. A. (1963). Studies on the initiation of growth and moulting in *Locusta migratoria migratorioides* R. and F.—III. The role of the frontal ganglion. *J. Insect Physiol.* **9**, 411-421.

Connolly, K. J. (1966). Locomotor activity in *Drosophila* as a function of food deprivation. *Nature, Lond.* **209**, 224.

Connolly, K. (1967). Locomotor activity in *Drosophila.* III. A distinction between activity and reactivity. *Anim. Behav.* **15**, 149-152.

Dadd, R. H. (1960). Observations on the palatability and utilization of food by locusts, with particular reference to the interpretation of performances in growth trials using synthetic diets. *Entomologia exp. appl.* **3**, 283-304.

Dadd, R. H. (1961). Evidence for humoral regulation of digestive secretion in the beetle, *Tenebrio molitor. J. exp. Biol.* **38**, 259-266.

Davey, K. G. (1962a). The release by feeding of a pharmacologically active factor from the corpus cardiacum of *Periplaneta americana. J. Insect Physiol.* **8**, 205–208.

Davey, K. G. (1962b). The nervous pathway involved in the release by feeding of a pharmacologically active factor from the corpus cardiacum of *Periplaneta. J. Insect Physiol.* **8**, 579-583.

Davey, K. G. and Treherne, J. E. (1963). Studies on crop function in the cockroach (*Periplaneta americana* L.). II. The nervous control of crop-emptying. *J. exp. Biol.* **40**, 775-780.

Davey, P. M. (1954). Quantities of food eaten by the desert locust, *Schistocerca gregaria* (Forsk.), in relation to growth. *Bull. ent. Res.* **45**, 539-551.

David, W. A. L. and Gardiner, B. O. C. (1961). Feeding behaviour of adults of *Pieris brassicae* (L.) in a laboratory culture. *Bull. ent. Res.* **52**, 741-762.

David, W. A. L. and Gardiner, B. O. C. (1966). The effect of sinigrin on the feeding of *Pieris brassicae* L. larvae transferred from various diets. *Entomologia exp. appl.* **9**, 95-98.

Davies, D. M. and Lall, S. B. (1970). Differences in gustatory responses of newly emerged and field-caught females of *Hybomitra lasiophthalma* (Diptera: Tabanidae) to sucrose solutions. *Ann. ent. Soc. Am.* **63**, 1192-1193.

Delvi, M. R. and Pandian, T. J. (1972). Rates of feeding and assimilation in the grasshopper *Poecilocerus pictus. J. Insect Physiol.* **18**, 1829-1843.

Dethier, V. G. (1952). Adaptation to chemical stimulation of the tarsal receptors of the blowfly. *Biol. Bull. mar. biol. Lab., Woods Hole,* **103**, 178-189.

Dethier, V. G. (1957). Communication by insects: physiology of dancing. *Science, Wash.* **125**, 331-336.

Dethier, V. G. (1961). Behavioural aspects of protein ingestion by the blowfly *Phormia regina* Meigen. *Biol. Bull. mar. biol. Lab., Woods Hole,* **121**, 456-470.

Dethier, V. G. (1966). Feeding behaviour. *In* "Insect Behaviour" (Ed. P. T. Haskell). *Symp. R. ent. Soc. Lond.* **3**, 46-58.

Dethier, V. G. (1967). Feeding and drinking behaviour of invertebrates. *In* "Handbook of Physiology", Section 6, Alimentary Canal (Ed. C. F. Code), Vol. 1, pp. 79-96. Am. Physiol. Soc., Washington, D.C.

Dethier, V. G. (1969). Feeding behaviour of the blowfly. *In* "Advances in the Study of Behaviour" (Eds D. Lehrman, R. A. Hinde and E. Shaw), Vol. 2, pp. 111-266. Academic Press, New York and London.

Dethier, V. G. (1974). Sensory input and the inconstant fly. *In* "Experimental Analysis of Insect Behaviour" (Ed. L. Barton-Browne), pp. 21-31. Springer-Verlag, Berlin, Heidelberg, New York.

Dethier, V. G. and Bodenstein, D. (1958). Hunger in the blowfly. *Z. Tierpsychol.* **15**, 129-140.

Dethier, V. G. and Evans, D. R. (1961). The physiological control of water ingestion in the blowfly. *Biol. Bull. mar. biol. Lab., Woods Hole,* **121**, 108-116.

Dethier, V. G. and Gelperin, A. (1967). Hyperphagia in the blowfly. *J. exp. Biol.* **47**, 191-200.

Dethier, V. G. and Hanson, F. E. (1965). Taste papillae of the blowfly. *J. cell. comp. Physiol.* **65**, 93-100.

Dethier, V. G. and Rhoades, M. V. (1954). Sugar preference–aversion functions for the blowfly. *J. exp. Zool.* **126**, 177-204.

Dethier, V. G., Evans, D. R. and Rhoades, M. V. (1956). Some factors controlling the ingestion of carbohydrates by the blowfly. *Biol. Bull. mar. biol. Lab., Woods Hole,* **111**, 204-222.

Dethier, V. G., Solomon, R. L. and Turner, L. H. (1965). Sensory input and central excitation and inhibition in the blowfly. *J. comp. Physiol. Psychol.* **60**, 303-313.

Dethier, V. G., Solomon, R. L. and Turner, L. H. (1968). Central inhibition in the blowfly. *J. comp. Physiol. Psychol.* **66**, 144-150.

de Wilde, J., Hille Ris Lambers-Suverkropp, K. and van Tol, A. (1969). Responses to air flow and airborne plant odour in the Colorado beetle. *Neth. J. Pl. Path.* **75**, 53-57.

Dingle, H. (1965). The relation between age and flight activity in the milkweed bug, *Oncopeltus. J. exp. Biol.* **42**, 269-283.

Dingle, H. (1968). The influence of environment and heredity on flight activity in the milkweed bug, *Oncopeltus. J. exp. Biol.* **48**, 175-184.

Dixon, A. F. G. (1959). An experimental study of the searching behaviour of the predatory coccinellid beetle *Adalia decempunctata* (L.). *J. Anim. Ecol.* **28**, 259-281.

Dodds, S. E. and Ewer, D. W. (1952). Effect of desiccation on the humidity response of *Tenebrio. Nature, Lond.* **170**, 758.

Ellis, P. E. (1951). The marching behaviour of hoppers of the African migratory locust (*Locusta migratoria migratorioides* R. and F.) in the laboratory. *Anti-Locust Bull.* **7**, 48 pp.

Ellis, P. E. and Hoyle, G. (1954). A physiological interpretation of the marching of hoppers of the African migratory locust (*Locusta migratoria migratorioides* R. and F.). *J. exp. Biol.* **31**, 271-279.

Evans, D. R. (1961). Control of the responsiveness of the blowfly to water. *Nature, Lond.* **190**, 1132-1133.

Evans, D. R. and Barton Browne, L. (1960). The physiology of hunger in the blowfly. *Am. Midl. Nat.* **64**, 282-300.

Evans, D. R. and Dethier, V. G. (1957). The regulation of taste thresholds for sugars in the blowfly. *J. Insect Physiol.* **1**, 3-17.

Feir, D. and Beck, S. D. (1963). Feeding behaviour of the large milkweed bug, *Oncopeltus fasciatus. Ann. ent. Soc. Am.* **56**, 224-229.

Friend, W. G. and Smith, J. J. B. (1971). Feeding in *Rhodnius prolixus*: potencies of nucleoside phosphates in initiating gorging. *J. Insect Physiol.* **17**, 1315-1320.

Gallopín, G. C. and Kitching, R. L. (1972). Studies on the process of ingestion in the predatory bug *Podisus maculiventris* (Hemiptera: Pentatomidae). *Can. Ent.* **104**, 231-237.

Galun, R. and Margalit, J. (1969). Adenine nucleotides as feeding stimulants of the tsetse fly *Glossina austeni* Newst. *Nature, Lond.* **222**, 583-584.

Gatehouse, A. G. (1967). Synergistic effect of two stimulants to induce probing in *Stomoxys calcitrans* (L.). *Nature, Lond.* **216**, 794-795.

Gelperin, A. (1966a). Control of crop emptying in the blowfly. *J. Insect Physiol.* **12**, 331-345.

Gelperin, A. (1966b). Investigations of a foregut receptor essential to taste threshold regulation in the blowfly. *J. Insect Physiol.* **12**, 829-841.

Gelperin, A. (1967). Stretch receptors in the foregut of the blowfly. *Science, Wash.* **157**, 208-210.

Gelperin, A. (1971a). Regulation of feeding. *A. Rev. Ent.* **16**, 365-378.

Gelperin, A. (1971b). Abdominal sensory neurons providing negative feedback to the feeding behaviour of the blowfly. *Z. vergl. Physiol.* **72**, 17-31.

Gelperin, A. (1972). Neural control systems underlying insect feeding behaviour. *Am. Zool.* **12**, 489-496.

Gelperin, A. and Dethier, V. G. (1967). Long-term regulation of sugar intake by the blowfly. *Physiol. Zool.* **40**, 218-228.

Getting, P. A. (1971). The sensory control of motor output in fly proboscis extension. *Z. vergl. Physiol.* **74**, 103-120.

Getting, P. A. and Steinhardt, R. A. (1972). The interaction of external and internal receptors on the feeding behaviour of the blowfly, *Phormia regina*. *J. Insect Physiol.* **18**, 1673-1681.

Gillary, H. L. (1966). Stimulation of the salt receptor of the blowfly. I. NaCl. *J. Gen. Physiol.* **50**, 337-350.

Gordon, H. T. (1968). Intake rates of various solid carbohydrates by male German cockroaches. *J. Insect Physiol.* **14**, 41-52.

Grabowski, C. T. and Dethier, V. G. (1954). The structure of the tarsal chemoreceptors of the blowfly, *Phormia regina* Meigen. *J. Morph.* **94**, 1-20.

Green, G. W. (1954). Some laboratory investigations of the light reactions of the larvae of *Neodiprion americanus banksianae* Roh. and *N. lecontei* (Fitch) (Hymenoptera: Diprionidae). *Can. Ent.* **86**, 207-222.

Green, G. W. (1964a). The control of spontaneous locomotor activity in *Phormia regina* Meigen—I. Locomotor activity patterns in intact flies. *J. Insect Physiol.* **10**, 711-726.

Green, G. W. (1964b). The control of spontaneous locomotor activity in *Phormia regina* Meigen—II. Experiments to determine the mechanism involved. *J. Insect Physiol.* **10**, 727-752.

Gwadz, R. W. (1969). Regulation of blood meal size in the mosquito. *J. Insect Physiol.* **15**, 2039-2044.

Hans, H. and Thorsteinson, A. J. (1961). The influence of physical factors and host plant odour on the induction and termination of dispersal flights in *Sitona cylindricollis* Fahr. *Entomologia exp. appl.* **4**, 165-177.

Hardee, D. D., Mitchell, E. B. and Huddleston, P. M. (1966). Effect of age, nutrition, sex, and time of day on response of boll weevils to an attractant from cotton. *Ann. ent. Soc. Am.* 59, 1024-1025.

Harrewijn, P. and Noordink, J. P. W. (1971). Taste perception of *Myzus persicae* in relation to food uptake and developmental process. *Entomologia exp. appl.* 14, 413-419.

Haskell, P. T. and Mordue (Luntz), A. J. (1969). The role of mouthpart receptors in the feeding behaviour of *Schistocerca gregaria. Entomologia exp. appl.* 12, 591-610.

Haskell, P. T. and Schoonhoven, L. M. (1969). The function of certain mouthpart receptors in relation to feeding in *Schistocerca gregaria* and *Locusta migratoria migratorioides. Entomologia exp. appl.* 12, 423-440.

Haskell, P. T., Paskin, M. W. J. and Moorhouse, J. E. (1962). Laboratory observations on factors affecting the movements of hoppers of the desert locust. *J. Insect Physiol.* 8, 53-78.

Haslinger, E. (1935). Über den Geschmackssinn von *Calliphora erythrocephala* Meigen und über die Verwertung von Zuckern und Zuckeralkoholen durch diese Fliege. *Z. vergl. Physiol.* 22, 614-639.

Hassett, C. C., Dethier, V. G. and Gans, J. (1950). A comparison of nutritive values and taste thresholds to carbohydrates for the blowfly. *Biol. Bull. mar. biol. Lab., Woods Hole,* 99, 446-453.

Highnam, K. C., Hill, L. and Mordue, W. (1966). The endocrine system and öocyte growth in *Schistocerca* in relation to starvation and frontal ganglionectomy. *J. Insect Physiol.* 12, 977-994.

Hill, L. and Goldsworthy, G. J. (1968). Growth, feeding activity, and the utilization of reserves in larvae of *Locusta. J. Insect Physiol.* 14, 1085-1098.

Hill, L., Luntz, A. J. and Steele, P. A. (1968). The relationships between somatic growth, ovarian growth, and feeding activity in the adult desert locust. *J. Insect Physiol.* 14, 1-20.

Holling C. S. (1966). The functional response of invertebrate predators to prey density. *Mem. ent. Soc. Can.* 48, 86 pp.

Hopkins, B. A. (1964). The probing response of *Stomoxys calcitrans* (L.) (the stable fly) to vapours. *Anim. Behav.* 12, 513-524.

Hori, K. (1971). Studies on the feeding habits of *Lygus disponsi* Linnavuori (Hemiptera: Miridae) and the injury to its host plant. II. Frequency, duration and quantity of the feeding. *Appl. Ent. Zool.* 6, 119-125.

Hosoi, T. (1959). Identification of blood components which induce gorging in the mosquito. *J. Insect Physiol.* 3, 191-218.

House, H. L. (1965). Effects of low levels of the nutrient content of a food and of nutrient imbalance on the feeding and nutrition of a phytophagous larva, *Celerio euphorbiae. Can. Ent.* 97, 62-68.

Hoyle, G. (1954). Changes in the blood potassium concentration of the African migratory locust (*Locusta migratoria migratorioides* R. and F.) during food deprivation, and the effect on neuromuscular activity. *J. exp. Biol.* 31, 260-270.

Hudson, A. (1958). The effect of flight on the taste threshold and carbohydrate utilization of *Phormia regina* Meigen. *J. Insect Physiol.* 1, 293-304.

Husain, M. A., Mathur, C. B. and Roonwal, M. L. (1946). Studies on *Schistocerca gregaria* (Forskål)—XIII. Food and feeding habits of the desert locust. *Indian J. Ent.* **8**, 141-163.

Jermy, T., Hanson, F. E. and Dethier, V. G. (1968). Induction of specific food preference in lepidopterous larvae. *Entomologia exp. appl.* **11**, 211-230.

Judson, C. L. (1967). Feeding and oviposition behaviour in the mosquito *Aedes aegypti* (L.). I. Preliminary studies of physiological control mechanisms. *Biol. Bull. mar. biol. Lab., Woods Hole*, **133**, 369-377.

Kennedy, J. S. (1965). Co-ordination of successive activities in an aphid. Reciprocal effects of settling on flight. *J. exp. Biol.* **43**, 489-509.

Kennedy, J. S. (1966). The balance between antagonistic induction and depression of flight activity in *Aphis fabae* Scopoli. *J. exp. Biol.* **45**, 215-228.

Kennedy, J. S. (1974). Changes in responsiveness in the patterning of behavioural sequences. *In* "Experimental Analysis of Insect Behaviour" (Ed. L. Barton Browne), pp. 1-6. Springer-Verlag, Berlin, Heidelberg, New York.

Kennedy, J. S. and Booth, C. O. (1963). Co-ordination of successive activities in an aphid. The effect of flight on the settling responses. *J. exp. Biol.* **40**, 351-369.

Kennedy, J. S. and Booth, C. O. (1964). Co-ordination of successive activities in an aphid. Depression of settling after flight. *J. exp. Biol.* **41**, 805-824.

Kennedy, J. S. and Moorhouse, J. E. (1969). Laboratory observations on locust responses to wind-borne grass odour. *Entomologia exp. appl.* **12**, 487-503.

Khan, A. A. and Maibach, H. I. (1970). A study of the probing response of *Aedes aegypti*. 1. Effect of nutrition on probing. *J. econ. Ent.* **63**, 974-976.

Khan, A. A. and Maibach, H. I. (1971). A study of the probing response of *Aedes aegypti*. 2. Effect of desiccation and blood feeding on probing to skin and an artificial target. *J. econ. Ent.* **64**, 439-442.

Knight, M. R., Sr. (1962). Rhythmic activities of the alimentary canal of the black blowfly, *Phormia regina* (Diptera: Calliphoridae). *Ann. ent. Soc. Am.* **55**, 380-382.

Kusano, T. and Adachi, H. (1969a). Proboscis extending time on distilled water, sugars and salts and their nutritive value in the cabbage butterfly (*Pieris rapae crucivora*). *Kontyû*, **36**, 427-436.

Kusano, T. and Adachi, H. (1969b). Relation between proboscis extending time and ingested amount of sugars, and the regulation of sucking of sucrose in the white cabbage butterfly, *Pieris rapae crucivora* Boisduval (Lepidoptera: Pieridae). *Appl. Ent. Zool.* **4**, 161-170.

Lall, S. B. (1969). Effect of starvation on the sensitivity of tarsal and labellar contact chemoreceptors in *Hybomitra lasiophthalma* (Diptera: Tabanidae). *Ann. ent. Soc. Am.* **62**, 1217.

Latheef, M. A. and Harcourt, D. G. (1972). A quantitative study of food consumption, assimilation, and growth in *Leptinotarsa decemlineata* (Coleoptera: Crysomelidae) on two host plants. *Can. Ent.* **104**, 1271-1276.

Lavoipierre, M. M. J. (1958). Biting behaviour of mated and unmated females of an African strain of *Aedes aegypti*. *Nature, Lond.* **181**, 1781-1782.

Leonard, D. E. (1970). Effects of starvation on behaviour, number of larval instars, and developmental rate of *Porthetria dispar*. *J. Insect Physiol.* **16**, 25-31.

Ma, W-C. (1972). Dynamics of feeding responses in *Pieris brassicae* Linn. as a function of chemosensory input: a behavioural, ultrastructural and electrophysiological study. *Med. Landbouwhogeschool Wageningen,* 72-11, 162 pp.

Maddrell, S. H. P. (1963). Control of ingestion in *Rhodnius prolixus* Stål. *Nature, Lond.* **198**, 210.

Madge, D. S. (1964). The light reactions and feeding activity of larvae of the cutworm *Tryphaena pronuba* L. (Lepidoptera: Noctuidae). I. Laboratory investigations. *Entomologia exp. appl.* **7**, 47-61.

Manjra, A. A. (1971). Regulation of threshold to sucrose in a mosquito, *Culiseta inornata* (Williston). *Mosquito News*, **31**, 387-390.

McCutchan, M. C. (1969). Responses of tarsal chemoreceptive hairs of the blowfly, *Phormia regina. J. Insect Physiol.* **15**, 2059-2068.

McGinnis, A. J. and Kasting, R. (1967). Dietary cellulose: effect on food consumption and growth of a grasshopper. *Can. J. Zool.* **45**, 365-367.

McLean, P. L. and Kinsey, M. G. (1969). Probing behavior of the pea aphid, *Acyrthosiphon pisum.* IV. Effects of starvation on certain probing activities. *Ann. ent. Soc. Am.* **62**, 987-994.

Mehta, R. C. and Saxena, K. N. (1973). Growth of the cotton spotted bollworm, *Earias fabia* (Lepidoptera: Noctuidae) in relation to consumption, nutritive value and utilization of food from various plants. *Entomologia exp. appl.* **16**, 20-30.

Meisner, J., Ascher, K. R. S. and Flowers, H. M. (1972). The feeding response of the larva of the Egyptian cotton leafworm, *Spodoptera littoralis* Boisd., to sugars and related compounds—I. Phagostimulatory and deterrent effects. *Comp. Biochem. Physiol.* **42A**, 899-914.

Mellanby, K. and French, R. A. (1958). The importance of drinking water to larval insects. *Entomologia exp. appl.* **1**, 116-124.

Miller, P. L. (1974). Rhythmic activities and the insect nervous system. *In* "Experimental Analysis of Insect Behaviour" (Ed. L. Barton Browne), pp. 114-138. Springer-Verlag, Berlin, Heidelberg, New York.

Minnich, D. E. (1922). A quantitative study of tarsal sensitivity to solutions of saccharose in the red admiral butterfly, *Pyrameis atalanta* L. *J. exp. Zool.* **36**, 445-457.

Minnich, D. E. (1929). The chemical sensitivity of the legs of the blowfly, *Calliphora vomitoria* Linn., to various sugars. *Z. vergl. Physiol.* **11**, 1-5.

Minnich, D. E. (1931). The sensitivity of the oral lobes of the proboscis of the blowfly, *Calliphora vomitoria* Linn., to various sugars. *J. exp. Zool.* **60**, 121-139.

Mittler, T. E. (1967). Effect of amino acid and sugar concentrations on the food uptake of the aphid *Myzus persicae. Entomologia exp. appl.* **10**, 39-51.

Mittler, T. E. (1970). Effects of dietary amino acid on the feeding rate of the aphid *Myzus persicae. Entomologia exp. appl.* **13**, 432-437.

Moloo, S. K. and Kutuza, S. B. (1970). Feeding and crop emptying in *Glossina brevipalpis. Acta Tropica*, **27**, 356-377.

Moorhouse, J. E. (1969). Locomotor activity and orientation in locusts. Ph.D. Thesis, University of London.

Moorhouse, J. E. (1971). Experimental analysis of the locomotor behaviour of *Schistocerca gregaria* induced by odour. *J. Insect Physiol.* 17, 913-920.

Moorhouse, J. E., Barton Browne, L. and van Gerwen, A. C. M. (In preparation). Factors affecting the rate of ingestion of fluids by the Australian plague locust, *Chortoicetes terminifera.*

Mordue, W. (1969). Hormonal control of malpighian tube and rectal function in the desert locust, *Schistocerca gregaria. J. Insect Physiol.* 15, 273-285.

Nadel, D. J. and Peleg, B. A. (1965). The attraction of fed and starved males and females of the Mediterranean fruit fly, *Ceratitis capitata* Wied., to "Trimedlure". *Israel J. agric. Res.* 15, 83-86.

Norris, D. M. (1970). Quinol stimulation and quinone deterrency of gustation by *Scolytus multistriatus* (Coleoptera: Scolytidae). *Ann. ent. Soc. Am.* 63, 476-478.

Norris, D. M. and Baker, J. M. (1969). Nutrition of *Xyleborus ferrugineus.* I. Ethanol in diets as a tunneling (feeding) stimulant. *Ann. ent. Soc. Am.* 62, 592-594.

Norris, M. J. (1961). Group effects on feeding in adult males of the desert locust, *Schistocerca gregaria* (Forsk.), in relation to sexual maturation. *Bull. ent. Res.* 51, 731-753.

Omand, E. (1971). A peripheral sensory basis for behavioural regulation. *Comp. Biochem. Physiol.* 38A, 265-278.

Omand, E. and Dethier, V. G. (1969). An electrophysiological analysis of the action of carbohydrates on the sugar receptor of the blowfly. *Proc. natn. Acad. Sci. U.S.A.* 62, 136-143.

Pflumm, W. (1970). Lokalisierendes Schmecken bei Insekten. *Z. vergl. Physiol.* 68, 49-59.

Pienkowski, R. L. and Golik, Z. (1969). Kinetic orientation behavior of the alfalfa weevil to its host plant. *Ann. ent. Soc. Am.* 62, 1241-1245.

Reynierse, J. H., Manning, A. and Cafferty, D. (1972). The effects of hunger and thirst on body weight and activity in the cockroach (*Nauphoeta cinerea*). *Anim. Behav.* 20, 751-757.

Rice, M. J. (1970). Cibarial stretch receptors in the tsetse fly (*Glossina austeni*) and the blowfly (*Calliphora erythrocephala*). *J. Insect Physiol.* 16, 277-289.

Rice, M. J. (1973). Cibarial sense organs of the blowfly, *Calliphora erythrocephala* (Meigen) (Diptera: Calliphoridae). *Int. J. Insect Morph. Embriol.* 2, 109-116.

Riegert, P. W. (1958). Humidity reactions of *Melanoplus bivittatus* (Say) and *Camnula pellucida* (Scudd.) (Orthoptera: Acrididae): Reactions of starved and of moulting grasshoppers. *Can. Ent.* 90, 680-684.

Roberts, J. A. and Kitching, R. L. (1974). Ingestion of sugar, protein and water by adult *Lucilia cuprina* (Wiedemann) (Diptera: Calliphoridae). *Bull. ent. Res.* 64, 81-88.

Roth, L. M. and Willis, E. R. (1951). The effects of desiccation and starvation on the humidity behaviour and water balance of *Tribolium confusum* and *Tribolium castaneum. J. exp. Zool.* 118, 337-361.

Rowell, C. H. F. (1963). A method for chronically implanting stimulating electrodes into the brains of locust, and some results of stimulation. *J. exp. Biol.* 40, 271-284.

Salama, H. S. (1969). Factors affecting taste sensitivity in *Aedes aegypti* L. (Diptera: Culicidae). *Bull. Soc. ent. Égypte,* 51, 343-346.

Saxena, K. N. (1963). Mode of ingestion in a heteropterous insect *Dysdercus koenigii* (F.) (Pyrrhocoridae). *J. Insect Physiol.* **9**, 47-71.

Saxena, K. N. (1967). Some factors governing olfactory and gustatory responses of insects. *In* "Olfaction and Taste—II" (Ed. T. Hayashi), pp. 799-819. Pergamon Press, Oxford.

Schoonhoven, L. M. (1967). Loss of host plant specificity by *Manduca sexta* after rearing on an artificial diet. *Entomologia exp. appl.* **10**, 270-272.

Schoonhoven, L. M. (1969). Sensitivity changes in some insect chemoreceptors and their effect on food selection behaviour. *Koninkl. Ned. Akad, Wet., Proc.* Series C, **72**, 491-498.

Sinoir, Y. (1968). Étude de quelques facteurs conditionnant la prise de nourriture chez les larves du criquet migrateur, *Locusta migratoria migratorioides* (Orthoptera, Acrididae). II. Facteurs internes. *Entomologia exp. appl.* **11**, 443-449.

Shrihari, T. (1970). Étude quantitative de la consommation et de l'utilisation de la nourriture au cours de la croissance larvaire de *Pieris brassicae* (Lep. Pieridae). *Annls Soc. ent. Fr.* **6**, 1003-1014.

Srivastava, P. N. and Auclair, J. L. (1971). Influence of sucrose concentration on diet uptake and performance by the pea aphid, *Acyrthosiphon pisum. Ann. ent. Soc. Am.* **64**, 739-743.

Stoffolano, J. G. (1974). Control of feeding and drinking in diapausing insects. *In* "Experimental Analysis of Insect Behaviour" (Ed. L. Barton Browne), pp. 32-47. Springer-Verlag, Berlin, Heidelberg, New York.

Strangways-Dixon, J. (1961a). The relationship between nutrition, hormones and reproduction in the blowfly *Calliphora erythrocephala* (Meig.). I. Selective feeding in relation to the reproductive cycle, the corpus allatum volume and fertilization. *J. exp. Biol.* **38**, 225-235.

Strangways-Dixon, J. (1961b). The relationships between nutrition, hormones and reproduction in the blowfly *Calliphora erythrocephala* (Meig.). II. The effect of removing the ovaries, the corpus allatum and the median neurosecretory cells upon selective feeding, and the demonstration of the corpus allatum cycle. *J. exp. Biol.* **38**, 637-646.

Stürckow, B., Holbert, P. E. and Adams, J. R. (1967). Fine structure of the tip of chemosensitive hairs in two blow flies and the stable fly. *Experientia*, **23**, 780-782.

Syrjämäki, J. (1962). Humidity perception in *Drosophila melanogaster. Ann. Zool. Soc. "Vanamo"*, **23**, 1-74.

Tobe, S. S. and Davey, K. G. (1972). Volume relationships during the pregnancy cycle of the tsetse fly *Glossina austeni. Can. J. Zool.* **50**, 999-1010.

Treherne, J. E. (1957). Glucose absorption in the cockroach. *J. exp. Biol.* **34**, 478-485.

van Herrewege, C. (1971). Consommation alimentaire chez les mâles adultes de *Blatella germanica* (L.): influence de l'âge, de la nourriture larvaire et du jeûne. *Arch Sci. Physiol.* **25**, 401-413.

Walker, P. R., Hill, L. and Bailey, E. (1970). Feeding activity, respiration, and lipid and carbohydrate content of the male desert locust during adult development. *J. Insect Physiol.* **16**, 1001-1015.

Wellington, W. G. (1948). The light reactions of the spruce budworm, *Choristoneura fumiferana* Clemens (Lepidoptera, Tortricidae). *Can. Ent.* **80**, 56-82.

Wensler, R. J. (1971). Locomotor and feeding activity of larvae of the scarabaeid *Sericesthis geminata* (Coleoptera). *Entomologia exp. appl.* **14**, 270-282.

Wensler, R. J. D. (1972). The effect of odors on the behavior of adult *Aedes aegypti* and some factors limiting responsiveness. *Can. J. Zool.* **50**, 415-420.

Wilczek, M. (1967). The distribution and neuroanatomy of the labellar sense organs of the blowfly *Phormia regina* Meigen. *J. Morph.* **122**, 175-202.

Wilkens, J. L. (1968). The endocrine and nutritional control of egg maturation in the fleshfly *Sarcophaga bullata*. *J. Insect Physiol.* **14**, 927-943.

Yinon, U. (1969). Food consumption of the armored scale lady-beetle *Chilocorus bipustulatus* (Coccinellidae). *Entomologia exp. appl.* **12**, 139-146.

Youdeowei, A. (1967). The reactions of *Dysdercus intermedius* (Heteroptera, Pyrrhocoridae) to moisture, with special reference to aggregation. *Entomologia exp. appl.* **10**, 194-210.

The Cytophysiology of Insect Blood

A. Clive Crossley

School of Biological Sciences, University of Sydney, Australia

1 Introduction

The blood cells of insects were first described by Schwammerdam (1758) who reported (p. 31) that when he opened the abdomen of a louse, blood flowed out containing "transparent globules". He compared insect blood cells with blood cells from man, reporting that human blood "has ruddy globules swimming in clear liquor". His observation, which implied a lack of haemoglobin in insect blood cells, has been supported by subsequent observers, although a few insect species do possess haemoglobin in solution in the blood plasma (Buck, 1953). It is probably the presence of a tracheal rather than a corpuscular respiratory system which explains the paucity of cells in insect blood by comparison with vertebrate blood. Thus human blood contains about 5 million cells in a microlitre, and these occupy nearly 50 per cent of the blood volume, whereas insect blood usually contains less than one hundred thousand cells in a microlitre (Brady, 1967).

Nevertheless blood cells can occupy up to 10 per cent of the blood volume in adult *Periplaneta* (Wheeler, 1963), and an even higher percentage in insects preparing for metamorphosis. The important contribution of the cellular component to the physiology of insect blood is the subject of this review.

Insect blood cells have been the subject of several earlier reviews. Those of Rooseboom (1937) and Ermin (1939) summarize the early literature. The burgeoning literature on blood cell populations and classification was reviewed by Jones (1962, 1970), and Wigglesworth (1939, 1959) has provided general reviews on blood cell function. The current offering is an attempt to integrate recent cytophysiological literature on insect haemocytes with earlier work, and beyond this with diverse general observations on cells.

2 The fine structure of insect blood cells

The blood cells, or haemocytes from representative species of six orders of insects, have now been examined in detail in the electron microscope. These orders are Blattaria, Orthoptera, Hemiptera, Lepidoptera, Diptera, and Coleoptera. With these reports in mind it is possible to outline the general ultrastructural features of insect haemocytes, as well as remark on certain specializations.

The haemocyte boundary is a unit membrane of typical trilaminar fine-structure about 8 nm across (Fig. 3), but this is not enveloped by a visible extracellular basement membrane. The absence of an extracellular sheath facilitates recognition of haemocytes (Locke and Huie, 1972), with the qualification that certain embryonic cells which migrate in the haemolymph during metamorphosis (for example myoblasts; Crossley, 1972a) also lack a conspicuous basement membrane. The boundaries of insect cells bathed in haemolymph, such as epidermal cells (Locke, 1969a), fat body cells (Locke and Collins, 1965), pericardial and garland cells (Crossley, 1972b), normally have a sheath of basement membrane external to the plasma membrane. The plasma membrane of haemocytes is seldom

Fig. 1. A phagocytic haemocyte from a *Calliphora* larva injected 8 h earlier with 2.4 per cent colloidal thorium dioxide stabilized with dextrin. Numerous coated vesicles (arrowed) are present, and granules of electron-dense thorium derivatives appear in them, both at the cell surface and within the cytoplasm. The vesicles contribute their contents to the lysosomal vacuoles (ly) where thorium dioxide can be seen to have accumulated. The plasma membrane is extended (e) to form filopodia or folds. Golgi centres (g) and microtubules (t) are numerous throughout the cell. Electron micrograph x8000.

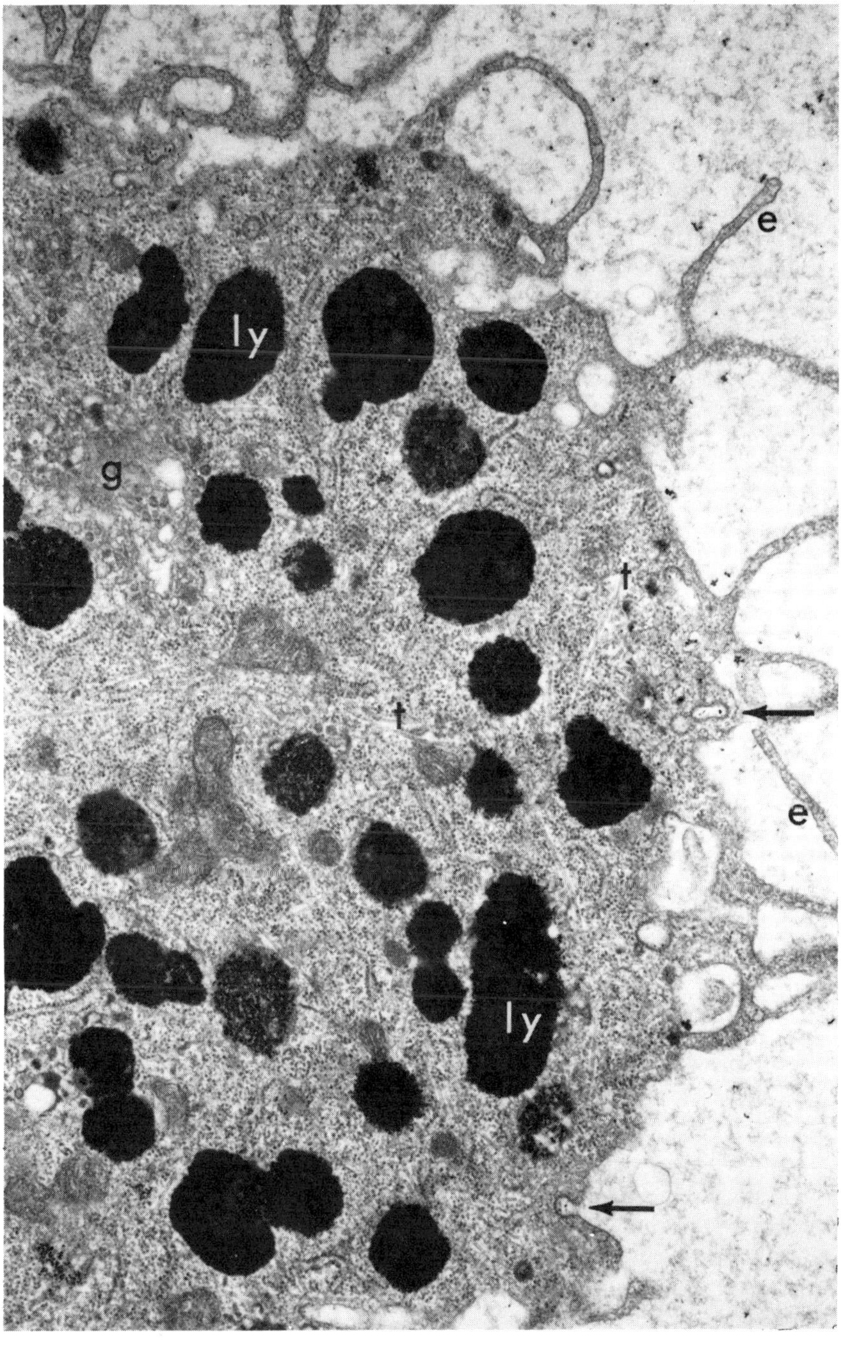

Fig. 1.

smoothly contoured, but is often found extended into finger-like projections (filopodia) or lamellar extensions (lamellopodia) (Figs 1 and 3). Indentations are confined to small surface infoldings and pinocytotic vesicles of various kinds, but extensive internal channels such as those seen on oenocytes (Locke, 1969a) or pericardial cells (Crossley, 1972b) are not present, except in the thrombocytoids of Diptera (Zachary and Hoffmann, 1973).

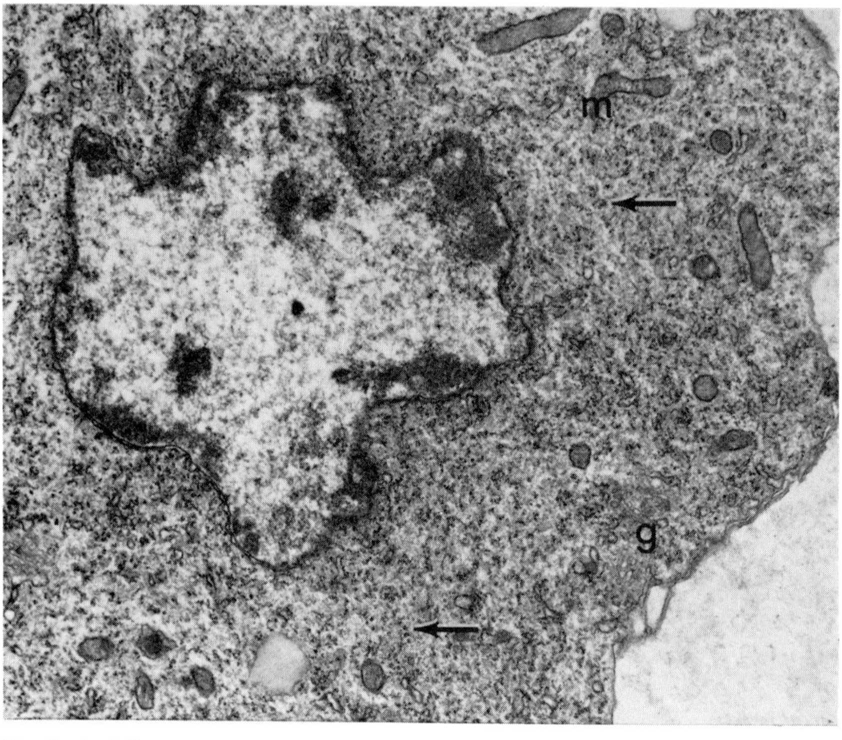

Fig. 2. A differentiating haemocyte from a *Calliphora* larva, replete with polysomes (arrowed) and mitochondria (m). Golgi centres (g) are present but endoplasmic reticulum is sparse, and the cell is probably developing into a type without vacuoles, such as that shown in Fig. 8. Electron micrograph x8000.

The nucleus of haemocytes is surrounded by an envelope composed of inner and outer unit membranes, and the space between these is frequently seen to be continuous with cisternae of endoplasmic reticulum (Smith, 1968; Scharrer, 1972). The nuclear envelope is frequently invaginated to form deep clefts (e.g. in *Ephestia:* Grimstone *et al.*, 1967). The nuclear profile appears to be a rather consistent structural feature, and may prove an aid in identification of cell types, as in the case of vertebrate neutrophils and monocytes (Huhn, 1967). One danger in this is the possibility that the shape of the nucleus may change when the blood clots, as reported for *Limulus* (Dumont *et al.*, 1966). The nucleus of some haemocyte types is consistently eccentric (e.g. in the oenocytoids of Hollande, 1920, and of Hoffmann, 1966a). Binucleate haemocytes have been reported (Hollande, 1920), as have multinucleate haemocytes (Tauber and Griffiths, 1943; Whitten, 1964). Multinucleate haemocytes in *Chironomus* reportedly contain more than 12 nuclei and reach giant size (10 μm) (Maier, 1969). The ultrastructure of multinucleate haemocytes has yet to be described.

The overall shape of haemocytes appears to be largely determined by the disposition of microtubules, as is the shape of many other cells (see review by Tilney, 1969). In *Ephestia* microtubules 21 nm in diameter are particularly abundant just below the plasma membrane and often extend the length of the cell. In flattened haemocytes engaged in encapsulation of foreign objects, the microtubules lie parallel to the long axis of the cell (Grimstone *et al.*, 1967). "Discoid haemocytes" in *Periplaneta* (Baerwald and Bousch, 1970), and "granular haemocytes" of elliptical profile in *Leucophaea* (Hagopian, 1971), have marginal bundles of about 30 microtubules, each about 25 nm in diameter. Similar bundles of microtubules have been described in *Limulus* haemocytes (Dumont *et al.*, 1966) and in vertebrate blood platelets and discoid erythrocytes (Behnke, 1970). Figure 3 shows microtubules in *Calliphora* haemocytes adjacent to the plasma membrane. These microtubules have an average diameter of 26 nm, i.e. within the size range of 24-27 nm reported for cells in other invertebrates and vertebrates.

Fusiform haemocytes (spherule cells) of *Melolontha* also have microtubules in bundles at the periphery of the cell (Devauchelle, 1971). Fusiform myoblasts migrating in the haemolymph of *Calliphora* pupae have bundles of oriented microtubules extending into elongated poles of the cell (Crossley, 1972a; Fig. 17), and these cells collapse to a spherical shape when 10^{-4} M colchicine is introduced into the haemolymph, with the object of inhibiting microtubule polymerization, indicating the importance of microtubules as cytoskeletal elements in insect cells. Honey bee haemocytes that initially had fusiform shapes quickly assumed a circular profile on removal from the insect to a glass slide (Yeager and Knight, 1933). Other factors controlling or changing haemocyte shape are discussed in connection with locomotion below.

The cytoplasm of haemocytes is well provided with mitochondria, but these are usually small and elongated, and have electron-dense matrices (Cassier and Fain-Maurel, 1968; Scharrer, 1972). There has, as yet, been no assessment of the activities of mitochondria in haemocytes. In addition to

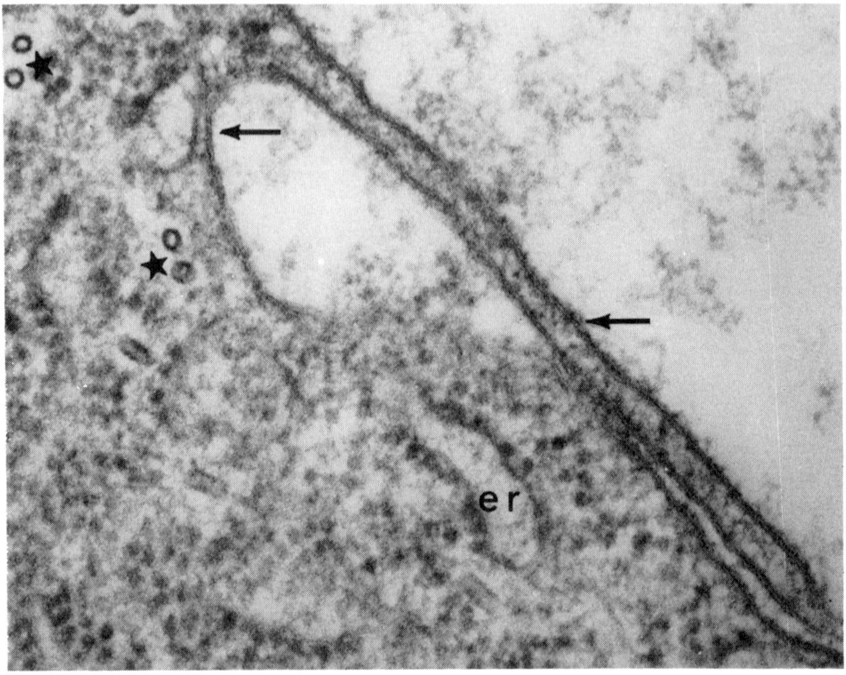

Fig. 3. An area of folded plasma membrane showing the trilaminar unit membrane structure (arrowed), but the absence of an extracellular basement membrane, on this *Calliphora* haemocyte. Microtubules (asterisks) 24 nm in diameter are ubiquitous. er, ribosome-studded endoplasmic reticulum. Electron micrograph x115 000.

oxidative phosphorylation, these organelles may be involved in synthetic pathways (e.g. of lipids) as is suggested by observations of mitochondria with electron-dense matrices in other kinds of insect cells (Crossley and Waterhouse, 1969).

Haemocytes are rich in ribosomes, and these are frequently seen in polysome clusters (Fig. 2), a probable indication of active protein synthesis, by analogy with vertebrate lymphocytes (Johnson *et al.*, 1966). Some insect haemocytes contain an enormous number of ribosome-like granules without any indication of polysome formations (Fig. 6). Rough surfaced (ribosome studded) endoplasmic reticulum cisternae occur in sinuous profiles and in stacks, and are often very numerous in haemocytes (Fig. 8). Smooth-surfaced cisternae are frequently encountered in haemocytes (Scharrer, 1972). Although glycogen is well known as a component of insect haemocytes as a result of histochemical studies, ultrastructural methods for glycogen (e.g. Gupta *et al.*, 1969; Vye and Fischman, 1971) have not been applied to haemocytes. Images highly suggestive of glycogen have been obtained using normal ultrastructural stains (e.g. for *Rhodnius*: Lai-Fook 1968; for *Spodoptera*: Harpaz *et al.*, 1969; for *Blaberus*: Moran, 1971, Fig. 1; and for *Gromphadorina*: Scharrer, 1972, Fig. 1), Fig. 7, but many granules in electron micrographs of insect haemocytes could be either glycogen or ribosomes.

Many haemocytes contain a well-developed "vacuolar apparatus" (term from de Duve and Wattiaux, 1966), consisting of lysosomes, phagosomes and residual bodies of several kinds. Lysosomes are not always easy to identify in electron micrographs of haemocytes, although such identification is facilitated by histochemical methods for acid phosphatase (Grimstone *et al.*, 1967; Crossley, 1968; Harpaz *et al.*, 1969). In haemocytes primary lysosomes are single-membrane bound vacuoles containing rather electron-dense homogeneous or finely granular contents, and are 0.1-3 μm in diameter. Since Golgi centres are implicated in lysosome formation, nearby small homogeneous, very electron dense vacuoles are often presumed to be lysosomes in insect cells, including haemocytes (Crossley, 1972b, Fig. 10; Scharrer, 1972, Fig. 2) (Fig. 9). In *Calliphora* haemocytes, fusion of primary lysosomes with larger vacuoles, some of which may contain redundant cellular organelles or ingested debris from the haemolymph, leads to the formation of vacuoles of heterogeneous structure. These are moderately electron dense and contain active acid phosphatase (Crossley, 1968, Fig. 14) (Fig. 4). *Leucophaea* haemocyte lysosomes contain active thiamine pyrophosphatase as well as acid phosphatase (Scharrer, 1972). The involvement of these organelles in phagocytosis is discussed in section 8.

Although microbodies containing catalase and urate oxidase have been identified in insect fat body cells (Locke and McMahon, 1971), their occurrence in haemocytes has yet to be confirmed by histochemistry. Microbody-like elements have been described in *Leucophaea* haemocytes by Hagopian (1971).

The insect haemocyte lysosome is apparently analogous to the azurophil granule of the vertebrate polymorphonuclear leucocyte. Both are single membrane bound, electron-dense, homogeneous vacuoles derived from the Golgi apparatus, containing lysosomal enzymes of acid pH optima, which are released into phagocytic vacuoles. Polymorphonuclear leucocytes also contain specific granules of low electron density, derived from the Golgi apparatus. These contain alkaline phosphatase, lysozyme, and lactoferrin (Bainton and Farquhar, 1966; Bainton, 1973). Vacuoles of similar fine structure are present in many insect haemocytes, but as yet no application

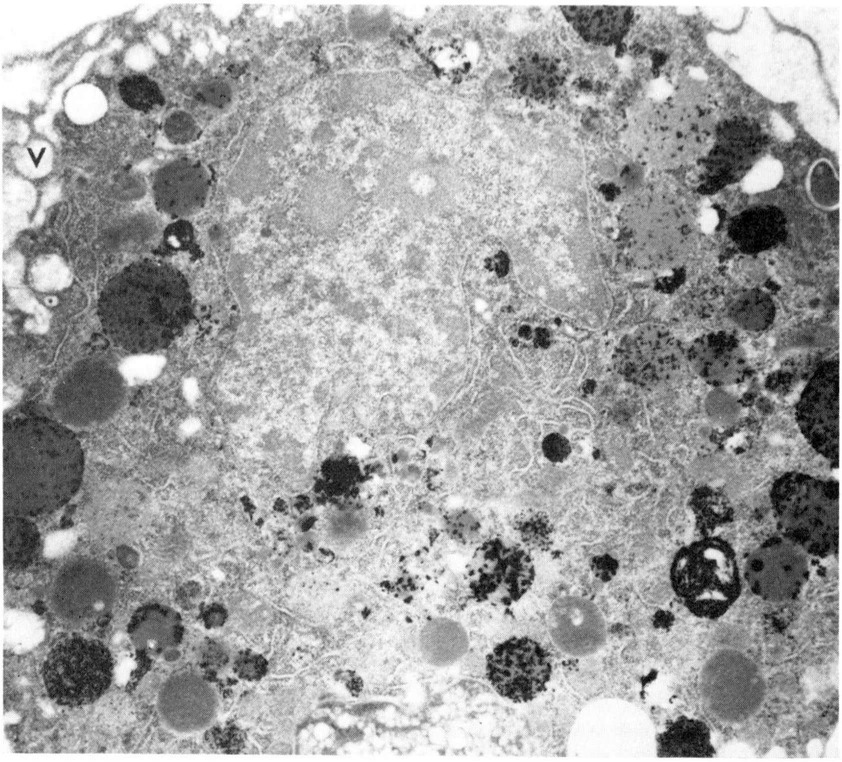

Fig. 4. Acid-phosphatase activity in a larval *Calliphora* haemocyte is here demonstrated by Gomori's technique. Lead deposits indicating acid-phosphatase activity appear in many electron-dense vacuoles, which are members of the lysosome series, but not in the cytoplasm. The plasma membrane is extensively folded, and many vacuoles (v) are probably the result of endocytosis. Electron micrograph ×18 000.

to insect haemocytes of the well-developed histochemical techniques for specific granule enzymes has been reported. The diversity of vacuole morphology reported in insect haemocytes is considerable and, at the risk of biochemical over-simplification, ultrastructural reports have been tabulated for convenience under six headings in Table 1.

Membrane-limited vacuoles containing flocculent material similar in structure to the precipitated haemolymph seen outside the cell are in all probability pinocytotic vesicles, or even indentations of the plasma membrane (Table 1, column 1) (Figs 5 and 8). Some authors provide a series of profiles indicating possible ingestion stages (e.g. Grimstone *et al.*, 1967; Smith, 1968). The difficulty of interpreting the static electron microscope image in a directional sense intrudes at this point, since a number of

TABLE 1

Vacuoles in Insect Blood Cells. Ultrastructural descriptions of vacuole contents have been tabulated under six headings, which are discussed in the text. The numbers refer to classification terms used to describe parent haemocytes by the authors

Insect	Authors	Date	1 Flocculent heterogeneous low electron density	2 Homogeneous high electron density	3 Containing membranous fragments as sheets, cylinders or microvesicles	4 Containing tubular elements (diameter nm)	5 Microfibrillar or fibrous	6 Granular
Periplaneta	Baerwald *et al.*	1970		0	0		0	0
Blaberus	Moran	1971		0	0	0 (34 nm)		
Leucophaea	Hagopian	1971		4	4	4 (25 nm)		
Blattaria spp.	Scharrer	1972	0	0	0	0 (34 nm)		9
Locusta	Hoffmann	1966a		8		8 (32 nm)		
Locusta	Hoffmann	1966b	4	4	4	4 (10-15 nm)		
Locusta	Hoffmann *et al.*	1968		6	2	8	6	
Locusta	Cassier *et al.*	1968		8		8 (30-40 nm)		
Rhodnius	Lai-Fook	1968		0	0			
Rhodnius	Wigglesworth	1973		2/3,8	2/3	2/3 (15 nm)	2/3	
Ephestia	Grimestone *et al.*	1967		0				
Antheraea	Beaulaton	1968				5 (17 nm)	5	
Calliphora	Crossley	1964, 1968	3	3	3			
Melolontha	Devauchelle	1971	4,5,6	1,2,4,5,6,7,9	2,7	5 (18 nm)	4 or 6	7
Tenebrio	Stang-Voss	1970	1,9,6			9,6,3 (16 nm)		

Cell type as stated by author: 0, unclassified; 1, prohaemocyte; 2, plasmatocyte; 3, phagocyte; 4, granular haemocyte; 5, adipohaemocyte; 6, coagulocyte; 7, spherule cell; 8, oenocytoid; 9, other specified type.

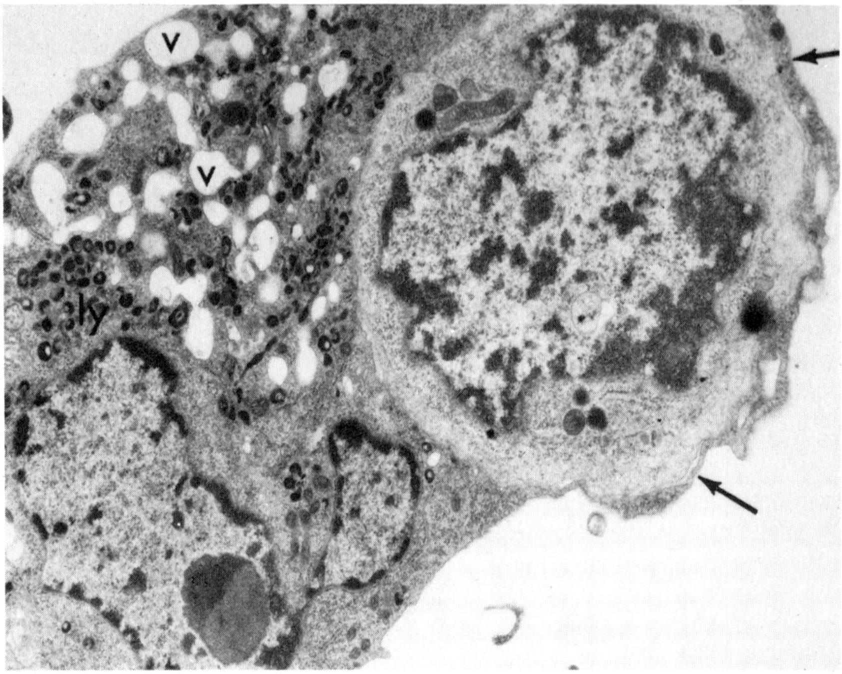

Fig. 5. A phagocytic haemocyte from a newly emerged adult *Lucilia*, in the process of engulfing another haemocyte. Folds of phagocyte plasma membrane (arrowed) wrap around the engulfed haemocyte. The cytoplasm of the phagocyte is packed with small electron-dense vacuoles interpreted as lysosomes (ly). Evidence of extensive endocytosis of blood plasma is provided by the electron-lucent vacuoles (v). Electron micrograph x14 300.

authors suggest that haemocytes secrete material by reverse pinocytosis (exocytosis). Examples of plausible exocytosis include certain vacuoles in *Gromphadorina* (Scharrer, 1972), and of plasmatocytes in *Rhodnius* (Wigglesworth, 1973), all of which probably contain mucopolysaccharides or mucoprotein, and the microfibrillar surface vacuoles of *Melolontha* (Devauchelle, 1971). In order to establish whether a vacuole is pinocytotic or exocytotic, electron-dense tracers can be introduced into the haemolymph before fixation (e.g. Thorium dioxide, ferritin, lanthanum: Crossley, 1968; 1972b; or saccharated iron oxide: Scharrer, 1972) (Fig. 1).

Electron-dense vacuoles of homogeneous content, or containing recognizable organelle fragments (Table 1, columns 2 and 3) (Fig. 16) are commonly primary or secondary lysosomes, and this presumption can be confirmed by histochemistry, as discussed above. The lysosome series also contains multivesicular bodies, which have been described in *Locusta*

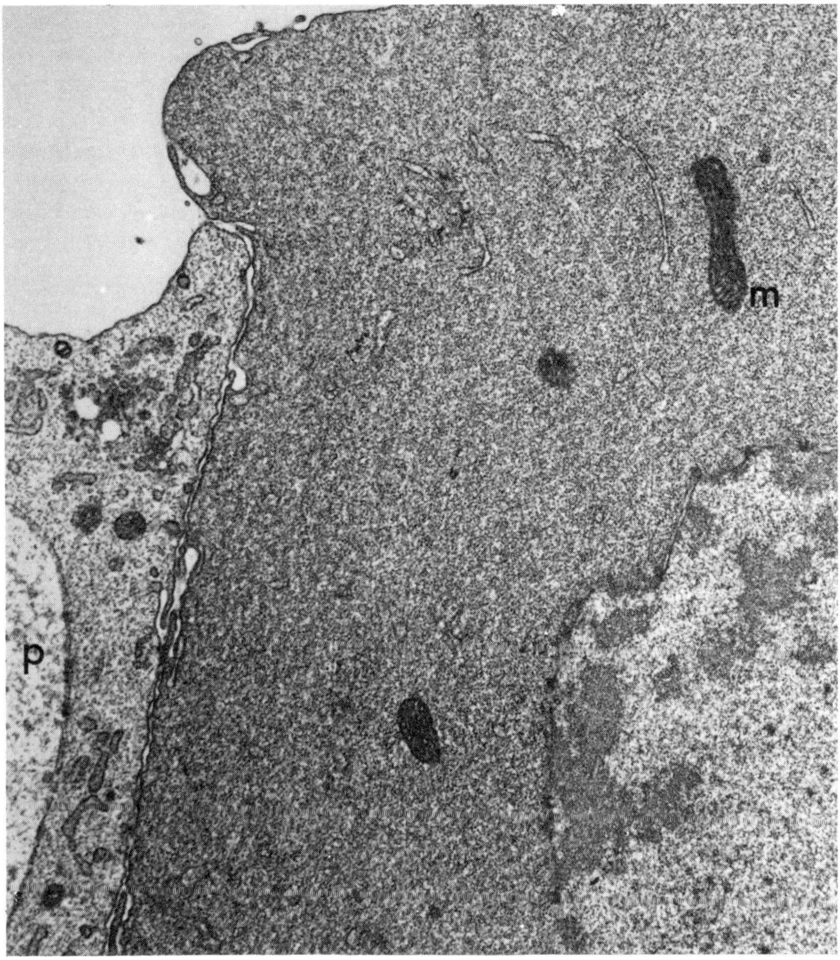

Fig. 6. A small undifferentiated prohaemocyte (p) or stem-cell lies next to a large electron-dense haemocyte, in blood from a *Calliphora* 1st instar larva. The electron-density of the cytoplasm is largely due to the affinity of the amorphous ground-plasm for heavy metals, although 20 nm granules are also stained. The latter could either be ribosomes or glycogen (see text). m, small electron dense mitochondrion. Electron micrograph x 17 000.

haemocytes by Hoffmann (1966b), and in *Leucophaea* haemocytes by Scharrer (1972).

The hypothesis proposed by Hagopian (1971) that electron-dense inclusions, both homogeneous and granular, may be melanosomes or premelanosomes finds little support at the anatomical level and needs histochemical verification. Vertebrate premelanosomes are polymorphic, but patterned areas of electron-dense protein matrix, frequently of zig-zag or helical conformation are typical (Birbeck, 1963; Maul, 1969), and these patterns differ from those in haemocyte vacuoles.

Many insect cells do contain extremely electron-dense vacuoles that are acid phosphatase negative (e.g. in *Lucilia*: Seligman *et al*., 1974), and these could contain pigment, but they require further investigation. Other homogeneous electron-dense vacuoles (e.g. in *Rhodnius* oenocytoids) apparently contain liquid (Wigglesworth, 1973). Scharrer (1972) has detected "masked" microtubular elements in occasional electron-dense vacuoles of certain *Gromphadorina* haemocytes, and suggests that in these cells a close relationship exists between tubule-containing vacuoles and homogeneous electron-dense vacuoles.

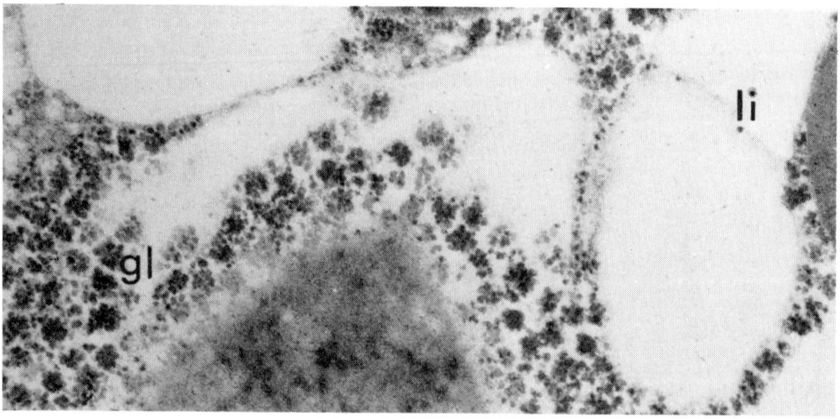

Fig. 7. β-glycogen (gl) and lipid droplets (li) in a phagocytic haemocyte from a *Calliphora* puparium. Electron micrograph ×12 000.

An inclusion which at the present time appears to be confined to insect haemocytes is composed of closely packed tubular elements. Such inclusions were first described in cockroaches by Scharrer (1966) and in *Locusta* by Hoffmann (1966b) but have subsequently been reported in all insect orders examined, except the Diptera (Table 1, column 4). Each tubule is a cylinder of electron-dense material surrounding an electron-lucent core, but a lipid bilayer does not enter into the tubule structure. In some insects the tubules are apparently free in the cytoplasm (Hoffmann, 1966a) but in most they are contained in membrane-limited spaces variously described as endoplasmic reticulum, nuclear envelope, or vacuoles (Cassier and Fain-Maurel, 1968; Granados *et al.,* 1968; Moran, 1971; Hagopian, 1971). The high resolution micrographs of Hagopian (1971) for *Leucophaea* and of

Moran (1971) for *Blaberus* show that the tubules are contained within a unit membrane. Scharrer (1972) believes that the membrane-bound tubular elements in cockroaches originate in Golgi-derived granular vacuoles.

The tubular elements of many haemocyte inclusions do not have the dimensions of typical cytoplasmic microtubules (24-27 nm diameter), nor have they been demonstrated to be sensitive to colchicine or vinblastine, nor yet composed of tubulin subunits. It is thus premature at this stage to call them "microtubules" without qualification. Three distinct size groups

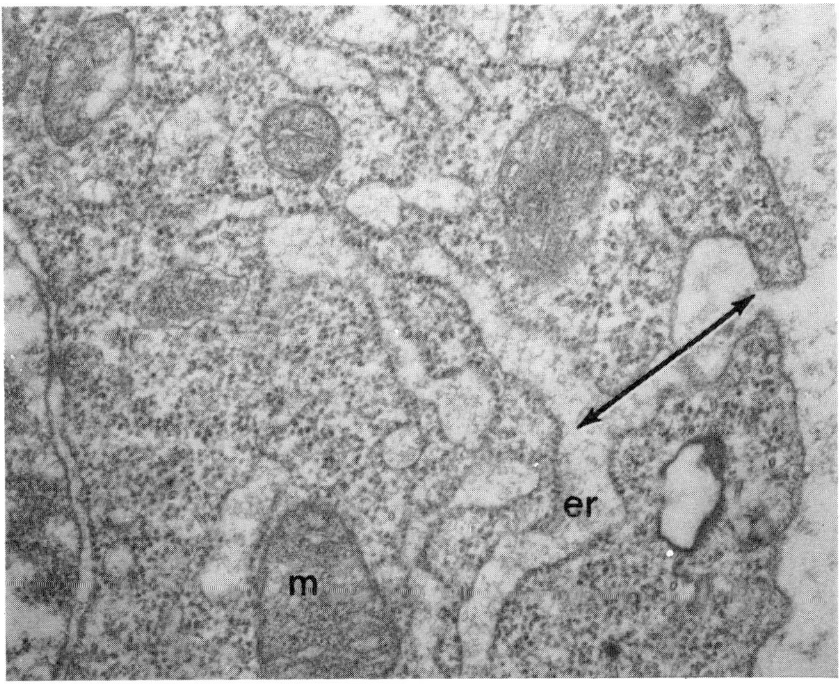

Fig. 8. In this *Calliphora* haemocyte the ribosome-studded endoplasmic reticulum (er) is in the form of expanded cisternae containing amorphous material. Similar material is seen in the haemolymph. Continuity of the exoplasmic compartment of the vacuome and the haemolymph (arrow) provides circumstantial evidence for exchange of material between cell and haemolymph. m, mitochondrion. Electron micrograph ×54 000.

have been reported (Table 1) and each has its counterpart in vertebrate cells. The tubules in *Blaberus* and *Locusta* oenocytoids are 32-40 nm in diameter, like the tubules in canine Kupffer cells (Boler, 1969). The tubules in *Antheraea, Rhodnius, Bombyx, Melolantha, Tenebrio* and *Locusta* granular haemocytes have maximum dimensions between 15-18 nm, like those in cytoplasmic bodies of arterial endothelia (Weibel and Palade, 1964). Only in *Leucophaea* are the dimensions of inclusion tubules (25 nm) comparable to cytoplasmic microtubules, but even in *Leucophaea* the inclusion tubules have atypical substructure, consisting of a central 5-nm dense area and eight radial spokes inside the tubule. The function of these organelles is discussed below in connection with clotting and basement membrane formation.

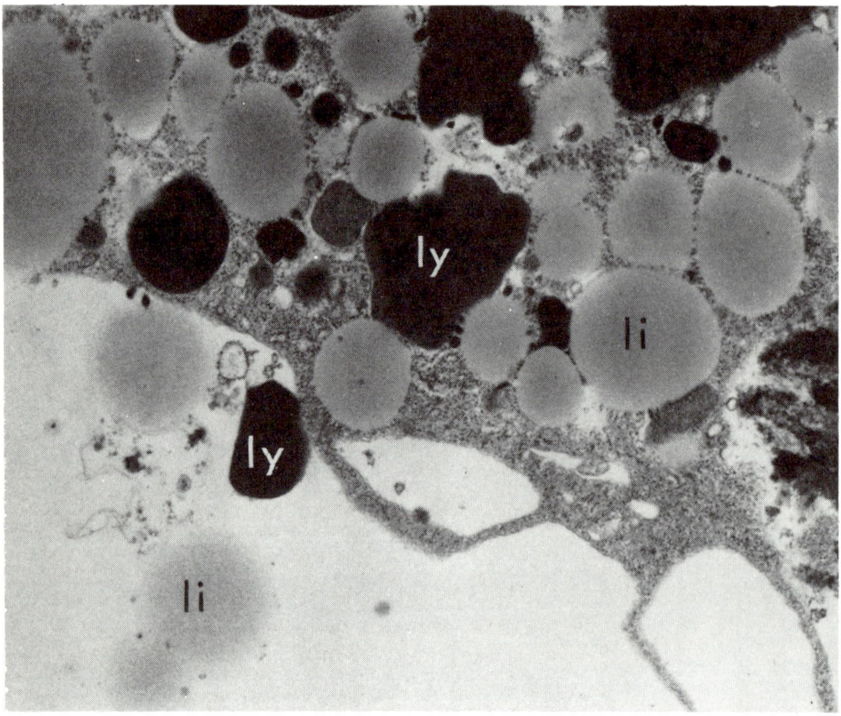

Fig. 9. During the later stages of metamorphosis in *Calliphora* the haemolymph becomes filled with lipid droplets (li) and lysosomal residual granules (ly) many of which appear to be released from phagocytic haemocytes such as this one. Note the similarity of ultrastructure of elements in cell and in haemolymph. Electron micrograph x14 000.

Vacuoles containing fibrous material (Table 1, column 5) close to the plasma membrane have been implicated in basement membrane formation in *Rhodnius* (Wigglesworth, 1973), and may contain connective tissue precursors in other species (Scharrer, 1972, Fig. 2). The significance of unit membrane stacks or cylinders (Table 1, column 3) and of granular vacuoles (Table 1, column 6) in haemocytes is not known.

Homogeneous electron lucent or "grey" droplets of smooth rounded profile (Figs 7 and 9) are commonly seen in haemocytes and are usually interpreted as lipid (e.g., in *Ephestia*: Grimstone *et al.*, 1967; Smith, 1968; in *Calliphora*: Crossley, 1968, Zachary *et al.*, 1973).

Some vacuoles seen in haemocytes may be postmortem artifacts, as it is well known that some haemocytes are very labile and become granulated when observed *in vitro* under the light microscope. This lability is, of course, a very general observation of cell behaviour. If, for example, starfish cells are crushed, the protoplasm becomes filled with large vacuoles (Heilbrunn, 1961). The fixation of insect haemocytes for electron microscopy has been discussed by Wittig (1969).

3 Insect blood cell diversity

Study of insect blood cells with the electron microscope affords not only cytophysiological comparisons with other cell types, but also a potential unification of blood cell classification using the terminology of fine structure. The immediate difficulty is to correlate our electron images with the existing classification, which the electron microscopist is liable to find "complex and confusing" (Moran, 1971). One trend is to abandon classification altogether because "the many transitional features of haemocyte morphology favour the concept of functional flexibility of one basic cell type, rather than a strict classification into distinctly separate cellular types" (Scharrer, 1972). This policy also has inherent justification because the process of differentiation leads from unspecialized to specialized cells through intermediate forms. But there is little doubt that when an extremely specialized cell presents itself, it can be distinguished from its undifferentiated precursors by a variety of techniques, morphological, physiological or chemical. Furthermore, the number of cells in a particular state of differentiation at a given time may be a highly significant factor in the physiology of the insect. It is largely the need for quantitative information about haemocyte populations that generates the need to classify individual cells. Other goals of a classification include haemocyte comparative physiology and the understanding of haemocyte differentiation pathways. Is it true, for instance, that every blood cell passes through all types of specialization, as has been suggested for *Sialis* by Selman (1962)?

The complexity and ambiguity of existing blood cell classification derives largely from the admixture of three kinds of terminology, each having different objectives:

a. *Shape terminology* seeks to describe the form or morphology of the cell as seen under the microscope. Examples of shape terminology are given in the diagram opposite, which is based largely on Yeager's (1945) descriptions of blood cells in *Prodenia*. Shape terminology implies no functional connotations, but the probability that some functions are associated with some shapes has led to widespread use of shape terminology in the literature.
b. *Population terminology* seeks to circumscribe the range of morphological and physiological variation of blood cell types in order to facilitate differential counting, with the understanding of blood cell population dynamics as the primary objective. It is reviewed below.
c. *Cytophysiological terminology* seeks to describe what individual blood cells do in the life of an insect. It relates to physiological parameters of the cell and to its fine structure, but not necessarily to its shape. Progress with this terminology is discussed below.

The three kinds of terminology overlap, and each has its place in current classifications.

It is necessary to consider the evolution of blood cell classification in order to understand the origins of terms in common use today. The first attempt at a general classification of insect haemocytes was that of Hollande (1911) which was based on study of five orders of insects, and on a distillation of even earlier work, particularly that of Cuénot (1897) and Kollmann (1908). Hollande (1911) provides us with definitions of the terms proleucocyte, phagocyte, granular leucocyte, adipoleucocyte, oenocytoid, and spherule cell, as follows:

Proleucocytes—are rounded cells with heavily-staining nuclei occupying nearly all of the cell. The cells are frequently seen in mitosis. The cytoplasm is slightly basophilic.
Phagocytes—are cells of variable shape—rounded, amoeboid or fusiform, with one or two lightly staining nuclei. Mitotic activity is infrequently seen. The cytoplasm is homogeneous or vacuolated, and the vacuoles have heterogeneous staining reaction. The cells are capable of phagocytosis.
Granular leucocytes—are elongated or oval cells with cytoplasm filled with either acidophil or basophil granules. The cells are occasionally capable of phagocytosis.
Spherule cells—are cells filled with large spherical inclusions of rather uniform size, which may be acidophil or basophil. The nucleus is small and mitotic activity is seen, even when spherules have formed. The cells are occasionally capable of phagocytosis. The cells multiply particularly at the approach of pupation.
Oenocytoids—are spherical to oval, never elongated, with a small eccentrically located nucleus. The cytoplasm is uniform and dense and strongly acidophilic. These cells are non-phagocytic.

SHAPE TERMINOLOGY APPLIED TO HAEMOCYTES

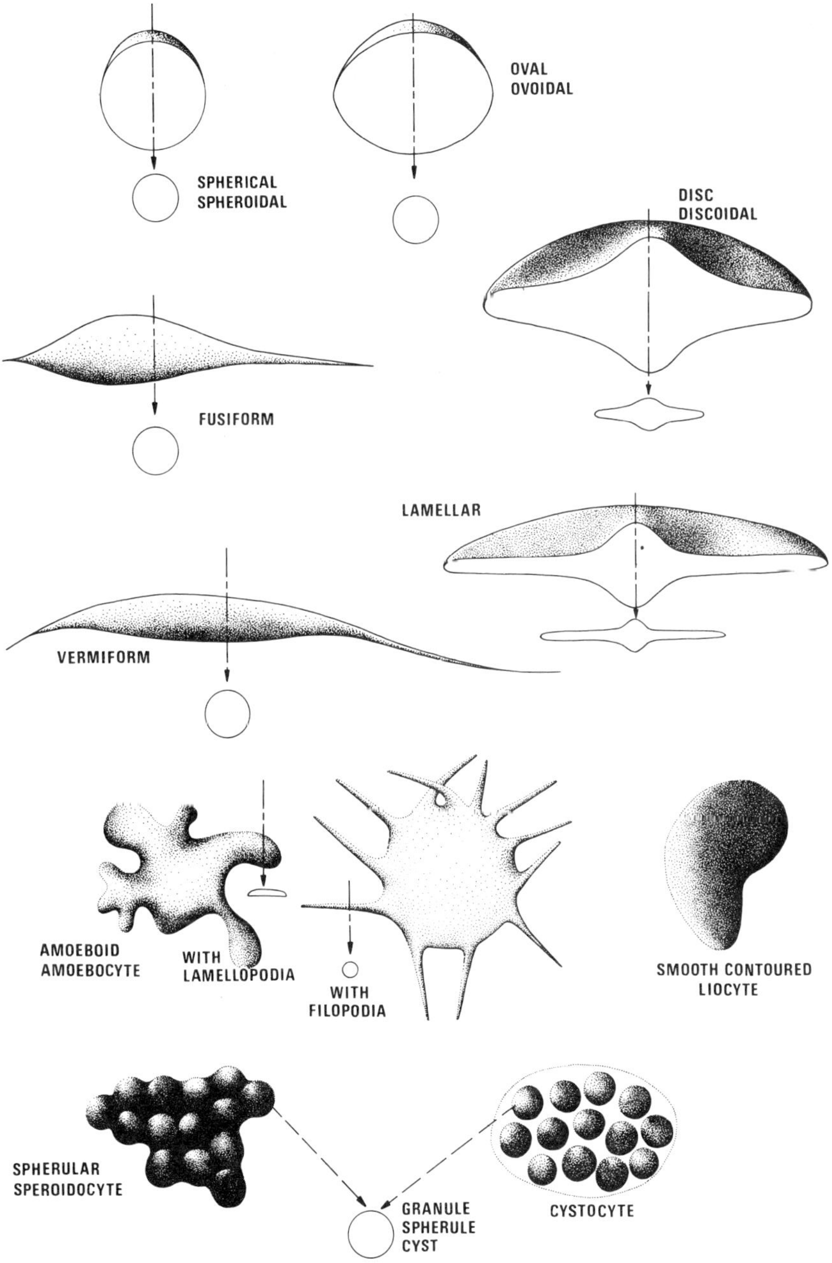

Adipoleucocytes—are spherical cells with large heterogeneous nuclei. The cytoplasm is basophilic at first but this changes on ageing and the cell becomes filled with small fat droplets, between which occur acidophilic granules. Green pigmented droplets are also seen. These cells are slightly phagocytic.

Wigglesworth's (1939) review provided a general classification of insect blood cells which included each of Hollande's categories. Phagocytes were divided into two classes on a basis of nuclear size following Paillot and Noel (1928). Wigglesworth pointed out that both granular leucocytes and spherule cells appeared often to be phagocytes laden with granules.

The introduction of phase contrast microscopy to haematology (Gregoire and Florkin, 1950a, 1950b) led to numerous studies of unstained insect blood cells examined *in vitro*. The work of Jones, summarized in his first review (Jones, 1962), pioneered the study of blood cell population dynamics and attempted to relate the new information obtained for living cells with earlier work on fixed stained cells. Jones based his classification, which was essentially designed for population studies, on that of Hollande, but he introduced notable changes. He abandoned the term "leucocyte" which was used in earlier papers to mean "white cell", in favour of the term "haemocyte" meaning "blood cell", a more appropriate term for cells circulating in a haemocoel. Jones (1962) preferred to avoid physiological terms, such as "phagocyte", "amoebocyte", "trephocyte", "coagulocyte", on the grounds that they "tend to be too specific for cells with multiple functions, or for structurally different cells which may perform the same function". However, it is hard to avoid physiological criteria in the description of cell types, even when these are primarily examples of population terminology. Thus Jones (1962) adopts the term "plasmatocyte" (Yeager, 1945) defined as "basophilic polymorphic haemocytes which do not rapidly disintegrate *in vitro*, but tend either to send out many pseudopodia or to round up, both *in vivo* and *in vitro*". Jones (1962) also adopted two of Yeager's (1945) shape terms: "podocyte" and "vermiform cell". He narrowed Hollande's (1911) definition of granular haemocyte to exclude cells with basophilic granules, but noted that the acidophilic granules might be coloured green, blue or yellow in the living cell (i.e. they are heterogeneous). A new form of granular haemocyte, the "cystocyte", was also recognized which also contained acidophilic inclusions, but which was unstable *in vitro*, tending to eject material into the haemolymph and become hyaline. The original definition of "oenocytoid" provided by Hollande (1911, 1920) was also greatly modified by Jones and this term will be discussed below.

The burgeoning literature on insect haematology has led to increasing difficulty with the evolved terminology, because of the looseness of the original definitions; nevertheless, Gupta (1969) produced an attempt at synonomy of haemocyte terminology. He adopted Jones' classification but included "prohaemocytes" in the general term "plasmatocyte" because of reports (e.g. Clarke and Harvey, 1965) that the two types could not be distinguished either on a basis of appearance or behaviour.

The use of an electron microscope imposes a large measure of standardization on cytological technique, and has led to a terminology common to the entire spectrum of cells. It is possible to distinguish between cells of different size, profile, surface specialization, organelle content, and storage vacuole structure. It is also possible to carry out cytochemistry on cells thus distinguished, and to comment on their enzymology. Detection of phagocytic capacity, and of secretory capacity, is possible using tracer substances. A comprehensive cytophysiological classification of insect haemocytes is not practicable on the basis of existing information, but we can begin by recognizing four cell specializations, realizing that more than one may exist in a given cell:

a. *Specialization for multiplication*

This is evidenced by the capacity for frequent mitosis and the corresponding phase of DNA synthesis. It is the specialization of stem cells or prohaemocytes or undifferentiated plasmatocytes. In cytophysiological terms it usually excludes the formation of specialized organelles and secretions. Prohaemocytes are usually smaller than 10 μm in diameter and have a nucleocytoplasmic volume ratio of 0.5 or more. The cytoplasm is well provided with ribosomes and mitochondria but has sparse endoplasmic reticulum and few Golgi zones. Large vacuoles or inclusions are absent. The cell is nonphagocytic, and is capable of mitosis. A centriole is present. Such stem cells are found in circulation, on tissue surfaces, and in haemocytopoetic centres. Representative micrographs include: for *Locusta* –Hoffmann *et al.*, 1968a, their Fig. 1, and Devauchelle (1971), Fig. 1, for *Melolontha.* The general term prohaemocyte is adopted to describe this specialization in the present review.

b. *Specialization for phagocytosis*

Cells with a capacity for phagocytosis have a well-developed vacuolar apparatus composed of lysosomes and phagosomes. These are best identified by histochemistry for acid hydrolases, but are seen following conventional fixation as electron-dense vacuoles ranging from 0.1-3.0 μm in diameter, limited by a single unit membrane. The Golgi apparatus and endoplasmic reticulum are extensive, as a result of involvement in lysosome formation. The plasma membrane frequently bears micropapillae, filopodia or other irregular processes, as well as pinocytotic or coated vesicular invaginations. Representative micrographs include: for *Calliphora*—Crossley (1964), his Fig. 7, (1968), his Figs 13-14; for *Locusta*—Hoffman Stoekel *et al.* (1968), his Fig. 1; for *Blaberus*—Moran (1971), his Fig. 2 (Figs 1, 4 and 5). In the literature phagocytic cells have been termed leucocytes, amoebocytes, plasmatocytes, micronucleocytes, spherule cells or adipohaemocytes, but the differences between the cells described by these terms are not sharp. The term phagocyte is adopted to describe this specialization in the present review.

c. *Specialization for secretion or storage*

Such cells are provided with organelles for synthesis of molecules to be eventually released from the cell, sometimes after a period of storage. The cells contain numerous membrane-limited vacuoles, some of which contain stored material *other than* hydrolytic enzymes or the products of cellular autophagy. Profiles suggesting release of vacuoles from the cell are seen. These cells contain well-developed rough endoplasmic reticulum or Golgi centres when they are engaged in protein, mucoprotein, or membrane synthesis. Lipid or cholesterol metabolism may be evidenced by hypertrophy of smooth membranes and of mitochondria. Representative micrographs include for *Locusta*—Hoffmann 1966a, his Plate 1; for *Melolontha*—Devauchelle, 1971, his Figs 25-26. In the literature such cells have been termed trephocytes, plasmatocytes, amoebocytes, adipohaemocytes, oenocytoids, granular haemocytes, spherule cells etc. The term "trephocyte" is adopted to describe this specialization in the present review.

d. *Specialization of haemostasis or injury metabolism*

These cells change their metabolism in response to foreign surfaces, and are often recognizably labile. In the light microscope the cells appear to be hyaline because of the sparse cytoplasmic organelle complement, which is composed mainly of ribosomes and small mitochondria. Few vacuoles are present. These cells are often reported to contain phenol-oxidizing enzymes (see section 9). Representative micrographs include: for *Locusta*—Hoffmann and Stoekel (1968); for *Melolontha*—Devauchelle (1971, Fig. 33); see also Figs 10, 12 and 15. In the literature these cells have been termed hyaline haemocytes, phenol-oxidizing haemocytes, coagulocytes, cystocytes, or oenocytoids.

In *Calliphora* a group of labile haemocytes which fragment on contact with foreign surfaces is present, and these have been termed "thrombocytoids" (Zachary *et al.*, 1973). These cells differ in ultrastructure from other cells concerned with haemostatis. They contain well-developed endoplasmic reticulum and Golgi centres, as well as numerous electron-dense granules and occasional autophagic vacuoles. They also have intricate invaginations of the plasma membrane, and this character alone serves to distinguish them from other kinds of haemocyte. The functions and nomenclature of this diverse group of cells will be considered below in sections concerned with blood clotting, defense reactions, and phenol metabolism.

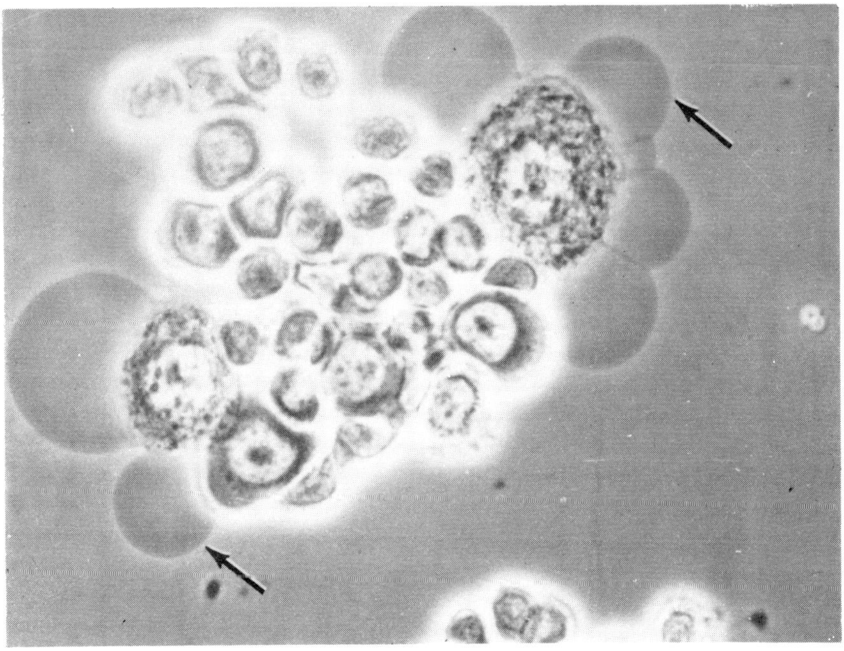

Fig. 10. In this light micrograph of a group of *Calliphora* larval haemocytes alterations to the membranes of two labile cells are evidenced by blistering (arrows) (cf. Fig. 12). Clotting by cell fragmentation has been inhibited by the addition of 5×10^{-5} M Colcemid. No plasma gelation occurs in this species. Light micrograph ×200.

The origins of the cytophysiological classification outlined above are apparent in a number of ultrastructural studies of insect blood cells. Stang-Voss (1970) recognizes four classes of haemocytes in *Tenebrio*: cells of embryonic nature (*ursprungszellen*) phagocytes (*phagocytare amoebocyten*), clotting cells (*gerinungzellen*) and spindle-shaped cells (*spindleformig amoebocyten*). The shape terminology of the last class is related to cells which have storage granules and are presumably trephocytes. Marschall (1966) places *Tenebrio* haemocytes into two physiological groups: the first a category of labile cells involved in blood clotting and mucopolysaccharide synthesis, the second a heterogeneous group concerned with phagocytosis and secretion. Scharrer (1972) recognizes a division of labour as reflected in the diverse ultrastructural appearances of haemocytes, but favours the concept of functional flexibility of one basic cell type. Most other authors on haemocyte fine structure assign their electron micrographs to categories erected on a basis of light microscopy. These categories have the advantage of wide acceptance, but at least three, adipohaemocytes, spherule cells, and coagulocytes, are far from satisfactorily defined.

"Adipohaemocyte", is a term derived from the "adipoleucocyte" of Hollande (1911), suggesting that demonstrable lipid is a major component of the cell, and has been used with little justification in the literature. Ultrastructural studies of so-called adipohaemocytes include cells which contain no reported lipid (Devauchelle, 1971, Figs 18-22), lipid of doubtful authenticity (Pipa and Woolever, 1965) or material believed to be mucoprotein or mucopolysaccharide (Beaulaton, 1968, Fig. 7). The "adipo" prefix seems inappropriate for these cells, even though they may contain some lipid as do all cells. The adipohaemocytes described for *Galleria* appear to be phagocytes (Pipa and Woolever, 1965), whilst those described for *Melolontha*, *Antheraea* and *Bombyx* (Beaulaton, 1968; Devauchelle, 1971) are perhaps best considered as trephocytes.

"Spherule cells" are morula-like cells containing large acidophilic or basophilic inclusions of rather uniform size. The term was introduced by Perez (1910) as *sphéres des granules,* and by Hollande (1911) as *cellule à sphérules,* and has been adopted in two different senses in the literature. The first sense is that of a phagocyte with a well-developed lysosomal vacuolar apparatus, sometimes gorged with engulfed cellular-debris. It was in this sense that the term was used for *Calliphora* by Perez (1910), by Akesson (1953), by Crossley (1964, 1968), and by Barritt and Birt (1971) for *Lucilia.* In a second sense the term "spherule cell" is used to describe haemocytes containing large storage vacuoles of diverse nature, which may be released by cell lysis *in vitro.* As Jones (1962) has pointed out, these cells are probably a "specialised kind of granular haemocyte". In *Antheraea* spherule cells are elongated, fusiform or flattened cells containing vacuoles 2-3 μm in diameter filled with flocculent *electronlucent* material and small microvesicles (Beaulaton, 1968). In *Melolontha* spherule cells are elongated fusiform cells with microtubules in marginal bundles, containing vacuoles 1.5-3.5 μm diameter, filled with *electron-dense* material which is granular in the smaller vacuoles (Devauchelle, 1971). To the present author the distinction between spherules and granules appears arbitrary, and these cells are perhaps best considered as trephocytes of diverse nature and unknown function.

The term "coagulocyte" was introduced by Gregoire and Florkin (1950b) to describe *Gryllulus* type III cells which were spherical and of large size but had small sharply outlined nuclei. The cytoplasm was hyaline and contained a few small dark granules in active motion. The cell was labile, and within seconds after contacting a glass surface the cell underwent intense modifications characterized by internal movement or bubbling of the cytoplasm. In some cells the plasma membrane appeared to break, allowing part of the cytoplasm to escape. Changes in the surrounding haemolymph followed and a granular precipitate formed around the cell, which was interpreted as an initial stage in plasma coagulation. Very similar cells had been described earlier in Crustacea as "explosive corpuscles" by Hardy (1892). Furthermore it had been shown that the explosive crustacean cells liberated phenolase as the blood clotted (Pinhey, 1930; Bhagvat and Richter, 1938). An eruptive cell type termed a "rhegmatocytoid" was also described in *Prodenia* by Yeager (1945). He reported that these very unstable cells had basophilic cytoplasm with acidophilic (eosinophilic) inclusions and sometimes also granules at the periphery. Erupted cells showed release of eosinophilic material into the nearby plasma. Yeager (1945) also introduced the term "cystocyte", simply meaning a cell with cyst-like inclusions. The cystocyte also had basophilic cytoplasm containing eosinophilic or colourless inclusions, and a definite ectoplasmic region in the active state. He did not, however, consider cystocytes to be unstable. Jones (1950) again uses the term "cystocyte" to describe "coarsely granular haemocytes", but later Jones (1962) uses "cystocyte" to describe "highly unstable haemocytes all of which rapidly turn into brightly hyaline forms, ejecting granules or droplets into the plasma". The inclusions in these cells are eosinophilic. Thus Jones does not recognize the distinction made by Yeager (1945) between stable (cystocyte) and unstable (rhegmatocytoid) eosinophilic haemocytes. Furthermore he considers instability as a poor criterion of cell identity since a number of morphological types are, on occasion, unstable (Jones, 1962, p. 216). The evidence that the coagulocyte is involved in plasma coagulation is still circumstantial (see section 6), so the terminology surrounding the eruptive cells described as coagulocytes in many species of insects by Gregoire is very unsatisfactory.

The possibility that the presence of phenol-oxidizing enzymes is a feature of the coagulocyte type offers new hope in this terminological maze. These enzymes are present in explosive cells of Crustacea, and in the explosive type B cells of *Calliphora* (Crossley, 1964), and the explosive coagulocytes of *Locusta* (Hoffmann *et al.*, 1970). An electron micrograph of such an unstable, phenol-oxidizing haemocyte from *Calliphora* is shown in Fig. 12. There is, however, a difficulty. This arises from the use of the term "oenocytoid" to describe phenol-oxidizing cells in *Sarcophaga* by Dennell (1947), and the suggestion by Gregoire (1955a) that hyaline haemocytes (coagulocytes) were oenocytoids. The question arises as to whether there is a useful distinction between coagulocyte and oenocytoid, and to decide this we shall next have to consider the lineage of the term oenocytoid.

The term oenocytoid was introduced by Poyarkoff (1910), but Hollande (1920) gave the first useful criteria for recognizing oenocytoids. They are cells of circular or oval form which do not produce external protoplasmic filaments. Their nucleus is relatively small, often eccentric and occasionally paired. *In vivo* the cytoplasm is shining and homogeneous, and on staining is eosinophilic. The cell is non-phagocytic (Hollande, 1920). Wigglesworth (1959) gives a similar interpretation and describes oenocytes as rounded disc-shaped cells which always have a smooth outline, with no trace of pseudopodial filaments. They have a clear homogeneous cytoplasm and minute vacuoles in contact with the cell surface are visible in fresh or osmium-fixed light microscope preparations. The cells are eosinophilic.

The definitions given by Hollande (1920) and Wigglesworth (1959) allow the electron microscopist to identify an oenocytoid with reasonable certainty. The first electron micrographs of oenocytoids identified as such are those for *Melolontha* (Devauchelle, 1971, Figs 32 and 33). The oenocytoid of *Calliphora* is extremely similar (Zachary *et al.*, 1973, Fig. 9)—also see Figs 10 and 12—but contains phenol-oxidizing enzymes (Crossley, 1974). The oenocytoid is a large spherical, oval, or fusiform cell with a smooth outline lacking pseudopodia or filopodia. The cytoplasm is poorly provided with organelles, but ribosomes and mitochondria are conspicuous. The endoplasmic reticulum is sparse and is in the form of elongated cisternae. Small Golgi centres are present, but lysosomes are absent. Small spherical electron-dense vacuoles are confined to the periphery of the cell. Oenocytoids are non-phagocytic, as evidenced by the absence of a vacuolar apparatus, and by numerous observations in the literature.

Cells which come within the category oenocytoid as defined above include the cells described as "coagulocytes" for *Locusta* by Hoffman and Stoekel (1968). These cells are rich in free ribosomes, and contain long and sinuous ER cisternae, often dilated with flocculent contents. Other organelles include numerous small mitochondria and Golgi centres, and homogeneous, dense peripheral vacuoles, about 1 μm in diameter. These *Locusta* cells differ from the oenocytoids of *Melolontha* and of *Calliphora* in having vacuoles containing finely fibrillar material in bundles, but in both *Calliphora* and *Locusta* such cells are highly labile, and undergo changes to the endoplasmic reticulum, mitochondria, vacuoles and nucleus as the blood coagulates.

The oenocytoids of *Rhodnius* have uniformly dense cytoplasm with a few small mitochondria and very few ribosomes. The electron-dense peripheral inclusions are always perfectly spherical; their contents are presumably liquid. There is circumstantial evidence for *Rhodnius* that oenocytoids discharge their inclusions at the time when cuticle formation is taking place (Wigglesworth, 1973). Cells described in the literature as oenocytoids which are not readily included on the definitions given above include the oenocytoids of *Locusta* described by Hoffmann (1966a) and by Hoffmann *et al.* (1968), which contain large cytoplasmic inclusions composed of bundles of tubular elements. Similar cells were described by Cassier and Fain-Maurel (1968) who also used the term oenocytoid, and suggested that the bundles of tubular elements were composed of acid mucopolysaccharide. Haemocytes containing tubular elements have been described by many authors but there is no agreement as to the type of cell involved (see Table 1), although they could all be considered as trephocytes.

4 Humoral control of insect blood cell populations

It has repeatedly been observed that the number of haemocytes circulating in the haemolymph relates to the physiological state and developmental stage of the insect, but nevertheless it is hard to obtain quantitative data. Profound variations in the total haemocyte count (THC = cells per mm^3 haemolymph) may be recorded without any change in the actual numbers of haemocytes present in the insect. These variations may be the result either of difficulties in making cell counts, or of changes in the blood volume. Wheeler (1963) distinguishes between the calculated number of haemocytes in the entire insect, which he terms the "absolute count", and the total haemocyte count (THC) which will vary according to haemolymph volume. When in *Periplaneta*, for example, the blood volume was determined by an amaranth dye dilution method, modified after Yeager and Munson (1950), the absolute number of circulating haemocytes did not increase prior to ecdysis even though the number of haemocytes per cubic millimetre increased. This relative increase in haemocytes was due to a decrease in haemolymph volume. Haemolymph volume is expressed as a percentage of body wet weight ($\mu l\ mg^{-1} \times 100$), and in the last moulting cycle of *Periplaneta* varies from 14 to 21 per cent, with a large transient increase at ecdysis (Wheeler, 1963). With the adoption of SI units

1 μl = 1 mm^3 exactly, so the absolute or volume-corrected haemocyte count obtained by multiplying THC in mm^3 with haemolymph volume in μl is a useful standard parameter. (It should of course be remembered that this parameter is sensitive to variations in insect weight.) The haematocrit, which is the wet volume ratio of cells to plasma, can also be expressed as the percentage of haemocytes in whole haemolymph.

A proportion of haemocytes adhere to tissue surfaces, as pointed out by Yeager *et al.* (1932). Wigglesworth (1956, 1959) has suggested that very large numbers of haemocytes reside in loose accumulations on tissue surfaces in general, and Scharrer (1972) provides electron micrographs of such sessile haemocytes. Jones (1970, p. 26) has suggested that nonproliferating accumulations of haemocytes in insects be referred to as "haemocytic reservoirs", to distinguish them from proliferation and differentiation centres. Since the total haemocyte count is based on the number of circulating cells it is liable to be a variable parameter, usually lower than the actual number of haemocytes present in the insect. Formerly sessile haemocytes are released by certain sampling pretreatments, the best known of which is heating (Yeager, 1938). Direct comparison of numbers of cells obtained inTHCs before and after heating the insect has been made for cockroaches by Wheeler (1963). He found that the mean THC of 59 400 cells per mm^3 obtained for untreated insects when blood was drawn into 2 per cent EDTA, became 103 200 cells per mm^3 when the insects were heated to 60° C for 1 min before sampling. The difference between these two THCs is apparently partly due to blood cell clumping, since Wheeler found that 2 per cent EDTA was not consistently effective in preventing haemocyte clumping, whereas clumping was not encountered in heat-treated samples. Rosenberger and Jones (1960) had earlier pointed out for *Prodenia* that if effective inhibition of coagulation is not obtained, the availability of haemocytes, and hence the THC, will decrease as a result of cell clumping. However, in the bug *Halys*, there is no coagulation of the blood *in vitro*; nevertheless by holding the insects at 60° C for 1 min in water before withdrawal of a haemolymph sample, the THC can be abruptly increased in comparison with unheated control samples, for reasons that are not understood (Bahadur and Pathak, 1971). It appears likely that cell to tissue adhesion, as well as coagulation, is influenced by heating.

By carefully standardizing sampling techniques it is possible to obtain useful comparative counts for a given insect, which detect real physiological changes in numbers of circulating haemocytes, but interspecific comparisons are less meaningful. Changed numbers of circulating haemocytes are the result of a dynamic balance between four factors: (a) mitosis of circulating haemocytes; (b) death or fragmentation of circulating haemocytes; (c) release or retention of haemocytes at haemocytic reservoirs; (d) release of sessile haemocytes from haemocytopoietic tissues.

There are many indications, both natural and experimental, that humoral control systems regulating haemocyte numbers exist in insects. Haemorrhage has been extensively employed as a haematological stress, and can be shown to entrain corrective increases in haemocyte numbers. Cuénot (1896) found that increased mitosis of haemocytes of several species of insects followed haemorrhage. Bogajavlensky (1932) found a great increase in the number of dividing haemocytes five to six days after haemorrhage in *Carausius*. Cameron (1934) and Dakhnoff (1938) noted that mitoses were frequent in *Galleria* after severe haemorrhaging.

Looking more closely at the effect of haemorrhage on *Galleria*, Shristava and Richards (1965) noted an increase in mitotic activity and an approximate doubling of THC, but reported that a small haemorrhage produced a comparable effect to a large haemorrhage. This suggests that mitotic rate is not coupled directly to blood volume or total circulating haemocyte number. In diapausing saturniids an increase in haemocyte numbers, mainly due to augmentation of plasmatocytes, may be induced by injury with minimal haemorrhage (Harvey and Williams, 1961). Wigglesworth (1937) reported that the intensity of the response of haemocytes to burns was greater than the response to cuts. Feir and McClain (1968b) compared the effect of haemorrhage and of injury directly and their results showed that a severe cut produced an increased mitotic index comparable to haemorrhage, but a burn produced a greater effect than haemorrhage, raising the mitotic index to nearly 6 per cent. When the experimental treatment was applied in the day of the moult, the peak of mitotic activity was 48 h later, and it was suggested that a hormone may set the time at which the haemocytes indulge in mitotic activity. There are two reports that cockroach haemocyte numbers are not altered by haemorrhage (Yeager and Tauber, 1933; Schlumberger, 1952).

Crossley (1964) reports that in *Calliphora* the mean mitotic index in posterior sessile haemocyte masses, which is 1.2 per cent in unoperated larvae, rises to 1.45 per cent following massive haemorrhage resulting in 18-22 per cent loss of body weight. This difference in mitotic index is significant at the 99 per cent level. In *Locusta* fifth instar larvae bled so as to lose approximately 80 per cent of their normal blood volume, the THC increases immediately. Differential haemocyte counts show that this increase is due to the sudden release of poorly differentiated blood cells, the cytological characteristics of which are identical with those of "reticular cells" of the dorsal diaphragm (Hoffmann, 1969). Thus in representatives of hemi- and holometabolous insects there is circumstantial evidence implicating consistently located foci of sessile cells in replacement of haemocytes lost by haemorrhage.

Beard's (1949) observations on the effect of haemorrhage on *Popillia* show a significant change in THC. The low $[K^+]$ 7 h after bleeding reported by Beard could also be attributed to a reduced cell fraction in the blood (Brady, 1967), since blood liquid is replaced rapidly, at the expense of other fluids following bleeding. Hoffmann (1969) found for *Locusta* that whereas blood volume was almost entirely restored within 24 h of bleeding, the total and differential counts could not be considered normal until 48 h after the loss of blood. Injections of distilled water, many chemicals, metabolites and bacteria also increase the mitotic index in haemocytes of various insects, and experimental reports are reviewed by Jones (1970, p. 52).

There is an extensive literature on haemocyte population dynamics in growth and development. In 1933 Wigglesworth noticed an increase in haemocyte mitosis and in numbers of sessile haemocytes following feeding in 4th instar *Rhodnius* nymphs. Jones (1967b) also working on 4th instar *Rhodnius* nymphs found the average mitotic index of circulating cells was about 0.3 per cent, with a maximum index of 0.9 per cent, in the first six days following a feed. Thereafter Jones reported a faster decline in mitotic rate for circulating cells than had earlier been observed for sessile cells by Wigglesworth. In *Melanoplus* Avery (1963, cited by Jones, 1967) estimated that a haemocyte population increased 5.1-fold (from 1.5 to 7.7 million) in the last nymphal stage, whilst in *Sarcophaga* the increase is 4.5-fold in the last larval stage (Jones, 1967). Tauber (1937) found that the number of mitotically dividing haemocytes in *Blatta* fell before and during moulting, but rose again following the moult. The mean mitotic index of circulating haemocytes of nymphs and adults was 0.2 per cent, but rose to about 1.0 per cent at moulting (Tauber, 1936, 1937). Increases in mitotic index following a moult have also been reported for *Oncopeltus* haemocytes, where a value of 4.06 per cent was recorded 30 h after the moult to the 5th instar (Feir and McClain, 1959). However, the contrary situation, a decrease in mitotic index, has also been reported to follow moulting in some insects, e.g. *Bombyx* (Bogojavlensky, 1932); *Rhodnius* (Jones, 1967); *Halys* (Bahadur and Pathak, 1971). It should be noted in passing that mitotic index data are not subject to errors due to changes in haemolymph volume (Wheeler, 1963). The decline in mitotic index as pupation is approached which was reported for *Calliphora* by Crossley (1964) and for *Galleria* by Jones and Liu (1968) may be accounted for by the presence of increasing numbers of differentiated cell types, which do not divide. However, a consistent picture cannot be painted since in some insects haemocyte mitotic rates increase as pupation is approached (e.g. *Bombyx*: Bogojavlensky, 1932; *Leptinotarsa*: Patay, 1939).

In insects with a strong diurnal rhythm of activity the haemocyte mitotic index may be subject to diurnal variation. Rooseboom (1937) found the greatest mitotic activity between 0300-0900 hours in *Carausius*, the period which included ecdyses, but persistent diurnal rhythm was not detected in hourly mitotic counts of *Oncopeltus* haemocytes made by Feir and McClain (1968a). Jones and Liu (1968) also sought diurnal numerical rhythms in *Galleria* haemocytes, but encountered large individual variations and were unable to obtain significant differences.

Synchronized changes in haemocyte mitotic index and population size are circumstantial indications of the operation of humoral controls, but the changes in haemocyte population structure, as reflected in differential haemocyte counts, provide even stronger evidence. Perhaps the first direct evidence of humoral control over haemocyte activity is provided by the work of Rizki (1957, 1958, 1962) on *Drosophila*. He reported that specific changes in cell morphology, such as the sending out of filiform pseudopodia, and later general flattening of the cell, only occurred at certain stages of the life cycle. The transformation to lamellocyte form normally occurs in the *ore-R* strain as puparium formation approaches, whilst in the *tumor W* mutant it is advanced to the time of the second moult. Experimental exclusion of the ring gland and brain was found to result in earlier appearance of lamellocytes, whilst reintroduction of ring gland normalizes the situation. The only disquieting feature of these experiments is the observation that distilled water injected into larvae lacking brain and ring gland also leads to a massive increase in the lamellocyte fraction. Apart from this, the experiments are interpreted as arguing for humoral control over haemocyte morphology, but probing more deeply, there are indications that the changes in haemocytes are secondary to earlier changes in other tissues. In particular, the experiments on posterior fat masses in the *tumor W* mutant indicate that the fat body here becomes necrotic as a result of genetic miscoding and that haemocytes respond to necrosis by assuming an encapsulating (or phagocytic) lamellocyte form (Rizki, 1958, 1962). Changes in haemocyte populations just before pupariation, and on experimental injection of distilled water, could equally be in response to the onset of necrosis. Several other authors, including Selman (1962) who worked on *Sialis* and Gupta and Sutherland (1966) who worked on *Sarcophaga* have remarked that haemocyte types may transform from one into another, and the question of endocrine control again arises. In some cases the actual occurrence of transformation is itself questioned; for example, in the case of *Sarcophaga*, according to Jones (1967) the numerical data does not support the notion of transformation, and plasmatocytes, granular haemocytes and spherule cells supposedly constitute distinct cell lines. In any case Gupta and Sutherland (1966) doubt whether humoral influences are involved because the transformations are not synchronized. On the other hand Shrivastava and Richards (1965) report the transfer of ^{3}H-thymidine label from prohaemocytes, through plasmatocytes to adipohaemocytes, indicating a differentiation sequence occupying about 3 days at 35° C, and a haemocyte life expectancy of six days. It would be very interesting to see this type of experiment quantified and extended to other species.

There is now direct experimental evidence that haemocyte populations are controlled by hormone levels. Crossley (1964, 1968) showed for *Calliphora* that the proportion of phagocytic haemocytes, as estimated by acid phosphatase histochemistry for lysosomes, increased sharply as the larva ceased feeding and prepared for pupariation. If larvae were injected with 2 Calliphora units of β-ecdysone, whilst they were still feeding, the haemocyte population structure shifted prematurely, as reflected by a significant increase in the proportion of circulating phagocytic cells. Furthermore, larvae which had been ligatured so as to confine ring glands to the anterior portion of the body not only failed to pupariate in the posterior section, but also maintained an abnormally low proportion of phagocytic cells in the posterior section. This low proportion could be experimentally increased by injection of β-ecdysone. Comparable results were obtained by Bohm (1968) for *Calliphora*, also using the ligature technique, but with Schering synthetic ecdysone. He found only 20 per cent phagocytic (spherule) cells in larval zones of the ligatured insect, but 90 per cent phagocytic cells in pupal zones of the same insect. Jones (1967) showed that if *Rhodnius* was ligatured behind the head immediately after a blood meal, the changes that would normally occur in the blood picture were modified. Jones proposed that the ligature prevents hormone circulation, and that the cyclical modifications in the blood picture are under humoral control. Jones and Liu (1969) carried out total haemocyte counts on *Galleria* larvae ligatured behind the head or behind the metathoracic segment. The thoracic ligature resulted in an average THC 2.2 times lower than normal in the posterior region, but the head ligature was without significant effect. There are strikingly fewer mitoses in the posterior halves of thoracic ligature subjects, but the possible influence of humoral centres on these results was not considered. Jones (1967) reported that if fifth instar *Rhodnius* larvae were ligatured *before* moulting hormones were secreted, the amount of haemolymph from a severed appendage remained abnormally high. The haemocyte count, particularly that of prohaemocytes and plasmatocytes, also remained high instead of declining towards the end of the instar, whilst other haemocytic changes were abnormal. If the head is ligatured *after* the secretion of moulting hormones, the amount of haemolymph from an appendage is not affected and the haemocytes mostly respond like those of normal insects. It was suggested that hormones, in part, regulate the amount of haemolymph available from an appendage and also the cyclic changes in the types and numbers of haemocytes.

The observations of Judy and Marks (1971) on *Manduca* presumptive haemocytes in tissue culture are also relevant here, since the introduction of β-ecdysone into the medium apparently produced an increase in migratory activity. The increased activity of migratory cells began 10-24 h after application of hormone, and the cells were ultimately moving at 5 to 10 times their rate before the application of hormone. There was also reportedly a general tendency for cells to disassociate, but no mitoses were seen in migrant cells during the period of hormone response. The authors suggested that haemocyte mobility may depend directly on the blood ecdysone titre, and that the ecdysone titre may thus control the concentration of free-circulating haemocytes. The inactive isomer 22-isoecdysone produced no response, indicating a high degree of stereospecificity for the 22-hydroxy position by the ecdysone receptor site. Arnold (1961) remarks that for *Blaberus* the presence of a humoral factor stimulating movement and extension of pseudopodia is indicated by the increased movement and pseudopodial extension in haemocytes from older adults, with maximal activity in moribund adults.

Hoffmann (1970) presents experimental evidence to show extensive involvement of endocrine glands in the differentiation of haemocytes. In *Locusta*, the ventral prothoracic glands stimulate the production and differentiation of haemocytes during the last larval instar, in an analogous fashion to the ring gland of Diptera. Although it is difficult to entirely remove the ventral gland in *Locusta*, partial ablation apparently reduces the titre of prothoracic gland hormone sufficiently to drastically curtail the increase in both coagulocyte and granulocyte numbers which normally occurs in 5th stadium insects prior to the adult moult. Furthermore, in *Locusta*, the corpora allata and corpora cardiaca can of course be surgically removed independently of each other. When the corpora allata of adults are removed in either male or female insects the production of differentiated haemocytes is slowed. Introduction of supernumary corpora allata into males raises the THC, and particularly the plasmatocyte count, well above normal for a given age. Electrocoagulation of the pars intercerebralis in males produces an increase of 235 per cent in the volume of the haemolymph, but does not alter the relative number of each type of haemocyte. In females the total haemocyte count jumps significantly, the plasmatocytes showing the greatest proportional increase. Thus destruction of the pars intercerebralis in females in general parallels the effect of introducing supernumary corpora allata into males. If we ignore sex differences, we can presume that these results indicate a moderating control by the median neurosecretory cells of the pars intercerebralis on the corpora allata. Removal of the corpus cardiacum of male adult locusts does not significantly alter the THC, but does cause a relative increase in the number of plasmatocytes. However, the changes induced in tissues other than blood by removal of the corpus cardiacum are so extensive that the effects on the blood itself may be secondary (Hoffmann, 1970).

We have remarked above that haemocyte numbers are the result of a balance between four factors, and it would be useful to know which factors are under endocrine control.

We can firstly speculate whether the population changes can be accounted for by mitosis of circulating haemocytes, or whether release of cells from storage or haemocytopoietic areas is necessary to explain the observations. Unfortunately data on the cell cycle in insect haemocytes is scanty and conflicting (Jones, 1970, p. 22). Lawrence (1968) has measured mitotic phase and interphase lengths for insect epidermal cells by a tritiated thymidine labelling procedure, and finds a minimum interphase of approximately 16 h. Labelled epidermal cells do not divide again until about 24 h after a previous mitosis. Data from insect tissue culture confirms this general timing (Schmidtt and Williams, 1953; Mitsuhashi, 1966). In mammalian cells in culture the total cell cycle time is 18-20 h with the M-phase taking about 1 h (Mitchison, 1971, p. 60), so these insect cycle times are comparable. The very short division times of *c.* 3 min obtained by Rizki for insect plasmatocytes (Jones, 1970, p. 22) thus need confirmation. Equations for estimating the proportion of the cell cycle spent in mitosis normally assume that all cells in the population are growing, that growth is exponential, and that there is no synchrony (Mitchison, 1971, p. 18). These assumptions are almost certainly not valid for insect haemocyte populations, which contain many non-dividing differentiated cells, and may show synchrony related to diurnal rhythms (see above).

There is an obvious need for tritiated thymidine labelling experiments on insect haemocyte cell populations, both pulse labelling experiments to measure the total cell cycle time, and label experiments spanning one cell cycle period to measure the proportion of dividing cells in the population. Only when these data are available will we be able to calculate the mitotic contribution to population increases with precision. For the present, we can estimate the percentage of cells dividing by measuring the total population of cells and its increase with time, and by estimating the cell cycle time. Thus for the data of Jones (1967) for *Sarcophaga* larvae, there is an increase of THC from 514 155 to 2 335 500 cells in five days. Assuming mitosis is confined to a single group of cells with a mitotic cycle of one day, this increase could be accounted for by the mitotic activity of about 10 per cent of the circulating cells. The increase may, however, be the result wholly or partly of release into the circulation of sessile cells, in which case the actual proportion of actively dividing cells in the population will be lower.

We can now examine the role of haemocytopoietic centres in endocrine regulation of cell numbers. There is strong circumstantial evidence that haemocytopoeic centres exist in several insect species. In Diptera Lange (1932) describes, under the name "phagocytic organ" accumulations of haemocytes in the anal region of larvae which disappear at the time of puparium formation with the liberation of phagocytic haemocytes. Arvy (1953) reported haemocyte accumulations in the postero-dorsal region of *Chironomus*, *Calliphora*, *Musca*, and *Phaonia*. In *Musca* and *Phaonia* the accumulations abut the hypoderm in the vicinity of the posterior spiracles, and grow in size during succeeding larval instars. From a group of only 24 stem cells in larvae 3 h post eclosion, the accumulations grow into a large mass of differentiated haemocytes by the end of the third instar, but disappear entirely at pupariation with the release of many phagocytic haemocytes. In *Calliphora* Arvy (1953) reports haemocyte accumulations in the vicinity of the posterior spiracles, which disappear at the end of the first larval instar. This release was incorrectly reported as taking place at the end of the third larval instar in Crossley (1964, p. 387). However, Crossley (1964) reports that small haemocyte masses indeed do exist in second and third instar *Calliphora* larvae, and have a similar disposition and fate to those of *Musca* and *Phaonia*.

In view of this coherent picture, for cyclorrhaphous Diptera, of release of cells from posterior accumulations at pupariation, it seems highly probable that release of sessile haemocytes also contributes to the increased THC reported by Jones in *Sarcophaga*. Furthermore it is likely that the disappearance of over 10^6 haemocytes from circulation in a period of three hours at pupariation in *Sarcophaga* (Jones, 1967) is mainly the result of enhanced adhesion to tissues, as is observed in histological preparations, rather than of actual changes in cell numbers.

In both *Locusta* and *Gryllus*, Hoffmann *et al.* (1968a, 1968b) have reported not only that consistent foci of differentiating haemocytes exist in the vicinity of the dorsal diaphragm but also that, following experimental haemorrhage, cells similar to those present in presumptive haemocytopoietic foci appear in circulation. Such haemocytes are slightly differentiated stem cells, and differ from cells normally circulating in the haemolymph. In *Locusta* such cells are released immediately after haemorrhage, and differentiate rapidly, contributing to the restoration of the normal blood picture (Hoffmann, 1969). Jones (1970), in a detailed review of insect haemocytopoiesis, demands seven minimum criteria of a normal haemocytopoietic tissue: (1) the histological discovery of a compact tissue of haemocytes, (2) the occurrence and (3) the quantitation of mitotic divisions in the tissue, and/or (4) the presence of various stages of differentiation of various types of cells within such tissue, but also most critically (5) specific correlations between the numbers of cells being produced in and (6) being released from such organs, and (7) statistically valid increases in the number of circulating cells. These requirements are certainly hard to meet experimentally, and may be too stringent. Given the consistent anatomical observation of a grouping of haemocytes in particular foci, perhaps the crucial property of a haemocytopoietic tissue is that it supports a greater rate of multiplication of haemocytes than occurs in the circulation. If the multiplication rate in the group of cells is the same as in the circulation, then the group may simply be sessile haemocytes. If, however, the multiplication rate in the group is greater than in the circulation as a whole, then the group may be said to be specialized for multiplication and thus a haemocytopoietic centre. Evidence for the existence of such a centre was provided ten years ago for the posterior haemocyte masses of *Calliphora* (Crossley, 1964). It was reported that consistent accumulations of haemocytes made up primarily of poorly differentiated (Type Z) haemocytes increased in size during larval life. Furthermore it was shown that following experimental haemorrhage the mitotic index for the dorsal areas was significantly greater ($p > 0.1$) than for extreme lateral areas, which contained cells multiplying at rates not significantly different from those obtaining in the haemolymph. The postero-dorsal accumulations thus represent areas specialized for haemocyte multiplication, i.e. are haemocytopoietic centres. Evidence is also presented, although without strong statistical support, that the centre of haemocytopoiesis is at the posterior of the heart in the vicinity of the spiracles. This centre is surrounded by a spatial gradient of declining cell multiplication activity, passing forwards and sideways (Crossley, 1964).

With an entirely different experimental technique Hoffmann (1972) demonstrated a focus of haemocytopoiesis in *Locusta*. He irradiated locusts latero-dorsally with a single 25 000 R X-ray dose, thus inhibiting division in dorsal haemocyte accumulations of likely haemocytopoietic potential, and in cells of the dorsal vessel, pericardial cells, and parts of other widely distributed tissues such as fat body. Control irradiations of equal dose and over an equal area were made on other insects in the latero-ventral region. The result was a dramatic decrease (of 64 per cent in 5th instar larvae) in the number of haemocytes 24 h after irradiation of the dorsal region, but no significant effect after irradiation of the ventral region. There was also a differential effect on haemocyte types, with the coagulocytes showing the greatest decline following dorsal irradiation. This is suggestive of an important role of the dorsal accumulation in haemocyte production, but the rapidity of disappearance of cells from circulation suggests an extraordinarily high rate of haemocyte turnover and replacement, if really only haemocytopoiesis is affected by the experiments. The possibility remains open that endocrine controls, either over haemocyte production, or more probably over haemocyte adhesion to tissues, are also disturbed by dorsal but not ventral irradiation. It is well known that sub-lethal doses of X-irradiation may decrease the rate of development and cause morphological abnormalities in insects (Obrien and Wolfe, 1964). Harshbarger and Moore (1966) using 12 000 R doses of X-irradiation (lower doses than those used by Hoffmann) report numerous morphological changes, including the development of melanized lesions in *Galleria*. There is no evidence at present for, or against, involvement of heart tissues and associated neurosecretory centres, or pericardial cells, in haemocyte endocrine control, so the significance of the irradiation of these cells for the experimental results is unknown. Zachary and Hoffmann (1973) later applied a similar X-irradiation technique to *Calliphora* larvae, and confirmed the existence of the haemocytopoietic centres located by Crossley (1964).

5 Insect blood cell locomotion and social behaviour

Living haemocytes have been examined as they circulate in the wing veins of the cockroach *Blaberus*, using bright field or phase contrast illumination projected through the transparent cuticle (Arnold, 1959a, 1959b). Small spherical prohaemocytes in which the nuclear area exceeds that of the cytoplasm, and larger phagocytic cells termed plasmatocytes were distinguished. The plasmatocytes of young adults were smooth-surfaced cells, 60 per cent being fusiform, whilst nearly all the rest were disc-shaped. Although the cells were flexible, and became deformed and bent by circulation currents as they passed obstructions, they resumed their original shapes as soon as possible, indicating that a particular shape was favoured, and presumably maintained, by the cell. Furthermore, cells showed an inherent symmetry, since the development of the main pseudopodia on the cell tended to be confined to particular sites into which they could be absorbed and from which they would later re-emerge (Arnold, 1961). Changes were noticeable in both the morphology and behaviour of living haemocytes as the adult aged. The disc-shaped cells came to predominate, and granulation or vacuolation within them became pronounced. Staining indicated that many of the granules were eosinophilic. The cells also became more adhesive and tended to occlude small wing veins, stopping blood circulation. Haemocytes then degenerated into bizarre or necrotic forms, and in some cases the veins reopened as the haemocytes disintegrated. When these haemocytes were studied with the aid of time-lapse cinematography (Arnold, 1961) it became apparent that they were capable of locomotion in the absence of blood flow. Some cells showed streaming of the cytoplasmic mass into amorphous pseudopodia, analogous to amoeboid movement. Protoplasmic turbulence was observed immediately prior to the start of migration, or to a reversal of direction of movement. Unfortunately the rates of cell movement given (57 μm min^{-1} at 25° C) are not fully quantitative, since they were made in partially occluded veins, without monitoring blood circulation rate. It was observed that the more granular the cell, the more active its amoeboid movement. Hyaline cells showed periods of prolonged quiescence. Other cells moved by the formation of lamellar hyaline pseudopodia with scalloped edges, and in these the rate of movement was 3.5 μm min^{-1} at 25° C (Arnold, 1959b). The lamellar areas showed enhanced adherence to the substrate and, in this and other particulars, the lamellar pseudopodia described by Arnold (1959b, 1961) for *Blaberus* resemble the leading lamellae, or "lamellopodia" of vertebrate fibroblasts moving over glass in tissue culture (Abercrombie *et al.*, 1970b, 1971). The extent to which lamellopodia are formed is correlated with the total amount of cell movement. However, the net forward displacement of the vertebrate fibroblast mainly results from the greater proportion of time spent in forward movement (30 per cent) than in backward movement (20 per cent). An average fibroblast speed is about 1.8 μm min^{-1} (Abercrombie *et al.,* 1970a).

Another quite distinct form of locomotion is mediated by long filiform pseudopodia, which can both oscillate slowly through an arc of 15°, and also retract rapidly into the cell body (Arnold, 1959b). In a later paper Arnold (1961) describes how sequential adhesion and release is related to locomotion. At first the posterior of the cell remains attached whilst the cell body moves freely forward. Next the anterior end adheres to the substrate, and this is followed by release of the posterior attachment and the movement of the bulk of the cell body towards the anterior attachment point. Cells can perform two or three successive jumps, moving the cell over the substrate. The formation of oscillating exploratory and adhesive filopodia has also been described in *Calliphora* pupal myoblasts in tissue culture medium, and here the cell has been examined in the electron microscope (Crossley, 1972a). Pupal myoblasts are spindle-shaped cells with two long filiform pseudopodia, which owe their shape to bundles of oriented microtubules. The polymerization of the microtubules can be disrupted by introducing 10^{-5} M colchicine into the environs of the cell, whereupon it rounds up with withdrawal of all filopodia. As in the *Blaberus* haemocyte system described by Arnold, the *Calliphora* myoblast can attach to, or detach from, the substrate at the extremities of the filopodia, where are situated minute leading lamellae. The main body of the cell is not attached to the substrate. Sequential attachment-detachment cycles coupled with contractile filopodia here constitute a locomotory system. A similar type of movement has been observed by the author in spindle-shaped *Calliphora* haemocytes, and in mouse myoblasts in tissue culture (Crossley, unpublished observations). It is not clear whether this form of locomotion can be explained solely by reference to microtubule subunit polymerization (c.f. Tilney, 1968), or whether an actomysin or other contractile system is present. Further examination of living insect haemocytes in wing veins, using modern interference microscopy, with the literature on vertebrate cell locomotion in mind, should be highly profitable. The techniques used by Du Praw (1965) for the study of microtubules and amoeboid activity in honey bee embryonic cells could also be usefully applied to haemocytes. Microtubules also appear to play an important part in the production of cell asymmetries and movement in grasshopper embryos (Kessel and Eichler, 1966).

Arnold (1959b) reports that in certain *Blaberus* haemocytes, which are often hyaline and vacuolated, the filopodia are both branched and adhesive. These adhesive filopodia appear to contribute to cellular agglutination, and may be analogous to certain of Gregoire's (1955a) coagulocyte types. It should also be borne in mind that insect phagocytic haemocytes are almost invariably bristling with attenuated filopodia (e.g. Rizki (1957) for *Drosophila*; Crossley (1964) for *Calliphora*; Marschall (1966) for *Tenebrio*) and these may play an important part in adhesion to foreign particles prior to phagocytosis (Fig. 1). In a later paper Arnold and Salkeld (1967) were able to recognize four types of haemocytes on a basis of morphology of fixed stained preparations, but did not correlate these types with particular forms of locomotion. They did note that granular haemocytes and spherule cells, which both contain neutral mucopolysaccharide, were nonmotile, suggesting that motility is confined to prohaemocytes and plasmatocytes.

The living haemocytes of *Drosophila* larvae have been observed in capillary tubes, and such tubes can be arranged either to connect two larvae in parabiosis, or sealed at one or both ends. Individual cells in glass tubes were observed to send out filopodia (termed the "podocyte transformation") and later to flatten out on the glass (termed the "lamellocyte transformation"), all within a period of 30 min (Rizki, 1962). The flattening to lamellocyte form appears to be correlated with increased adhesiveness and is apparently the response of the haemocyte to a foreign surface. This explanation could apply both to the changes on glass surfaces and those at necrotic lesions induced by genetic factors in *tumor W* mutants. Changes in the lamellocyte fraction of the population can be induced, probably indirectly, by humoral factors and also most strikingly by the injection of distilled water (discussed elsewhere) (Rizki, 1962, Fig. 3). Changed adhesiveness of lamellocytes is also implied by the results of experiments in which *Drosophila* were fed glucosamine hydrochloride (Rizki, 1961).

The movement of silkmoth phagocytic plasmatocytes has been studied in a culture chamber which permits the cells to respire through a film of polyethylene (Walters and Williams, 1966; Walters, 1970). These authors report that formation of ruffled membranes (leading lamellae) occurs during cell movement at 25° C at rates up to $2\mu m\ min^{-1}$. Locomotion requires contemporaneous formation of ATP (since 2-4 DNP inhibits movement), and possibly the presence of calcium (since the cells are promptly immobilized by EDTA). Plasmatocytes are said to exhibit "contact inhibition", but no nuclear overlap data are given. Contact inhibition of moving haemocytes is also indicated by the data of Clark and Harvey (1965) and by the observation of Arnold (1961) that cells moving towards each other in a wing vein "reversed direction on contact". In the culture chambers of Walters and Williams the cells became attached to the substrate by small fan-shaped extremities and, as they move, a fine filament of cytoplasm up to 15 nm long extends behind them. This filament is contractile and can be withdrawn in a few seconds. The cytoplasmic filaments also serve to connect cells by the formation of adhesion zones at their extremities, not only between plasmatocytes but also between plasmatocytes and fat body cells. However, granulocytes do not form adhesion zones, and merely become entrapped in the meshwork formed between the other cells. A differential recognition process is thus involved in this cellular behaviour, and a clue to its nature is provided by experiments in which Sephadex ion-exchange resins are used as foreign surfaces in the haemocyte environment. The haemocytes reportedly adhere strongly to resins which bear a positive ionic charge (e.g. DEAE-Sephadex), but much less strongly to negatively charged resins (e.g. SE-Sephadex), and hardly at all to uncharged resins (e.g. G-Sephadex). The interesting experiments reported above can be criticized on two grounds, firstly that phenylthiourea (a potent inhibitor of copper oxidases and other enzymes, e.g. peroxidase) (see section 9) is present in the medium to retard darkening, and secondly because it is not clear to what extent the clotting mechanisms of haemocytes are stimulated to give rise to the adhesions described.

Blood cell contacts have been studied in *Ephestia* using the electron microscope by Grimstone, Rotherham and Salt (1967). During capsule formation haemocytes adhere closely to each other and often show membranous interdigitations, but nevertheless they retain their individuality and do not form syncytia. Two types of specialized contact zone are formed; tight junctions of the "zona occludens" type, and striated contact regions reminiscent of septate desmosomes but less prominent and less regular. Fine intracellular fibrils are present between closely apposed haemocytes, and these fibrils may be comparable to the terminal-web fibrils of vertebrate cells. The fine structural aspects of cell communication in insects have recently been reviewed by Satir and Gilula (1973).

Gupta and Sutherland (1966) have observed the behaviour of *Periplaneta, Galleria,* and *Tenebrio* haemocytes *in vitro*, in either a simple saline containing 2 per cent EDTA, or in the tissue culture medium of Martignoni. A wide variety of changes occur in the cells, but many of these may relate to incipient blood clotting and cell death, since the suitability of even the tissue culture medium for maintaining healthy haemocytes is in doubt. The rapidity with which haemocytes die in unfavourable tissue culture medium is shown by Feir and Pantle (1971). They found that cell death, as evidenced by Trypan blue uptake, occurred in 1 per cent of the cells after 30 min, 75 per cent of the cells in 8 h and 100 per cent of the cells in 12 h. The changes in *Periplaneta* haemocytes during clotting and cell death have been described by Yeager, Shull and Farrar (1932). They report that "the blood cells lose their original fusiform or discoid shape, round up, become more refractive, form pseudopodia, agglutinate into a number of clumps, spread out on supporting surfaces, and seemingly disintegrate".

Several workers have nevertheless reported success with haemocyte tissue culture. Mitsuhashi (1966, 1967) reported successful primary cultivation of prohaemocytes from *Chilo*, and subsequently established a line of these cells, in spite of the fact that the cells which attached to glass degenerated. Ritter and Bray (1968) and Ritter and Blissit (1969) reported that cockroach haemocytes could be cultured for a year in Grace's medium, where they were motile, and in some cases synthesized crescent-shaped inclusion bodies. However, the population of cells was heterogeneous, and certainly included some epithelial cells.

The tissues of the tobacco hornworm *Manduca* have been cultured by Judy and Marks (1971) in Yunkers' *et al.* (1967) modification of Grace's medium, which contains no insect serum. Cells identified tentatively as plasmatocytes lived for up to 3 months, and moved in a "gliding manner" over the glass, with continual changes in morphology, but without evidence of mitosis. Granular haemocytes were reported to rotate on the glass surface, whilst maintaining their shape. Cells that moved from explant tissue out onto the coverglass only covered a few millimetres and then died (Judy and Marks, 1971). None of the cells illustrated by the latter authors are clearly identifiable as haemocytes. If the cells are haemocytes then the reported increase in migratory activity induced by β-ecdysone may well relate to events of physiological significance, as discussed in the section concerned with humoral control of haemocytes. Judy and Marks also observed that pupal fat body is caused to dissociate into single cells by ecdysone, and the dissociated cells then strongly resembled adipohaemocytes.

Kurtii and Brooks (1970) cultured haemocytes drawn from severed prolegs of *Trichoplusia, Chorista*, and *Malacosoma* (Lepidoptera) for up to 100 days in Grace's medium. Good proliferation by mitosis was observed in fibroblast-like cells interpreted as prohaemocytes. Plasmatocytes were also fibroblast-like, but appeared to grow primarily by increase in cell size rather than by increase in number. Many of them exceed 10 nm in diameter and are equipped with large lobulated nuclei. Granular haemocytes were never seen dividing, and spherule cells did not attach to the culture vessel. The cells were not successfully subcultured. Sohi (1971) also cultured prohaemocytes of *Malacosoma* in Grace's medium but succeeded in maintaining growth for 16 months over 35 subculturings. These cells were naturally infected with a microsporidian parasite. Blisters and vesicles believed to result from pinocytosis were very common in primary cultures.

Possible chemotaxis by insect haemocytes has been reported by Nappi and Stoffolano (1972), but the evidence is not strong. One way to approach definitive evidence would be to use defined chemotactic fields generated on microscope slides, following the techniques developed for vertebrate leucocytes (Grimes and Barnes, 1973).

By making time-lapse cinematographic records of unharmed haemocytes in living pupae of *Tenebrio*, Marschall (1966) found that conventional methods of obtaining haemolymph extensively changed the shape and behaviour of blood cells.

Arnold and Salkeld (1967) have quantitatively monitored the changes in haemocytes occurring during fixation, by comparing populations of living *Blaberus* cells within the insect with similar cells fixed in various ways. They found that the full range of variability seen in living cells was not present in fixed cells, whilst on the other hand some changes were induced by the fixation procedure. Granular haemocytes, for example, shrank measurably during fixation. Rizki (1957) carried out an analysis of variance of plasmatocyte size as obtained by three different haematological techniques: cell suspension in immersion oil, wet smears, and fixed stained smears. The variability between fixed stained samples and fresh samples was significantly greater than that within a single sample, whilst cell suspension in immersion oil gave the least variability.

6 Insect blood clotting

The involvement of haemocytes in haemostasis in insects has been the subject of numerous reviews (Wyatt, 1961; Heilbrunn, 1961; Gregoire and

Tagnon, 1962; Gregoire, 1964, 1970, 1971). There appear to be many parallels between insect and crustacean blood clotting mechanisms, and advances in our knowledge can often be traced to pioneer work on crustacean material. Thus seventy years ago Loeb (1903) reported that in crustaceans clotting proceeded in two stages: firstly he observed a *cellular agglutination* in which haemocytes became a system of sticky threads that later retracted; secondly he observed a *plasma coagulation* involving substances previously in solution which precipitated. The plasma coagulation could be inhibited independently of the cellular agglutination (Loeb, 1903). These two physiological phenomena occur widely, possibly universally, in the arthropoda, although in different degrees of dominance, wherever a haemostatic mechanism exists at all. Thus in different crustaceans Tait (1911) recognized three permutations. In Type A only cellular agglutination occurs. In Type B agglutination of the cells is followed subsequently by plasma coagulation. In Type C cell agglutination is relatively insignificant, but coagulation of the plasma occurs in two stages. Initially localized clots form in immediate relation to special blood corpuscles, the "explosive corpuscles" of Hardy (1892), but later the entire plasma coagulates.

Muttkowski (1924) observed, for insect material, that agglutination and coagulation can occur independently of each other. He described how clotting amoebocytes spread out fibrillar or lappet-like pseudopodia which interlaced with other cells to form a living meshwork. After examining forty-seven species of insects Yeager and Knight (1933) found cellular agglutination without plasma coagulation (Tait Type A) in some species, e.g. *Periplaneta*. In other species cellular agglutination and plasma coagulation were accompanied by haemocyte disintegration (Tait Type C), e.g. *Gryllus*. In a few species no clotting at all was detected (e.g. *Apis* larvae). Beard (1948) found that in *Galleria* the coagulum formed by cellular agglutination with only incidental plasma gelation; but in *Popillia* plasma gelation occurs, the cells being entrapped passively in the coagulum.

The introduction of the phase-contrast microscope to insect haematology by Gregoire and Florkin (1950b) greatly facilitated study of the cytology of clotting. In a wealth of publications Gregoire and his co-workers eventually examined 1600 species of arthropods (Gregoire, 1971). It proved possible to place them in four categories, largely on the basis of the relative significance and comparative morphology of cellular agglutination and plasma coagulation phenomena. The categories are given in Gregoire (1970) as follows:

I. Selective alterations in a category of fragile hyaline haemocytes result in exudation without cell rupture, or in explosive discharge of cell material (granules) in to the surrounding fluid. Coagulation of the plasma starts in the form of circular clouds of granular consistency around altered cells. Further the fluid plasma in the channels which separate the islands clots into a granular substance progressively organized into networks of granular fibrils.

[In this type of haemostasis then, coagulation affects the plasma only, and follows observed changes in a particular form of hyaline haemocyte. It occurs initially only around that haemocyte, but later spreads to the entire plasma.]

II. On contacting glass, fragile hyaline corpuscles extrude straight threadlike expansions carrying along cytoplasmic granules, and highly adhesive to solid particles, physical interfaces (bubbles) and to other categories of haemocytes. These expansions form pseudopodial meshworks within which the plasma clots in the form of transparent elastic veils without forming distinct islands as in pattern I.

[In this form of haemostasis cellular agglutination, apparently initiated by changes in a class of fragile haemocyte, is followed by a plasma coagulation of a different form to pattern I.]

III. Hyaline haemocytes form meshworks as in pattern II. The plasma clots as in pattern I (islands with a hyaline corpuscle in the centre) and as in pattern II (veils).

[In this form of haemostasis cellular agglutination is apparently initiated by changes in a class of fragile haemocyte, and is also followed by plasma coagulation of different form to pattern I.

IV. Hyaline haemocytes resembling the unstable corpuscles involved in other patterns do not alter or undergo clarification after ejection of cell substance in the surrounding fluid. Under the phase contrast microscope no change can be detected in the consistency of the plasma in the vicinity of these inert or altered hyaline haemocytes.

[In this class, there is neither cellular agglutination nor coagulation of the plasma in the vicinity of hyaline haemocytes, although these may show morphological alterations.]

According to Gregoire (1970): "The other categories of haemocytes do not take part in the process of coagulation. In contrast with the fragile corpuscles these haemocytes remained unaltered or underwent slow modifications without cytolysis. They were passively entrapped at random in the plasma clots, in the veils, in both formations, or they gathered along the highly adhesive expansions already developed by the fragile corpuscles."

A number of concepts arise from the observations and classification developed by Gregoire. The first concept is that in insects, as in crustaceans, changes in a particular class of haemocyte precede the alterations in the haemolymph that occur as the blood clots. This concept arose from the observations, made for many species of insects, that changes in hyaline haemocytes occurred in all species where the blood was observed to clot, but these haemocytes did not exhibit any important alteration in insects where blood clotting did not occur (Gregoire, 1951, p. 1191). Thus a strong circumstantial connection, but no causal relationship, was established between transformed hyaline cells and blood clotting. This was further supported by a report of parallelism between the number of active coagulocytes and plasma coagulability in insects irradiated by X-rays (Gregoire, 1955b; Hoffmann, 1972), and by the report of Wheeler (1963) for *Periplaneta* and Brehelin (1971) for *Locusta* that there is a correlation between the percentage of hyaline haemocytes (cystocytes) and the coagulability of the haemolymph. It is still possible that changes in labile cells may be the effect, rather than the cause, of coagulation of the plasma, as has been suggested by Gupta and Sutherland (1966). A significant difficulty is posed by insects which fall into Gregoire's category IV. These contain hyaline haemocytes which undergo moderate to profound alteration on contacting a foreign surface, but which do not produce a detectable change in the nearby plasma (Gregoire, 1955a, p. 105; Gregoire, 1971, Figs 7 and 8). Lea and Gilbert (1961) report for *Hyalophora* that a class of haemocyte termed an "oenocytoid" rapidly transforms *in vitro* into a hyaline form, and that this transformation is accompanied by a visible discharge of fluid material into the plasma. However, the expulsion of cytoplasm does not stimulate any visible reaction in the neighbouring haemolymph. Similarly, in *Calliphora* larvae a large haemocyte (Type B, Fig. 4(a-d) of Crossley, 1964) rapidly swells and becomes hyaline, but fails to induce local plasma precipitation, or to form filamentous extensions (Fig. 12). Gregoire (1971, p. 178) suggests that transforming hyaline haemocytes which do not induce clotting are "relics of a formerly functional process". There is also a possibility that they are involved in an altogether different aspect of wound healing, namely, the production and release of bacteriostatic substances, and the evidence for this is reviewed in section 9.

The second important concept to arise from Gregoire's work is that blood clotting in insects is "initiated by alterations taking place in contact with foreign surfaces in a *single* category of highly fragile haemocytes" (Gregoire, 1971, p. 172). However, it has been reported by Marschall (1966) that in *Tenebrio* one type of haemocyte would induce islands of coagulation in the plasma whilst another type would send out pseudopodial threads resulting in cellular agglutination. In *Calliphora* larvae the formation of pseudopodial threads occurs without apparent involvement of hyaline haemocytes analogous to those described by Gregoire, Zachary and Hoffmann, 1973 (see Fig. 11). Gregoire himself (1971, p. 176) points out that "it is still unknown if the fragile haemocytes including coagulocytes, belong to the same or different category of haemocytes".

The term "coagulocyte" was introduced by Gregoire (1950a) to describe a cell in *Gryllulus* that was spherical, had a small nucleus and pale hyaline cytoplasm with a few small dense granules. The hyaline coagulocytes

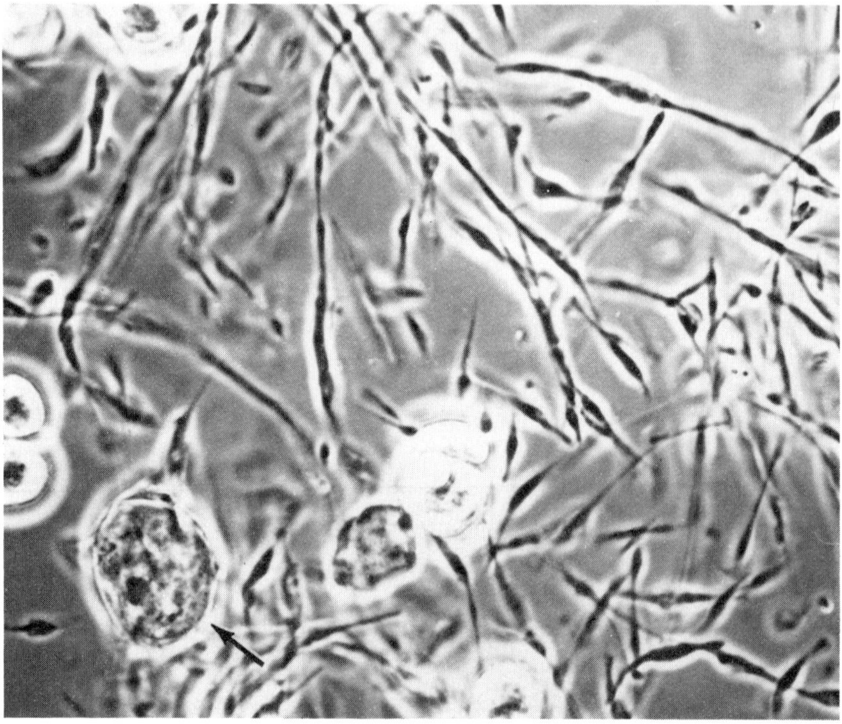

Fig. 11. Intense clotting by cell fragmentation and fragment elongation is seen in this light micrograph of *Calliphora* cells treated with 10^{-3} M EGTA indicating that calcium is apparently not required for cell fragmentation. After fragmentation residual nuclei with very little cytoplasm (arrowed) are abundant. Light micrograph ×525.

rapidly underwent cytolysis on glass, releasing cytoplasmic material into the blood, forming an island of granular material around the cell. The *same* class of cell was said to send out filiform pseudopodia which were intensely thigmotropic. In later papers Gregoire uses the terms hyaline haemocyte and coagulocyte independently, and in Gregoire (1955a, p. 133) a coagulocyte is defined simply as "a hyaline haemocyte efficient in the inception of the process of coagulation". Presumably then, many different forms of haemocyte could be coagulocytes. Jones (1954, 1962, 1964) identifies

Fig. 12. Shows one of a number of hyaline haemocytes congregating at a puncture wound made in the cuticle of *Calliphora* one hour earlier. The plasma membrane of the cell is unstable and is releasing cytoplasm in the vicinity of the wound (asterisk). The cytoplasm reacts positively with amino-benzidine reagent for peroxidase. Ribosome-studded endoplasmic reticulum (arrow) is prominent. n, nucleus. Electron micrograph ×10 000.

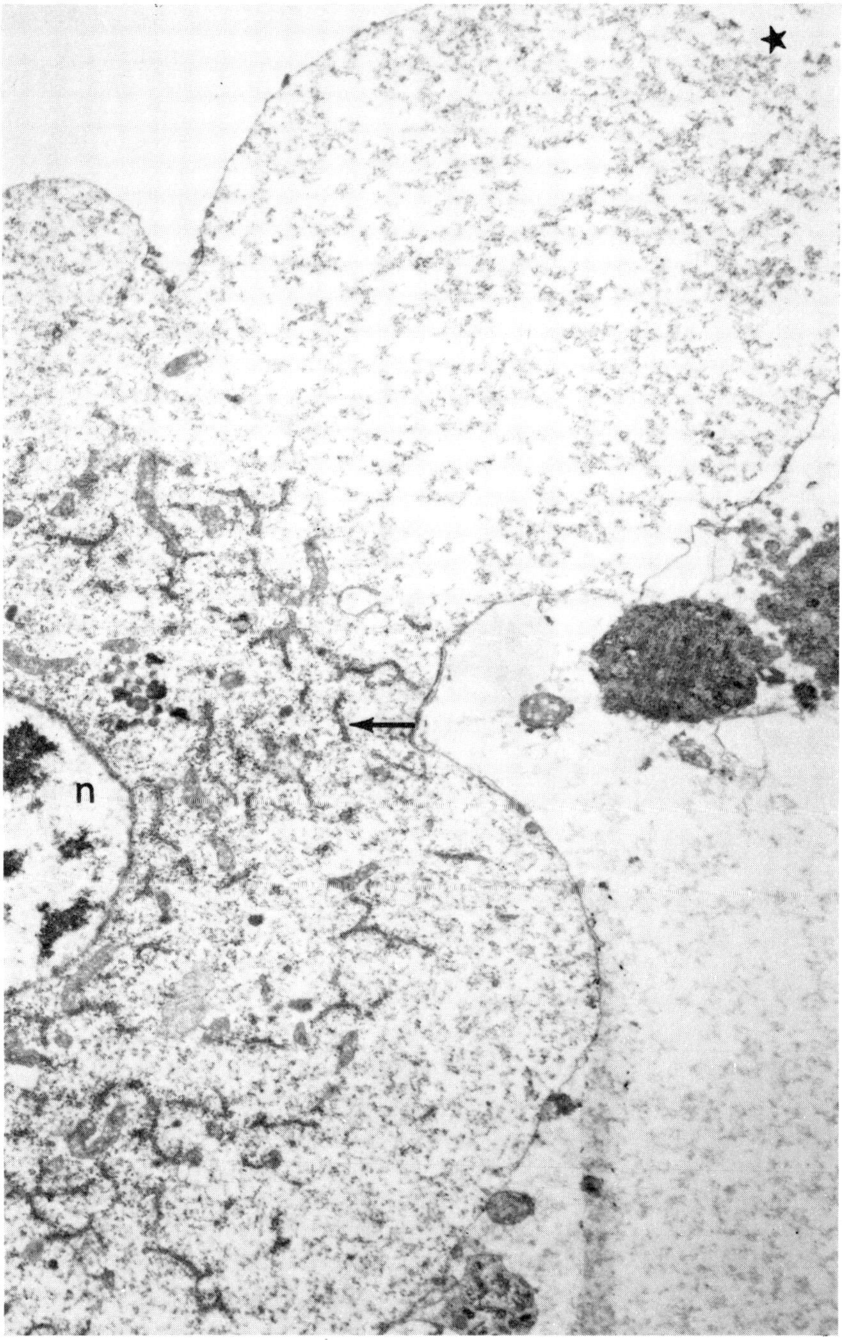

Fig. 12.

coagulocytes with his cystocytes, but the latter is itself an imprecise category.

The oenocytoids described by Poyarkoff (1910) and Hollande (1920) have many similarities to coagulocytes, including a labile acidophilic cytoplasm and relatively small, often eccentric, nuclei. Gregoire (1955a, p. 131) distinguishes between oenocytoids and coagulocytes on the basis of the tendency for coagulocytes but not oenocytoids to develop pseudopodia.

The nature of the materials released from unstable hyaline haemocytes is not yet known, nor is it certain whether the action of these cells is the release of a pharmacological signal which induces the plasma to coagulate, or the release of the coagulation precursors themselves, or both. Release of vacuolar contents by haemocytes in contact with a foreign surface, soon after removal from the animal, has been reported for *Limulus* and spiders (Gregoire, 1971), for *Locusta* (Hoffmann and Stoekel, 1968), and for cockroaches (Scharrer, 1972). At the outset it should be borne in mind that released materials may not simply be coagulins but may be components of an anti-bacterial mechanism. The example here is the well-known degranulation of vertebrate polymorphonuclear leucocytes, where two different types of vacuole, each with a characteristic complement of enzymes, are released sequentially as part of an anti-bacterial, rather than a haemostatic, mechanism (Bainton, 1973). The vacuoles released from insect haemocytes in contact with a foreign surface are of diverse morphology. In *Locusta* Hoffmann and Stoekel (1968) have studied the change in "coagulocytes" on clotting with the aid of the electron microscope. These cells are rich in free ribosomes, and have well-developed sinuous endoplasmic reticulum cisternae which dilate and subsequently fragment at the onset of clotting. Other cytological changes at clotting include a swelling of the nucleus accompanied by pycnosis, and disappearance of free ribosomes and of vacuoles containing fibrillar material. However, the cell plasma membranes were not seen to rupture and there was no massive liberation of granular material. To these authors the changes appeared to be secondary manifestations of a changed cellular permeability.

It has recently been suggested that the hyaline haemocyte (coagulocyte) undergoes a rapid change in its membrane on removal from the insect, as evidenced by the appearance of hydrophobic groups detected with 8-anilino-1-naphthalene sulphonic acid (Belden and Cowden, 1971). Moran (1971) reports that disappearance of tubule-containing bodies is associated with clotting in *Blaberus*. Furthermore, the presence of many such bodies in haemocytes correlates with high coagulability of the haemolymph. Scharrer (1972) provides profiles suggestive of release of vacuolar material with tubular substructure (her Fig. 8) and granular substructure (her Fig. 18) and believes that the stepwise transformations of cytoplasmic inclusions culminate in their release of their content into the haemolymph, and play a decisive role in the clotting process. However, as reported above, Wigglesworth (1973) provides similar electron microscope images for cells implicated in connective tissue formation in *Rhodnius*. Similar molecular mechanisms and similar haemocytes might be involved in the two processes, and also in the formation of the plug at wound sites. The gelatinous coagulate in clotted *Popillia* blood and the cellular agglutinate in clotted *Galleria* blood both give histochemical indication of a mucopolysaccharide nature (Beard, 1950).

In *Limulus* the tubular and granular vacuoles released from haemocytes of clotting blood are considered to be sequential forms of a single type of a transforming vacuole (Dumont *et al.*, 1966). In *Limulus* there is evidence that the clottable protein of the blood is located entirely in the cells and released during coagulation. This protein has been partially characterized (Solum, 1970). In several crustaceans it has been shown that a PAS-positive muco- or glycoprotein is released from haemocytes as blood clots. The material is released initially from PAS-positive hyaline haemocytes but a similar activity of PAS-positive granular haemocytes follows (Wood *et al.*, 1971). PAS-positive contents have repeatedly been observed in insect haemocytes (Wigglesworth, 1956; Ashhurst and Richards, 1964b; Scharrer, 1972) but as noted above, this reaction (histochemistry) is by no means specific for coagulins.

In Crustaceans, cellular agglutination leading to the formation of a cytoplasmic meshwork of fibrillar strands is also part of the clotting reaction (Wood *et al.*, 1971). In crabs the agglutination involves the appearance of membrane-limited projections which make contact with neighbouring cells. Furthermore, the agglutination is accelerated by addition of a crude extract of crab haemocytes (Bang, 1971). Similar accelerating factors may exist in insects, but have not been reported to date.

In 1891 Cuénot (p. 375) suggested that some coagulative enzyme like rennin or thrombin might be involved in insect blood clotting. This idea was revived by Heilbrunn (1961) who believed that the mechanisms of protoplasmic clotting, of invertebrate blood clotting and of vertebrate blood clotting showed basic similarities. He suggested that whenever protoplasm flowed out of a cell it flowed out unimpeded unless there was calcium in the environment. If calcium was present the protoplasm coagulated by what was termed "surface precipitation reaction" (SPR). However, when vertebrate blood clotted a substance was formed that could induce clotting of other blood in the absence of appreciable amounts of calcium. This substance has been called thrombin, a proteolytic clotting agent. Heilbrunn suggested that an equivalent thromboplastic agent could be released from all injured cells including the coagulocytes of insects. This hypothesis raises two questions. Firstly, in general terms, does haemostasis in vertebrates and invertebrates involve similar chemistry? Secondly, in particular, is insect haemostasis controlled by calcium levels?

At the outset we should notice that crustacean and vertebrate haemostatic mechanisms involve different chemistry. Thrombin fails to induce a clot in crayfish blood, and prothrombin and partial thromboplastin tests using human reagents yield negative results. Gilliam and Shimanuki (1970) report that the thrombin clot time for larval *Apis* blood is 2 h, compared to 10.8 s for human blood. Their conclusion that "the clotting system for honey bee haemolymph is similar to man" lacks justification from the evidence presented. Crayfish haemolymph contains six times as much calcium as vertebrate blood, and 15 per cent sodium citrate is needed to prevent clotting (Wood and Karpawich, 1972). For insects Gregoire (1953) reports that reagents that sequester calcium ions (e.g. 0.2 per cent potassium oxalate) prevent disintegration of coagulocytes, and prevent blood clotting in *Periplaneta* or *Gryllotalpa*, where coagulation islands are normally formed. However, in *Calliphora*, which shows clotting by cellular agglutination but does not develop coagulation islands, 5×10^{-3} M EGTA does not prevent clotting (Fig. 11). It is therefore important in studies in insect haemostasis to distinguish between the effects of inhibitors on cellular agglutination and on plasma coagulation. This distinction was made by Beard (1950) in his study of *Popillia* and *Galleria*, but his results with respect to calcium levels are difficult to interpret. There is no consensus as to the effect of sequestration agents on invertebrate haemostasis. Several authors indicate that insect blood can clot in the absence of calcium (Muttkowski, 1924; Beard, 1950), but the effects of citrate and oxalate are pH-dependent (Franke, 1960a, 1960b). In this connection it is interesting to remember that Tsuji (1909) cited by Wyatt (1961), and Levenbook and Hollis (1961), found extremely high levels of citrate in normal insect blood. A value of 32 mM citrate was recorded for *Bombyx* blood.

The insect blood clotting mechanisms are extraordinarily insensitive to high concentrations of metabolic inhibitors, e.g. cyanide and azide (Shull *et al.*, 1932; Beard, 1950) and to antithrombins, e.g. heparin and hirudin (Beard, 1950). The clotting by haemocyte agglutination in *Calliphora* is not inhibited in the presence of final concentrations of 10^{-3} M cyanide, 10^{-2} M azide, 5×10^{-3} M thiourea, 5×10^{-3} M caffeine, 10^{-4} M cycloheximide, 5×10^{-3} M EDTA, or saturated aqueous cytochalasin B, when these are introduced into the haemolymph either 30, or 2, minutes before a blood sample is taken for clotting assay (Crossley, unpublished observations). A number of anticoagulant treatments effective in certain insects have been reported including acetic acid vapour (Shull and Rice, 1933), X-irradiation

(Gregoire, 1955b), sodium hydrosulphite, 0.1 M ascorbic acid, freezing followed by thawing, and addition of frozen blood (Beard, 1950). In *Popillia* and *Galleria* Beard (1948) found that freezing, heating to temperatures between 55° C and 70° C, and ultrasonic waves (400 KC) inhibited coagulation where exposure was made *in vivo*, but not when made *ex vivo*. The haemolymph failed to coagulate when exposed *ex vivo* to ultrasonic waves, provided it was in contact with body tissues, or with previously inhibited haemolymph. Inhibited haemolymph from *Galleria* could not be substituted for inhibited *Popillia* haemolymph in mediating the inhibition of fresh haemolymph of *Popillia* upon exposure to ultrasonic waves. Dilution with Ringer's solution *in vivo* prevents coagulation of *Galleria* but not in *Popillia* (Beard, 1948). These experiments suggest that clotting is inhibited by generalized tissue disruption, and that this disruption may release an anticoagulant, with some species specificity, from tissue other than the blood itself. Dilution with veronal buffer inhibits coagulation in *Schistocerca* (Bowen and Kilby, 1953), and in general dilution appears to be the inhibition technique best suited to separation of cells from haemolymph for biochemical analysis. In spite of the report that over half the nondialysable components, and some protein, are removed from *Schistocerca* haemolymph on clotting (Bowen *et al.*, 1953), very little is known about the detailed chemistry of the clotting process in insects. Siakotos (1960a, 1960b) compared the unclotted plasma with clotted serum from *Periplaneta* by electrophoresis. He reported that on clotting two new lipoproteins are formed, one of which is the relatively insoluble coagulum network, but the relative contribution of blood cells and plasma cannot be obtained from his results. Brehelın (1972), also using electrophoretic techniques, reported that only minor changes in plasma proteins occurred on clotting in *Locusta*, but that haemocytes were essential for normal coagulation.

The insect blood clot has been examined in the electron microscope by Gregoire *et al.* (1949). The *Dytiscus* and *Gryllus* clots were found to be composed of granular fibrils with a parallel fibrillar substructure. However, no periodicity was seen in the fibrillae, which thus differ from fibrin.

There are close analogies between cellular clotting in invertebrates and platelet clumping in vertebrates (Bang, 1971; Gregoire, 1971). Mammalian platelets are membrane-bound cytoplasmic fragments about 3 μm long, which lack nuclei, although they are derived from nucleated megakaryocytes of the bone marrow. In circulation they maintain a biconcave shape, but at wound sites they become both rounded and adhesive (viscous transformation). The change in shape is irreversible, and associated with the formation or loss of microtubules (White and Krivit, 1967). It is of interest therefore to note that the cellular fragments of *Calliphora* blood contain numerous microtubules (Fig. 15), and even more importantly, the formation of a cellular meshwork of fragments and cellular pseudopodia on clotting can be entirely inhibited by the addition of 10^{-4} M colchicine or 10^{-5} M vinblastine sulphate (Fig. 14). These reagents are anti-mitotic agents, which prevent the normal assembly of the tubulin subunits of microtubules (Tilney, 1968; Bensch *et al.*, 1969). It appears that in *Calliphora* at least, the clot meshwork consists largely of cytoplasm rich in microtubule subunits, which polymerize to form microtubules and generate elongated meshworks in the clot reaction. Microtubules have also been implicated in the formation of cell clots in the hermit crab *Eupagurus*, where they appear in the pseudopodial processes of blood cells within 3 min of the onset of clotting (Bang, 1970). The fragments in *Calliphora* blood remain membrane-limited, since they can be caused to swell and become spherical in hypotonic solutions. There is a further analogy with vertebrate platelet physiology in the origin of cellular fragments in insect haemolymph. During development, megakaryocytes of the vertebrate bone marrow undergo multiple nuclear divisions, and are at times multinucleate, although nuclear fusion leads to giant multinucleolate nuclei. Platelet demarcation membranes form by coalescence of membranous profiles, which eventually extend to the edge of the cell, allowing platelets to disperse. A thin residual layer of cytoplasm bounded by an intact plasma membrane is left surrounding the nucleus (Bloom and Fawcett, 1968). Remarkably similar events occur during the clotting of blood in certain insects. It is known that insect haemocytes can fragment, losing portions of their cytoplasm to the haemolymph (Gupta and Sutherland, 1966). Scharrer (1972) pointed out that this fragmentation was particularly marked in coagulating blood, and she provides electron micrographs of fragmentation in *Leucophaea*. In *Calliphora* larvae Zachary and Hoffmann (1973) have shown that certain haemocytes, which they term "thrombocytoids", fragment by forming invaginations of the plasma membrane which eventually dissect the peripheral cytoplasm. The cytoplasm then disperses, leaving nuclei surrounded by only a thin layer of cytoplasm (Fig. 15). The analogy with vertebrate megakaryocytes is striking.

In *Calliphora* the haemocytes undergoing fragmentation include multinucleate cells, sometimes containing as many as ten nuclei (Fig. 13). These multinucleate forms are not seen often in normal microscopic preparations because they rapidly disintegrate as the clot forms. Perhaps the multinucleate haemocytes reported in other insects are comparable. In *Chironomus* the fragmentation of large multinucleate plasmatocytes to form smaller spindle-shaped cells has been reported, although the process is not related to clotting phenomena (Maier, 1969).

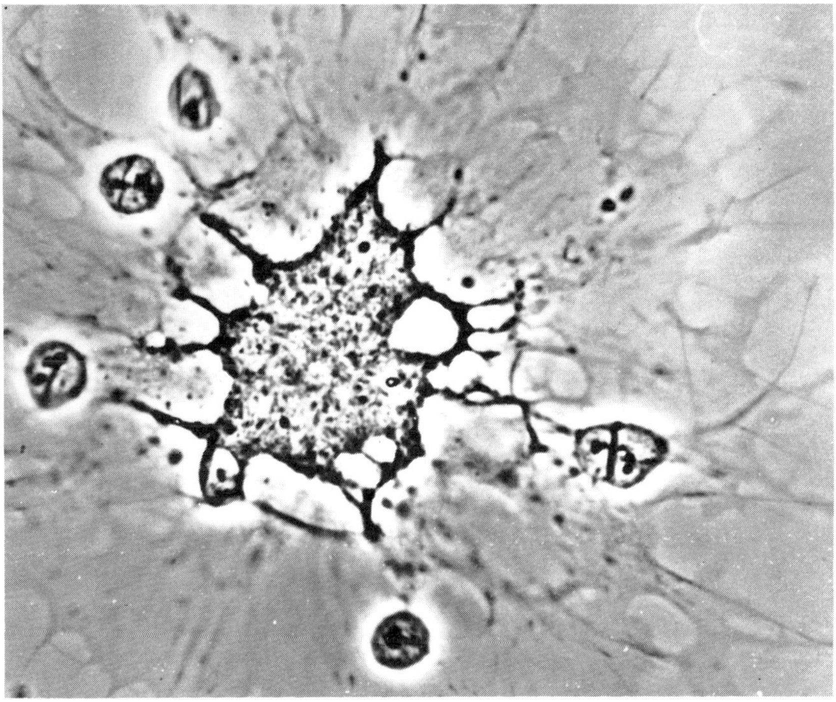

Fig. 13. A light micrograph of *Calliphora* larval haemolymph 5 min after removal from the insect and the onset of clotting. A multinucleate haemocyte is seen expelling its 5 nuclei. Cytoplasmic fragmentation followed. Light micrograph x525.

In insects cell fragmentation is not confined to haemocytes, since it occurs at eclosion in wing hypodermis, where it results in the release of membrane-bound fragments into the blood (Seligman and Doy, 1972, 1973; Seligman *et al.*, 1974). Cellular fragmentation may be considered to be a widespread phenomenon in animals, and has been termed "apoptosis" (Kerr *et al.*, 1972).

A situation intermediate between insect haemocyte fragmentation and vertebrate platelet formation is seen in the protochordata, where cellular fragments, budded off from cytoplasmic appendages of nucleated blood cells, become strongly adhesive (Andrew, 1961). Gregoire and Tagnon (1962) point out that insect coagulocytes do not have, *in vitro*, the property of autoagglutinability that characterizes vertebrate platelets. However, these particular haemocytes may not be concerned in cellular agglutination, which appears to be the domain of a distinct form of labile

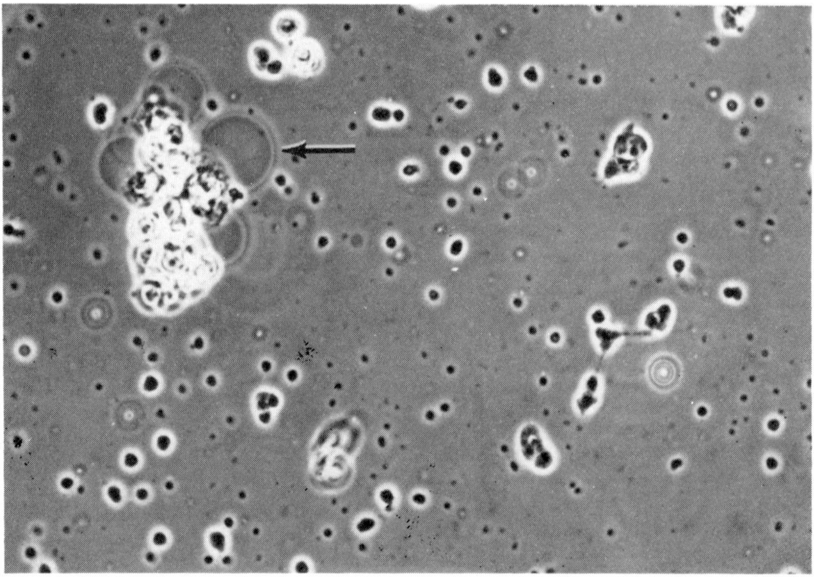

Fig. 14. In this preparation of *Calliphora* blood addition of vinblastine sulphate (5×10^{-6} M) 30 s after removal of blood and onset of fragmentation of labile haemocytes, has caused rounding-up of fragments, presumably due to failure of microtubule polymerization. Normal changes in labile haemocytes occur (arrow). Light micrograph ×500.

haemocyte, at least in *Calliphora*. Calcium has a controlling influence on platelet function, being required for fibrogenesis and thrombinogenesis, and viscous transformation. It is obvious therefore, that further information on the calcium relations of insect haemocytes is urgently required.

In *Rhodnius*, where the blood does not clot, any blood that escapes to the surface of a wound dries into a brownish scab, which is slowly converted into an amber-coloured, highly insoluble substance. The scab is not soluble in strong hydrochloric or nitric acids, but dissolves in hot nitric acid saturated with potassium chlorate (cerinic acid test) and in hot saturated potassium chloride. A similarity to cuticulin is suggested (Wigglesworth, 1937).

In summary we can state that insect blood clotting involves two distinct phenomena, cellular agglutination and plasma coagulation, which vary in their contribution to haemostasis in a given species. Plasma coagulation may involve substances released from labile hyaline haemocytes. The chemistry of this coagulation awaits detailed investigation, but correlations with events in crustacean blood may be expected. The cellular agglutination phenomenon involves viscous transformation of certain haemocytes, some-

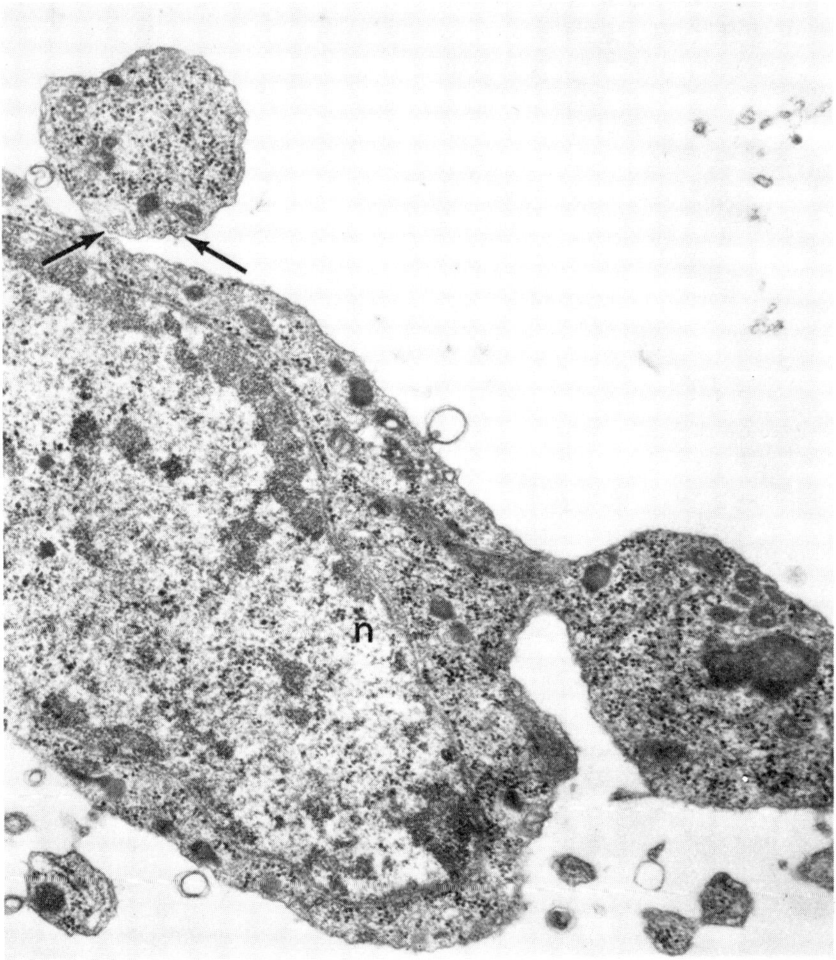

Fig. 15. Fragmentation of a *Calliphora* haemocyte in clotting blood is seen in this electron micrograph. One fragment is held by a tenuous cytoplasmic bridge, others are already detached and contain numerous microtubules (arrows). The residual structure consists of a nucleus (n) surrounded by a thin layer of cytoplasm. Electron micrograph x30 000.

times other than coagulocytes, and the disintegration of certain haemocytes to form membrane-bound fragments which also undergo a viscous transformation. The development of elongated cellular meshworks in an insect depends on the formation of microtubules, and can be inhibited by colchicine. The detailed nature of the signals and intermediaries in these changes remain to be investigated.

7 Insect blood cells in defence reactions

Insects live in a world in which potentially pathogenic bacteria and other organisms are abundant and menacing, yet they are remarkably resistant to bacterial attack (Burnet, 1968). The wax-moth *Galleria* has been challenged by injection with large doses of each of 32 species of bacteria, representing most of the major bacterial families, but only seven were pathogenic (Chadwick, 1967, table 2). The contribution of haemocytes to defence reactions may be through their capacity for sequestration of foreign materials, or through release of anti-microbial secretions. Sequestration by phagocytosis is discussed in section 8, whilst sequestration by encapsulation has recently been comprehensively reviewed by Salt (1970). We shall begin here by considering the evidence for the involvement of haemocytes in the production of internal defensive secretions, either generated in the vicinity of an invading organism, or released as humoral factors.

For fifty years immunologists have searched for a serological basis to the insect cellular defence response, but their results when critically examined have all been negative. Any response has been found to be nonspecific, and has not involved immunoglobulins or complement fixation. Nevertheless several investigators have shown that insects can acquire resistance to pathogenic bacteria or toxins, i.e. that the first dose is more harmful than similar later doses. Many different immunological techniques have been applied to insects, but throughout the copious early literature on insect immunity neither precipitin activity nor complement fixation have been described (Chadwick, 1967). The spontaneous precipitation which occurs in haemolymph proteins of *Galleria* and *Periplaneta* diffused against each other in phosphate-buffered agar gels, producing bands not unlike vertebrate antigen-antibody precipitates, is of unknown significance (Bullock and Steinhauer, 1970). There is no evidence that circulating antibodies are formed in insects (Briggs, 1958; Stephens, 1959, 1962a; Kamon and Shulov, 1965). Even when soluble antigens and sensitive fluorescent detection methods are used, antibodies are not detected (Feir and Schmidt, 1968). It has been pointed out that lymphoid tissue and lymphoid cells are absent from invertebrates (Good and Papermaster, 1964), and although insect haemocytes often contain well-developed endoplasmic reticulum it does not commonly occur as stacks of flattened cisternae as it does in vertebrate plasma cells. The technique of immunization with ferritin, coupled with localization of developing antibody cells by electron microscopy, so successful with vertebrate material (de Petris and Karlsbad, 1965) has yet to be applied to insect blood cells.

Chadwick (1967) has remarked that the only significant demonstrable immune response in insects is bacteriostatic, bacteriolytic, or bacteriocidal activity. The experimental evidence that leads to these conclusions on the insect immune response has been reviewed by Carton (1969). Nearly all the bacteriocidal factors described from insects are heat stable and nonspecific (Zernoff, 1931; Briggs, 1958; Gingrich, 1964), although some are believed to be associated with blood proteins (Rehm, 1948; Gingrich, 1964).

Chadwick (1967) suggests that substances which produce an acquired immune response in insects be termed "insect immunogens". Her work has revealed that if heat-killed vaccine of *Pseudomonas aeruginosa* is injected into *Galleria*, a bacteriocidal factor appears in the haemolymph within 24 h. This factor is dialysable, heat stable, unaffected by trypsin, of low molecular weight and is not mediated by complement fixation. The factor is described as "specific" because immune sera prepared against some bacterial strains were not active against others (Stephens, 1962a; Stephens and Marshall, 1962). However, the possibility exists that some strains are simply less sensitive to a nonspecific antibacterial agent, just as some bacteria are resistant to antibiotics. *Galleria* maintains its haemolymph free of bacteria and curiously sometimes shows an inhibition of normal melanization reactions during the period of resistance (Stephens, 1962b; Chadwick, 1967). It seems possible that the immune factor that develops in *Galleria* blood in the experiments of Stephens is a phenolic metabolite. Many intermediary compounds on the pathway to melanin, particularly the quinones, are bacteriostatic and could be generated by phenolases or peroxidases in the blood or blood cells (see section 9). Such oxidation need not proceed to melanin, but rather might be directed by other enzymic pathways to a noncoloured antibacterial compound, giving the inhibition of melanization described by Stephens.

There are indications that the endotoxins produced by bacteria such as *Pseudomonas* may be the active insect immunogens (Chadwick and Vilk, 1969). Both formolized vaccine and homologous endotoxin of *Pseudomonas* significantly increase the LD_{50} on challenge in *Galleria*, but for *Proteus* only the vaccine is an effective immunogen (Chadwick, 1971). When a ^{51}Cr-labelled lipopolysaccharide derived from *Shigella* was used as an immunogen in *Galleria*, both the development of immune response and the fate of the immunogen could be followed. A large proportion of label was rapidly removed from the haemolymph and appeared in melanized cysts throughout the haemocoel, but final deposition of the label appeared to be in the cuticle. The haemocytes never contained more than 15 per cent of the endotoxin at one time, although the apparent haemocyte uptake roughly coincided with clearance from the serum, and haemocyte encapsulation of immunogen is postulated. The immunogen also causes a large increase in the number of circulating haemocytes, as monitored by haematocrit (Schwalbe and Boush, 1971). Certain venoms have been reported to act as insect immunogens. In *Locusta* an injection of scorpion venom gives a degree of acquired immunity within 48 h (Kamon and Shulov, 1961). Three sublethal doses of venom give maximal protection, but no precipitin or agglutinin activity was detected, nor did the immune haemolymph inhibit toxicity of the venom *in vitro* (Kamon and Shulov, 1965). Similar results were obtained by Bettini (1965) for *Musca* injected with spider venom, and again response was maximal within 48 h. The injection of serum albumin into *Locusta* also provided some active immunity towards venom, again arguing for a nonspecific system unlike that regulating production of vertebrate antibodies (Kamon and Shulov, 1965). Bakula (1970) has isolated an antibacterial factor from *Drosophila* pupae by preparation of acid extracts. These extracts were effective in killing Gram-positive, but not Gram-negative, bacteria tested *in vitro*. Bakula suggests that the factor is a basic protein, either a histone or of lysosomal origin. Starting with homogenized whole larvae Bakula (1971) later isolated an acid soluble antibacterial factor by freeze-drying and dialysis. Although this originates from a cellular fraction which includes lysosomes, neither the organelles of origin, nor the cells of origin, are identified. The factor does not appear to be lysozyme, but could originate from the lysosome complex. It would be interesting to know if the factor can be isolated from haemocytes. The inducible antibacterial defence system described for *Drosophila* by Boman, Nilsson and Rasmuson (1972) is also nonspecific, since although only one of the three bacterial species tested (*Aerobacter cloacae*) is an insect immunogen, at least three bacterial species including *E. coli*, are susceptible to the factor.

The enzyme lysozyme (E.C. 3.2.1.17) is present in insect haemolymph and its titre increases in *Galleria* haemolymph on injection of a bacterial pathogen (Mohrig and Messner, 1968). However, in *Calliphora* larvae injected with bacteria, the rising lysozyme titre has been attributed to release from pericardial cells rather than from the haemocytes (Crossley, 1972b). In uninfected *Galleria,* lysozyme is present in the haemolymph at about 700-800 $\mu g\ ml^{-1}$, but after injection the lysozyme level rises to 2000-5000 $\mu g\ ml^{-1}$. Although the time course of the rise in lysozyme activity parallels the rise in protective immunity, the decrease in immunity is already apparent at 36 h after infection, at which time the lysozyme concentration is at its peak (Chadwick, 1970). It thus appears that lysozyme is one of several agents with antibacterial activity contributing to the observed immunity.

The enzyme chitinase (E.C. 3.2.1.14) isolated from cockroaches has lysozyme-like properties since both are 1-4 acetyl hexosaminidases. Although cockroach chitinase solubilizes *Micrococcus lysodeikticus* cell walls, it is less active with this substrate than with chitin (Powning and Irzykiewicz, 1967).

Changes in the haemocyte picture during immunization of an insect have been studied by Hink and Briggs (1968) and Hink (1970). They report a decreased percentage of plasmatocytes but an increased percentage of oenocytoids and vacuolated cells in immune *Galleria* haemolymph, suggesting that the haemocytes are themselves involved in the immune response. However, haemocytes from immunized donor larvae did not confer passive immunity on recipient larvae, as do antigenically stimulated lymph node and spleen cells when transferred syngeneically in vertebrates. Nor were attempts to isolate antibacterial factors from *Galleria* haemocytes successful. Nevertheless, oenocytoids are rarely phagocytic, and their increased numbers during infection need further investigation.

Although introduction of metazoan parasites into an insect usually elicits an encapsulation reaction, this is not invariably the case, and in some instances such parasites appear to succumb to humoral factors. Bess (1939) showed that in mealy bugs resistant to hymenopterous parasites, immunity was not necessarily accompanied by phagocytosis or encapsulation. Some aphids (e.g. *Myzus*), resistant to a hymenopterous parasite (*Monoctonus*), slow and eventually arrest its development, without the apparent intervention of haemocytes. Only in later stages is the already degenerating parasite attacked by phagocytes and no capsule is formed. In other aphid species (e.g. *Aulacorthum*) haemocytes reportedly secrete a melanized capsule around the same parasite (Griffiths, 1961). According to Lewis and Vinson (1968) a humoral reaction is implicated in the resistance of *Heliothis zea* toward a braconid parasite (*Cardiochiles*) which apparently involves alteration of the parasite egg to make it susceptible to encapsulation by haemocytes. *Heliothis virescens* lacks this humoral reaction and can only encapsulate the braconid egg if this has been made susceptible by prior implantation in *Heliothis zea*. This experiment does not rule out damage to, or contamination of, the parasite in potentiation of the response in *H. virescens*. Interestingly, *H. zea* larvae which had been previously exposed to the parasite egg could encapsulate eggs implanted later significantly faster, demonstrating acquired immunity, and strengthening the case for a humoral factor. Lewis and Vinson's attempts to identify the humoral factor by electrophoretic, Ouchterlony diffusion and histochemical techniques were, however, unsuccessful. These authors use of the terminology of vertebrate serology is unfortunate, as pointed out by Salt (1970, p. 91). In another host-parasite system, previous infection does not seem to affect the success of a superimposed infection (Lackie, 1972).

Nappi and Stoffolano (1971) followed the changes in blood cell differential count in *Drosophila* following infection with a hymenopterous parasite that dies, but is not encapsulated, in the host fly. They found that marked changes (predominantly decreases) occurred in the number of oenocytoids detected in circulating blood following parasitization, suggesting some sort of humoral "alert". In *Drosophila* oenocytoids are capable of turning into hyaline forms and lysing. In other dipterans the oenocytoids or their analogues are involved in phenolic oxidation (Dennell, 1947; Crossley, 1964) as discussed in section 9, and are implicated in defensive reactions (Taylor, 1969b). It is therefore significant that Nappi (1972) reports that dead but unencapsulated parasites in *Drosophila* are melanized. Encapsulated parasites are also frequently melanized in other insects (Salt, 1970). But it has been shown that in *Strongylogaster* parasitized by *Mesoleius*, where the parasite egg is invariably encapsulated, the parasite is able to develop so long as it is not melanized (Adam, 1966). The experiments of Brewer and Vinson (1971) indicate that reagents which can interfere with melanin formation actually reduce encapsulation by haemocytes in *Heliothis* larvae. But their results show that a number of substances without known effect on melanization, or with very general toxic effects on insect tissues, also reduce encapsulation, highlighting the need for caution in the interpretation of this type of experiment. As pointed out in section 9, it may well be a function of labile haemocytes to release phenol-oxidizing enzymes, which in turn generate phenolic intermediates and eventually dark pigment. Recent chemical analysis has shown that the dark pigment in several insect capsules is indeed melanin (Misko, 1973).

As Burnet (1968) points out, invertebrates have a foreignness-recognition mechanism incorporated into the wandering phagocytic cells. The question is: What form does this recognition system take? Burnet suggests that some sort of proto-antibodies may be present amongst the globulins of invertebrate wandering cells. He suggests that the two developments needed to convert these into an adaptive immune system would be (a) increased flexibility in genetic coding for protoglobulins, and (b) contact of foreign pattern with recognition globulin under appropriate conditions inducing proliferation of the haemocyte concerned. His view of invertebrate wandering cells places them ancestral to the immunocyte, polymorphonuclear and macrophage cells of vertebrates.

We may well ask what evidence there is for the existence of globulins on insect cells. Haemagglutination tests have indicated that insect haemagglutinins are nonspecific (Bernheimer, 1952; Bernheimer *et al.*, 1955; Feir and Walz, 1964), and this may be true of all invertebrate haemagglutinins (Tripp, 1969).

The haemagglutinin from *Periplaneta* is a heat-labile, nondialysable, euglobulin-type protein (Scott, 1971a). In electrophoresis experiments it displays mobility comparable to the slow-moving α-β region of mouse serum. In fact γ-globulins are reportedly absent from cockroach blood (Siakotos, 1960a; Scott, 1971a), although proteins migrating in the γ- and β-globulin range have been observed for *Locusta* (Nelstrop *et al.*, 1970). Attempts to detect γ-globulins using specific antiserum would seem to be the next step. In *Periplaneta*, protein, lipid, saccharide and esterase activity are associated with the euglobulin material in the electrophoresis band. Scott points out that there appears to be a molecular similarity between the haemagglutinins of the oyster, the crayfish and the cockroach (Scott, 1972). It is interesting that crayfish haemagglutinin is very resistant to degradation by both trypsin and pronase, although it is sensitive to heat, and has a molecular weight greater than 150 000. It does not show detectable bacteriocidal activity (Miller *et al.*, 1972). Both the oyster and the crayfish haemagglutinins appear to function as opsonins, facilitating uptake of material by phagocytotic haemocytes (Tripp, 1966; McKay and Jenkin, 1970), but no such opsonic activity has been demonstrated for cockroach haemagglutinin (Scott, 1971a). Scott (1971b) has studied the adherence of sheep and chicken red blood cells to cockroach haemocytes, generating structures resembling the rosettes formed between mammalian macrophages and red cells in the presence of cytophilic antibody (Berken and Benacerraf, 1966). A proteinaceous surface receptor appears to be involved since prior treatment of the haemocytes with trypsin abrogates this effect. Pretreatment of the haemocytes with cockroach haemolymph did not increase the number of red cells adhering to the haemocytes whereas pretreatment of the haemocytes with rabbit anti-sheep red cell sera did; however, the results of the control experiment using normal rabbit serum were not reported. By analogy with the mammalian macrophage-cytophilic antibody system, it might be expected that pretreatment of the sheep red blood cells with the rabbit antisera would increase even further the adherence of the red blood cells to cockroach haemocytes, but no effect was observed by Scott.

It is not possible from these experiments to draw any conclusions as to the nature of the binding of red blood cells to cockroach haemocytes. It might be that some protoglobulin present in the cockroach haemolymph is a specific haemagglutinin which can also bind to the haemocytes. However, no specificity of the haemagglutinin has been demonstrated by such methods as competitive binding of sheep and chicken red blood cells. Burnet (1968) has suggested that insect physiologists should carry out experiments to see whether haemocytes can be divided into specific sub-populations, according to their capacity to adsorb specific particles such as bacteria or red cells. It should also be possible to use specific antisera to investigate the nature of the red cell receptor on cockroach haemocytes, and to determine whether the Fc portion of vertebrate immunoglobulin can bind to the haemocyte surface.

Burnet (1968) has presented evidence that suggests that cellular immunity may have developed first in an evolutionary sense. It is possible that the recognition and binding of foreign particles by insect haemocytes may represent the primitive beginnings of the cell mediated (T cell) immune system of the vertebrates. In this case opsonins and cytophilic antibodies would probably not be involved. Such a system has not been investigated in insects.

In his review of the cellular defence reactions of insects, Salt (1970) points out: "No zoologist traces the evolution of vertebrates through the insects; and the two groups so widely distinct in the middle Palaeozoic have had ample opportunity for development of different means of discrimination." He suggests (p. 71) that since the wide range of particles and foreign bodies to which insect blood cells can adhere can have no positive characteristics in common, they can be alike only in lacking something. It is necessary to envisage a general stimulus to adhere, together with a mechanism whereby in specific instances the stimulus is absent. Salt suggests that it is the connective tissue sheath, or basement membrane, which surrounds insect cells which the haemocyte associates with "self". When this sheath is absent the haemocyte initiates phagocytosis or encapsulation "nonself" reactions. Salt states (p. 67): "As long as an organ or foreign body is completely covered by a membrane of the appropriate connective tissue, blood cells make no concerted reaction to it. Stretching of the membrane, as in growth, initiates reaction of a few cells which replenish and repair it; perforation of the membrane leads to encapsulation, or to phagocytosis." Support for this view comes from examination of the objects that are projected from haemocytic reaction. The most important protected tissues are those of the insect itself. Intact organs of the same insect and even organs from unrelated insects (see below) are not subjected to haemocytic reaction. However, damage to the surface of organs elicits the attention of haemocytes, although experiments in which only the connective tissue sheath has been damaged have not been reported. Salt (1960, 1965, 1966) has shown that if the surfaces of organ transplants or of a tolerated parasite were changed in various ways (e.g. by heating, by abrasion, or by fat solvents), they were encapsulated on implantation in the same species. It would be interesting to extend these treatments to include reagents specific for the mucopolysaccharides thought to make up the greater part of insect connective tissue. A particular difficulty encountered by the hypothesis is the fact that haemocytes themselves do not have a visible connective tissue sheath or basement membrane, yet they show relatively little tendency to engage in cellular battles. Furthermore, there are times in the life of some insects when connective tissues, including basement membranes, are enormously reduced or actually absent from an organ, but haemocytes show no tendency to adhere to the organ. The pupae of some Diptera are characterized by the absence of investing membranes (Whitten, 1962), but many such "naked" tissues avoid the attentions of phagocytes. Metamorphosing muscles in *Calliphora* for

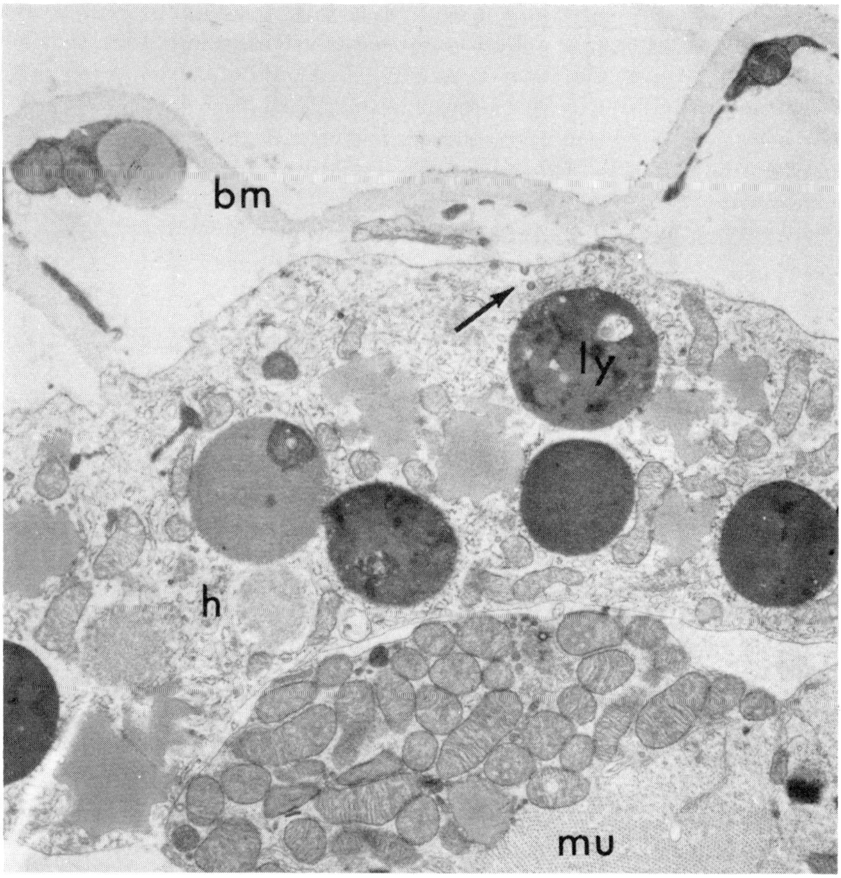

Fig. 16. Penetration of a phagocytic haemocyte beneath the basement membrane (bm) of a *Calliphora* intersegmental muscle (mu) during phagocytosis in the puparium. Coated vesicles are present near the haemocyte surface (see Fig. 17) and heterogeneous lysosomal vacuoles are prominent within the cell (ly). Electron micrograph x12 000.

example, pass without being engulfed through a period when basement membranes are absent, as evidenced by electron microscopy (Crossley, 1972a, cf. Figs 2 and 9). Even when phagocytosis is demonstrable in *Calliphora*, the invasion of haemocytes takes place through a substantially intact basement membrane (Fig. 16), although it is always possible that some areas are damaged, and solicit entry of haemocytes. It seems to the author that we must remain sceptical of any involvment of basement membranes in cellular recognition phenomena, and continue our search for the identification factors to which phagocytes respond. The extent of species or generic specificity of the cellular recognition factors is of interest, and is accessible to experimental investigation. Plagge (1936) transplanted testes of caterpillars between members of different genera, and even between different families. For example, testes from *Plodia, Galleria, Carpocapsa*, *Ptycha* and *Plusia* were transplanted into *Ephestia*, where they not only survived but developed normally in the foreign host. The success of transplantations depended on the size and consistency of the graft, rather than on the closeness of its taxonomic relation to the host. Scott (1971c) has recently carried out transplantation studies using *Periplaneta* as host organism, and has reported that nerve cords from conspecific cockroaches are not encapsulated, but interspecific nerve cords, and *Calliphora* or mouse brains are encapsulated. Since insect physiologists routinely carry out interspecific grafts in the course of their experiments, the report of Scott needs confirmation and extension.

An entirely different mechanism of haemocyte stimulation has been postulated to occur at wound sites. It is suggested that an injury factor is released by damage to cells, or a change in their metabolism, which solicits haemocyte activity. Wigglesworth (1937) suggested that epidermal cells and haemocytes probably respond to the same stimulus or injury factor which he believes are protein degradation products. Interestingly, very superficial incisions which do not penetrate the basement membrane fail to induce accumulation of haemocytes. Nor are the epidermal cells dependent for their activation in wound healing on substances produced by the haemocytes. It may be profitable to consider both injury factors and foreignness recognition (nonself) factors together here, because three changes in the haemocyte population are induced by both: (a) mobilization of formerly sessile cells; (b) multiplication by mitosis; (c) attachment at localized sites. The underlying mechanisms of response in the two situations, injury or nonself, may well be similar or identical. We need not consider the wound-healing responses of the epidermal cells here, which are reviewed by Wigglesworth (1937) and Lai-Fook (1966), but we must consider the haemocyte responses.

There are good reasons for supposing that in most insects haemocytes are predominantly sessile cells, attached to or moving over the surfaces of organs (Wigglesworth, 1959). This is evidenced by the fact that wounding dramatically increases the number of circulating haemocytes (Harvey and Williams, 1961; Lea and Gilbert, 1961). In diapausing pupae of *Hyalophora cecropia* the increase in circulating cells is about one hundred-fold and the dissemination of injury response is shown by parabiosis experiments to be humoral. Furthermore the ancillary effects of injury include an increase in heart rate, which should increase the rate of circulation of haemocytes, and hence the probability of a haemocyte encountering the wound site (Harvey and Williams, 1961). The spread of the injury factor appears to be decremental. The magnitude of the production of the injury factor as monitored by QO_2 is related to the size of the wound, and perhaps specifically to the circumference of the wound, indicating that the injury factor is generated at the margin of the wound. However, the extent of involvement of haemocytes is related to the area of wound surface, since introduction of a sponge greatly increases both the injury metabolism and the number of haemocytes taking part in wound closure (Harvey and Williams, 1961). The differential between injury and normal metabolic rates is greatly amplified in diapausing pupae, but there are reports of increased haemocyte counts following injury in insects with a normal metabolic rate (Jones, 1967; Hoffmann, 1969). That the increase in circulating blood cells in wounded cecropia pupae is due to cell mobilization rather than to mitosis is shown by Clarke and Harvey (1965), who find that mitotic figures are not seen until 18-24 h after wounding, whereas substantial increases in cell numbers are observed by this time. However, mitosis does contribute to later increases in cell numbers, the reported increase in mitotic index following wounding being 0.67 ± 0.13 per cent. It is of interest to note that these experiments involved the introduction of about 18 mg of phenylthiourea per insect, which could severely reduce the activities of a large variety of copper- and iron-containing enzymes, and suggests indirectly that such enzymes, e.g. tyrosinase, are probably not directly involved in production of "injury factor". In cecropia the selective adhesion of special classes of haemocyte described as "plasmatocytes" and as "hyaline cells" is reported, and these compose about 60 per cent of the total haemocyte population. The hyaline cells here may relate to the hyaline haemocytes involved in blood clotting (q.v.) although this is not discussed by the authors. The behaviour of the cells changes on arrival at a wound site. A flattening is observed first, followed by formation of leading lamella and random movement akin to that seen in fibroblasts by Ambrose (1961). Since encounters could result in permanent attachments between cells, a meshwork of cytoplasm was formed, which gradually occluded the wound (Clarke and Harvey, 1965). In spite of the fact that enhancement of attachment or increased stickiness of haemocytes at wound sites is reported by numerous workers (Lazarenko, 1925; Wigglesworth, 1937; Harvey and Williams, 1961; Clarke and Harvey, 1965), it is hard to quantify (see section 5).

In an ultrastructural study of haemocyte-wound interaction in *Rhodnius*

Lai-Fook (1970) showed that a cell type (termed a plasmatocyte) with dense homogeneous granules is most active in repair of wounds. These cells accumulate at incisions in the integument within 12 h, and on attachment display changes in the morphology of contained granules culminating in the appearance within these of 15 nm diameter "microtubules." By analogy with the work of Beaulaton (1968) on *Antheraea* and *Bombyx* we might suppose the granules to be released from haemocytes as part of the clotting or, perhaps more generally, the foreignness recognition system. Although, as Lai-Fook rightly emphasizes, the granules of these haemocytes represent a strange organelle and their density changes could be either associated with formation or breakdown. Haemocytes clustering around wound sites can form cell junctions characteristic of tissues. *Zonulae adherens* junctions occur first and are most numerous, but *Zonulae occludens* and septate desmosomes are also seen connecting haemocytes (Lai-Fook, 1970, plate 6).

The injury factor may thus be envisaged as emanating directly from an injured or changed cell, or indirectly from a specialized haemocyte which is sensitive to foreign molecules. There is ample evidence that labile haemocytes sensitive to foreign surfaces exist in insects. Such haemocytes are called coagulocytes, oenocytoids, hyaline haemocytes, or simply labile haemocytes by various authors and are postulated to play a part in clotting and bacteriostatic mechanisms (see above). It is entirely possible that they are also involved in wound healing and encapsulation reactions, by activation of similar foreignness recognition mechanisms. It is also possible that released substances potentiate phagocytosis or encapsulation by other haemocytes, i.e. act as opsonins. What is required is a renewed examination of the consequences for the insect of disruption of labile haemocytes, with attention to changes other than blood clotting.

Salt (1970, p. 75) argues against the involvement of a chemical stimulus in encapsulation because inert objects such as glass or polyfluorocarbon are subject to haemocyte encapsulation. However, a localized chemical stimulus could arise indirectly, from a labile haemocyte sensitive to foreign surfaces, which mediated in the encapsulation reaction. Such a haemocyte might, or might not, itself form part of the capsule. As Salt points out, the diffusion of a stimulus would wane as it passed through successive layers of congregating haemocytes, and could explain why blood cells eventually cease to add themselves to a capsule. Crossley (1975) has recently obtained evidence that disrupted labile haemocytes accumulate at wound sites, locally releasing their cytoplasmic contents which are rich in phenoloxidizing enzymes (Fig. 12). This local release, coupled with availability of substrate in the haemolymph and oxygen at the wound site, leads to local melanization. There seems little reason to doubt that a similar sequence of events occurs on parasites that are recognizably foreign to the labile haemocytes. Furthermore the release of a foreignness recognition signal by certain haemocytes could also explain the speedy aggregation of blood cells. As Salt (1970) puts it in analogy: "A chance passerby might notice a fire and sound the alarm." The evidence for haemocyte chemotaxis is discussed by Nappi and Stoffolano (1972) but is not strong at present, and further work is needed to support or refute a chemotaxis hypothesis.

8 Endocytosis by insect blood cells

Endocytosis is the process by which the plasma membrane of a cell is infolded, leading to vesiculation and transfer of membrane and enclosed substances into the cell without actual permeation of the membrane. The vesiculation theory of cellular ingestion was first clearly stated by Bennett (1956) and the terms "cytosis" (Novikoff, 1961) and "endocytosis" (de Duve, 1963) were later coined to stress the conceptual unity of phagocytosis and pinocytosis. Studies on amoebae suggested that both molecules in solution and particles in suspension are taken up by the same mechanism, and that materials taken up form a continuum. At one extreme, large insoluble particles are said to be phagocytozed, and at the other extreme, solute molecules are said to be pinocytozed (Brandt and Pappas, 1960). Subsequently it became apparent that phagocytosis and pinocytosis are comparable in mechanism of absorption, kinetics of uptake, and in the fate of the bodies after engulfment. Engulfed substances have access only to that semi-external phase of the cell which corresponds to the contents of phagosomes, lysosomes, and secretory vacuoles, and which has therefore been christened the "exoplasmic" compartment of the vacuome (Jacques, 1969). Nevertheless insect cells are capable of selective uptake of particles. Thus *Calliphora* haemocytes ingest soluble precipitated particles of thorium dioxide and whole bacteria (Figs 1, 18, 19), as well as such soluble molecules as horseradish peroxidase. However, pericardial nephrocytes can ingest only small soluble molecules (Crossley, 1972b). Selective uptake of certain insoluble and soluble molecules is mediated in haemocytes, nephrocytes, and in many other cells by coated vesicles. Coated vesicles were described in insect oöcytes by Roth and Porter (1964) where they were shown to be concerned with uptake of yolk protein. They are especially numerous in insect cells concerned with endocytosis, such as pericardial cells (Bowers, 1964; Crossley, 1972b) and haemocytes (Crossley, 1968). In haemocytes coated vesicles are flask-shaped indentations of the plasma membrane about 120-165 nm in diameter. The indentations are coated on their cytoplasmic surface by alveolariform thickenings, which in sections appear as bristles projecting into the cytoplasm. Kaneseki and Kadota (1969) describe the structure of coated vesicles in mammalian cells, and suggest that the "basket work" on the membrane is exclusively an apparatus to control the infolding and fissioning mechanism of the membrane. Insect haemocytes sequester a variety of protein (e.g. ferritin) and nonprotein (e.g. lanthanum and thorium dioxides) tracer molecules in coated vesicles, although uptake of nonproteins may follow their attachment to proteins in the haemolymph. The vesicles pinch off from the plasma membrane and sink into the cytoplasm (Fig. 17). Since the sequestered material accumulates in membrane-bound vacuoles, it is to be presumed that the coated vesicles fuse with these vacuoles, contributing their contents. However, as is to be expected from the hypothesis of Kaneseki and Kadota (1969), the large vacuoles do not bear an alveolariform coating, and the fate of individual quanta of membrane is not clear. Since multiple

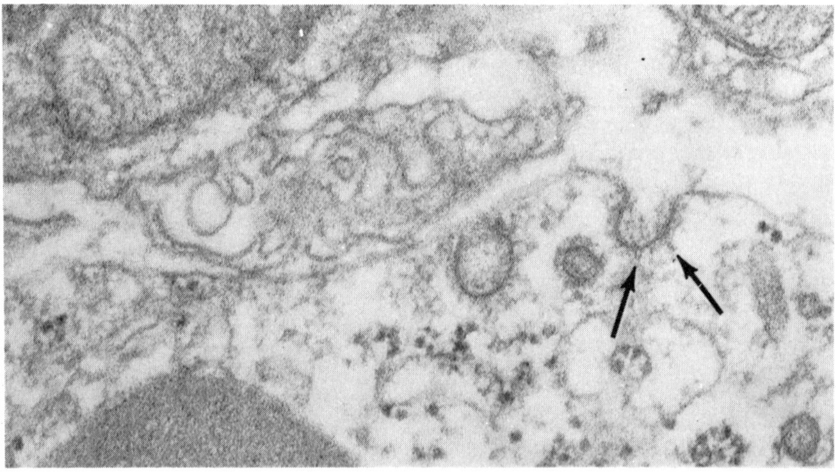

Fig. 17. Coated vesicles at the surface of a haemocyte such as that seen in Fig. 16. These vesicles detach and pass into the cell. Fuzzy material is present on the haemolymph aspect of these vesicles, whilst a structural coating appearing as "bristles" is present on the cytoplasmic aspect (arrows). Electron micrograph x88 000.

small vesicles have a greater surface area than a single large vesicle, some recycling would seem feasible, although this has not been proven. The large vacuoles which contain sequestered material can be demonstrated to contain acid phosphatase and are members of the lysosomal vacuolar apparatus discussed below.

The coated vesicle adsorption mechanism must include a molecular recognition component if it is to be at all selective. In insect pericardial cells a selective uptake of some molecules and a rejection of others by coated vesicles has been experimentally demonstrated (Crossley, 1972b), but the capabilities of the haemocyte-coated vesicles have yet to be investigated. Coated vesicles are numerous on the surface of haemocytes which approach cells undergoing phagocytosis (Crossley, 1968, Fig. 16), and are probably employed in selective adsorption complementary to the main phagocytic effort.

One possible complication in this interpretation should be interposed here. Friend and Farquhar (1967) have observed two size classes of coated vesicle in the rat vas deferens, and believe that only the larger vesicles are concerned with transport of adsorbed proteins to lysosomes. Smaller coated vesicles (< 75 nm) never took up peroxidase marker and were apparently involved in outward transport of molecules from the Golgi apparatus. Some small coated vesicles fuse with the cell membrane and could be involved with conveyance of enzymes or surface coat material to the plasmalemma, or could replace membrane lost from the cell surface during protein absorption. However, in insect haemocytes the coated vesicles observed by the author have been 100 nm or more in diameter and are presumably concerned with transport of adsorbed molecules.

Phagocytosis, the uptake of large insoluble particles by haemocytes, plays a part both in morphogenesis and in resistance to disease. In insects undergoing metamorphosis the phagocytic haemocytes are instrumental in the rapid breakdown and reutilization of cellular components. The phagocytosis of muscle by haemocytes provides an excellent system for study, because the high level of three-dimensional organization of muscle makes it easy to monitor structural integrity. The extensive literature on insect muscle phagocytosis is reviewed by Crossley (1965). Using the electron microscope it can be seen that invasion of muscle by phagocytes is preceded by changes in the muscles themselves, in which vacuoles appear and the sarcoplasmic reticulum begins to break down. All the visible changes which precede invasion are confined to the sarcoplasm and the myofilaments are eventually engulfed without disruption of hexagonal packing or loss of striation. Haemocytes have been observed attacking muscles enclosed in a full thickness of basement lamella, and rolls of finely granular lamella material could be distinguished in vacuoles within the haemocytes (Crossley, 1968). These observations suggest that disruption of the extracellular basement membrane is as likely to be a result of, rather than a signal for, the invasion of haemocytes, although, as we have seen, the underlying recognition phenomena are poorly understood at present (section 7).

The plasma membrane of phagocytic haemocytes is extended into attenuated pseudopodia and flattened folds, and these serve to infiltrate between organelles of the muscle. The cooperative activities of numerous haemocyte pseudopodia isolate fragments of muscle cytoplasm, and these fragments then sink into the haemocyte where they are sequestered within membrane-limited vacuoles (Crossley, 1968). Similar extensions are observed in vertebrate cells engaged in phagocytosis and have been termed "marginal folds" (Fawcett, 1965). In *Sarcophaga* Whitten (1964, 1969) describes the involvement of large multinucleate granular haemocytes in phagocytosis, but this observation has not been extended to other cyclorrhaphous Diptera such as *Calliphora* (Crossley, 1968) or *Lucilia* (Barritt and Birt, 1971).

Lockshin and Williams (1964, 1965a, 1965b, 1965d, 1969) show that phagocytic haemocytes do not enter the muscles of *Antheraea* until the muscle organelles are disrupted. They present evidence that a nervous signal, potentiated by hormonal changes, brings about disruption of lysosomes in apparently viable muscle. In *Calliphora*, however, the muscle cytoplasm is lacking lysosomal enzymes, as evidenced by Gomori's histochemical method for acid phosphatase, although invading haemocytes are rich in this enzyme (Crossley, 1968). Although there is abundant evidence that insect muscles can, in certain circumstances, break down without the intervention of haemocytes at all (Viallanes, 1882; Lockshin and Williams, 1964, 1965a; Lockshin, 1969b; Crossley, 1965, 1972a), it is by no means certain that such breakdown involves lysosomes. For autolysing intersegmental muscle in *Calliphora* pupae there is no evidence for enclosure of myofilaments in membrane-limited vacuoles of any sort during breakdown. Myofilament depolymerization appears to be mediated by unknown factors in the cytoplasm (Crossley, 1972a). In other *Calliphora* muscles, that were destroyed by haemocytic phagocytosis, it was shown that the availability of muscles for breakdown could be controlled by the hormone β-ecdysone, which also influenced the number of phagocytic haemocytes (Bohm, 1968; Crossley, 1968).

Metchnikoff (1905) clearly foreshadowed modern ideas of intracellular digestion, but he thought that lytic enzymes which he called "cytases" were confined within the blood cells unless these were damaged. It is only recently that evidence has accumulated from studies on vertebrate leucocytes that lysozyme, β-glucuronidase, and certain other lysosomal enzymes are released into the medium during phagocytosis. The release of enzymes is selective, and a nongranular enzyme, lactic dehydrogenase, is not released, showing that cell damage is not a necessary condition of enzyme release (Wright and Malawista, 1972). The release of lytic enzymes may assist the phagocytic action of the leucocyte, although the rapid inactivation of some enzymes released, such as α-naphthyl phosphatase and cathepsin, presumably serves to localize tissue damage and inflammation. The possibility that lytic enzymes are released from phagocytic insect haemocytes does not appear to have been investigated.

Insect phagocytic haemocytes have been found to ingest a variety of microorganisms and tracer materials (Graber, 1871; Hollande, 1930; see also the review of Jones, 1962) (Figs 18 and 19). Metalnikov (1924) injected 50 species of bacteria into *Galleria* and found 47 to be ingested and progressively broken down by haemocytes. Cameron (1934) reported that some bacteria retained their virulence after engulfment, but it is surely difficult to be certain that all organisms were engulfed. Salt (1970, table 3) has surveyed the wide range of microorganisms subject to insect phagocytosis. Wittig (1966) has discussed many pitfalls in the quantitative

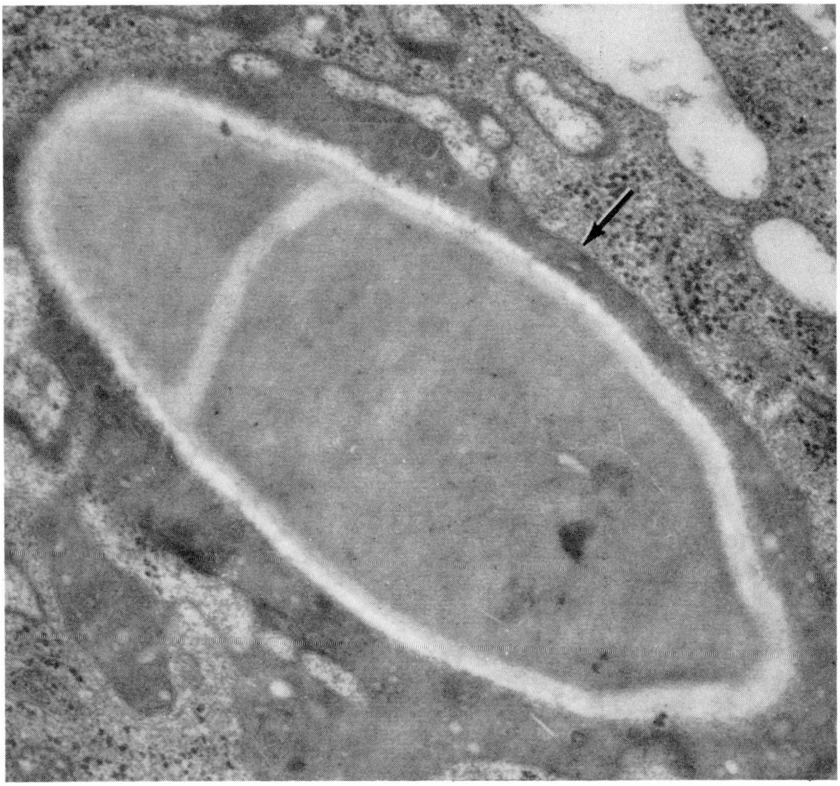

Fig. 18. *Bacillus cereus* following haemocyte endocytosis in *Calliphora.* The bacterium is enclosed within a lysosomal vacuole bounded by a single unit membrane (arrow) within the haemocyte. Electron micrograph x75 000.

experimental study of phagocytosis in insects. She points out that the reaction is compounded of reaction due to wounding, reaction derived from the volume and chemistry of diluent, and reaction due to pathogen. That each reaction is separately quantifiable was shown by examination of the total haemocyte count. Pricking the insect alone caused an increase in the THC significant at the 1 per cent level. If sufficient distilled water to increase the blood volume by 5 per cent was introduced, a further increase in the THC occurred, again significant at the 1 per cent level. However, Wittig's Fig. 1 (1966) graphically expresses the extreme fluctuation in haemocyte numbers in normal insects that undermines any quantitative study based on haemocyte numbers. In last instar *Pseudaletia*, even within a group of healthy caterpillars of the same physiological state, individual THC ranged from 75 to 140 per cent of the mean of the group. Wittig pointed out that poor techniques, both experimental and statistical, can further exacerbate this difficulty.

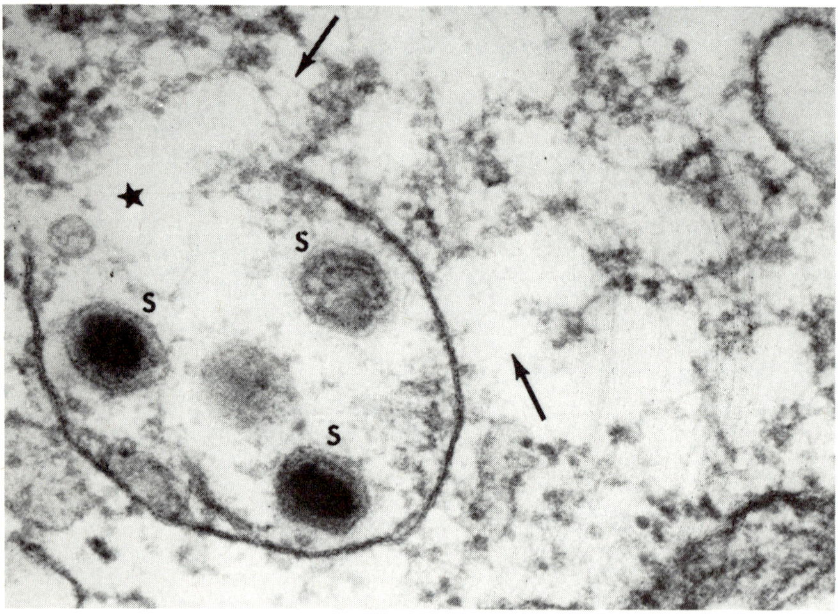

Fig. 19. Sericesthis iridescent virus (SIV) following endocytosis by a *Calliphora* haemocyte(s). Virus particles appear initially in membrane-limited vacuoles, but the membranes subsequently break down (asterisk) releasing virus into the cytoplasm. The cytoplasm contains unusual filamentous structures (arrowed) which stain with uranyl acetate, and may be M-RNA. Electron micrograph x107 000.

When phagocytic capacity of the haemocytes is studied, particle numbers, shapes, and sizes become decisive. Thus Indian ink is found to be an unsuitable test material because of gross variations in size and shape of the particles. Polystyrene latex particles were to be preferred as they could be obtained in several graded sizes, and were sequestered as rapidly, and by the same types of cells, as Indian ink. The phagocytic capacity of army worm blood was saturated by a dose of 66.3 million, 1.3 μm diameter, polystyrene particles per larva. If killed *Bacillus thuringiensis* were used in light doses, up to 280 000 could be picked up by haemocytes within 30 min of injection. Heavy doses (*c.* 20 million bacteria) exceeded the initial pickup capacity of the haemocytes and many particles remained free in the haemolymph for more than 12 h after injection (Wittig, 1966). In any insect the vast majority of haemocytes are potentially phagocytic. In *Drosophila* 90-95 per cent are potentially phagocytic plasmatocytes, in *Galleria* 95-98 per cent are potentially phagocytic plasmatocytes or adipohaemocytes (Werner and Jones, 1969).

Many authors report changes in the haemocyte count when foreign matter is introduced into insects. The total haemocyte count is usually increased (e.g. Schwartz and Townshend, 1968) but occasionally decreased (e.g. Vinson, 1971). Increased haemocyte numbers are sometimes thought to be due to an increase in the mitotic rate, for example in the cockroach *Blatta*, infection with *Staphylococcus* sends the mitotic index from the normal value of 0.5 to 3.1 per cent (Tauber, 1940). On the other hand, increased numbers of haemocytes which were observed following introduction of foreign particulate material in another cockroach, *Periplaneta*, were reportedly not the result of enhanced mitosis. Furthermore, the injection of foreign particles also increases the haemolymph volume (Ryan and Nicholas, 1972). Thus it is reasonable to suppose that here mobilization of sessile haemocytes is largely responsible for the change in numbers. There are also reports of changes in the differential haemocyte count when foreign objects are introduced into insects (Shapiro, 1968; Werner and Jones, 1969; Vinson, 1971; Maier, 1973). The difficulty that is inescapable is that differential counts depend on visual anatomical, and near instantaneous, classification of individual cells as the count proceeds, and each worker has his own subjective terms of reference. Within their terms of reference Werner and Jones (1969) report that spherule cells and oenocytoids in *Galleria* show no phagocytic ability and that phagocytic adipohaemocytes and plasmatocytes show correlated changes in numbers following injection of foreign matter. However, Vinson (1971) working on *Heliothis* reports that, in his terms of reference, prohaemocytes, plasmatocytes, granular haemocytes, and oenocytoids are phagocytic, whereas spherule cells are nonphagocytic. Since the act of wounding, with minimal introduction of foreign material can itself increase the haemocyte density (e.g. produce an increase of four times in cockroaches as evidenced by heat-fixed pellets: Brady, 1967), and in view of the work of Wittig (1966) cited above, and of Walters (1970) cited elsewhere, it is obvious that the controls used in injection experiments need careful evaluation.

Phagocytic haemocytes ingest other haemocytes during the development of some insects. In *Rhodnius* nonphagocytic "oenocytoids" are taken up by phagocytic haemocytes, and undergo cytolysis within them, prior to a moult (Wigglesworth, 1933). A similar phenomenon in *Lucilia* is shown in Fig. 5.

Insect haemocytes respond by encapsulation and phagocytosis towards transplants of cell culture lines of vertebrates. *Leucophaea* is well able to defend itself against vertebrate cells of either normal or neoplastic origin (Dawe *et al.*, 1967). One point of interest is that in granulomas of mammals the usual rule is to find flattening of macrophages and transitions towards fibrocytes at the periphery, rather than flattening in the inner layer of cells as is the case in this and most other insect capsules. Interactions of various erythrocyte preparations with phagocytes of *Galleria* and of the mouse were compared *in vivo* and *in vitro* by Rabinovitch and de Stefano (1970). Haemocytes ingested *in vivo* a broader range of particles than *in vitro*, but it was not shown whether haemolymph recognition factors could explain the discrepancy. Extensive damage to sheep erythrocytes, for example by formaldehyde treatment, was needed for their interaction *in vitro* with either insect or mouse phagocytes. It is suggested by these authors that the similar patterns of particle recognition by insect and mouse phagocytes may be used as evidence for a comparable foreignness recognition mechanism, at least when haemolymph or serum factors are excluded from the reaction system. Insect cellular recognition phenomena are discussed in section 7.

The phagocytosis of virus is not always effective in preventing infection. The ingestion of SIV virus by haemocytes of *Galleria* initially leads to virus sequestration within phagocytic vesicles. However, the virus disappears within 3 h, and viral DNA synthesis in haemocyte cytoplasm is detected $4\frac{1}{2}$ h after inoculation. Viral progeny are detected 12 h after inoculation (Leutenegger, 1967) (Fig. 19). Wound tumour virus (WTV) is also able to multiply in insect haemocytes, and haemocytes may also transport virions to other tissues. The plasmatocytes are liable to become so heavily infected that almost the entire cytoplasm is replaced by WTV. Granular haemocytes, containing vacuoles with periodic tubular elements as substructure, were never seen to ingest virus, and did not become infected (Granados *et al.*, 1968). Nuclear polyhedrosis virus can also multiply in *Spodoptera* plasmatocytoids, which are again a phagocytic haemocyte type (Kislev *et al.*, 1969). An entompox virus also replicates in granular haemocytes following initial phagocytosis and sequestration in phagocytic vesicles. Release of the virus into the haemolymph is apparently by exocytosis (Devauchelle *et al.*, 1971). A *Rickettsia* species has also been reported to undergo phagocytosis followed by successful multiplication within haemocytes, following transformation to a Rickettsia-like particle (Devauchelle *et al.*, 1972). Mention should also be made of an interesting trypanosome, *Trypanosome rangeli*, which multiplies in the haemocytes of *Rhodnius prolixus*. Crithidial forms enter the plasmatocytes where they divide. *T. rangeli* is, so far, unique in that it is lethal to its invertebrate, rather than to its vertebrate, host (Tobie, 1970).

9 Phenol metabolism in insect blood cells

Phenol metabolism extends into many areas of insect physiology and the blood and blood cells, as well as the integument, are implicated in this metabolism. Haemocytes containing enzymes capable of accelerating the oxidation of phenolic compounds have been described in *Sarcophaga* (Dennell, 1947; Jones 1956); in *Calliphora* (Crossley, 1964, 1975); in *Periplaneta* (Mills *et al.*, 1968); in *Antheraea* (Evans, 1968); in *Locusta* (Hoffmann *et al.*, 1970); and in *Chironomus* (Maier, 1973). It has also been established that insect blood plasma usually contains substrates oxidizable by phenol-oxidizing enzymes. In *Calliphora*, for example, the haemolymph (plasma) contains tyrosine (Finlayson and Hamer, 1949) and dihydroxy-phenylalanine (dopa) (Pryor, 1955b). The paradoxical failure of the blood to darken as a result of phenol oxidation *in vivo* has elicited a number of hypothetical explanations in the literature. Dennell (1947) suggested that dehydrogenase activity maintained the redox potential of the blood too low for phenol-oxidase activity.

Karlson and Wecker (1955) suggested that a permanent induction period was maintained for phenol oxidizing enzymes by the oxidation of substrate dopa by the cytochrome system.

However, Pryor (1955a) pointed out that a bubble of air introduced into an insect does not cause darkening, and suggested that physical separation of enzyme and substrate was maintained by tissue organization. This idea was further developed by Mills and Whitehead (1970) who showed that the permeability of haemocytes in *Periplaneta* toward tyrosine changed at ecdysis, and might be under humoral control. Thus bursicon could allow access of tyrosine to the enzymes which convert it to dopa and dopamine. Experimental support for this idea also comes from the demonstration of *in vitro* conversion of tyrosine in the haemolymph of *Pieris*, which is induced by bursicon (Post, 1972). However, as Mills and Whitehead (1970) point out for their own experiments, overall blood cell permeability could be increased by hormone, with the result that swelling and eventual lysis of cells liberates tyrosine metabolizing enzymes. Since the phenol oxidizing haemocytes are particularly labile, as discussed below, this possibility must be taken seriously.

The hypothesis that has gained widest experimental support proposes that the blood contains an inert pre-enzyme (or pro-enzyme) which requires activation. This hypothesis arose from the observations of Bodine and Allen (1938) and Bodine (1945) on *Melanoplus* eggs, where tyrosinase pro-enzyme could be activated by a lipid also derived from the eggs. Ohnishi (1953) described a protyrosinase system in the haemolymph of *Drosophila* larvae, and showed that the activating principle was a protein. A crystalline *O*-diphenol oxidase was isolated from whole *Calliphora* by Karlson and Liebau (1961) and later Karlson *et al.* (1964) extracted an activator of this enzyme from the integument of *Calliphora.* The activator itself appeared to be an enzyme, and produced *in vitro* activation of an inert *O*-diphenol oxidase percursor from haemolymph. A number of proteolytic enzymes could mimic the natural activator, and activation was postulated to involve limited proteolysis (Schweiger and Karlson, 1962; Karlson *et al.*, 1964).

Prophenolases and their protein or lipid activators have subsequently been described for *Musca* (Inaba and Funatsu, 1964) *Tenebrio* (Heyneman, 1965) and *Bombyx* (Ashida, 1971). In *Antheraea* concentrated blood cell homogenates were found to have dramatic activating influence on extracted *Antheraea* haemocyte prophenoloxidase. The activator component of the homogenate was mooted, on a basis of dialysis and heating experiments, to be a protein (Evans, 1968). It is not known whether activator and enzyme occupy the same blood cells, but in any case there is evidence that the activator is itself formed from an inactive percursor in *Antheraea*, as originally suggested for *Calliphora* pro-phenolase activator by Schweiger and Karlson (1962).

In larval *Calliphora*, activation of an inert pro-enzyme is not the only limiting factor in phenol oxidation, since a potent *O*-diphenolase inhibitor is released, at times, from the salivary glands (Thomson and Sin, 1970). A phenoloxidase inhibitor was also reported for *Antheraea* by Evans (1968).

Integument and blood each contain a variety of phenol-oxidizing enzymes. Lai-Fook (1966) showed that the cuticle of *Sarcophaga* contained not only a phenolase involved in puparium formation, but also a prophenolase which is activated by injuries. The work of Hackman and Goldberg (1967) on *Lucilia* distinguishes between a water-soluble active *O*-diphenoloxidase derived from the cuticle, and a pro-*O*-phenoloxidase derived from centrifuged haemolymph. The latter, once activated, had a wider range of acceptable substrates. Freezing and thawing of *Lucilia* larvae brought about release of a further active dopamine-oxidizing enzyme absent from normal plasma, and perhaps derived from cells.

There is evidence that the phenol-oxidizing enzymes in plasma are formed and released by haemocytes. This was first demonstrated for certain crustacean haemocytes, which release phenolase during explosive contact interaction with foreign surfaces (Pinhey, 1930; Bhagvat and Richter, 1938). Later Dennell (1947) found phenol-oxidizing enzyme in insect

(*Sarcophaga*) haemocytes. Crossley (1964) found that analogous haemocytes in *Calliphora* which contained both phenol-oxidase and peroxidase were labile cells ("Type B"), which became blistered as the blood clotted, suggesting a changed plasma-membrane permeability. In *Locusta* Hoffmann, Porte and Joly (1970) have also reported that phenol-oxidizing enzymes are present in a labile class of haemocytes (coagulocytes). When *Galleria* larvae were injected with bacteria (*Serratia* or *Enterobacter*) the percentage of haemocytes containing polyphenol oxidase increased when compared with that of control larvae injected with saline. The identity of enzyme(s) oxidizing the pyrogallol substrate was not established. It was suggested that the "polyphenol oxidase is related to the initial appearance of antibacterial factors in insect haemolymph associated with immunization" (Pye and Yendol, 1972).

In *Calliphora* haemocytes the phenol oxidizing enzymes have a greater oxidative repertoire than is normal for *O*-diphenolases (= tyrosinase = *O*-diphenol: O_2 oxidoreductase), and it is probable that a peroxidase (Donor: H_2O_2 oxidoreductase) is also present (Crossley, 1975). Since peroxidases from a number of different plant and human cells can oxidize tyrosine to melanin (Patel *et al.*, 1971; Okun *et al.*, 1971), it is quite possible that a biochemical pathway to melanin, independent of oxidase enzymes, also exists in insects. The *Calliphora* haemocytes containing both phenol oxidase and peroxidase are strikingly labile, and appear to release the enzymes on contact with foreign surfaces (Fig. 10). They also accumulate at wound sites (Fig. 10) (Crossley, 1975). It is thus possible to argue that these cells are involved in bacteriostasis and wound healing, particularly since vertebrate leucocytes releasing peroxidase have been shown to be involved in bacteriostasis (Klebanoff, 1968).

Karlson and Sekeris (1964) suggested that during puparium formation in *Calliphora* the phenolase pro-enzyme moves from the haemolymph into the cuticle, where it is activated and brings about sclerotization of the cuticle by conversion of *N*-acetyl dopamine to quinone. Support for this notion comes from the observations that in *Calliphora* the differential count of phenol-oxidizing haemocytes indicates a reduction of numbers at the time of puparium formation, whilst at the same time phenol-oxidizing enzymes appear in the plasma, as revealed by spectrophotometric assay (Crossley, 1975). To be convinced of the reality of transfer of sclerotizing or darkening enzymes from blood cells to integument, we also need evidence of uptake from the blood in to the integument. At present this evidence is lacking and not easy to obtain (Locke and Krishnan, 1971; Koeppe and Gilbert, 1973).

With the exception of the work by Evans (1968) on the *Antheraea* system, the natural activators of haemocyte phenol-oxidizing enzymes have neither been identified nor located. In *Calliphora*, artificial activators, such as alcohol or anionic detergents, are effective on the haemocyte pre-enzymes, but the natural activator extracted from integument is only effective on the pre-enzymes already present in the blood plasma (Crossley, 1975). It would presumably be in the interests of the insect to limit release of activator to localized sites at specific times, such as wounds, capsules, or the cuticle at sclerotization, to avoid general melanization of the blood. However, it may well be the generalized release of the activators that leads to blackening of insects heavily infected with bacteria.

L-Tyrosine has been identified as a precursor of protocatechuic acid found in oöthecae of *Periplaneta* (Brunet, 1963), but an enzyme catalysing the β-hydroxylation of tyramine has been demonstrated in *Periplaneta* haemolymph (Lake *et al.*, 1970). The product of the reaction, *p*-hydroxyphenyl ethanolamine, is a precursor of noradrenalin and could also be an intermediate in a pathway leading to the oöthecal hardening agent protocatechuic acid (see diagram). It is not known whether the tyramine hydroxylase involved is in the cells or plasma. However, another enzyme, monoamine oxidase, which may initiate an alternative pathway of protocatechuic acid formation (see diagram) has been reported in the haemocytes of *Periplaneta* by Whitehead (1970b). It has been suggested that the control of protocatechuic acid synthesis is exerted in haemocytes associated with colleterial glands. Protocatechuic acid cannot be synthesized by the isolated gland. The results of an experiment with allatectomized cockroaches "pulsed" with (U-^{14}C)-tyrosine suggest that the pathway in the colleterial system is controlled by juvenile hormone and inhibited by α-ecdysone. Isolated haemocytes treated with juvenile hormone and (U-^{14}C) tyrosine showed an enhanced ability to deaminate tyrosine (Whitehead, 1970b). Tyrosine decarboxylase and possibly dopa decarboxylase are also said to be present in *Glossina* haemocytes, but here synthesis from (U-^{14}C) tyrosine results *in vitro* in the production of neutral substances unaffected by xanthine oxidase, which could be *n*-acetyl-dopamine, and *N*-acetyl tyramine (Whitehead, 1973). The contribution of various haemocyte phenol metabolism pathways need investigation in a range of insect forms to enable us to construct a coherent scheme, but the importance of haemocytes in the physiology of hardening and darkening is increasingly obvious.

10 Insect blood cells in connective tissue formation

Insect tissues are swathed in a variety of extracellular matrix and fibrous components of diverse chemistry and ultrastructure, which are known collectively as connective tissue. Connective tissue in the form of strands

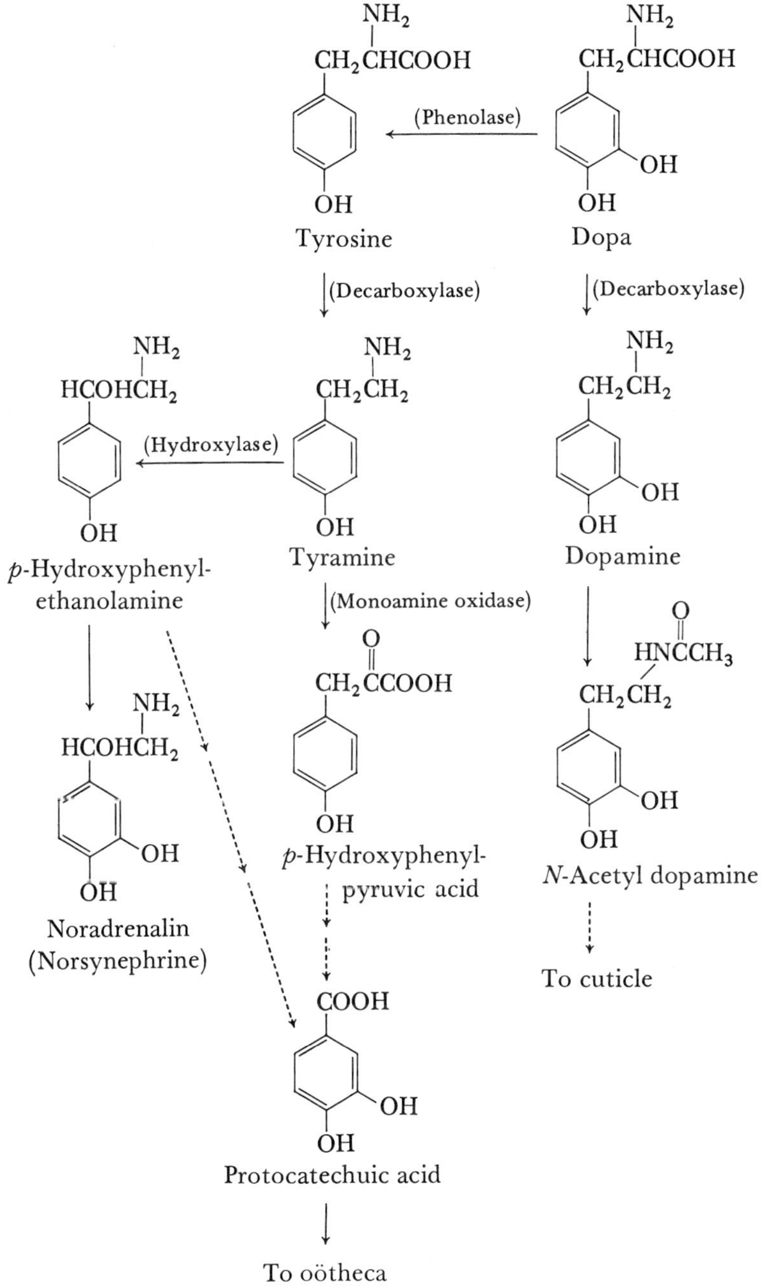

METABOLISM OF TYROSINE IN HAEMOCYTES

holding internal organs in their correct position is present, but not conspicuous, in insects because this structural function has largely been taken over by tracheae (Waterhouse, 1950). Connective tissue in the form of thin layers of extracellular material applied to cell membranes is most aptly described as basement membrane. Basement membranes are for the most part under 0.5 μm thick, and thus barely visible under the light microscope (Ashhurst, 1968). Secretion of basement membrane by wandering mesenchyme cells was postulated by Lazarenko (1925b) as a result of observations on capsule formation in *Oryctes*. Wigglesworth (1933) noted that, in *Rhodnius*, four types of haemocyte (prohaemocytes, amoebocytes, oenocytoids and adipohaemocytes) congregate at the basement membrane at the time of moulting. Wigglesworth (1956) later showed that amoebocytes contained periodic acid-Schiff (PAS) positive inclusions and appeared to contribute these by discharge to the basement membrane, which itself appeared to be PAS-positive, and was presumably mucopolysaccharide. Amoebocyte inclusions and connective tissue also shared three other identical staining reactions; they both stained blue-black in chrome haematoxylin, deep red in fuchsin-paraldehyde, and intense green in Masson's trichrome. The two former reactions are also given by neurosecretory cells, but these are PAS-negative. Basement membranes were reported to form with the mediation of haemocytes around thoracic muscles, and around groups of phagocytic haemocytes that had been experimentally caused to ingest Indian ink. However, the PAS-positive basement membranes of the gut, and of nervous tissue (perilemma) were not formed in this way. Forty years after his first observations on the subject, Wigglesworth (1973) further showed that the brief period of congregation of haemocytes below the epidermis of the abdomen, six days after feeding in the 4th instar larvae, was precisely the time at which the basement membrane increased greatly in thickness (from 0.15-0.22 μm at six days to 0.36-0.45 μm at seven days). He also demonstrated that the amoebocyte (plasmatocyte of some authors) inclusions stained also with the Sudan B hypochlorite technique for bound lipid and free triglycerides (Wigglesworth, 1971). He concluded that in *Rhodnius* both amoebocyte inclusions and basement membrane contain protein, carbohydrate and lipid.

Since the connective tissues surrounding insect nerve cells are of particular physiological interest, stemming from their contribution to ionic regulation, they have been intensively studied. The composition of these specialized connective tissues, and the terminology applied to them, have been discussed by Ashhurst (1968). The extracellular neural lamella is composed of acid or neutral mucopolysaccharide and contains a fibrous component in most insects. It has been shown by electron microscopy, and by the detection of hydroxyproline in hydrolysates by chromatographic analysis, that collagen is present in the neural lamella of several species. The neural lamella is apparently formed by subjacent perineurial (glial) cells in cockroaches (Scharrer, 1939, 1972) in *Rhodnius* (Wigglesworth, 1956) and in *Schistocerca* (Ashhurst, 1965). In *Galleria*, haemocytes which are termed adipohaemocytes (the plasmatocytes of some authors) are intimately applied to the neural lamella, but these appear to be engaged in phagocytosis rather than secretion (Ashhurst and Richards, 1964a; Pipa and Woolever, 1965), and the same conclusion can be drawn from the autoradiographic studies on *Galleria* of Shrivastava and Richards (1965). McLaughlin (1974) observes that in *Manduca* also the neural lamella degenerates during metamorphosis and is phagocytosed by adipohaemocytes. However, these haemocytes remain in the vicinity of the newly forming adult neural lamella and could also be involved in its formation. The composition of the neural lamella and associated connective tissue in cockroaches and locusts and in *Carausius*, as revealed by histochemistry, is somewhat variable. Chondroitin, dermatan and keratan sulphates are present, but the glial lacunar system contains only hyaluronic acid (Ashhurst and Costin, 1971b, 1971c). In the case of the collagenous reproductive connective tissue (basal layer) of the accessory glands ejaculatory duct in *Locusta*, which is not a typical basement membrane similar sulphated glycosaminoglycans are present, as also is neutral glycoprotein (Odhiambo, 1970; Ashhurst and Costin, 1971a). Unfortunately no histochemical analysis in comparable depth is available for typical basement membrane known to be secreted with haemocyte involvement. Application of the Alcian Blue 8GX techniques developed by Ashhurst and Costin to such membranes would be interesting. The positive PAS reaction observed after the standard procedure is attributed to reaction with polysaccharide groups of collagen molecules, as well as with neutral glycoproteins of the matrix (Ashhurst and Costin, 1971a). It should also be practicable to distinguish between collagenous fibres and elastic fibres, since the latter, but not the former, are stained by spirit-blue and aldehyde fuchsin techniques. Elastic fibres similar to vertebrate oxytalan fibres are present in many invertebrates, but insects have not been widely investigated (Elder, 1973). Elastic fibres have been reported in the orthopteran dorsal diaphragm by Nutting (1951), and occur in the pericardium of *Calliphora* (Crossley, unpublished observation).

The histochemistry of haemocyte inclusions has often indicated a carbohydrate content. Arnold and Salkeld (1967) report that in *Blaberus* both granules of granular haemocytes and spherules of spherule cells contain azurophilic, diastase-stable, PAS-positive content, and the cell types are considered to differ only in size and shape. Staining cut-off below pH 6.2 in the methylene blue procedure suggests the presence of a neutral mucopolysaccharide in these cells. Gupta and Sutherland (1967) examined the histochemistry of the spherule cells of *Periplaneta, Blaberus, Leucophaea*, and other orthopteroids, and concluded that the spherules contained acid mucopolysaccharide. They found the spherules to be PAS-negative, but the very brief incubation times used could have contributed to this negative reaction. Although the positive results obtained by Gupta *et al.* for spherules treated with Alcian Blue in 0.3 per cent acetic acid certainly suggest that mucopolysaccharide is present, this is not a specific test since under these conditions it is bound by phosphate groups of nucleic acids and proteins, as well as by the carboxyl and sulphate ester groups of glycosoaminoglycans (Ashhurst and Costin, 1971a).

The electron microscope provides another view of the involvement of haemocytes in connective tissue formation. Beaulaton (1968) presents ultrastructural evidence for the involvement of two types of haemocyte, termed adipohaemocytes and spherule cells, in tunica propria deposition around the prothoracic gland in *Leucophaea* and *Bombyx*. Haemocytes are figured infiltrating the tunica propria and extruding vacuolar contents, composed of 2 nm microfibrils, very similar to the material making up the tunica propria. This association occurs in *Bombyx* at the time when the tunica propria increases in diameter from 0.1 to 3.0 μm. Furthermore the adipohaemocytes contain material giving similar reactions to the tunica propria in histochemical tests; both are PAS-positive, lead tetra-acetate Schiff positive, Bauer negative, and Sudan negative. Vacuoles replete with tubular elements 15 nm in diameter arranged in bundles are proposed as an intermediate step in the synthesis of tunica propria microfibrillae. The Golgi apparatus is implicated in the packaging of this secretion in a fashion comparable with that of vertebrate cells secreting mucopolysaccharide, such as the goblet cells investigated by Neutra and Leblond (1966). The tunica propria is a form of basement membrane, but it is continuous with connective tissue strands. The latter contain large (30-35 nm diameter) electron-dense fibrils, lacking periodicity and of unknown origin (see below).

In *Rhodnius* Wigglesworth (1973) described how plasmatocytes settled on the surface of tissues and began active pinocytosis by formation of coated vesicles, which fused to form larger vesicles. Certain larger inclusions in the settled haemocytes contained parallel rods or tubules 15 nm in diameter. Such inclusions with tubular contents are figured during discharge from the plasmatocyte onto the surface of basement lamellae. A series of micrographs shows how the tubular structure of inclusions could transform into loose fibrous material comparable with the basement membrane itself. The tubular substructure in granules of haemocytes has been described repeatedly (see section 2) but there is little consensus about the size of the tubules or the physiological involvement of the parent cells. Hoffmann (1966a) has applied histochemistry to the cells (oenocytoids) of *Locusta* which contain granules with 32 nm tubular substructure. He finds them to be feebly PAS-positive, ninhydrin-Schiff positive, Sudan B negative, and to colour rose in Malachite pyronin, a colour stable to ribonuclease digestion.

Scharrer (1972) believes that tubular inclusion bodies of haemocytes are represented at the light microscope level by aldehyde-fuchsin and PAS-positive granules, which are highly refractile, and contain neutral mucopolysaccharide and mucoprotein comparable to that of the extracellular stroma. However, she also suggests that opaque bodies and tubule-containing bodies in haemocytes are parts of a developmental series with other intermediate forms, and that the development of the series culminates with the release of the vacuole contents into the haemolymph, "a process whose major effect seems to be the initiation of the clotting process". However, as suggested in discussion of haemostasis in section 6, the production of coagulins and of connective tissue may have biochemical features in common. At the present time our histochemical information does not enable us to distinguish between the cells involved in these activities. Wigglesworth (1973) points out that in *Drosophila* the mucoprotein glue produced by salivary glands (and used by the insect to cement the puparium to the substrate) also contains parallel rods of opaque material (Rizki, 1967).

Several authors have suggested that basement membrane is secreted by the enveloped tissue itself, rather than by wandering cells. Edwards *et al.* (1958) suggest that all cells are probably capable of secreting outside membranes. From a study of connective tissue in cyclorrhaphous diptera, Whitten (1962) concludes that "circumstantial evidence suggests active participation on the part of component cells of the tissue or organ in relation to which the connective tissue is secreted". However, in the case of *Sarcophaga* Whitten (1969, p. 773) adds that haemocytes which have undergone phases of phagocytosis and intracellular digestion subsequently produce connective tissue strands, but she does not provide evidence for this. Lai-Fook (1968) suggests that the epidermal cells of *Rhodnius* secrete basement membrane, perhaps with the involvement of hemidesmosomes, but without the participation of blood cells.

The production of collagen is another problem. Scharrer (1972) provides electron micrographs of *Leucophaea* haemocytes containing recognizable collagen substructure, both within vacuoles and at the interface of the haemocyte plasma membrane with cells of the prothoracic gland. Scharrer believes that insect collagen may be assembled intracellularly by haemocytes, probably from precursors formed in the excessively distended and interconnected cisternae of rough endoplasmic reticulum prominent in these cells. There is a precedent for this notion in the observation of Gupta that collagen of the rectal papillae in *Calliphora* originates within the endoplasmic reticulum of medullary cells (Ashhurst, 1968). Recent ultrastructural studies of connective tissue around the nerve cord and heart of *Calpodes* (Locke and Huie, 1972) have shown that in addition to a matrix material, and to 13-nm wide collagen fibrils with a macroperiod of about 66 nm, two other fibrous components are present. Fibrils less than 6 nm diameter, and broad 40-nm diameter fibres, are illustrated. The 40-nm fibres stain with phosphotungstic acid and react selectively for peroxidase, but do not show a repeat banding pattern. They are most abundant in connective tissue that is elastic. The origin of insect connective tissue fibres other than collagen is not known, but it may be significant that, in several species, haemocyte vacuoles contain fibrous or tubular components of comparable sizes (see section 2), and haemocytes could be involved in the formation of the fibrous as well as the matrix components of connective tissue.

It is abundantly clear that in insects the haemocytes perform not only some of the roles of vertebrate leucocytes, but also some of the roles of vertebrate fibroblasts.

11 Insect blood cells in synthesis, secretion and plasma homeostasis

Crustacean blood cells are involved in both metabolism and storage of polysaccharides. Electron micrographs of *Carcinus* blood cells reveal both α and β forms of glycogen, and preparations of similar cells for the light microscope give a PAS reaction that is sensitive to diastase. If crustacean blood is centrifuged to remove the cells, then hydrolysed, the plasma hydrolysate lacks glucose, indicating that glycogen is absent from the plasma. An analogy with vertebrate liver can be drawn, since crustacean blood cells contain a glucose-6-phosphatase, probably involved in glucose release from stored glycogen (Johnston *et al.*, 1971). In the insect *Bombyx* a hexose-1-phosphatase has been reported in the blood, but its distribution between cells and plasma was not determined. This enzyme forms part of a pathway from glycogen to blood glucose in insects. The essential difference from glycogen breakdown in mammalian liver is that phosphoglucomutase is apparently not involved, since all attempts to demonstrate this enzyme in silkworm blood have given negative results (Faulkner, 1955). Blood cell glycogen not only represents a store of glucose as an energy reserve, but also a store of hexose units for chitin synthesis (Johnston and Davies, 1972). *Drosophila* larvae can regulate the osmotic pressure and glucose concentration of their haemolymph when placed in hypo- or hypertonic media, but the role of haemocytes in this presumptive homeostasis is not known (Zwicky, 1954). Certain insect haemocytes contain glycogen at some stages of development, as evidenced by diastase-sensitive PAS reaction (e.g. oenocytoids of *Galleria*; Ashhurst and Richards, 1964b), and by electron microscopy (Fig. 7). Babers (1941) assayed the distribution of glycogen between haemocytes and plasma in *Prodenia* larvae fed on a glucose-starch paste. He found 16.5 mg glycogen per 100 ml blood in the cell fraction, but only 0.92 mg glycogen per 100 ml blood in the plasma, and concluded that most, if not all, the glycogen came from the haemocytes.

Since fluctuating demand on blood sugars is a feature of insect physiology, a homeostatic mechanism would appear to be mandatory to maintain osmotic equilibrium. In flies carbohydrate appears to serve as the primary flight fuel (Sacktor, 1965), whilst haemolymph glucose is known to be incorporated into the chitin of growing cuticle in several insect orders (Bade and Wyatt, 1962). Insect blood cells may well cooperate with fixed tissues such as fat body in blood carbohydrate regulation, if the pointers provided by work on Crustacea are significant.

Fat-droplets are formed in *Prodenia* haemocytes as glycogen disappears from the cells (Munson and Yeager, 1944), and there is evidence for many insects that lipids accumulate in certain haemocytes during growth. In *Ephestia* neutral fats accumulate in spheroidocytes just before pupation, and the same cells increase in number during starvation and bacterial infection when energy reserves are being depleted, indicating a haemocyte role in fat metabolism (Arnold, 1952). In *Galleria* similar haemocytes (adipo-haemocytes) also gradually fill with lipid during the prepupal stage (Ashhurst and Richards, 1964b). In both *Ephestia* and *Galleria* the cells involved are phagocytic, and the lipids are probably formed in the lysosomal vacuolar apparatus.

Although Cuénot (1891, p. 316) remarked that phagocytic haemocytes (in *Gryllus*) are absolutely incapable of digesting fat-droplets, these do disappear from blood cells. Much of the lipid formed in haemocytes appears to be released directly into the plasma (Fig. 9), although tracer studies are lacking. It has been suggested that lipid could be released from blood cells amongst fixed tissues, where a direct nutritional effect is postulated (Zeller, 1938). Whitten (1969) describes the life-history of phagocytes trapped amongst fixed tissues in *Sarcophaga* puparia. Digestion of fragments of larval cells generates a large number of osmiophilic globules in these blood cells. The globules subsequently become reduced in number whilst adjacent epidermal cells accumulate similar globules, providing circumstantial evidence of direct nutrient transfer from haemocyte to epidermis. In *Rhodnius* larvae, when the basement membrane is complete and deposition of new cuticle is beginning, oenocytoids come into intimate contact with the plasma membrane of epidermal cells and enlarge. The inclusions of these haemocytes appear to contain liquid, and to be in the process of liberation, although their nature is not known (Wigglesworth, 1973).

Insect blood contains very high levels of free amino acids. In third instar *Phormia* larvae, 60 per cent of the total free amino acids is localized in the haemolymph and 40 per cent in other tissues. The distribution of individual amino acids is very uneven, but in general glutamate is concentrated in tissues whilst nearly all glutamine is in the haemolymph (Levenbook, 1966). The distribution of amino acids between cells and plasma was not investigated by earlier workers, but has recently assumed importance as a result of the proposal that glutamate acts as a transmitter at the insect neuromuscular junction. It has been suggested that the measured levels of glutamate in the haemolymph of the locust would be sufficient to activate excitatory synapses (Miller *et al.*, 1973). Thus it has been postulated that the neuromuscular junctions of insects are protected in some way from high levels of free amino acids. Enclosure of free amino acids within haemocytes would partition these away from the post-synaptic membrane, but measurements on cockroaches have indicated that only 12 per cent of the total free amino acid pool was bound to haemocytes in this insect (Holden, 1973). In experiments on *Calliphora* when particular precautions were taken to prevent haemocyte disruption during preparation of samples, over 60 per cent of the dicarboxylic amino acids glutamate and aspartate appeared to be sequestered into the haemocyte fraction. When glutamate was injected directly into *Calliphora* haemolymph, the concentration of this amino acid within haemocytes rapidly increased, indicating that haemocytes were capable of selectively accumulating and retaining amino acids against a concentration gradient. This observation suggests that insect haemocytes may have a homeostatic function with respect to plasma amino acids (Evans and Crossley, 1974).

In some insects, haemolymph amino acid homeostasis can, however, be overcome. Extirpation of lepidopteran silk glands causes accumulation in the haemolymph of massive amounts of several amino acids that are constituents of silk or their metabolic precursors (e.g. glutamine, threonine, glycine). Free amino acids of haemolymph are major precursors for silk synthesis and a delicate balance exists between secretion of amino acids into the haemolymph and withdrawal from it (Wyatt, 1964).

Haemolymph amino acids are presumably available for blood-protein synthesis in cells, particularly of the fat body, but there is little concrete evidence of protein synthesis in haemocytes, even though this was envisaged by Cuénot (1891). Coles (1965) found for *Rhodnius* that most haemolymph proteins corresponded to fat body proteins in electrophoresis gels. However, Coles suggested that some of the minor protein bands, which were not detectable in fat body but which increased after injury of the insect, could be formed in haemocytes. However, injury may activate nonhaemocytic defence reactions involving protein synthesis and stronger evidence is needed.

Haemocytes have been implicated in the metabolism of moulting hormones. The initial indication came from the observation of Wigglesworth (1955) that loading of phagocytic haemocytes by the injection of Indian ink or iron saccharate into 4th stage larvae of *Rhodnius* caused a delay in moulting of one to four weeks. Similar injections caused much smaller delays if made later than the third day after feeding, indicating that they did not interfere with ordinary processes of intermediary metabolism. Furthermore moulting induced by injection of ecdysone is not delayed by loading haemocytes. Wigglesworth concluded that haemocytes in *Rhodnius* intervened in some way between the brain and the site of ecdysone synthesis. Further support for this notion came from Wigglesworth's (1956a) report that plasmatocytes of *Rhodnius* show definite signs of secretory activity at the time when the hormone of the prothoracic gland is being produced. However, Jones (1965) was unable to detect a change in the circulating haemocytes of *Rhodnius* at the time when moulting hormone is produced.

Classical histochemical methods for neurosecretory products such as the paraldehyde-fuchsin, paraldehyde-thionin, or performic acid-Victoria-blue techniques give positive indications in some haemocytes (Dogra, 1970). As these methods probably stain the associated carrier proteins rather than hormones (Scharrer and Scharrer, 1954), and the specificity of the protein staining can be questioned, it must be concluded that evidence for haemocyte involvement in hormone metabolism remains unsatisfactory.

Wigglesworth (1963) has reported that the juvenile hormone analogue farnesyl methyl ether passes through haemocytes following experimental injection in oil into *Rhodnius* haemolymph, but this may simply represent endocytosis of a foreign substance by haemocytes. It is now believed that the water-insoluble juvenile hormone is carried on a haemolymph lipoprotein (Emmerich, 1973) but the cytophysiology of this carrier has yet to be explored.

Acknowledgements

The author would like to thank Dr Darcy Gilmour, Mr Robin Callard, and Mr Philip Peake for critically reading parts of the manuscript. The author gratefully acknowledges the financial support of the Australian Research Grants Committee, and the hospitality of the Department of Zoology in The University of Cambridge, during the preparation of this review.

References

Abercrombie, M., Heaysman, J. E. M. and Pegrum, S. M. (1970a). The locomotion of fibroblasts in culture. I. Movements of the leading edge. *Expl. Cell. Res.* **59**, 393-398.

Abercrombie, M., Heaysman, J. E. M. and Pegrum, S. M. (1970b). The locomotion of fibroblasts in culture. II. "Ruffling". *Expl. Cell. Res.* **60**, 437-444.

Abercrombie, M., Heaysman, J. E. M. and Pegrum, S. M. (1971). The locomotion of fibroblasts in culture. IV. Electron microscopy of the leading lamella. *Expl. Cell. Res.* **67**, 359-367.

Adam, H. (1966). Die hämocytären Abwehrreaktionen des Blutes von *Strongylogaster xanthoceros* (Stephens) und *Strongylogaster lineata* (Christ) gegen die endoparasitische Ichneumonide *Mesolius niger* (Gravenhorst). *Beitr. Ent.* **15**, 893-965.

Åkesson, B. (1953). Observations on the haemocytes during the metamorphosis of *Calliphora erythrocephala. Ark. Zool.* **6**, 203-211.

Ambrose, E. J. (1961). The movements of fibrocytes. *Expl. Cell. Res.* **8**, 54-72.

Andrew, W. (1961). Phase microscope studies of living blood cells of the Tunicates under normal and experimental conditions, with a description of a new type of motile cell appendage. *Q. Jl. microsc. Sci.* **102**, 89-105.

Arnold, J. W. (1952). Effects of certain fumigants on haemocytes of the Mediterranean flour moth, *Ephestia kuhniella* Zell. (Lepidoptera: Pyralididae). *Can. J. Zool.* **30**, 365-374.

Arnold, J. W. (1959a). Observations on living haemocytes in wing veins of the cockroach *Blaberus giganteus* (L) (Orthoptera: Blattidae) *Ann. ent. Soc. Am.* **52**, 229-235.

Arnold, J. W. (1959b). Observations on amoeboid motion of living haemocytes in the wing veins of *Blaberus giganteus* (L) (Orthoptera: Blattidae). *Can. J. Zool.* **37**, 371-375.

Arnold, J. W. (1961). Further observations on amoeboid haemocytes in *Blaberus giganteus* (L) (Orthoptera: Blattidae). *Can. J. Zool.* **39**, 756-765.

Arnold, J. W. and Salkeld, E. H. (1967). Morphology of the haemocytes of the giant cockroach *Blaberus giganteus*, with histochemical tests. *Can. Ent.* **99**, 1138-1145.

Arvy, L. (1953). Contribution a l'étude de la leucopoiese chez quelques Diptères. *Bull. Soc. zool. Fr.* **78**, 158-171.

Ashhurst, D. E. (1965). The connective tissue sheath of the locust nervous system: its development in the embryo. *Q. Jl. microsc. Sci.* **106**, 61-73.

Ashhurst, D. E. (1968). The connective tissues of insects. *Ann. Rev. Ent.* **13**, 45-74.

Ashhurst, D. E. and Costin, N. M. (1971a). Insect mucosubstances. I. The mucosubstances of developing connective tissue in the locust *Locusta migratoria. Histochem. J.* **3**, 279-295.

Ashhurst, D. E. and Costin, N. M. (1971b). Insect mucosubstances. II. The mucosubstances of the central nervous system. *Histochem. J.* **3**, 297-310.

Ashhurst, D. E. and Costin, N. M. (1971c). Insect mucosubstances. III. Some mucosubstances of the nervous system of the wax-moth (*Galleria mellonella*) and the stick insect *Carausius morosus. Histochem. J.* 3, 379-387.

Ashhurst, D. E. and Richards, A. G. (1964a). A study of the changes occurring in the connective tissue associated with the nervous system during the pupal stage of the wax-moth, *Galleria mellonella. J. Morph.* 114, 225-236.

Ashhurst, D. E. and Richards, A. G. (1964b). Some histochemical observations on the blood cells of the wax moth *Galleria mellonella.* L. *J. Morph.* 114, 247-254.

Ashida, M. (1971). Purification and characterization of pre-phenoloxidase from hemolymph of the silkworm *Bombyx mori. Archs. Biochem. Biophys.* 144, 749-762.

Babers, F. H. (1941). Glycogen in *Prodenia eridania*, with special reference to the ingestion of glucose. *J. Agric. Res.* 62, 509-530.

Bade, M. L. and Wyatt, G. R. (1962). Metabolic conversions during pupation of the Cecropia silkworm. I. Deposition and utilization of nutrient reserves. *Biochem. J.* 83, 470-477.

Baerwald, R. J. and Boush, G. M. (1970). Fine-structure of the haemocytes of *Periplaneta americana* (Orthoptera; Blattidae) with particular reference to marginal bundles. *J. Ultrastruct. Res.* 31, 151-161.

Bahadur, J. and Pathak, J. P. M. (1971). Changes in the total haemocyte counts of the bug *Halys dentata*, under certain specific conditions. *J. Insect Physiol.* 17, 329-334.

Bainton, D. F. (1973). Sequential degranulation of the two types of polymorphonuclear leucocyte granules during phagocytosis of microorganisms. *J. Cell Biol.* 58, 249-264.

Bainton, D. F. and Farquahar, M. G. (1966). Origin of granules in polymorphonuclear leucocytes: Two types derived from opposite faces of the Golgi complex in developing granulocytes. *J. Cell Biol.* 28, 277-301.

Bakula, M. (1970). Antibacterial compounds in the cell-free haemolymph of *Drosophila melanogaster. J. Insect Physiol.* 16, 185-197.

Bakula, M. (1971). The isolation of intracellular antibacterial activity from *Drosophila melanogaster* larvae. *J. Insect Physiol.* 17, 313-320.

Bang, F. B. (1970). Cellular aspects of blood clotting in the sea star and the hermit crab. *J. Reticuloendoth. Soc.* 7, 161-172.

Bang, F. B. (1971). A factor in crab amebocytes which stimulates *in-vitro* clotting of crab blood. *J. Invert. Pathol.* 18, 280-283.

Barritt, L. C. and Birt, L. M. (1971). Development of *Lucilia cuprina:* correlation of biochemical and morphological events. *J. Insect Physiol.* 17, 1169-1183.

Beard, R. L. (1948). Blood coagulation in an insect (*Galleria mellonella*) Lepid. *Anat. Rec.* 101, 736-773.

Beard, R. L. (1949). Physiological effects of induced haemorrhage in Japanese Beetle Larvae. *J. N.Y. ent. Soc.* 57, 79-91.

Beard, R. L. (1950). Experimental observations on coagulation of insect haemolymph. *Physiol. Zoöl.* 23, 47-57.

Beaulaton, J. (1968). Étude ultrastructurale et cytochimique des glandes prothoraciques de vers à soie aux quatrième et cinquième âges larvaires. I. La tunica propria et ses relations avec les fibres conjunctives et les hémocytes. *J. Ultrastruct. Res.* 23, 474-498.

Behnke, O. (1970). A comparative study of microtubules of disk-shaped blood cells. *J. Ultrastruct. Res.* 31, 61-75.

Belden, D. A. and Cowden, R. R. (1971). Detection of early changes in cockroach haemocytes during coagulation with 8-Anilino-1-Naphthalene sulfonic acid. *Experientia*, 27/4, 448-449.

Bennett, H. S. (1956). The concepts of membrane flow and membrane vesiculation as mechanisms for active transport and pumping. *J. Biophysic. biochem. Cytol.* 2, 99-103.

Bensch, K., Marantz, R., Wisiewski, H. and Shelanski, M. (1969). Induction *in vitro* of microtubular crystals by *Vinca* alkaloids. *Science, N.Y.* 165, 495-496.

Berken, A. and Benacerraf, F. (1966). Properties of antibodies cytophilic for macrophages. *J. exp. Med.* 123, 119-127.

Bernheimer, A. W., Caspari, E. and Kaiser, A. D. (1955). Studies of antibody formation in caterpillars. *J. exp. Zool.* 119, 23-35.

Bernheimer, A. W. (1952). Hemagglutinins in caterpillar bloods. *Science,* 115, 150-151.

Bess, H. A. (1939). Investigations on the resistance of mealy-bugs (Homoptera) to parasitization by internal hymenopterous parasites, with special reference to phagocytosis. *Ann. ent. Soc. Am.* 32, 189-226.

Bettini, S. (1965). Acquired immune response of the house fly *Musca domestica* (L) to injected venom of the spider *Latrodectus mactans tredecimguttatus* (Rossi). *J. Invert. Path.* 7, 378-383.

Bhagvat, K. and Richter, D. (1938). Animal phenolases and adrenaline. *Biochem. J.* 32, 1397-1406.

Birbeck, M. S. C. (1963). Electron microscopy of melanocytes: the fine structure of hair-bulb premelanosomes. *Ann. N.Y. Acad. Sci.* 100, 540-547.

Bloom, W. and Fawcett, D. W. (1968). "A Textbook of Histology". Saunders, Philadelphia.

Bodine, J. H. (1945). Tyrosinase and phenols. Action of diversely activated tyrosinase on monohydric and *o*-dihydric phenols. *Proc. Soc. exp. Biol. Med. N.Y.* 58, 205-209.

Bodine, J. H. and Allen, T. H. (1938). Enzymes in ontogenesis (Orthoptera). IV. Natural and artificial conditions governing the action of tyrosinase. *J. cell. comp. Physiol.* 11, 409-423.

Bogojavlensky, K. S. (1932). The formed elements of the blood of insects. *Arch. Russes. Anat. Hist. Embryol.* 11, 361-387.

Bohm, G. A. (1968). Ecdysontiter und Blutbild bei *Calliphora vicina* R.D. unter normalen und experimentellen Bedingungen. *Mitt. D. Ent. Gesellschaft.* 27, 52-59.

Boler, R. K. (1969). Fine structure of canine Kupffer cells and their microtubule-containing cytosomes. *Anat. Rec.* 163, 483-496.

Boman, H. G. Nilsson, I. and Rasmuson, B. (1972). Inducible antibacterial defence system in *Drosophila. Nature,* 237, 232-234.

Born, G. V. R. (1970). Observations on the change in shape of blood platelets brought about by adenosine diphosphate. *J. Physiol., Lond.* 209, 487-511.

Bowen, T. J. and Kilby, B. A. (1953). Electrophoresis of Locust haemolymph. *Archs Intern. Physiol.* 61, 413-416.

Bowers, B. (1964). Coated vesicles in the pericardial cells of the aphid (*Myzus persicae* sulz.) *Protoplasma,* **59**, 351-367.

Brady, J. (1967). The relationship between blood ions and blood cell density in insects. *J. exp. Biol.* **47**, 313-326.

Brandt, P. W. and Pappas, G. D. (1960). An electron microscopic study on pinocytosis in ameba. I. The surface attachment phase. *J. biophys. biochem. Cytol.* **8**, 675-687.

Brehélin, M. M. (1971). La coagulation de l'hémolymphe et ses modalités chez les larves de l'Orthoptère *Locusta migratoria* (phase grégaire). *C.r. hebd. Séanc. Acad. Sci., Paris,* **273**, 1598-1601.

Brehélin, M. M. (1972). Étude du mécanisme de la coagulation de l'hémolymphe d'un acridien; *Locusta migratoria migratorioides* (R and F). *Acrida,* **1**, 167-175.

Brewer, F. D. and Vinson, S. B. (1971). Chemicals affecting the encapsulation of foreign material in an insect. *J. Invert. Pathol.* **18**, 287-289.

Briggs, J. D. (1958). Humoral immunity in Lepidopterous larvae. *J. exp. Zool.* **138**, 155-188.

Brunet, P. C. J. (1963). Tyrosine metabolism in insects. *Ann. N.Y. Acad. Sci.* **100**, 1020-1034.

Buck, J. (1953). Physical properties and chemical composition of insect blood. *In* "Insect Physiology" (Ed. K. D. Roeder), pp. 147-190. John Wiley, New York.

Bullock, H. R. and Steinhauer, A. L. (1970). Natural precipitating proteins in insect haemolymph. *J. Insect Physiol.* **16**, 595-600.

Burnet, F. M. (1968). Evolution of the immune process in vertebrates. *Nature, Lond.* **218**, 426-430.

Cameron, G. R. (1934). Inflammation in the caterpillars of Lepidoptera. *J. Path. Bact.* **38**, 441-446.

Carter, S. B. (1966). Principles of cell motility: the direction of cell movement and cancer invasion. *Nature, Lond.* **208**, 1183-1187.

Carton, Y. (1969). Données récents sur les phénomènes d'immunité humorale acquise chez les invertébrés. *Année Biologique,* **8**, 657-682.

Caspari, E. (1933). Über die Wirkung eines pleiotropen Gens bei der Mehlmotte *Ephestia kühniella* (Zeller). *Arch. EntwMech. Org.* **130**, 358-381.

Cassier, P. and Fain-Maurel, M. A. (1968). Sur la présence de microtubules dans l'ergastoplasm et l'espace périnucléaire des oenocytoids du criquet migrateur *Locusta migratoria migratorioides* (Ret F.) *C.r. hebd. Séanc. Acad. Sci., Paris,* **266**, 686-689.

Chadwick, J. S. (1967). Serological responses of insects. *Fedn Proc. Fedn Am. Socs exp. Biol.* **26**, 1675-1679.

Chadwick, J. S. and Vilk, E. (1969). Endotoxins from several bacterial species as immunizing agents against *Pseudomonas aeruginosa* in *Galleria mellonella. J. Invert. Pathol.* **13**, 410-415.

Chadwick, J. S. (1970). Relation of lysosome concentration to acquired immunity against *Pseudomonas aeruginosa* in *Galleria mellonella.* J. Invert. Pathol. **15**, 455-456.

Chadwick, J. S. (1971). Effects of immunization on the LD_{50} of two pathogens of *Galleria mellonella. J. Invert Pathol.* **18**, 117-120.

Coles, G. C. (1965). The haemolymph and moulting in *Rhodnius prolixus* Stål. *J. Insect Physiol.* **11**, 1317-1323.

Crossley, A. C. S. (1964). An experimental analysis of the origins and physiology of haemocytes in the blue blow-fly *Calliphora erythrocephala* (Meig.) *J. exp. Zool.* **157**, 375-398.

Crossley, A. C. S. (1965). Transformations in the abdominal muscles of the blue blowfly, *Calliphora erythrocephala* (Meig.) during metamorphosis. *J. Embryol. exp. Morph.* **14**, 89-110.

Crossley, A. C. S. and Waterhouse, D. F. (1969). The ultrastructure of the osmeterium and the nature of its secretion in *Papilio* larvae (Lepidoptera). *Tissue and Cell*, **1**, 525-545.

Crossley, A. C. S. (1968). The fine structure and mechanism of breakdown of larval intersegmental muscles in the blowfly *Calliphora erythrocephala* (Meig.) *J. Insect. Physiol.* **14**, 1389-1407.

Crossley, A. C. S. (1972a). Ultrastructural changes during transition of larval to adult intersegmental muscle at metamorphosis in the blowfly *Calliphora erythrocephala.* I. Dedifferentiation and myoblast fusion. *J. Embryol. Exp. Morph.* 27, 43-74.

Crossley, A. C. S. (1972b). The ultrastructure and function of pericardial cells and other nephrocytes in an insect: *Calliphora erythrocephala. Tissue and Cell*, **4**, 529-560.

Crossley, A. C. S. (1975). The histochemistry, ultrastructure, population dynamics and function of phenol-oxidising haemocytes in an insect: *Calliphora.* In press.

Cuénot, L. (1891). Etudes sur le sang et les glandes Lymphatiques dans le série animale. *Archs. Zool. exp. gén.* 2es, 9, 365-475 and 592-670.

Cuénot, L. (1896). Études physiologiques sur les Orthoptères. *Archs. Biol.* **XIV**, 293-341.

Cuénot, L. (1897). Les globules sanguins et les organes lymphoides des invertébrés. *Archs Anat. micros. Morph. exp.* **1**, 153-192.

Dakhnoff, A. (1938). La régeneration du sang chez *Galleria mellonella* après une pris de sang. *C.r. Séanc. Soc. Biol.* **128**, 520-523.

Dawe, C. J., Morgan, W. D. and Slatick, M. S. (1967). Cellular response of a cockroach *Leucophaea maderae* to transplants of cell culture lines of vertebrates. *Fedn Proc. Fedn Am. Socs exp. Biol.* **26**, 1698-1706.

De Duve, C. (1963). "CIBA Symposium on Lysosomes" (Eds. A. de Reuck and M. Cameron), p. 126. Churchill, London.

De Duve, C. and Wattiaux, R. Functions of lysosomes. *A. Rev. Physiol.* **28**, 435-

Dennell, R. (1947). A study of the insect cuticle: the formation of the puparium of *Sarcophaga falculata. Proc. R. Soc. Lond. B.* **134**, 79-110.

De Petris, S. and Karlsbad (Gioanna) (1965). Localization of antibodies by electron microscopy in developing antibody producing cells. *J. Cell Biol.* **26**, 759-778.

Devauchelle, G. (1971). Étude ultrastructurale des hémocytes du Coléoptère *Melolontha melolontha* (L). *J. Ultrastruct. Res.* **34**, 492-516.

Devauchelle, G., Bergoin, M. and Vago, C. (1971). Étude ultrastructurale du cycle de replication d'un entomopoxvirus dans les hémocytes de son hôte. *J. Ultrastruct. Res.* 37, 301-321.

Devauchelle, G., Meynadier, G. and Vago, C. (1972). Étude ultrastructurale du cycle de multiplication de *Ricketsiella melolonthae* (Krieg), Philip. dans les hemocytes de son hôte. *J. Ultrastruct. Res.* **38**, 134-148.

Dogra, G. S. (1970). Functional significance of the neurosecretory material in the haemocytes of the adult female *Gryllotalpa africana*, Beauvois (Orthoptera: Gryllotalpidae). *Anat. Anz.* **126**, 355-362.

Dumont, J. N., Anderson, E. and Winner, G. (1966). Some cytologic characteristics of the hemocytes of *Limulus* during clotting. *J. Morph.* **119**, 181-208.

Dupraw, E. J. (1965). The organization of honey-bee embryonic cells. *Dev. Biol.* **12**, 53-71.

Edwards, G. A., Ruska, H. and De Harven, E. (1958). Electron microscopy of peripheral nerves and neuromuscular junctions in the wasp leg. *J. Biophys. biochem. Cytol.* **4**, 107-114.

Elder, H. Y. (1973). Distribution and functions of elastic fibers in the invertebrates. *Biol. Bull. mar. biol. Lab., Woods Hole,* **144**, 43-63.

Emmerich, H. and Hartmann, R. (1973). A carrier lipoprotein for juvenile hormone in the haemolymph of *Locusta migratoria. J. Insect physiol.* **19**, 1663-1677.

Ermin, R. (1939). Über bau und function der Lymphocyten bei Insekten *Periplaneta americana. Z. Zellforsch. mikrosk. Anat.* **29**, 613-669.

Evans, J. J. T. (1968). The distribution of prophenoloxidase and its activator in pupae of the Chinese oak silkmoth *Antheraea pernyi. J. Insect Physiol.* **14**, 107-119.

Evans, P. D. and Crossley, A. C. (1974). Free amino acids in the haemocytes and plasma of the larva of *Calliphora vicina. J. Exp. Biol.* **61**, 463-472.

Faulkner, P. (1955). A hexose-1-phosphatase in silkworm blood. *Biochem. J.* **60**, 590-596.

Fawcett, D. W. (1965). Surface specializations of absorbing cells. *J. Histochem. Cytochem.* **13**, 75-91.

Feir, D. and McClaine, E. (1968a). Mitotic activity of the circulating hemocytes of the large milkweed bug *Oncopeltus fasciatus. Ann. ent. Soc. Am.* **61**, 413-416.

Feir, D. and McClaine, E. (1968b). Induced changes in the mitotic activity of hemocytes of the large milkweed bug *Oncopeltus fasciatus. Ann. ent. Soc. Am.* **61**, 416-421.

Feir, D. and Pantle, C. R. (1971). *In vitro* studies of insect haemocytes. *J. Insect Physiol.* **17**, 733-773.

Feir, D. and Schmidt, G. T. (1968). Detection of antibody and antigen in the milkweed bug *Oncopeltus fasciatus. Experientia,* **24**, 1217-1218.

Feir, D. and Walz, M. A. (1964). An agglutinating factor in insect hemolymph. *Ann. ent. Soc. Am.* **57**, 388.

Finlayson, L. H. and Hamer, D. (1949). Free amino-acids in the haemolymph of *Calliphora erythrocephala* Meigen. *Nature, Lond.* **163**, 843-844.

Florkin, M. and Jeuniaux, C. (1964). Hemolymph composition. *In* "The Physiology of Insecta" (Ed. M. Rockstein), Vol. 3, pp. 110-152. Academic Press, New York and London.

Franke, H. (1960a). Licht-und elektronenmikroskopische Untersuchungen über die Blutgerinnung bei *Periplaneta orientalis. Zool. Jb. Abf. Allg. Zool. Physiol. Tiere.* **68**, 499-518.

Franke, H. (1960b). Licht-und elektronenmikroskopische Untersuchungen über die Blutgerinnung bei *Periplaneta orientalis*. *Zool. Jb. Abf. Allg. Zool. Physiol. Tiere.* **69**, 131-132.

Friend, D. S. and Farquhar, M. G. (1967). Functions of coated vesicles during protein absorption in the rat vas deferens. *J. Cell. Biol.* **35**, 357-376.

Gingrich, R. E. (1964). Acquired humoral immune response of the large milkweed bug *Oncopeltus fasciatus* (Dallas), to injected materials. *J. Insect Physiol.* **10**, 179-194.

Gilliam, M. and Shimanuki, H. (1970). Coagulation of hemolymph of the larval honey-bee (*Apis mellifera* L). *Experientia,* **26**, 908-990.

Good, R. A. and Papermaster, B. W. (1964). Ontogeny and Phylogeny of Adaptive immunity. *In* "Advances in Immunology" (Eds F. J. Dixon and J. H. Humphrey), Vol. 4, pp. 1-115. Academic Press, New York and London.

Graber, V. (1871). Über die Blutkörperchen der Insekten. Sitzungber. *Akad. Wiss. Wien,* **64**, 9-44.

Granados, R. R., Ward, L. S. and Maramorosch, K. (1968). Insect viremia caused by plant-pathogenic virus: electron microscopy of vector haemocytes. *Virology*, **34**, 790-796.

Grimes, G. and Barnes, F. S. (1973). A technique for studying chemotaxis of leucocytes in well defined chemotactic fields. *Expl. Cell Res.* **79**, 375-385.

Grégoire, Ch. (1953). Blood coagulation in Arthropods. III. Reactions of insect haemolymph to coagulation inhibitors of vertebrate blood. *Biol. Bull. mar. biol. Lab., Woods Hole,* **104**, 372-393.

Grégoire, Ch. (1955a). Blood coagulation in Arthropods. V. Studies on haemolymph coagulation in 420 species of insects. *Archs. Biol.* **66**, 103-143.

Grégoire, Ch. (1955b). Coagulation de l'hémolymph chez les insectes irradiés par les rayons X. *Archs. int. Physiol. Biochim.* **68**, 246-248.

Grégoire, Ch. (1964). Haemolymph coagulation. *In* "Physiology of Insecta" (Ed. M. Rockstein). Academic Press, New York and London.

Grégoire, Ch. (1970). Haemolymph coagulation in Arthropods. *Symp. zool. Soc. Lond.* 27, 45-74.

Grégoire, Ch. (1971). Haemolymph coagulation in Arthropods. *In* "Chemical Zoology" (Ed. M. Florkin), Vol. 6, pp. 145-186. Academic Press, New York and London.

Grégoire, Ch., Duchateau, Gh. and Florkin, M. (1949). La coagulation de l'hemolymph d'invertébré etudiée a l'aide du microscope électronique et du microscope a contraste de phases démonstration de micrographies. *Arch. Int. Physiol.* **57**, 117-119.

Grégoire, Ch. and Florkin, M. (1950a). Étude au microscope à contraste de phase du coagulocyte, du nuage granulaire et de la coagulation plasmatique dans la sang des Insectes. *Experientia,* **6**, 297-298.

Grégoire, Ch. and Florkin, M. (1950b). Blood coagulation in Arthropods. I. The coagulation of insect blood, as studied with the phase contrast microscope. *Physiol. Comp. Oecol.* **2**, 126-139.

Grégoire, Ch. and Tagnon, H. J. (1962). Blood coagulation. *In* "Comparative Biochemistry" (Eds Florkin and Mason), Vol. 4, pp. 453-482. Academic Press, New York and London.

Griffiths, D. C. (1961). The development of *Monoctonus paludum* Marshall (Hymenoptera: Braconidae) in *Nasonovia ribisnigri* on lettuce and immunity reactions in other lettuce aphids. *Bull. ent. Res.* 52, 147-163.

Grimes, G. and Barnes, F. S. (1973). A technique for studying chemotaxis of leucocytes in well defined chemotactic fields. *Exp. Cell Res.* 79, 375-385.

Grimstone, A. V., Rotherham, S. and Salt, G. (1967). An electron microscope study of capsule formation by insect blood cells. *J. Cell Sci.* 2, 281-292.

Gupta, A. P. (1969). Studies of the blood of Meloidae (Coleoptera). I. The haemocytes of *Epicauta cinerea* (Forster), and a synonymy of haemocyte terminologies. *Cytologia*, 34, 300-344.

Gupta, B. L., Mellon, D. and Treherne, J. E. (1969). The organisation of the central nervous connectives in *Anodonta cygnea* (L) (Mollusca: Eulamellabranchia). *Tissue and Cell*, 1, 1-30.

Gupta, A. P. and Sutherland, D. J. (1966). *In vitro* transformations of the insect plasmatocyte in some insects. *J. Insect Physiol.* 12, 1369-1375.

Gupta, A. P. and Sutherland, D. J. (1967). Phase contrast and histochemical studies of spherule cells in cockroaches (Dictyoptera). *Ann. ent. Soc. Am.* 60, 557-565.

Hackman, R. H. and Goldberg, M. (1967). The *O*-diphenoloxidases of fly larvae. *J. Insect Physiol.* 13, 531-544.

Hagopian, M. (1971). Unique structures in the insect granular haemocytes. *J. Ultrastruct. Res.* 36, 646-658.

Hardy, W. B. (1892). The blood corpuscles of the Crustacea, together with a suggestion as to the origin of the crustacean fibrin-ferment. *J. Physiol.* 13, 165-190.

Harshbarger, J. C. and Moore, B. S. (1966). Some effects of X-irradiation on larvae of *Galleria mellonella. J. Invert. Path.* 8, 277-279.

Harpax, I., Kislev, N. and Zelcer, A. (1969). Electron-microscope studies on haemocytes of the Egyptian cotton-worm *Spodoptera littoralis* (Boisduval) infected with a nuclear polyhedrosis virus, as compared to non-infected haemocytes. I. Non-infected haemocytes. *J. Invert. Path.* 14, 175-185.

Harvey, W. R. and Williams, C. M. (1961). The injury metabolism of the cecropia silkworm. I. Biological amplification of the effects of localized injury. *J. Insect Physiol.* 7, 81-99.

Heilbrunn, L. V. (1961). *In* "Functions of the Blood" (Eds MacFarlane and Robb-Smith), pp. 283-301. Academic Press, New York and London.

Heyneman, R. A. (1965). Final purification of a latent phenolase with mono- and diphenolase activity from *Tenebrio molitor. Biochem. biophys. Res. Commun.* 21, 162-169.

Hink, W. F. (1970). Immunity in insects. *Transplant Proc.* 2, 233-235.

Hink, W. F. and Briggs, J. D. (1968). Bactericidal factors in haemolymph from normal and immune wax moth larvae, *Galleria mellonella. J. Insect Physiol.* 14, 1025-1035.

Hoffmann, J. A. (1966a). Étude des oenocytoides chez *Locusta migratoria. J. Microscopie,* 5, 269-272.

Hoffmann, J. A. (1966b). Étude ultrastructurale de deux hémocytes à granules de *Locusta migratoria. C.r. hebd. Séanc. Acad. Sci., Paris,* 263, 521-524.

Hoffmann, J. A. (1969). Étude de la récupération hémocytaire après hémorragies expérimentales chez l'orthoptère *Locusta migratoria*. *J. Insect Physiol.* **15**, 1375-1384.

Hoffmann, J. A. (1970). Régulations endocrines de la production et de la différentiation des hémocytes chez un insect Orthoptère: *Locusta migratoria*. *Gen. comp. Endocr.* **15**, 198-219.

Hoffmann, J. A. (1972). Modifications of the haemogramme of larval and adult *Locusta migratoria* after selective X-irradiations of the haemocytopoietic tissue. *J. Insect Physiol.* **18**, 1639-1652.

Hoffmann, J. A., Porte, A. and Joly, P. (1968a). Présence d'un tissue hématopoiétique au niveau du diaphragme dorsal de *Locusta migratoria* (Orthoptère). *C.r. hebd. Séanc. Acad. Sci., Paris,* **266D**, 1882-1883.

Hoffmann, J. A., Porte, A. and Joly, P. (1968b). Sur la nature hématopoietique de "l'organe phagocytaire" (Cuenot) chez *Grillus bimaculatus* (Orthoptère: Ensifère). *C.r. hebd. Séanc. Acad. Sci., Paris,* **267D**, 776-777.

Hoffmann, J. A., Stoekel, M. E. Porte, A. and Joly, P. (1968). Ultrastructure des hemocytes de *Locusta migratoria* (Orthoptère). *C.r. hebd. Séanc. Acad. Sci., Paris,* **266**, 503-505.

Hoffmann, J. A., Porte, A. and Joly, P. (1970). On the localization of phenoloxidase activity in coagulation of *Locusta migratoria* (L. Orthoptera). *C.r. hebd. Séanc. Acad. Sci., Paris,* **270D**, 629-631.

Hoffmann, J. A. and Stoekel, M. E. (1968). Sur les modifications ultrastructurales des coagulacytes au cours de la coagulation de l'hémolymphe chez un insecte Orthopteroide: *Locusta migratoria. C.r. Séanc. Soc. Biol. Strasburg,* **162**, 2257-2259.

Hoffmann, J. A., Stoekel, M. E., Porte, A. and Joly, P. (1968). Ultrastructure des hémocytes de *Locusta migratoria* (Orthoptère). *C.r. hebd. Séanc. Acad. Sci., Paris,* **266**, 503-550.

Holden, J. S. (1973). Free amino acid levels in the cockroach *Periplaneta americana. J. Physiol.* 232, 61P-62P.

Hollande, A. C. (1911). Études histologiques comparée du sang des insectes a hémorrhée et des insectes sans hémorrhée. *Archs Zool. exp. gén.* 5es, **6**, 283-323.

Hollande, A. C. (1920). Oenocytoides et tératocytes du sang des chenilles. *C.r. hebd. Séanc. Acad. Sci., Paris,* **170**, 1341-1344.

Hollande, A. C. (1930). La digestion des Bacilles tuberculeux par les leucocytes du sang des chenilles. *Archs. Zool. Exp. gén.* **70**, 231-280.

Hollande, A. C. and Aghar, M. (1928). La phagocytose et la digestion de Bacille tuberculeux par les leucocytes du sang des chenilles autres que *Galleria mellonella. C.r. Séanc. Soc. Biol.* **99**, 120.

Huhn, D. (1967). Nuclear pockets in normal monocytes. *Nature, Lond.* **216**, 1240.

Inaba, T. and Funatsu, M. (1964). Studies on tyrosinase in the housefly. Pt. III. Activation of protyrosinase by natural activator. *Agric. Biol. Chem.* 28, 206-215.

Jacques, P. J. (1969). Endocytosis. *In* "Lysosomes in Biology and Pathology" (Eds Dingle and Fell), Vol. 2, pp. 395-420. North-Holland, Amsterdam.

Johnson, H. A., Schnappauf, H., Chanana, A. D. and Cronkite, E. P. (1966). Variability of ribosomal aggregation in lymphocytes. *Nature, Lond.* **211**, 420.

Johnston, M. A., Davies, P. S. and Elder, H. Y. (1971). Possible hepatic function for crustacean blood cells. *Nature, Lond.* **230**, 471-472.

Johnston, M. A. and Davies, P. S. (1972). Carbohydrates of the hepatopancreas and blood tissues of *Carcinus*. *Comp. Biochem. Physiol.* **41B**, 433-443.

Jones, J. C. (1950). The normal hemocyte picture of the yellow mealworm, *Tenebrio molitor* L. *Iowa St. J. Sci.* **24**, 355-361.

Jones, J. C. (1954a). A study of mealworm hemocytes with phase contrast microscopy. *Ann. ent. Soc. Am.* **47**, 308-315.

Jones, J. C. (1956). The hemocytes of *Sarcophaga bullata* (Parker). *J. Morph.* **99**, 233-257.

Jones, J. C. (1962). Current concepts concerning insect hemocytes. *Am. Zool.* **2**, 209-246.

Jones, J. C. (1964). The circulatory system of insects. *In* "Physiology of Insecta" (Ed. M. Rockstein). Academic Press, New York and London.

Jones, J. C. (1965). The hemocytes of *Rhodnius prolixus* Stål. *Biol. Bull. mar. biol. Lab., Woods Hole,* **129**, 282-294.

Jones, J. C. (1967). Normal differential counts of hemocytes in relation to ecdysis and feeding in *Rhodnius*. *J. Insect Physiol.* **13**, 1133-1141.

Jones, J. C. (1970). Hemocytopoiesis in insects. *In* "Regulation of Hematopoiesis" (Ed. A. S. Gordon), Vol. 1, pp. 7-65. Appleton-Century Crofts, New York.

Jones, J. C. and Liu, D. P. (1968). A quantitative study of mitotic divisions of hemocytes of *Galleria mellonella* larvae. *J. Insect Physiol.* **14**, 1055-1061.

Jones, J. C. and Liu, D. P. (1969). Effects of ligaturing on total hemocyte counts of *Galleria mellonella* larvae. *J. Insect Physiol.* **15**, 1703-1708.

Judy, K. J. and Marks, E. P. (1971). Effects of ecdysterone *in vitro* on hindgut and hemocytes of *Manduca sexta* (Lepidoptera). *Gen. comp. Endocr.* **17**, 351-359.

Kamon, E. and Shulov, A. (1961). Immunity response of African migratory locust to scorpion venom. *Bull. Res. Coun. Israel,* **9E**, 155-158.

Kamon, E. and Shulov, A. (1965). Immune response of locusts to venom of the scorpion. *J. Invert. Path.* **7**, 192-198.

Kaneseki, T. and Kadota, K. (1969). The "vesicle in a basket". A morphological study of the coated vesicle isolated from the nerve endings of the guinea pig brain, with special reference to the mechanism of membrane movements. *J. Cell Biol.* **42**, 202-220.

Karlson, P. and Liebau, H. (1961). Zum Tyrosinstoffwechsel der Insekten, V. Reindarstellung, Kristallisation und Substratspezifität der O-Diphenoloxydase aus *Calliphora erythrocephala. Hoppe-Seyler's Z. physiol. Chem.* **326**, 135-143.

Karlson, P., Mergen Hagen, D. and Sekeris, C. E. (1964). Zum tyrosinstoffwechsel der Insekten. XV. Weitore Untersuchungen über das O-diphenoloxidase-system von *Calliphora erythrocephala. Hoppe-Seyler's Z. physiol. Chem.* **338**, 42-50.

Karlson, P. and Sekeris, C. E. (1964). Comparative Biochemistry. Biochemistry of Insect Metamorphosis. Vol. 6, Chapter 4. Associated Press, New York.

Karlson, P. and Wecker, E. (1955). Die Tyrosinaseaktivät während der Pupariumbildung von *Calliphora erythrocephala. Hoppe-Seyler's Z. physiol. Chem.* **300**, 42-48.

Kessel, R. G. and Eichler, V. B. (1966). Microtubules in the microspikes and cortical cytoplasm of grasshopper embryonic cells. *J. Microscopie,* 5, 781-786.

Kerr, J. F. R., Wyllie, A. H. and Currie, A. R. (1972). Apoptosis: A basic biological phenomenon with wide ranging implications in tissue kinetics. *Br. J. Cancer,* **26**, 239-57.

Kislev, N., Harpaz, I. and Zelcer, A. (1969). Electron microscopic studies on hemocytes of the Egyptian cottonworm, *Spodoptera littoralis* (Boisduval) infected with a nuclear polyhedrosis virus as compared to non-infected hemocytes. II. Virus infected hemocytes. *J. Invert. Path.* **14**, 245-257.

Klebanoff, S. J. (1968). Myeloperoxidase–Halide–Hydrogen Peroxide antibacterial system. *J. Bact.* **95**, 2131-2138.

Koeppe, J. K. and Gilbert, L. I. (1973). Immunochemical evidence for the transport of haemolymph protein into the cuticle of *Manduca sexta. J. Insect Physiol.* **19**, 615-624.

Kollman, M. (1908). Recherches sur les leucocytes et le tissue lymphoide des invertebres. *Ann. Sci. nat. (Zool.)* **8**, 1-238.

Kurtti, T. J. and Brooks, M. A. (1970). Growth and differentiation of Lepidopteran myoblasts *in vitro. Expl. Cell Res.* **61**, 407-412.

Lackie, J. M. (1972). The course of infection and growth of *Moniliformis dubius* (Acanthocephala) in the intermediate host *Periplaneta americana, Parasitology,* **64**, 95-106.

Lai-Fook, J. (1966). The repair of wounds in the integument of insects. *J. Insect Physiol.* 12, 195-226.

Lai-Fook, J. (1968). The fine structure of wound repair in an insect *Rhodnius prolixus. J. Morph.* **124**, 37-78.

Lai-Fook, J. (1970). Haemocytes in the repair of wounds in an insect (*Rhodnius prolixus*). *J. Morph.* **130**, 297-314.

Lake, C. R., Mills, R. R. and Brunet, P. J. C. (1970). β-hydroxylation of tyramine by cockroach haemolymph. *Biochim. biophys. Acta,* **215**, 226-228.

Lange, H. H. (1932). Die Phagocytose bei Chironomiden. *Z. wiss. Biol.* B **16**, 753-805.

Lawrence, P. A. (1968). Mitosis and the cell cycle in the metamorphic moult of the milk-weed bug, *Oncopeltus fasciatus.* A Radioautographic Study. *J. Cell Sci.* 3, 391-404.

Lazarenko, T. (1925a). Histological observations on healing of integument wounds in insects. *Bull. Biol. Res. Perm. Univ.* 2, 287-398.

Lazarenko, T. (1925b). Beiträge zur vergleichenden Histologie des Blutes und des Bindegewebes. *Z. mikrosk.-anat. Forsch.* 3, 409-499.

Lea, M. S. and Gilbert, L. I. (1961). Cell division in diapausing silkworm pupae. *Am. Zool.* 1, 368-369.

Lea, M. S. and Gilbert, L. I. (1966). The hemocytes of *Hyalophora cecropia* (Lepidoptera). *J. Morph.* **118**, 197-215.

Leutenegger, R. (1967). Early events of SIV infection in hemocytes of *Galleria mellonella* (L). *Virology,* **32**, 109-116.

Levenbrook, L. (1966). Hemolymph amino acids and peptides during larval growth of the blowfly *Phormia regina. Comp. Biochim. Physiol.* 18, 341-351.

Levenbrook, L. and Hollis, V. W. (1961). Organic acid in insects. I. Citric Acid. *J. Insect Physiol.* 6, 52-61.

Lewis, W. J. and Vinson, S. B. (1968). Immunological relationships between the parasite *Cardiochiles nigriceps*; (Vierick) and certain *Heliothis* species. *J. Insect Physiol.* 14, 613-626.

Locke, M. (1969a). The structure of an epidermal cell during the development of the protein epicuticle and the uptake of the molting fluid in an insect. *J. Morph.* 127, 7-40.

Locke, M. (1969b). The ultrastructure of the oenocytes in the molt/intermolt cycle of an insect. *Tissue and Cell,* 1, 103-154.

Locke, M. and Collins, J. (1965). The structure and formation of protein granules in the fat-body of an insect. *J. Cell Biol.* 26, 857-884.

Locke, M. and Huie, P. (1972). The fiber components of insect connective tissue. *Tissue and Cell,* 4, 601-612.

Locke, M. and Krishnan, N. (1971). The distribution of phenoloxidases and polyphenols during cuticle formation. *Tissue and Cell,* 3, 103-126.

Locke, M. and McMahon, J. T. (1971). The origin and fate of microbodies in the fat body of an insect. *J. Cell Biol.* 48, 61-78.

Lockshin, R. A. (1969a). Programmed cell death. Activation of lysis by a mechanism involving the synthesis of protein. *J. Insect Physiol.* 15, 1505-1516.

Lockshin, R. A. (1969b). Lysosomes in insects. *In* "Lysosomes in Biology and Pathology" (Eds Dingle and Fell), Vol. 1, pp. 363-391. North-Holland, London.

Lockshin, R. A. and Williams, C. M. (1964). Programmed cell death. II. Endocrine potentiation of the breakdown of intersegmental muscles of silkmoths. *J. Insect Physiol.* 10, 643-649.

Lockshin, R. A. and Williams, C. M. (1965a). Programmed cell death. I. Cytology of degeneration of the intersegmental muscles of silkmoths. *J. Insect Physiol.* 11, 123-133.

Lockshin, R. A. and Williams, C. M. (1965b). Programmed cell death. III. Neural control of the breakdown of the intersegmental muscles of silkmoths. *J. Insect Physiol.* 11, 601-610.

Lockshin, R. A. and Williams, C. M. (1965c). Programmed cell death. IV. The influence of drugs on the breakdown of the intersegmental muscles of silkmoths. *J. Insect Physiol.* 11, 803-809.

Lockshin, R. A. and Williams, C. M. (1965d). Programmed cell death. V. Cytolytic enzymes in relation to the breakdown of the intersegmental muscles of silkmoths. *J. Insect Physiol.* 11, 831-844.

Loeb, L. (1903). On the coagulation of the blood of some Arthropods and on the influence of traction on the protoplasm of the blood cells of Arthropods. *Biol. Bull. mar. biol. Lab., Woods Hole,* 4, 301-318.

McKay, D. and Jenkin, C. R. (1970). Immunity in the invertebrates. The role of serum factors in phagocytosis of erythrocytes by haemocytes of the fresh water crayfish *Parachaeraps bicarinatus. Aust. J. exp. Biol. med. Sci.* 43, 139-150.

McLaughlin, B. J. (1974). The accessibility of a developing lepidopteran nervous system to Lanthanum and peroxidase. *J. Cell. Sci.* In press.

Maier, W. A. (1969). Die Hämocyten der Larve von *Chironomus thummi* (Dipt.). *Z. Zellforsch. mikrosk. Anat.* **99**, 54-63.

Maier, W. A. (1973). Die phenoloxidase von *Chironomus thummi* und ihre beeinflussyng durch parasitäre mermithiden. *J. Insect Physiol.* **19**, 85-95.

Marschall, K. J. (1966). Bau und Functionen der Blutzellen des Mehlkäfers *Tenebrio molitor* (L). *Z. Morph. Ökol. Tiere*, **58**, 182-246.

Metalnikov, S. (1924). Phagocytose et réactions des cellules dans l'immunité. *Annls Inst. Pasteur, Paris*, **38**, 787-826.

Metchnikoff, E. (1905). Immunity in Infective Diseases. Cambridge University Press, London.

Miller, V. H., Ballback, R. S., Pauley, G. B. and Krassner, S. M. (1972). A preliminary physicochemical characterisation of an agglutinin found in the hemolymph of the crayfish *Procambarus clarkii*, *J. Invert. Path.* **19**, 83-93.

Miller, R., Leaf, G. and Usherwood, P. N. R. (1973). Blood glutamate in Arthropods. *Comp. Biochem. Physiol.* **44A**, 991-996.

Mills, R. R. Androuny, S. and Fox, F. R. (1968). Correlation of phenoloxidase activity with ecdysis and tanning hormone release in the American cockroach. *J. Insect Physiol.* **14**, 603-611.

Mills, R. R. and Whitehead, D. L. (1970). Hormonal control of tanning in the American cockroach: changes in blood cell permeability during ecdysis. *J. Insect Physiol.* **16**, 331-340.

Misko, I. S. (1973). Thesis, Australian National University.

Mitchison, J. M. (1971). The Biology of the Cell Cycle. Cambridge University Press, London.

Mitsuhashi, J. (1966). Tissue culture of the rice stem borer *Clulo suppressalis*. Walker. (Lepid. Pyralidae): II. Morphology and in vitro cultivation of haemocytes. *Appl. Ent. Zool. (Japan)*, **1**, 5-20.

Mitsuhashi, J. (1967). Establishment of an insect cell strain persistently infected with an insect virus. *Nature, Lond.* **215**, 863-864.

Mohrig, W. and Messner, B. (1968). Immunreaktionen bei Insekten. I. Lysozym als grundlegender antibakterieller Faktor im humoralen Abwehrmechanismus der Insekten. *Biol. Zbl.* **87**, 439-470.

Moran, D. T. (1971). The fine structure of cockroach blood cells. *Tissue and Cell*, **3**, 413-422.

Munson, S. C. and Yeager, J. F. (1944). Fat inclusions in the blood cells of the southern armyworm (*Prodenia eridania*). *Ann. ent. Soc. Am.* 37, 396-400.

Muttkowski, R. A. (1924). Studies on the blood of insects. III. The coagulation and clotting of insect blood. *Bull. Brooklyn ent. Soc.* 19, 128-144.

Nappi, A. J. (1972). Defence reactions of *Drosophila euronotus* larvae against the hymenopterous parasite *Pseudeucoila bochei*. *J. Invert. Pathol.* 16, 408-418.

Nappi, A. J. and Stoffolano, J. G. (1971). *Heterotylenchus autumnalis* haemocytic reactions and capsule formation in the host, *Musca domestica*. *Expl. Parasit.* **29**, 116-125.

Nappi, A. J. and Stoffolano, J. G. (1972). Distribution of haemocytes in larvae of *Musca domestica* and *Musca autumnalis* and possible chemotaxis during parasitization. *J. Insect Physiol.* 18, 169-179.

Nelson, D. S. (1969). Macrophages and immunity. *In* "Frontiers of Biology" (Eds A. Neuberger and E. L. Tatum), Vol. 2. North Holland, Amsterdam.

Nelstrop, A. E., Taylor, G. and Collard, P. (1970). Comparative studies of serum and haemolymph proteins. *Comp. Biochem. Physiol.* 35, 191-199.

Neutra, M. and Leblond, C. P. (1966). Radioautographic comparison of the uptake of galactose H^3 in the golgi region of various cells secreting glycoproteins or mucopolysaccharides. *J. Cell. Biol.* **30**, 137-150.

Nittono, Y. (1960). Studies on the blood cells in the silkworm *Bombyx mori* (L.). *Bull. Sericul. Exp. Sta. Tokyo,* **16**, 261-266.

Novikoff, A. B. (1961). Lysosomes and related particles. *In* "The Cell" (Eds J. Brachet and A. Mirsky), Vol. 2, pp. 423-488. Academic Press, New York and London.

Nutting, W. L. (1951). A comparative anatomical study of the heart and structures of orthopteroid insects. *J. Morph.* 89, 501-598.

O'Brien, D. and Wolfe, L. S. (1964). "Radiation, Radioactivity and Insects". Academic Press, New York and London.

Odhiambo, T. R. (1970). The architecture of the accessory reproductive glands of the male desert locust. III. Components of the muscular wall. *Tissue and Cell,* 2, 233-248.

Ohnishi, E. (1953). Tyrosinase activity during puparium formation in *Drosophila melanogaster. Jap. J. Zool.* 11, 69-74.

Okun, M., Patel, R., Donnellan, B., Lever, W., Edelstein, L. and Epstein, D. (1971). DOPA compared with dihydroxy-fumarate as cofactor in peroxidase-mediated oxidation of tyrosine to melanin-histochemical studies with neutrophils, eosinophils, mast-cells, melanoma cells and neurons. *Histochemie,* 27, 331-338.

Paillot, A. and Noël, R. (1928). Recherches histologiques sur les cellules péricardiales et les éléments du sang des larves insectes. *Bull. Histol. Appl. Physiol. Path.* **5**, 1-128.

Patay, R. (1939). Contribution a l'étude d'un Coleoptere (*Leptinotarsa decemlineata* Say). Rennes, p. 132.

Patel, R. P., Okun, M. R., Edelstein, L. M. and Epstein, D. (1971). Biochemical studies of the peroxidase mediated oxidation of tyrosine to melanin; demonstration of the hydroxylation of tyrosine by plant and human peroxidases. *Nature, Lond.* 124, 439-441.

Perez, C. (1910). Recherches histologiques sur la metamorphose des Muscides. *Archs. Zool. exp. gén.* 5e, 4, 1-275.

Pinhey, K. G. (1930). Inhibitors of tyrosinase. *J. exp. Biol.* 32, 468-484.

Pipa, R. L. and Woolever, P. S. (1965). Insect neurometamorphosis. II. The fine structure of perineural connective tissue, adipohaemocytes, and the shortening ventral nerve cord of a moth *Galleria mellonella. Z. Zellforsch. mikrosk. Anat.* **68**, 80-101.

Plagge, E. (1936). Bewirkung der Augenausfarbung der rotaugigen Rasse von *Ephestia kuhniella* durch Implantation artfremder Hoden. *Nachr. Ges. Wiss. Gottingen Math-Phys. Kl. Nachr. a.d. Biol.*2, 251-256.

Post, L. C. (1972). Bursicon: its effect on tyrosine permeation into insect haemocytes. *Biochim. biophys. Acta,* **290**, 424-428.

Powning, R. F. and Irzykiewicz, H. (1967). Lysozyme-like action of enzymes from the cockroach *Periplaneta americana* and from some other sources. *J. Insect Physiol.* 13, 1293-1299.

Poyarkoff, E. (1910). Histology during moulting and metamorphosis. *Archs Anat. micros. Morph. exp.* 12, 333-474.

Pryor, M. G. M. (1955a). Inhibitors of tyrosinase. *J. exp. Biol.* 32, 468-484.

Pryor, M. G. M. (1955b). Tanning of blowfly puparia. *Nature, Lond.* 175, 600.

Pye, A. E. and Yendol, W. G. (1972). Haemocytes containing polyphenoloxidase in *Galleria* larvae after injections of bacteria. *J. Invert. Pathol.* 19, 166-170.

Rabin, H. (1970). Hemocytes, hemolymph and defense reactions in crustaceans. *J. Reticuloendothelial Soc.* 7, 195-207.

Rabinovitch, M. and De Stefano, M. J. (1970). Interactions of red cells with phagocytes of the wax-moth (*Galleria mellonella*), and mouse. *Expl. Cell Res.* 59, 272-282.

Rehm, E. (1948). "Insecticin", ein antibiotisch-bactericid wirkender stoff aus insekten. *Klin. Wschr.* 26, 120-121.

Ritter, H. and Bray M. (1968). Chitin synthesis in cultivated cockroach blood. *J. Insect Physiol.* 14, 361-366.

Ritter, H. and Blissit, L.(1969). Cell transformation in vitro: crescent cell origin in cockroach blood. *J. Cell Biol.* 43, 117A.

Rizki, M. T. M. (1957). Alterations in the haemocyte population of *Drosophila melanogaster. J. Morph.* 100, 437-458.

Rizki, M. T. M. (1961). The influence of glucosamine-hydrochloride on cellular adhesiveness in *Drosophila melanogaster. Expl. Cell. Res.* 24, 111-119.

Rizki, M. T. M. (1962). Experimental analysis of hemocyte morphology in insects. *Am. Zool.* 2, 247-256.

Rizki, M. T. M. (1967). Ultrastructure of the secretory inclusions of the salivary gland cell in Drosophila. *J. Cell Biol.* 32, 531-534.

Rooseboom, M. (1937). Contribution à l'étude de la cytologie du sang de certains insectes, avec quelques considérations générales. *Archs Néerl. Zool.* 2, 432-559.

Rosenberger, C. R. and Jones, J. C. (1960). Studies on the total blood cell counts of the southern armyworm larvae, *Prodenia eridania* (Lepidoptera). *Ann. ent. Soc. Am.* 44, 351-355.

Roth, T. F. and Porter, K. R. (1964). Yolk protein uptake in the oöcyte of the mosquito *Aedes aegypti. J. Cell Biol.* 20, 313-332.

Ryan, M. and Nicholas, W. L. (1972). The reaction of the cockroach *Periplaneta americana* to the injection of foreign particulate material. *J. Invert. Pathol.* 19, 299-307.

Sacktor, B. (1965). Energetics and respiratory metabolism of muscular contraction. *In* "The Physiology of Insecta" (Ed. M. Rockstein), Vol. 2, pp. 484-580. Academic Press, New York and London.

Salt, G. (1960). Surface of a parasite and the haemocytic reaction of its host. *Nature, Lond.* 188, 162-163.

Salt, G. (1965). Experimental studies on insect parasitism. XIII. The haemocytic reaction of a caterpillar to eggs of its habitual parasite. *Proc. R. Soc. Lond. B.* 162, 303-318.

Salt, G. (1966). Experimental studies in insect parasitism. XIV. The haemocytic reaction of a caterpillar to larvae of its habitual parasite. *Proc. R. Soc. Lond. B.* **165**, 155-178.

Salt, G. (1970). The cellular defence reactions of insects. Cambridge Monographs in Experimental Biology, No. 16, Cambridge University Press, London.

Satir, P. and Gilula, N. B. (1973). The fine structure of membranes and intercellular junctions in insects. *Ann. Rev. Ent. Am.* **18**, 143-161.

Scharrer, B. (1939). The differentiation between neuroglia and connective tissue sheath in the cockroach (*Periplaneta americana*). *J. comp. Neurol.* **70**, 77-88.

Scharrer, B. (1966). Ultrastructural study of the regressing prothoracic glands of Blattarian insects. *Z. Zellforsch. mikrosk. Anat.* **69**, 1-21.

Scharrer, B. (1972). Cytophysiological features of hemocytes in cockroaches. *Z. Zellforsch. mikrosk. Anat.* **129**, 301-319.

Scharrer, E. and Scharrer, B. (1954). Hormones produced by neurosecretory cells. *Recent Prog. Horm. Res.* **10**, 183-240.

Schlumberger, H. G. (1952). A comparative study of the reaction to injury; the cellular response to methylcholanthrene and to talc in the body cavity of the cockroach, *Periplaneta americana. Arch. Path.* **54**, 98-113.

Schmidt, E. L. and Williams, C. M (1953). Physiology of insect diapause. V. Assay of the growth and differentiation hormone of Lepidoptera by the method of tissue culture. *Biol. Bull. mar. biol. Lab., Woods Hole*, **105**, 174-187.

Schwalbe, C. P. and Boush, G. M. (1971). Clearance of ^{51}Cr-labelled endotoxin from hemolymph of actively immunized *Galleria mellonella. J. Invert. Path.* **18**, 85-88.

Schweiger, S. and Karlson, P. (1962). Zum tyrosinstoffwechsel der Insekten. X. Die aktivierung der Präphenoloxydase und das Activator-Enzym. *Hoppe-Seyler's Z. physiol. Chem.* **329**, 210-221.

Scott, M. T. (1971a). A naturally occurring haemagglutinin in the haemolymph of the American cockroach. *Archs. Zool. exp. gén.* **112**, 73-80.

Scott, M. T. (1971b). Recognition of foreignness in invertebrates. II. *In vitro* studies of cockroach phagocytic haemocytes. *Immunology*, **21**, 817-828.

Scott, M. T. (1971c). Recognition of foreignness in invertebrates: transplantation studies using the American cockroach (*Periplaneta americana*). *Transplantation*, **11**, 78-86.

Scott, M. T. (1972). Partial characterization of the hemagglutinating activity in haemolymph of the American cockroach *Periplaneta americana. J. Invert. Path.* **19**, 66-71.

Schwartz, P. H. and Townshend, B. G. (1968). Effects of milky disease on haemolymph coagulation and on the number of haemocytes in infected larvae of the Japanese beetle. *J. Invert. Path.* **12**, 288-293.

Seligman, I. M. and Doy, F. A. (1972). Studies in cyclic-AMP mediation of hormonally induced cytolysis of the alary hypodermal cells and of hormonally controlled DOPA synthesis in *Lucilia cuprina. Israel J. Ent.* **7**, 129-141.

Seligman, I. M. and Doy, F. A. (1973). Hormonal regulation of disaggregation of cellular fragments in the haemolymph of *Lucilia cuprina. J. Insect Physiol.* **19**, 125-135.

Seligman, I. M., Filshie, B. K., Crossley, A. C. and Doy, F. A. (1974). Hormonal control of morphogenetic cell death of the wing hypodermis in *Lucilia cuprina*. In press.

Selman, B. J. (1962). The fate of the blood-cells during the life history of *Sialis lutaria* (L.). *J. Insect Physiol.* **8**, 209-214.

Shapiro, M. J. (1968). Pathenogenic changes in the blood of the greater wax moth, *Galleria mellonella*, during the course of nucleopolyhedrosis and starvation. II. Differential haemocyte count. *J. Invert. Path.* **10**, 230-234.

Shapiro, M., Stock, R. D. and Ignoffo, C. M. (1969). Hemocyte changes in larvae of the bollworm, *Heliothis zea*, infected with nucleopolyhedrosis virus. *J. Invert. Path.* **14**, 28-30.

Shrivastava, S. C. and Richards, A. G. (1965). An autoradiographic study of the relation between haemocytes and connective tissue in the wax moth, *Galleria mellonella*. *Biol. Bull. mar. biol. Lab., Woods Hole*, **128**, 337-345.

Shull, W. E. and Rice, P. L. (1933). A method for temporary inhibition of blood coagulation in insects. *J. econ. Ent.* **26**, 1083-1089.

Shull, W. E., Riley, M. K. and Richardson, C. H. (1932). Some effects of certain toxic gases on the blood of the cockroach *Periplaneta orientalis* (L). *J. econ. Ent.* **25**, 1070-1072.

Siakotos, A. N. (1960a). The conjugated plasma proteins of the American cockroach. *J. Gen. Physiol.* **43**, 999-1013.

Siakotos, A. N. (1960b). The conjugated plasma proteins of the American cockroach. II. Changes during moulting and clotting processes. *J. gen. Physiol.* **43**, 1015-1030.

Smith, D. S. (1968). "Insect Cells: Their Structure and Function". Oliver and Boyd, Edinburgh.

Sohi, S. S. (1971). *In vitro* cultivation of hemocytes of *Malacosoma disstria* (Lepidoptera). *Can. J. Zool.* **49**, 1355-1358.

Solum, N. O. (1970). Coagulation in *Limulus*—some properties of the clottable protein of *Limulus polyphemus* blood cells. *Symp. zool. Soc. London.* **27**, 207-216.

Stang-Voss, C. (1970). Zur Ultrastructur der Blutzellen Wirbelloser Tiere. I. Über die Haemocyten der Larve des Mehlkäfers *Tenebrio molitor*. *Z. Zellforsch Mikrosk. Anat.* **103**, 589-605.

Stephens, J. M. (1959). Immune responses of some insects to some bacterial antigens. *Can. J. Microbiol.* **5**, 203-228.

Stephens, J. M. (1962a). Bactericidal activity of the blood of actively immunized wax moth larvae. *Can. J. Microbiol.* **8**, 491-499.

Stephens, J. M. (1962b). Influence of active immunization on melanization of the blood wax moth larvae. *Can. J. Microbiol.* **8**, 597-602.

Stephens, J. M. and Marshall, J. H. (1962). Some properties of an immune factor isolated from the blood of actively immunized wax moth larvae. *Can. J. Microbiol.* **8**, 719-725.

Swammerdam, J. (1758). The Book of Nature, or the History of Insects.

Tait, J. (1911). Types of Crustacean blood coagulation. *J. mar. Biol. Ass.* **9**, 191-198.

Tauber, O. E. (1936). Mitosis of circulating cells of *Blatta*. *Iowa St. J. Sci.* **10**, 431-439.

Tauber, O. E. (1937). Effects of ecdysis on number of mitotically dividing cells in *Blatta* haemolymph. *Ann. ent. Soc. Am.* **30**, 35-39.

Tauber, O. E. (1940). Mitotic response of roach haemocytes to certain pathogens. *Ann. ent. Soc. Am.* 33, 113-119.

Tauber, O. E. and Griffiths, J. T. (1943). Multinucleate haemocytes in roach *Blatta. Trans. Am. micros. Soc.* 62, 91.

Taylor, R. L. (1969a). Formation of tumorlike lesions in the cockroach *Leucophaea maderae* after nerve severance. *J. Invert. Path.* 13, 167-187.

Taylor, R. L. (1969b). A suggested role for the polyphenol-phenol-oxidase system in invertebrate immunity. *J. Invert. Path.* 14, 427-428.

Thomson, J. A. and Sin, Y. T. (1970). The control of prophenoloxidase activation in larval haemolymph of *Calliphora. J. Insect. Physiol.* 116, 2063-2074.

Tilney, L. G. (1968). The assembly of microtubules and their role in the development of cell form. *Devl. Biol. Suppl.* 2, 63-102.

Tobie, E. J. (1970). Observations on the development of *Trypanosoma rangeli* in the hemocoel of *Rhodnius prolixus. J. Invert. Path.* 15, 118-125.

Tripp, M. R. (1966). Hemagglutinin in the blood of the oyster *Crassostrea virginica. J. Invert Path.* 8, 478-484.

Tripp, M. R. (1969). General mechanisms and principles of invertebrate immunity. *In* "Immunity to Parasitic Animals" (Ed. Jackson *et al.*). Appleton, New York.

Viallanes, H. (1882). Recherches sur l'histologie des insectes, et sur les phénomènes histologiques qui accompagnent le dévelopment post-embryonnaire de ces animaux. *Ann. Sc. Nat. Zool.* (6) 14, 1-348.

Vinson, S. B. (1971). Defence reaction and hemocytic changes in *Heliothis virenscens* in response to its habitual parasitoid *Cardiochiles nigriceps. J. Invert. Path.* 18, 94-100.

Vye, M. V. and Fischman, D. A. (1971). A comparative study of three methods for the ultrastructural demonstration of glycogen in thin sections. *J. Cell Sci.* 9, 727-749.

Walters, D. R. and Williams, C. M. (1966). Reaggregation of insect cells as studied by a new method of tissue and organ culture. *Science*, 154, 516-576.

Walters, D. R. (1970). Hemocytes of Saturniid silkworms:— their behaviour in vivo and in vitro in response to diapause, development, and injury. *J. exp. Zool.* 174, 441-450.

Waterhouse, D. F. (1950). Connective tissue strands in blowfly larvae. *Aust. J. Sci.* 13, 25-26.

Weibel, E. R. and Palade, G. (1964). New cytoplasmic components in arterial endothelia. *J. Cell Biol.* 23, 101-112.

Werner, R. A. and Jones, J. C. (1969). Phagocytic haemocytes in unfixed *Galleria mellonella* larvae. *J. Insect Physiol.* 15, 425-437.

Wheeler, R. E. (1963). Studies on the total haemocyte count and haemolymph volume in *Periplaneta americana* (L) with special reference to the last moulting cycle. *J. Insect Physiol.* 9, 223-235.

White, J. and Krivit, W. (1967). An ultrastructural basis for the shape changes induced in platelets by chilling. *Blood*, 30, 625-635.

Whitehead, D. L. (1969). New evidence for the control mechanisms of sclerotization in insects. *Nature, Lond.* 224, 721-723.

Whitehead, D. L. (1970a). L-dopa decarboxylase in the haemocytes of diptera. *F.E.B.S. Letters*, 7, 263-266.

Whitehead, D. L. (1970b). The role of haemocytes in the biosynthesis of protocatechuate in the cockroach colleterial system. *Biochem. J.* **119**, No. 5, 65-66.

Whitehead, D. L. (1973). The events preceding formation of the puparium in *Glossina* larvae. *Trans. R. Soc. trop. Med. Hyg.* **67**, No. 2, 300-301.

Whitten, J. M. (1962). Breakdown and formation of connective tissue in the pupal stage of an insect. *Q. Jl. microsc. Sci.* **103**, 359-367.

Whitten, J. M. (1964). Connective tissue membranes and their apparent role in transporting neurosecretory and other secretory products in insects. *Gen. comp. Endocr.* **4**, 176-192.

Whitten, J. M. (1969). Haemocyte activity in relation to epidermal cell growth, cuticle secretion, and cell death in a metamorphosing cyclorrhaphan pupa. *J. Insect Physiol.* **15**, 763-778.

Wigglesworth, V. B. (1933). The physiology of the cuticle and of ecdysis in *Rhodnius proxilus* with special reference to the function of oenocytes and dermal glands. *Q. Jl. microsc. Sci.* **76**, 269-318.

Wigglesworth, V. B. (1937). Wound healing in an insect, *Rhodnius prolixus* (Hemiptera). *J. exp. Biol.* **14**, 364-381.

Wigglesworth, V. B. (1939). "Principles of Insect Physiology". Methuen, London.

Wigglesworth, V. B. (1955). The endocrine chain in an insect. *Nature, Lond.* **175**, 338.

Wigglesworth, V. B. (1956a). The function of the amoebocytes during moulting in *Rhodnius. Annls Sci. nat. (Zool.)* **18**, 139-144.

Wigglesworth, V. B. (1956b). The haemocytes and connective tissue formation in an insect, *Rhodnius prolixus* (Hemiptera). *Q. Jl microsc. Sci.* **97**, 89-98.

Wigglesworth, V. B. (1959). Insect blood cells. *A. Rev. Ent.* **4**, 1-16.

Wigglesworth, V. B. (1963). The juvenile hormone effect of Farnesol and some related compounds: quantitative experiments. *J. Insect Physiol.* **9**, 105-119.

Wigglesworth, V. B. (1971). Bound lipid in the tissues of mammal and insect: a new histochemical method. *J. Cell Sci.* **8**, 709-725.

Wigglesworth, V. B. (1973). Haemocytes and basement membrane formation in *Rhodnius. J. Insect Physiol.* **19**, 831-844.

Wittig, G. (1966). Phagocytosis by blood cells in healthy and diseased caterpillars. II. A consideration of the method of making haemocyte counts. *J. Invert. Path.* **8**, 461-477.

Wittig, G. (1969). Fixation of insect hemocytes for electron microscopy: a comment. *J. Invert Path.* **14**, 425-426.

Wood, P. J., Podlewski, J. and Shenk, T. (1971). Cytochemical observations of haemolymph cells during coagulation in the crayfish *Orconectes virilis. J. Morph.* **134**, 479-487.

Wood, P. J. and Karpawich, P. P. (1972). Studies of the coagulation process in the crayfish *Orconectes virilis* and attempts to identify a fibrinogen-like factor. *Comp. Biochem.* **42B**, 41-48.

Wright, D. G. and Malawista, S. E. (1972). The mobilization and extracellular release of granular enzymes from human leucocytes during phagocytosis. *J. Cell Biol.* **53**, 788-797.

Wyatt, G. R. (1961). The biochemistry of insect haemolymph. *A. Rev. Ent.* **6**, 75-102.

Yeager, J. F. (1938). A modified Wright's blood staining procedure for smears of heat-fixed insect blood. *Ann. ent. Soc. Am.* 31, 9-14.

Yeager, J. F. (1945). The blood picture of the southern armyworm. *J. Agric. Res.* 71, 1-40.

Yeager, J. F. and Knight, H. H. (1933). Microscopic observations on blood coagulation in several different species of insects. *Ann. ent. Soc. Am.* 26, 591-602.

Yeager, J. F., and Munson, S. C. (1950). Blood volume of the roach *Periplaneta americana* determined by several methods. *Arthropoda,* 1, 255-265.

Yeager, J. F., Shull, W. E. and Farrar, M. D. (1932). On the coagulation of blood from the cockroach *Periplaneta orientalis* (Linn.) with special reference to blood smears. *Iowa St. J. Sci.* 6, 325-345.

Yeager, J. F. and Tauber, O. E. (1933). On counting mitotically dividing cells in the blood of the cockroach, *Periplaneta orientalis* (Linn.). *Proc. Soc. exp. Biol. Med.* 30, 861-863.

Yunker, C., Vaughn, J. and Cory, J. (1967). Adaptation of an insect cell line (Grace's *Antheraea* cells) to a medium free of insect haemolymph. *Science,* 155, 1565-1566.

Zachary, D. and Hoffman, J. A. (1973). The haemocytes of *Calliphora erythrocephala* (Meig.)(Diptera). *Z. F. Zellforsch.* 141, 55-73.

Zeller, H. (1938). Blut und Fettkörper im Flugel der Mehlmotte *Ephestia kuhniella. Z. Morph. Ökol. Tiere,* 34, 663-738.

Zernoff, V. (1931). L'immunite et les anticorps non-specifiques chez les insectes (chenilles de *Galleria mellonella*). *C.r. Séanc. Soc. Biol.* 106, 151-153; 543-546.

Zwicky, K. T. (1954). Osmoregulatorische Reaktionen der Larve von *Drosophila melanogaster. Z. vergleich Physiol.* 36, 367-390.

Development and Physiology of the Oöcyte-Nurse Cell Syncytium[1]

William H. Telfer

Department of Biology, University of Pennsylvania, Philadelphia, Pennsylvania

[1] The literature search for this review was supported in part by NIH Grant 5-T01-HD00026, a training grant in Fertilization and Gamete Physiology, administered by C. B. Metz at the Marine Biological Laboratory, Woods Hole, Massachusetts, and in part by a grant from the National Science Foundation, GB.

1 Introduction

Egg formation is supported in many insects by ovarian nurse cells which provide the growing oöcyte with a major part of its cytoplasm. Intercellular bridges link the nurse cells to the oöcyte, and these facilitate the trophic functions of the system by providing channels across which synthetic products move. While the mechanism of transport through the bridges has not been adequately studied, there can be no question about the fact that movement in the system as a whole is strongly polarized. Thus, during their most active period of synthesis, the nurse cells in some species retain a relatively constant size, while the oöcyte undergoes a slow but steady growth. Even more striking are events at the termination of trophic function when a surge of residual nurse cell cytoplasm may flow into the oöcyte, just before the bridges finally break.

A recent study of the oöcyte-nurse cell complex in *Hyalophora cecropia* provided what may be a clue to the physiological basis of bridge polarity. A 10 mV potential gradient was found to be maintained between the nurse cells and the oöcyte during the several day period when cytoplasmic transfer occurs (Woodruff and Telfer, 1973). The direction of the gradient is consistent with its being a driving force behind the movement of negatively charged materials such as ribosomes, and other forms of oöcyte RNA that are known to be synthesized in the nurse cell. Evidence was presented that the potential gradient is in fact steep enough in the cytoplasmic bridges to effect the polarized movement of a micro-injected fluorescent protein.

Many questions remain to be answered about these observations. What generates the electrical potential gradient? Is it the cause or the result of the polarized cytoplasmic movement? When does it arise in the development of the complex? How is electrical polarity of the cytoplasm related to the differentiation of the nurse cell and oöcyte nuclei? Does the gradient extend far enough into the oöcyte to help establish the morphogenetic polarity that is built into the structure of the mature egg? It is clear from the last two questions in particular that the oöcyte-nurse cell complex provides an opportunity for the application of physiological techniques to some of the basic problems of cellular differentiation.

This review is designed to summarize the literature on the development and function of the oöcyte-nurse cell complex, with the particular hope of encouraging physiologically trained experimenters to examine the relations between these cells. The ideas presented come from the studies of S. B. Pollack, R. I. Woodruff, and I. Mandelbaum on *Hyalophora* follicles, and from the important background of contributions by Bier, R. C. King, Mahowald, E. Anderson, Urbani, and many others who pioneered the application of electron microscopy and autoradiography to nurse cell development and function. The results of their efforts have already been summarized by Bier (1967), King (1970), and Mahowald (1972), but are so basic to an understanding of nurse cell function that they are treated again in the special context of this presentation. Finally, much emphasis is placed on the histological and cytological literature, dating from the mid-19th century until about 1960, which provided many enduring principles of insect oögenesis. Ovarian function in insects is among the oldest subjects of cell biology and the literature accumulated has become too copious, too dispersed, and too multi-lingual to be readily mastered. In comparison with the decisiveness made possible by modern techniques, the early literature is uncritical on many issues of current interest, and this, added to its formidable proportions, has led many writers to be content with a treatment of recent findings. Yet many of the early papers contain perspective and insights that can still be extremely useful. The selection reviewed here proved to contain a number of unrecognized observations that add in surprisingly helpful ways to present concepts on the development and function of the nurse cells.

2 Morphological background

2.1 THE OVARIOLE AS THE FUNCTIONAL UNIT OF THE OVARY

At least as early as Malpighi's treatise on the anatomy of the silkworm in 1669, insect ovaries were recognized as being divided into morphologically separate ovarioles. Much of the diversity recognized in these structures at the present time was already portrayed in a review by Lubbock in 1859. The individual ovariole (Fig. 1) was conceptualized from the outset as a tube that generates new follicles in a germarium at its apex and releases fully grown and ovulated eggs to the oviduct at its base. Between the two extremes, in what is sometimes called the vitellarium of the ovariole, is a progression of follicles in graded stages of growth, yolk deposition, and chorion formation. In Fig. 1 nurse cells have been omitted for purposes of clarity but, as will be seen below, they too arise in the germarium and function primarily while the oöcyte is in the vitellarium. In some insects,

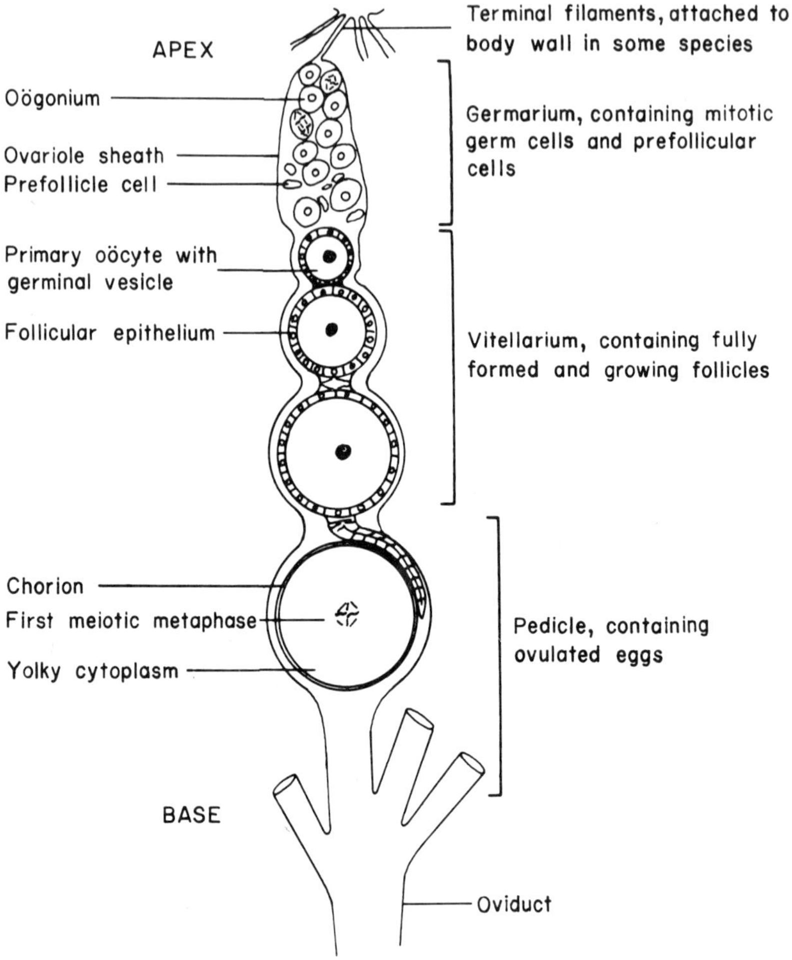

Fig. 1. The general plan of an insect ovariole. In steady-state systems mitosis of the oögonia and profollicle cells in the germarium compensates for the transfer of differentiated oöcytes and follicular epithelium to the vitellarium. In other insects the germarium functions only in immature stages and the vitellarium only in adults.

such as the Hemiptera and many Coleoptera and Lepidoptera, the germarium produces follicles only in nymphal or larval stages, and follicles reach maturity only in the adult. Thus complete developmental sequences are never present at any one time. In other insects all oögenetic stages, from actively dividing oögonial stem cells to ovulating eggs, are simultaneously present in the adult ovary (King, 1970). Grell and Chandley (1965), for instance, were able to show with thymidine labelling of steady-state, adult

Drosophila ovarioles that development from the S-phase of the primary oöcyte, which occurs in the germarium prior to follicle formation, until oviposition of the mature egg, requires an 8-9 day period of progress down the length of the ovariole.

In contrast to the ovariolar arrangement is the Collembolan ovary in which germarial tissue occurs in the wall of a single sac (Claypole, 1898). Follicles generated by the germaria are released to the lumen of the sac where they continue their development without being arranged in ordered rows. If an analogy is made to a glove with a germarium at the tip of each finger, the ovary in most insects is nearly all fingers, while in Collembola the fingers have fused or shortened so that the glove appears to be all palm.

The number of ovarioles and the ways in which they branch from the oviduct in various insects were summarized by Gross (1903) and Deejener (1928). There is at present no obvious relationship between these parameters and the way in which the individual ovariole functions. Its morphological isolation from neighbouring tissue evokes the speculation that each ovariole is a relatively independent unit, subject more to humoral influences reaching it through the hemolymph than to local control by other components of the ovary.

2.2 THE GERMARIUM

In principle the function of the germarium is clear. At or near its apex is a population of oögonia which grow and divide mitotically (Fig. 1). The progeny remaining at the apex continue to serve as oögonial stem cells, while those that are displaced in a basal direction eventually terminate mitosis and enter the first meiotic prophase. At the base of the germarium there is a second population of mitotic cells, the prefollicular cells (Fig. 1). These envelop the oöcyte in an epithelium one cell thick and thus complete the formation of a follicle. That the cells of the follicle all originate in the germarium, rather than being added to during their course down the vitellarium, has gone unchallenged since the question was analysed by Korshelt (1886). The early literature also paid much attention to the issues of whether oögonial and prefollicular cells arise from a common stem cell population, and whether they originate in the terminal filament that attaches the germarial apex to the body wall (Paulke, 1900). These two questions are commonly supposed to have been answered negatively, though the evidence is exclusively from descriptive morphology and confirmation by other methods would still be desirable.

In some ovaries the germarium functions in essence as diagrammed in Fig. 1, so that subsequent stages of oögenesis can be described primarily in terms of the growth and transformations of the oöcyte and its follicular epithelium. After analysing a germarium of this sort in *Periplaneta*, Brandt (1874) coined the adjective panoistic to designate the fact that all oögonial progeny become oöcytes, except the stem cells remaining at the germarial apex. In the majority of insects, by contrast, the oögonia generate not only

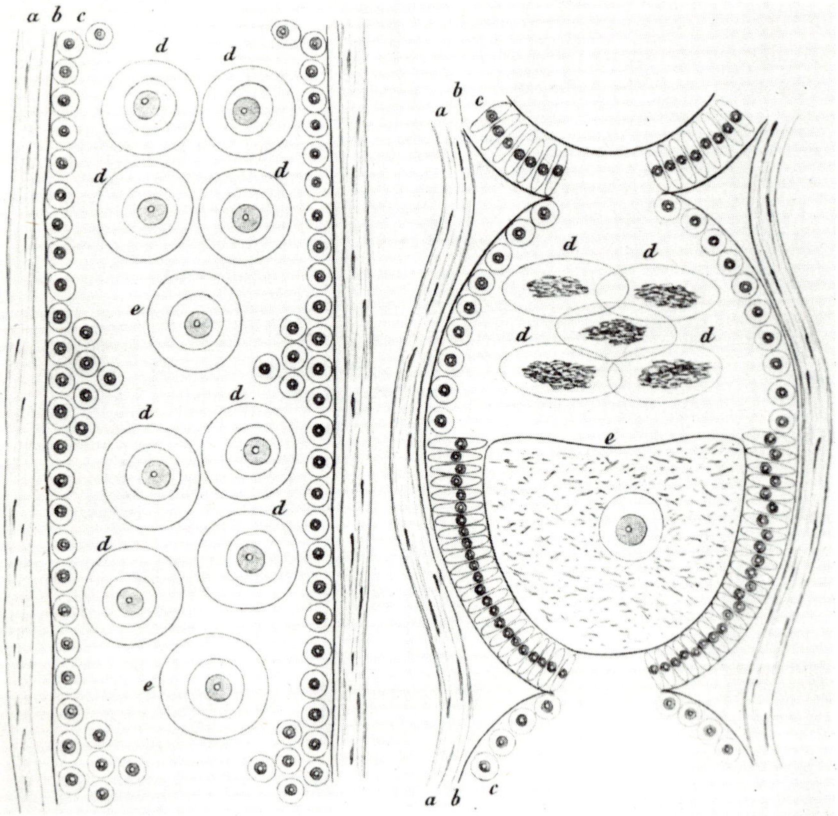

Fig. 2. Meyer (1849) described the differentiation of germ cells into nurse cells and oöcytes in the ovarioles of several Lepidoptera, including *Bombyx* and *Saturnia*. Because of their origin from oögonia, he called the nurse cells "abortive eggs". His drawings, shown here, were based on microscopic analysis of whole mounts. *a*, ovariole sheath; *b*, basement lamina; *c*, follicular epithelium; *d*, nurse cells; *e*, oöcyte. The left-hand drawing is an early stage of follicle formation typical of last instar caterpillars. The right-hand drawing is a follicle which is about to initiate yolk formation, and would occur in pupae or adults.

oöcytes, but also nurse cells. Impressed by the fact that not all oögonial products form eggs, Brandt coined the term meroistic for these ovarioles. The two terms are unfortunate in the obscurity of their meaning to biologists not trained in entomology, but they were endorsed and amplified by Gross (1903), and are now firmly established in the literature on insect oögenesis.

As early as 1859, Lubbock's review of ovarian structure recognized two distinct ways in which nurse tissues are organized. In one described by Meyer (1849) in Lepidoptera (Fig. 2), nurse cells and oöcytes alternate along the length of the ovariole. In the second, first seen in aphids (Huxley, 1858) the nurse cells remain confined to a chamber at the apex of the ovariole. To distinguish these two very different outcomes of germarial function, Gross (1903) introduced two additional adjectives, telotrophic for the hemipteran type ovary and polytrophic for the lepidopteran type. There are thus three patterns of germarial function (Fig. 3), each of which poses a unique set of developmental and physiological problems. This review necessarily makes frequent use of the terms panoistic (all germ cells form oöcytes), telotrophic (one nurse chamber at the apex of the ovariole), and polytrophic (a special nurse chamber for each oöcyte in the vitellarium), the latter two being sub-classes of meroistic ovaries.

In all three cases the fully established follicle contains a single meiotic prophase oöcyte surrounded by an epithelium one cell thick. While the follicles of panoistic and telotrophic ovarioles contain only these cells, those of polytrophic ovarioles include a third component, a cluster of nurse cells, which are mitotic siblings of the oöcyte (Figs 2 and 3). In the telotrophic ovariole a unique feature of the follicle is a cytoplasmic cord which arises from each oöcyte, and passes apically through the follicular epithelium (Fig. 3). The cord is long enough to attach at its apical end to the nurse chamber which, as will be seen below, is a transformation product of the larval or nymphal germarium.

The phylogenetic distribution of the three types of germarial function has been surveyed by Gross (1903), Deejener (1928), Bonhag (1958), and Zaffagnini (1969), and requires no further amplification here. All endopterygote orders have polyphagic ovaries, except the Siphonaptera which are panoistic, and the polytrophic Coleoptera which are telotrophic. Most exopterygotes are panoistic, but the Heteroptera are telotrophic, and several orders, among them the Dermaptera, are polytrophic. The Homoptera, which interested Huxley (1859), appear to be polytrophic at some seasons and to form something akin to a telotrophic ovariole at other seasons. Of the apterygotes, all orders are panoistic except the Collembola which, though lacking ovarioles, clearly possess polytrophic-type follicles (Claypole, 1898; Krzysztofowicz, 1971). Nurse cells are also known in several Crustacea, including the ostracods (Woltereck, 1898) and the branchiopods (Anteunis *et al.*, 1966). The seemingly whimsical occurrence of the meroistic ovary has made it impossible to say whether it is a primitive arthropodan trait that has been lost by some groups, or a recent innovation with such adaptive value that it has evolved independently several times.

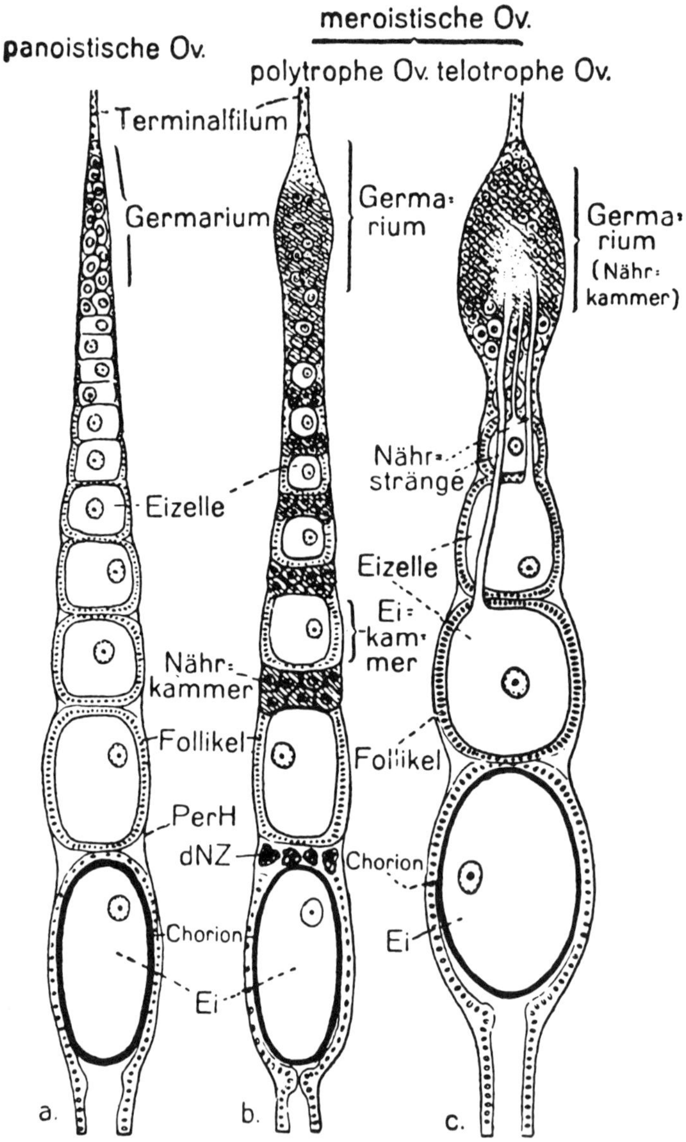

Fig. 3. A diagram of the three types of insect ovarioles. In most insects the oöcyte nucleus, unlike those shown here, forms the first meiotic metaphase when the chorion appears. dNZ, degenerating nurse cells; Per H, ovariole sheath. (After H. Weber, 1954. *Grundriss der Insektenkunde*. Fisher, Stuttgart.)

3 Germarial function in polytrophic ovarioles

In the follicles of polytrophic ovarioles, the nurse cells and the oöcyte are the mitotic progeny of a single germ cell which Brown and King (1964) termed a cystoblast. By their concept, which was developed from studies of *Drosophila* and was amplified by Koch and King (1966), oögonia are restricted to the apex of the germarium where they serve as stem cells. Mitosis of an oögonium yields two cells, one of which remains at the apex as a functional oögonium, while the other leaves the apex and becomes a cystoblast. The latter undergoes further mitosis and differentiation to generate a single oöcyte and in this species 15 nurse cells, all of which remain connected by intercellular bridges. The resulting sibling cluster is finally packaged in one follicle at the base of the germarium.

As early as 1901, Giardina recognized that the development of the sibling cluster is divided into two physiologically different phases. The first is a synchronous phase in which all progeny of one cystoblast undergo mitosis at precisely the same times. Chandley (1966) showed that the siblings also progress through the S-phase of the cell cycle together. Giardina proposed that synchrony results from these cells being linked in a syncytium by cytoplasmic bridges. That synchronizing cues are exchanged through intercellular bridges during the mitotic phase of sibling cluster formation is still a basic concept which is applicable not only to the polytrophic ovariole, but also to many cases of spermatogenesis (Fawcett, 1961, 1970).

In contrast to the synchrony of the early period is the more individualized behaviour of the cells in later stages of oögenesis. Thus, after completing the mitotic divisions, one sibling forms an oöcyte, while all others become nurse cells. As will be seen below, asynchrony also occurs between nurse cells in the timing of their endomitotic chromosome replications. Of central concern to this review is the fact that asynchrony occurs with the cytoplasmic bridges still intact, and therefore signals a fundamental change in the role of these structures. While the bridges assure that all cells behave alike early in development, they abet the differentiation of the system at later times by facilitating the transport of nurse cell products to the oöcyte. The discussion begins here with the intercellular bridges during the synchronous phase, and then turns to the question of differentiation and the physiology of asynchrony.

3.1 THE FORMATION OF INTERCELLULAR BRIDGES

Though intercellular bridges between germ cells were recognized as early as 1886 by Platner, precise information on their structure and formation required electron microscopy, and was not attained until Burgos and

Fawcett (1955) described them in spermatogenic cells of mammals. Meyer (1961) found a very similar configuration in *Drosophila* ovaries and testes, and since that time the intercellular bridges between oöcytes and nurse cells in all polytrophic insects examined have been found to conform to a general pattern (Fig. 4). The bridge is generally cylindrical in shape with a

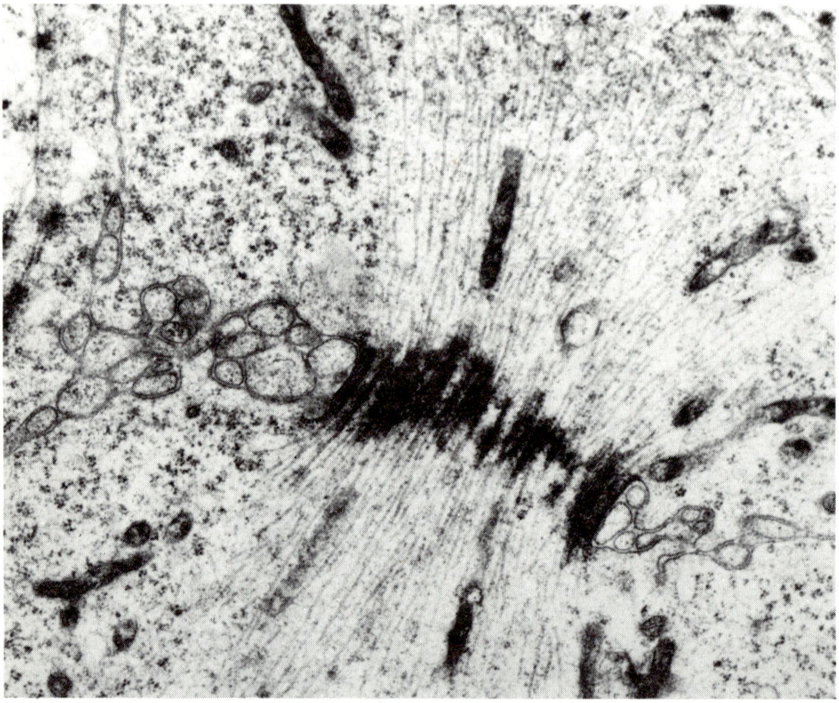

Fig. 4. A newly formed intercellular bridge in the germarium of *Hyalophora*. A mitotic spindle remnant, including longitudinally oriented microtubules and densely staining mid-body material, occupies the centre of the bridge. The cell membrane that lines the bridge is coated on its cytoplasmic side by a dense lining material. x 20 000. (Courtesy of I. Mandelbaum.)

diameter of 1 μm or less when first formed and a length of under 1 μm. Its lateral surfaces are covered by a unit membrane that is continuous with the cell membranes at either end. The membrane is invariably lined on the cytoplasmic side by a layer of densely staining material which has generally been supposed to stabilize the bridge by mechanically supporting its membrane.

At later stages of development, when the bridges have greatly enlarged, the dense material is visible by light microscopy as a darkly staining ring in paraffin sections that have been treated with hematoxylin (e.g. Giardina, 1901; Marshall, 1907; Gunthert, 1910; Maziarski, 1913; Hegner, 1915). Because of this appearance the intercellular bridges of insect ovaries are frequently termed ring canals. The term was applied to spermatogenic cells by King and Akai (1971), who prefer it to the more widely used intercellular bridge. An additional term, fusome, is also sometimes used, but for reasons that are detailed below this appears to be a misnomer.

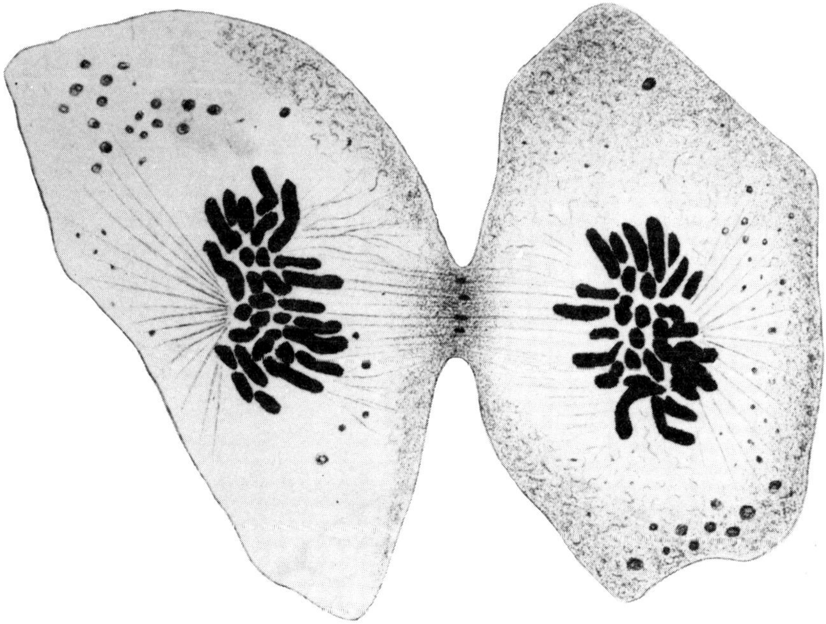

Fig. 5. A transitory intercellular bridge depicted by Flemming (1891) in lung epithelium. Flemming found such structures between newly divided cells in many tissues. They resemble that shown in Fig. 4 and in other germ cells, except that in somatic cells the lining material that coats the cell membrane of the bridge is apparently absent.

Fawcett *et al.* (1959) accepted the long-standing idea that an intercellular bridge is formed when the cleavage furrow is arrested upon contacting a mitotic spindle remnant late in cytokinesis. This would appear to be confirmed by the occurrence in the germaria of polytrophic ovaries of occasional bridges that contain a cluster of microtubules (Koch and King, 1969; Mahowald and Strassheim, 1970; Mahowald, 1971; Fig. 4).

Careful examination of cytokinesis in a number of somatic cells has indicated, as shown in Fig. 5, a transitory structure very similar to the intercellular bridges of germ cells (Flemming, 1891; Buck and Tisdale, 1962). Bridge formation and behaviour was particularly clearly recorded by cinemicrography in Hela cell cultures by Byers and Abramson (1968). There can be no comparable opportunity to observe bridge formation in germ cells until cystoblasts have been isolated in cell cultures, and it is therefore instructive to consider some of the details of this example. The bridge arises in Hela cells when the cleavage furrow has advanced centripetally until obstructed by the interzonal fibres of the mitotic spindle. Even then the furrow continues to advance, compressing the interzonal fibres until they form a tight bundle with a diameter of 1.5 μm. At this point cytokinesis is arrested with the two cells remaining connected by the resulting cytoplasmic bridge, and with the microtubules extending well into the cells on either side. Electron microscopy showed that the bridge contains not only microtubules, but also a densely staining mid-body on which the microtubules terminate (Paweletz, 1967). The similarity of this configuration to occasional bridges seen in the insect germarium suggests a similar mode of origin for those linking the members of a sibling cluster (Mahowald, 1971). It has also been suggested that the divisions giving rise to cytoplasmic bridges occur by vesicle fusion in the manner characterizing cytokinesis in plants, rather than by contractile ring furrowing (Koch and King, 1969; King, 1970).

In their subsequent development, Hela cell bridges do not develop the dense material that lines the membranes of the bridges in the germarium, and they have a much shorter life span, 3 h at the most. The completion of cleavage is not achieved by breaking the bridge at its centre, but by constricting it at one end so that the entire bridge goes to one of the two daughter cells. Constriction is preceded by waves of cytoplasm that initiate at the mid-body and move centrifugally along the surface of the bridge toward the cells on either side. The waves result in elongation of the bridge to a length as great as 3 μm, before it finally separates from one of the cells. Cystocyte bridges, by contrast, widen with time (Koch and King, 1969; Mahowald and Strassheim, 1970) without elongating during the synchronous phase of development. From this, as well as from their stability, it may be inferred that the centrifugal waves do not occur in the bridges produced during cystocyte division.

3.2 THE FUSOME AND ROSETTE FORMATION

Electron microscope studies of germaria have revealed sibling clusters in which a single bridge contains microtubules while other bridges to the same cell do not; this configuration suggests that the microtubules disappear from a newly formed bridge before the completion of the next cell cycle.

Of particular importance to this review is the frequently overlooked fact that the microtubules are replaced by a finely fibrillar material quite different in its texture and staining intensity from the cytoplasm elsewhere in the cell (Mahowald, 1971, for *Drosophila* and Fig. 6 for *Hyalophora*). This material usually excludes organelles the size of mitochondria from the bridges but frequently has ribosomes embedded in it. Its structure is

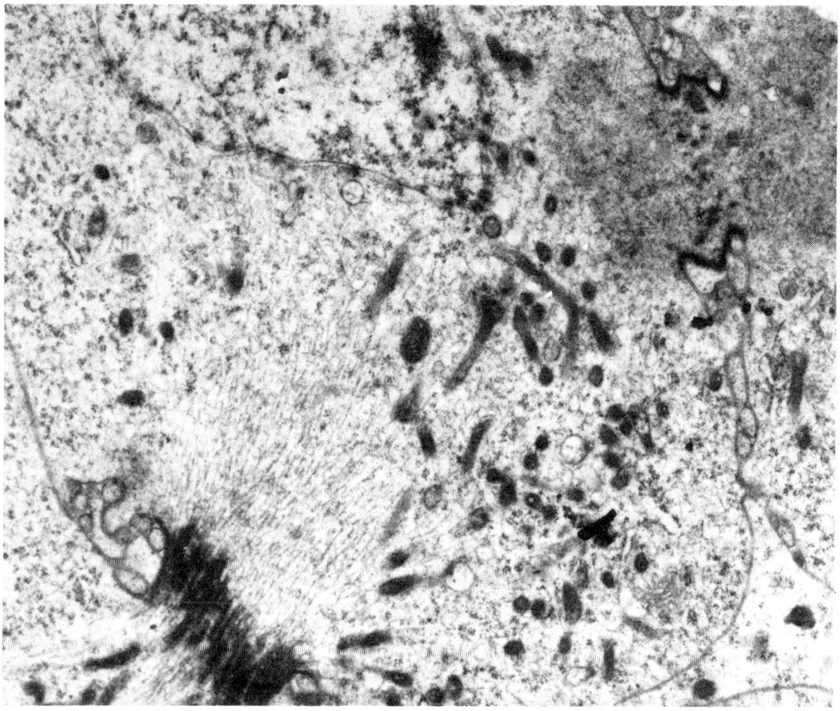

Fig. 6. A cystocyte from the germarium of *Hyalophora* with a newly formed bridge (lower left) and a stabilized bridge from a previous division (upper right). In the latter the microtubules and mid-body material have been replaced by a diffuse, finely fibrillar or granular material that fills the bridge and excludes organelles such as mitochondria. The material is designated here as the fusome. x 16 000. (Courtesy of I. Mandelbaum.)

particularly clear in glutaraldehyde-fixed specimens, though a special cytoplasm may be visible in the bridges in germaria fixed with osmium tetroxide as well (e.g. Fig. 10 in Meyer, 1961; Fig. 5 in Koch and King, 1966). The apparent replacement of microtubules by the fibrillar material raises what should be an experimentally answerable question. Are the microtubules themselves converted to a fibrillar form, or are they dissolved and replaced by a new material? In the former case they would represent a fibrillar phase of the protein tubulin, such as Weisenberg (1972) demonstrated in the meiotic interphase of *Spisula* eggs. There are at this time no observations that allow a distinction to be made between the two possibilities.

An additional feature of the fibrillar material is that it interconnects all bridges during the synchronous phase of sibling cluster development as a single, branching continuum. This relationship is shown in Fig. 7 for an 8-cell *Hyalophora* cluster, and has also been figured in *Drosophila* for a testicular cyst (Rasmussen, 1973) and for an osmium-fixed ovarian sibling cluster (Koch and King, 1966), though the latter two studies were concerned with other matters and so did not stress the finding. That the configuration has a more generalized distribution which has not yet been realized by electron microscopy is suggested by its similarity to a structure that was reported many times between 1886 and 1915 in both male and female sibling clusters of a wide variety of animals (Fig. 8). The technique utilized at that time generally included fixation with Carnoy's or Flemming's solution and staining of sections with iron hematoxylin. After this treatment a complex "intercellular ligament" could be seen, which stained more heavily than the rest of the cytoplasm, and which branched in such a way as to interconnect all members of a sibling cluster.

There is little agreement in the older literature about what to call the hematoxylin-staining structure. An abbreviated term, fusome, was introduced by Hirschler (1945), and this is arbitrarily adopted here. Hirschler applied the term not only to an explicit structure observed in germarial sibling clusters, but also to a generally distributed, hypothetical organelle, which he conceived of as governing spindle orientation and cleavage planes in all animal cells. His hypothetical extrapolation is not encompassed by the use of the term in this review. "Fusome" has recently been used as a synonym for "intercellular bridge" (e.g. Bier, 1963b; Bier *et al.*, 1967) but this is an unfortunate equating of two very different structures. A bridge is a tube of cytoplasm that connects two cells, while the fusome is a special material that sometimes occupies the bridge.

There is a substantial literature on the development and morphology of the fusome during sibling cluster formation, and this has gone largely unrecognized during the last two decades. Since the fibrillar material seen in Figs 6 and 7 is identical to the position and overall structure of the fusome described earlier, as well as in the stages of development when it is present, there can be little doubt that it is the electron microscope equivalent of the hematoxylin-staining material. A review of the early literature on the fusome can therefore add much to an understanding of the significance of the fibrillar material.

Platner (1886) first described what he termed a *Verbindungsbrücken*

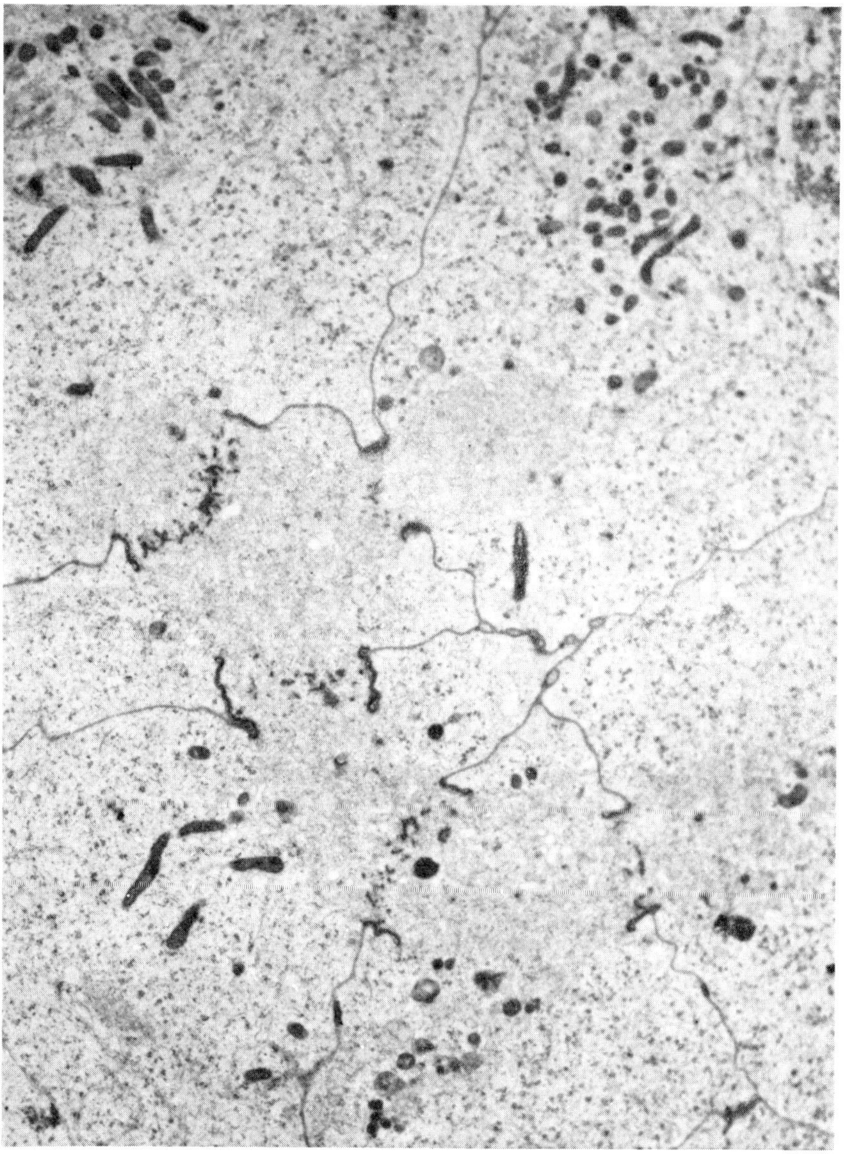

Fig. 7. A low-power electron micrograph showing five of the seven intercellular bridges in an eight-cell *Hyalophora* sibling cluster. Though not as densely staining as at younger stages (Fig. 6), the fusomal material can be seen here to form a single branching structure that traverses all five bridges. The fusome lies at the centre of the sibling cluster, and the cell bodies radiate from it like petals in a "rosette". x 11 600. (Courtesy of I. Mandelbaum.)

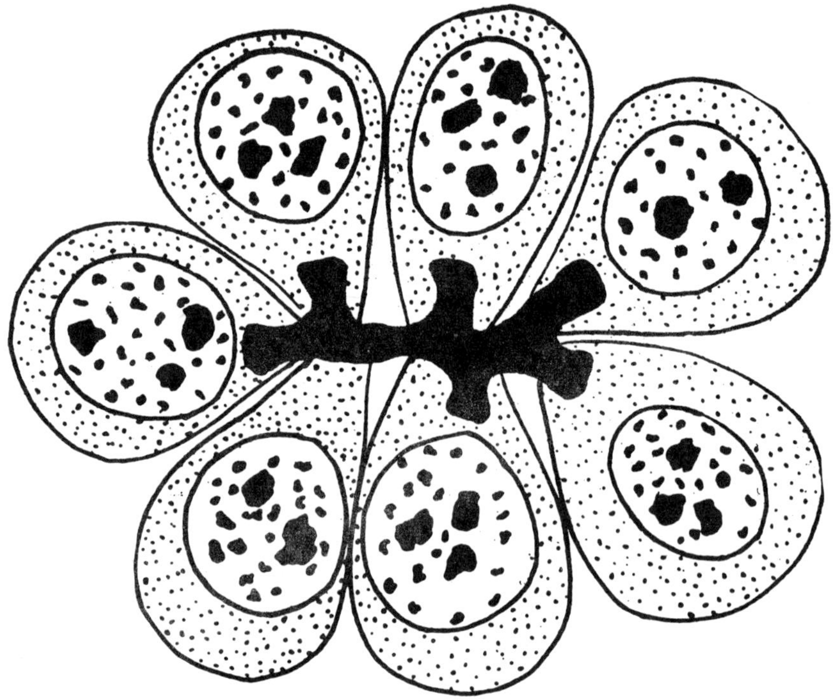

Fig. 8. A rosette of siblings from the germarium of a wasp, as drawn by Hegner (1915). The densely staining fusome sends branches into each cell at the centre of the rosette.

Fig. 9. Cystocyte divisions and the formation of the fusome in *Dytiscus.* These drawings were published by Giardina (1901) as water-colours, and though some of their elegance is lost by black and white reproduction, their clarity is unimpaired. (a) Two cystoblasts after completion of the last oögonial division. The dark mass in the cytoplasm is the *residue fusoriale*, a product of the last mitotic spindle and a precursor of what was later called the fusome. The nucleus contains a newly forming mass of chromatin which forms an extrachromosomal DNA body at the ensuing divisions. (b) Telophase of the first cystoblast division. The *residue fusoriale* and extrachromosomal DNA body both lie in the same cell. (c) Interphase after the first cystoblast division. The *residue fusoriale*, or fusome, has joined the mitotic spindle in a stabilized intercellular bridge. (d) Telophase of the second division. Again, a single cell retains the *residue fusoriale* and the DNA body. (e) Interphase following the second division. The two bridges attached to the cell with the DNA body have moved together and have been incorporated by the *residue fusoriale.* (f) Metaphase of the fourth and last division. The successive gathering of bridges at one pole of the cell with the DNA body, and the addition of their spindles to the *residue fusoriale* results in a rosette configuration.

between spermatocytes in *Sphinx* and *Pygaera*, and showed that it is distinct from the then recently discovered *Nebenkern*. Studies of spermatogenic cells in a variety of other animals soon confirmed that comparable structures are widely distributed (Prenant, 1888; Zimmerman, 1895; Lee, 1895; Meves, 1897; McGregor, 1899). Their formation in polytrophic insect ovaries was first described in detail by Giardina (1901),

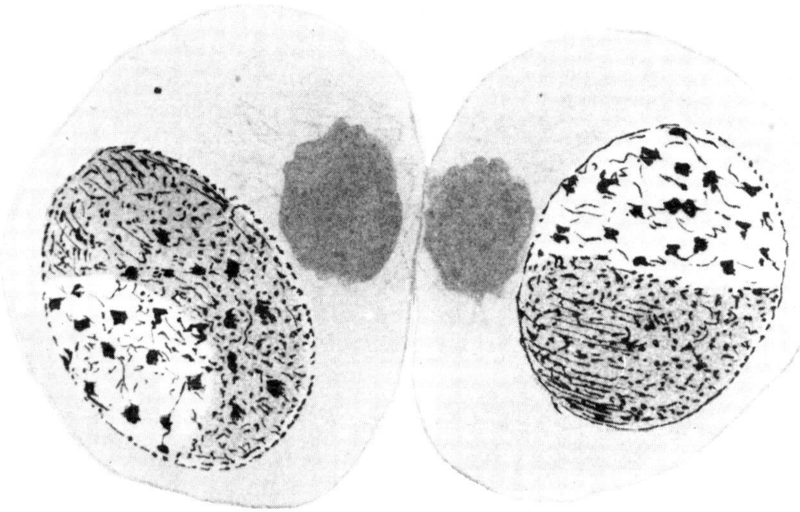

Fig. 9(a).

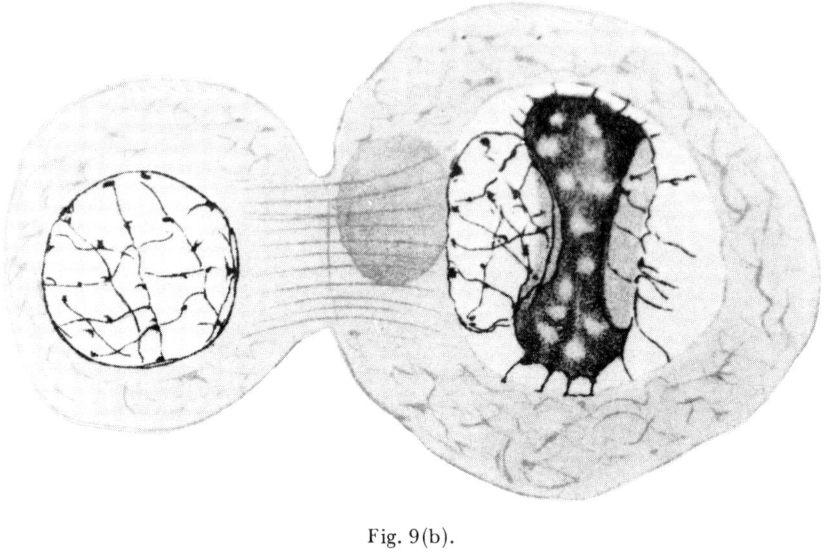

Fig. 9(b).

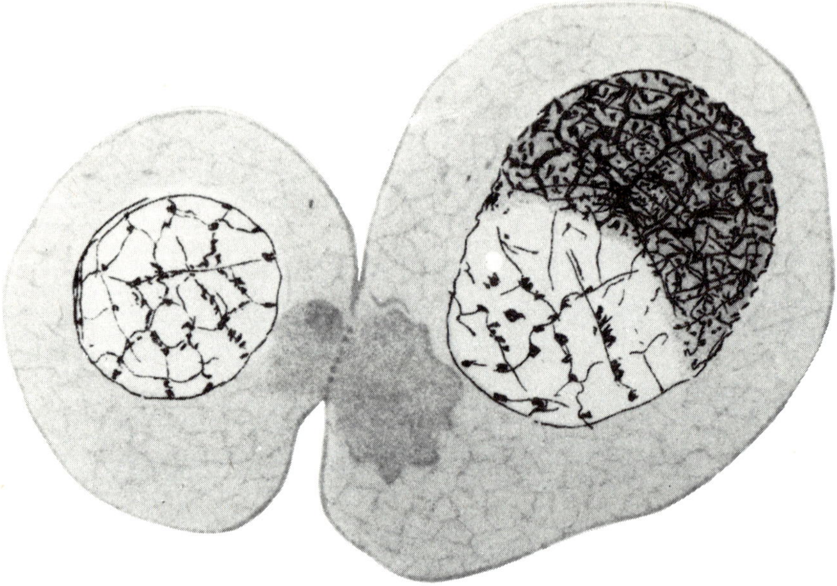

Fig. 9(c).

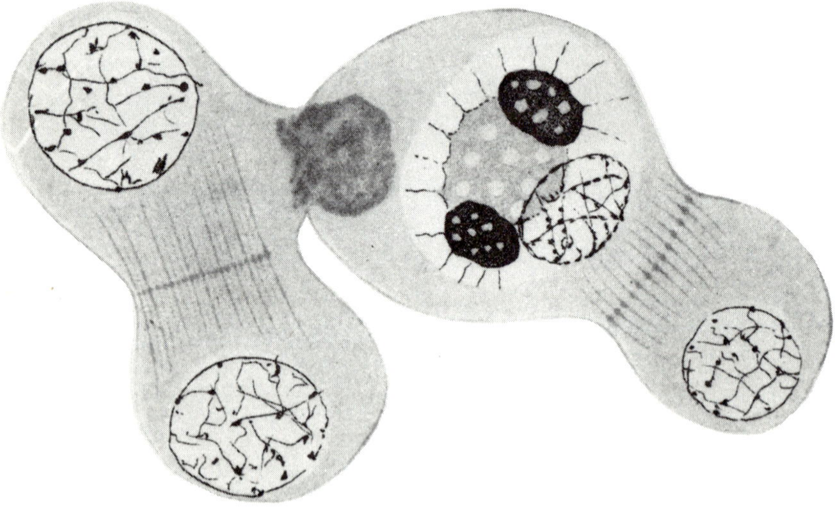

Fig. 9(d).

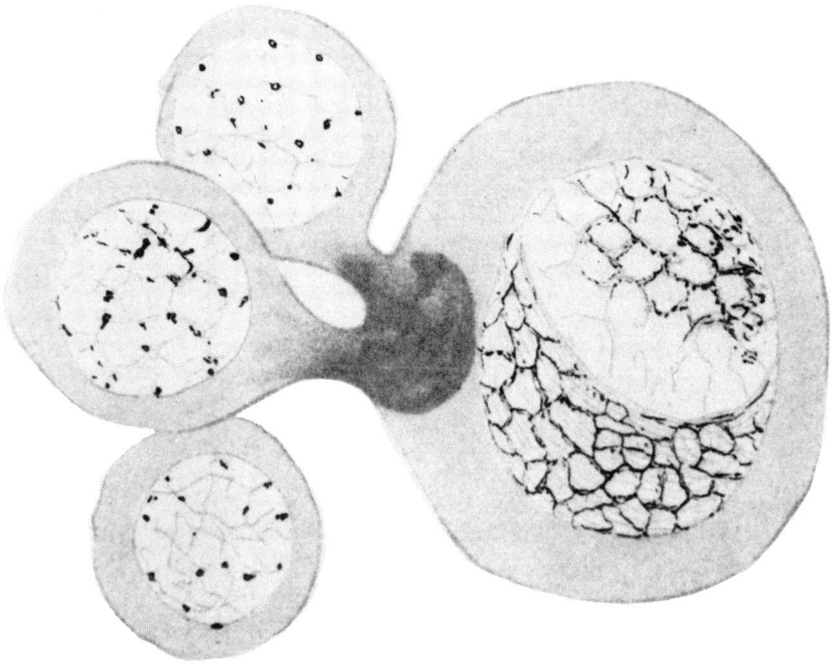

Fig. 9(e).

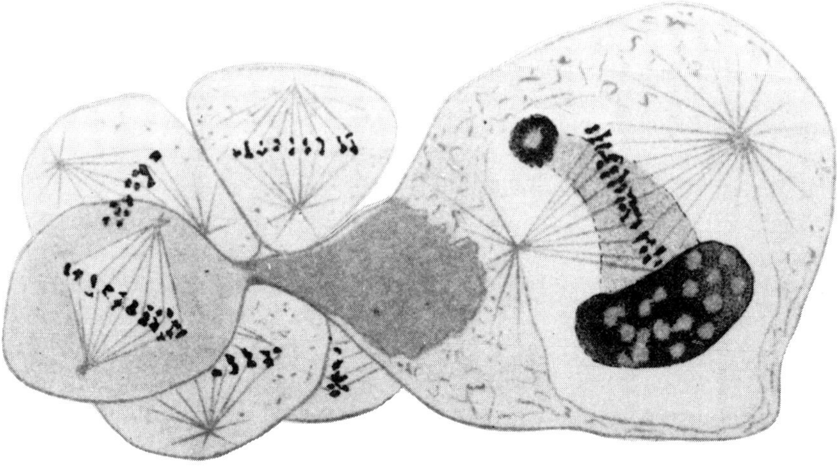

Fig. 9(f).

who agreed with the earlier findings on spermatogenic cells that the hematoxylin-staining material develops from a residue of the mitotic spindle, and who accordingly called it a *residue fusoriale*. While the structure is unquestionably a fusome in the sense defined here, it has some special features in the cystoblasts studied by Giardina, and it will be safer to use his terminology in discussing *Dytiscus*. His description of its origin and behaviour is so basic to several current questions about nurse cell development that a detailed summary is required.

The cystoblasts of *Dytiscus* undergo a sequence of four mitotic divisions, all entailing intercellular bridge formation; a cluster of sixteen siblings results, one of which will become an oöcyte while the remaining fifteen form nurse cells. A unique feature of this system is that the cystocyte giving rise to the oöcyte can be identified throughout sibling cluster formation by its large size and unique nuclear cytology. It was therefore possible for Giardina to detail bridge formation and the behaviour of the *residue fusoriale* in the particular cell line giving rise to the oöcyte. A stable *residue fusoriale* first arises during the last oögonial division, even though an intercellular bridge is not formed at this time (Fig. 9(a)). Its persistence in the cytoplasm during the subsequent interphase is the first visible sign that an oögonial division product has transformed into a cystoblast. During this interphase the cystoblast nucleus produces an extrachromosomal DNA body, and at the next division, which terminates with the formation of the first intercellular bridge, the DNA body and the *residue fusoriale* are both inherited by one daughter cystocyte (Fig. 9(b)). Giardina termed this a differential mitosis, and was able to show that the cell containing the two special structures would, after further divisions, give rise to the oöcyte, while its sibling would give rise only to nurse cells. Of particular concern to the present discussion is his observation that, after the first cystoblast cytokinesis has been arrested, the *residue fusoriale* moves to the newly formed intercellular bridge and fuses with its spindle residue (Fig. 9(c)).

At the next division, which generates four cells from two, the DNA body and the now enlarged fusome once more remain with one daughter cell (Fig. 9(d)). Two new bridges are produced, one between nurse cell progenitors, and one between the cell containing the DNA body, and its newly formed mitotic sibling. The position of the latter bridge becomes shifted during the subsequent interphase until its spindle remnant also combines with the fusome (Fig. 9(e)). These characteristics of the fusome, inheritance by only one daughter cell, and fusion with the spindle remnant in the new intercellular bridge, are repeated at each of the four differential divisions. As a consequence of bridge displacement, the four, eight, and sixteen cell clusters of *Dytiscus* assume the appearance of what Giardina described as a rosette during each successive interphase, with the cells radiating like petals from the centrally located fusome and the associated complex of intercellular bridges (Fig. 9(f)).

The prospective nurse cells in *Dytiscus* are much smaller than the oöcyte progenitor, and the formation of their bridges was not as clearly seen. Giardina described a rather different behaviour for them, with cytokinesis beginning at the periphery of the rosette, and proceeding centripetally, so that new bridges were initially formed adjacent to the older bridges, rather than being displaced to that position at a later time as they are in the cell containing the DNA body. In any case, all bridges and their spindle residues are gathered at the centre of the rosette, and it is easy to see that this behaviour, combined with the fusion of spindle remnants, would result in a continuum such as that shown by the fibrillar material seen in Figs 7 and 8.

Aside from Gunthert's (1910) reanalysis of *Dytiscus*, there has been no study of the fusome as thorough as that of Giardina. This is certainly due in part to the fact that the cystocytes of most other insects are not as large and as clearly marked as the oöcyte progenitor of *Dytiscus*. Many observations combine, however, to indicate that the phenomenon is widespread in polytrophic systems. A branching, hematoxylin-staining ligament that attaches the cells of the germarial sibling cluster to each other has been repeatedly seen in *Hymenoptera* (Marshall, 1907; Maziarski, 1913; and Hegner, 1915) and *Lepidoptera* (Dederer, 1915; Hirschler, 1942). Mandelbaum (1974) found that the fibrillar material disappears from the bridges of *Hyalophora* when the cells begin to differentiate from each other, and this may explain why electron microscopy has so infrequently revealed it. As has already been noted, Mahowald (1971) observed the fibrillar material in *Drosophila* bridges, Koch and King (1966) published a figure of a classical rosette configuration in this species, and Meyer (1961) observed similar material running between bridges in *Tipula*. Finally, as will be seen below, the intercellular bridges are seen to be distributed among the cells of mature sibling clusters in exactly the manner that would be predicted from Giardina's observations on the oöcyte progenitor of *Dytiscus*.

3.3 BRIDGE DISTRIBUTION

The formation of the sibling cluster in polytrophic ovarioles has been shown to entail a cleavage pattern that results in the bridges being distributed in a special manner. This was first observed by Knaben in 1934 for the eight-cell cluster of the moth *Tischeria*. As illustrated in Fig. 10(d), Knaben's map showed that four cells are attached to the mature complex by single bridges, two cells are each attached by two bridges, and two cells by three bridges. In every case examined the oöcyte had formed from one of the two cells with three bridges. The key feature resulting in this arrangement occurs at the third cystocyte division. In the four-cell cluster there are two cells that are each attached to two intercellular bridges (Fig. 10(c)). When these cells divide, both bridges are retained by a single daughter cell. With the addition of the newly formed bridge, these particular daughter cells thus come to be connected to a total of three bridges each. The elements of this behaviour were first observed in the oöcyte progenitor of *Dytiscus*. Electron microscope serial sections in which the bridges could be mapped in the 4-, 8- and 16-cell clusters showed that this behaviour is repeated at the fourth cystocyte division in *Drosophila* (Koch and King, 1966). While these are the only developmental analyses

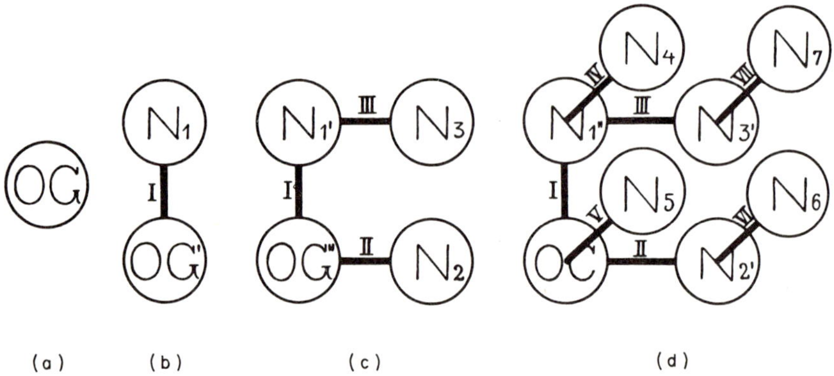

Fig. 10. Bridge distribution in the lepidopteran sibling cluster as proposed by Knaben (1934). The key feature is that when OG divides to form OC and N_5, bridges I and II are both retained by OC. A similar behaviour occurs in the division of $N_{1'}$. If this were not the case, a linear chain rather than a branched chain of cells would result.

showing bridge distribution at intermediate stages, mapping of the differentiated oöcyte-nurse cell complex in a number of additional species has confirmed that retention of pre-existing bridges by a single daughter cell must be a general characteristic of the cystocyte divisions. Thus, Knaben's findings were confirmed in two other Lepidoptera with 8-cell clusters, *Macrothylacia* (Hirschler, 1942) and *Hyalophora* (King and Aggarwal, 1965). *Drosophila* and *Dytiscus*, as we have seen, have 16-cell clusters which conform. Thirty-two cell clusters have not as yet been mapped, but Cassidy and King (1973) identified a cell with five bridges in a sibling cluster of *Habrobracon*. Engels (1968) described five bridges attached to the oöcyte of *Apis*, which has an anomalous number of 48 nurse cells.

What in fact is the mechanism assuring that pre-existing bridges are not separated from each other at cytokinesis? Rosette formation, with the resultant gathering of new bridges in the centre of the cluster after each set of divisions, would in itself tend to assure this result. It would still be necessary to assure, however, that cytokinesis divides each member of the rosette into a central cell and a peripheral one (Fig. 8). The plane of cytokinesis can be altered in several kinds of cells by displacing the spindle

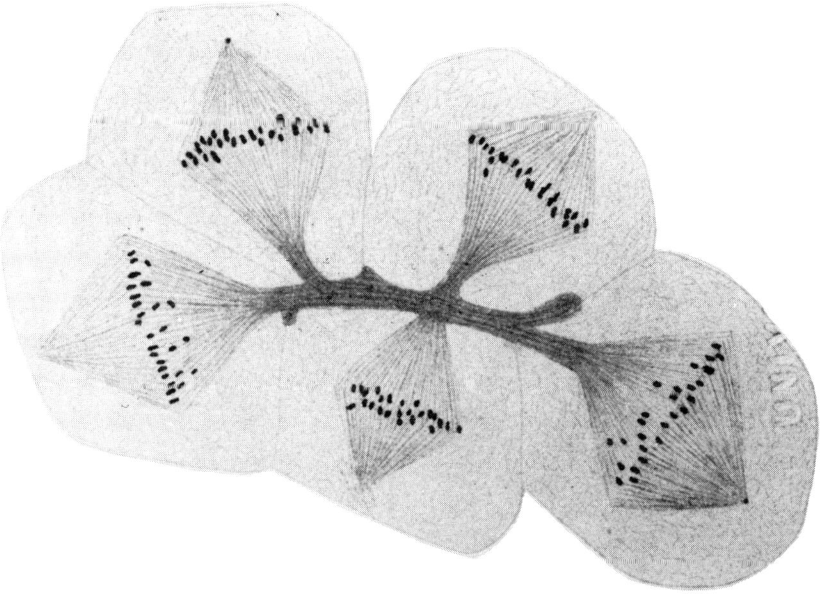

Fig. 11. The orientation of mitotic metaphase figures with relation to the fusome in the germarium of *Vespa*. This drawing suggests that one pole of each spindle is formed adjacent to or attached to a branch of the fusome. Equatorial cytokinesis in each cell would necessarily result in the retention of the fusome and all pre-existing bridges by one of the two daughter cells. This configuration thus appears to explain the results seen in Fig. 10. (From Maziarski, 1913.)

during or before metaphase (e.g. Kawamura, 1960). Cleavage always occurs in these instances through the equator of the spindle, and it is therefore reasonable to suppose that the plane of cytokinesis is governed by the position of the spindle. Assuming that this principle also holds for the cystocyte, what is needed in the rosette is a mechanism governing spindle orientation. One manner in which this might be achieved is suggested in Fig. 11, which was taken from a paper by Maziarski (1913). Here it can be seen that the mitotic spindles in a dividing sibling cluster are each arranged with one pole adjacent to the fusome. Whether or not the fusome itself is

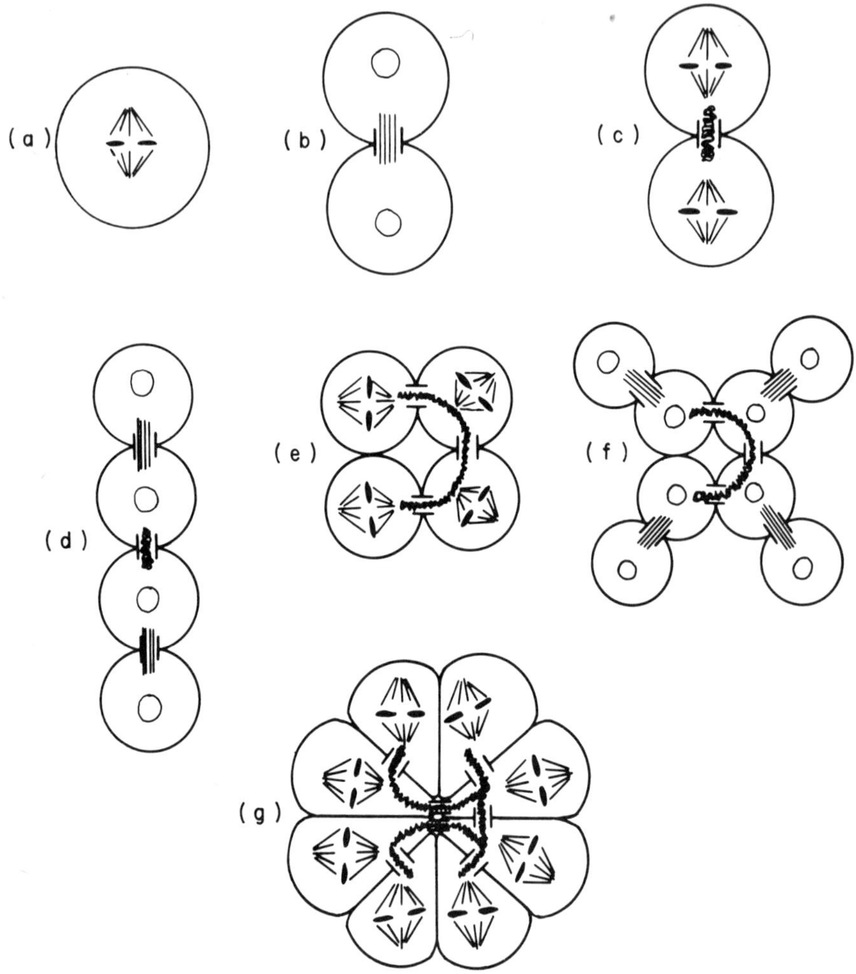

Fig. 12. A speculative synthesis of the information gathered in Figs 4 to 11. The result would be a distribution of bridges identical to that recognized in mature sibling clusters. (a) and (b) First cystoblast division, resulting in an intercellular bridge containing residual spindle fibres. (c) and (d) The second division, yielding four cells attached by three bridges. The spindle fibres in the first bridge have been replaced by a fusome, as in Fig. 6. (e) Metaphase of the third division. In going from (d) to (e), the three bridges have moved toward each other, and have become incorporated by one fusome. The spindles all have one pole oriented toward the fusome, as in Fig. 11. (f) As a consequence of the spindle orientation in (e), cytokinesis occurs so that the bridges are distributed as in Knaben's map (Fig. 10). (g) The centripetal gathering of bridges and their incorporation by the fusome has been repeated so that an 8-cell rosette is formed. If the cluster were to divide again, the spindles would be oriented as shown, and once more pre-existing bridges would remain with single daughter cells.

the orienting agent is not certain, but the arrangement would necessarily result in the retention of the fusome and all associated bridges by a single daughter cell.

Figure 12 summarizes fusome production from the mitotic spindle residues, the centripetal gathering of new bridges after each set of syncytial mitoses, and spindle orientation with one pole next to the fusome, and shows that these would result in the distribution of intercellular bridges found in mature sibling clusters. The diagram is a synthesis of old and new information, and, while speculative in nature, seems to be the most plausible way to account for what is presently known.

There have been two additional efforts to explain bridge distribution, one by Hirschler (1942, 1945, 1955), and one by Koch, Smith and King (1967), but neither of these utilize the process of rosette formation. Hirschler's is similar to that proposed here in attributing to the fusome an ability to orient the mitotic spindle. Beyond that, his proposal has proved too difficult to understand to allow a critical evaluation in this review.

Koch *et al.* proposed a model in which the bridges remain fixed at their site of formation, rather than being gathered centripetally after each division. Such bridges would be especially vulnerable to separation from each other at the next cytokinesis if it were not for a precise control of spindle orientation. They proposed that this would be achieved if the centriole, after each cystoblast division, should migrate to a new position 90° around the nucleus before dividing and then giving rise to a spindle (Fig. 13).

The trouble with this model is that it postulates a cell surface rigidity which is at variance with the early observations on rosette formation. Rosettes have been clearly figured in *Drosophila* (Koch and King, 1966), as well as in many other species, and the available evidence therefore points to the centripetal gathering of bridges as being of general occurrence. Mahowald (1971) observed that the adjacent surfaces of newly divided cells in the *Drosophila* sibling cluster are highly folded, and this too suggests a dynamic cell surface at the completion of bridge formation. A similar configuration is seen in Fig. 4.

Apart from the question of bridge position stability, the suggestions of Hirschler and of Koch and King have an important element in common with that proposed here, for all three emphasize factors that determine centriole location and spindle orientation in relation to pre-existing bridges. It is only necessary to propose that the centriole forming one of the spindle poles becomes anchored to the fusome to bring them into substantial agreement. In effect, Maziarski's illustration suggested that the fusome may be responsible for spindle orientation two decades before the necessity for such a mechanism became apparent.

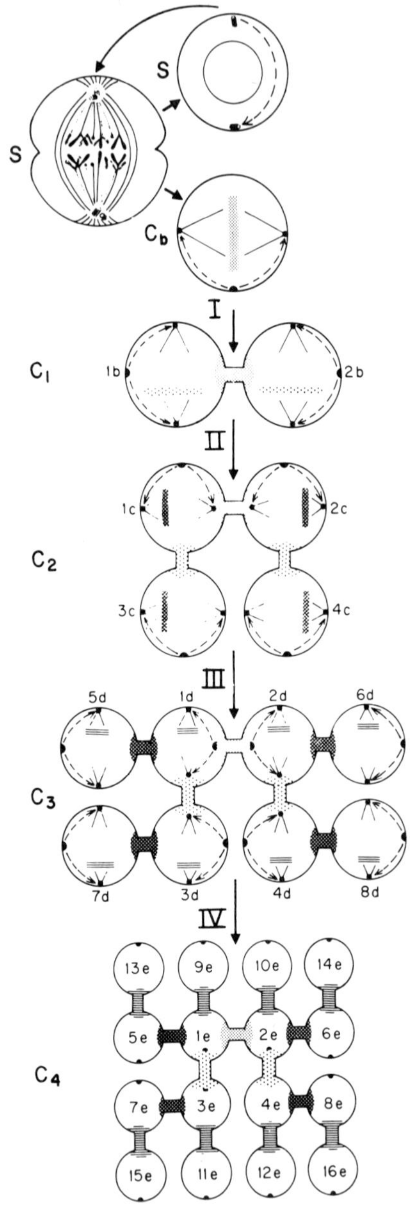
S
S
Cb
I
C1
1b
2b
II
1c
2c
C2
3c
4c
III
5d
1d
2d
6d
C3
7d
3d
4d
8d
IV
13e
9e
10e
14e
5e
1e
2e
6e
C4
7e
3e
4e
8e
15e
11e
12e
16e

Fig. 13.

3.4 SYNCHRONY, THE 2^n RULE, AND MITOTIC PROGRAMMING

A second aspect of developmental control in the sibling cluster is the number of mitotic divisions entailed. The number of nurse cells can be described by the term, $2^n - 1$, almost without exception. The value of n, which represents the number of times the cystocytes divide, is a constant for each species. Oöcyte formation from one of the siblings necessitates the subtraction of one. The principle was first published by Perez (1886), and Giardina (1901) attributed it to the mitotic synchrony resulting from the presence of intercellular bridges. There have been reports of deviations from the 2^n rule, including two nurse cells per oöcyte in *Panorpa* (Lubbock, 1859), five for Lepidoptera (Gross, 1903), and about 48 for *Apis* (Paulcke, 1900); but, as Hirschler (1945) pointed out, the first two resulted from casual impressions rather than from the serial section analysis that is required for greatest certainty. *Apis* does not fall into this category, but as will be seen later, may well be the exception that proves the rule. There are, in any case, no grounds at this time for doubting that the mitotic synchrony made possible by the intercellular bridges is basic to the 2^n rule.

Three properties of intercellular bridges are essential to their accounting for the 2^n rule. As in other syncytia, synchrony implies that cytoplasmic factors transmissible from nucleus to nucleus control the phases of the cell cycle, and it is clear that to be exchanged across the intercellular bridges these must be able to permeate the fusome. This matter is examined more fully in a later section. Second, the bridges must be stable enough to persist throughout the mitotic phase of sibling cluster development. Indeed, the bridges were seen at the outset to persist even into vitellogenesis as ring-shaped structures several microns wide, linking adjacent nurse cells and the oocyte (Giardina, 1901; Gunthert, 1910; Maziarski, 1913; Hegner, 1915). And third, once bridge formation commences it must occur at all subsequent divisions, so that all the mitotic progeny of one cystoblast remain bridged together. Thus, Johnson and King (1972) showed that in a female sterile mutant of *Drosophila* failure to form bridges at some divisions results in sibling clusters with cell numbers such as 3, 5, 6, 7, 9, 10 and 11.

Synchronous divisions and bridge formation normally continue until the

Fig. 13. A proposal on how the bridges become distributed in *Drosophila*, from Koch *et al.* (1967). As in Fig. 12, the plane of cytokinesis is equatorial to the spindle. The orientation of the spindle is determined by the centrioles. Before a spindle is formed, the centriole divides and each product migrates 90° in directions opposite to each other. As a result, each cytokinesis is oriented perpendicularly to its predecessor. Rosette formation and the incorporation of all bridges by one fusome are not utilized by this model.

establishment of a number of cells that is specific for the species and, in many cases, for the order of the insect. Lubbock (1859) figured one nurse cell for each oöcyte in the follicles of a dermapteran and this observation was subsequently confirmed for other species in the same order (e.g. Bonhag, 1956; Nath *et al.*, 1959; Engels, 1970; Zinmeister, 1973). *Panorpa* has three nurse cells per follicle (Ramamurty, 1967), and Knaben (1934) found that 11 of 12 species of Lepidoptera surveyed had 7 nurse cells, though one, *Ordenodes hexadactyla*, had only 3. Hirschler (1942), Colombo (1957) and King and Aggarwal (1965) described additional cases of 7 nurse cells in Lepidoptera. Species with 15 nurse cells occur among dytiscid beetles (Giardina, 1901; Gunthert, 1910), Diptera (Brown and King, 1964), and Hymenoptera (P. King and Richards, 1969). Other Hymenoptera contain 31 nurse cells (Meng, 1970; Cassidy and King, 1973), and, as has been noted, *Apis* contains approximately 48.

Intercellular bridges and synchronous mitosis are common among spermatogenic cells as well as in oögenesis, and here 2^n reaches values as high as 64, 128, and 256 (Phillips, 1970), again depending on the species. King and Akai (1971) point out that the 64 spermatids of the *Drosophila* sibling cluster would be formed by 4 mitotic and 2 meiotic divisions and that the 16-cell female sibling cluster would entail the 4 mitotic divisions alone. In Lepidoptera, by contrast, 3 mitotic divisions yield the 8-cell female cluster, while 6 mitotic and 2 meiotic divisions would be required in the male. It is clear that a species-specific and, in some cases, a sex-specific mechanism exists for determining the number of syncytial mitoses that follow the first cystoblast division.

While there are as yet no clues as to how n is controlled, there is evidence in *Drosophila* that a basic component of the mitotic apparatus is lost by the nurse cells with the onset of differentiation. Soon after the fourth set of mitoses, at which the cluster attains its final number of 16 cells, the centrioles begin to disappear from the nurse cells, and to accumulate in the oöcyte (Koch and King, 1969; Mahowald and Strassheim, 1970). Mahowald and Strassheim found 14 to 17 of the possible 32 centrioles in the oöcyte; any that remained in the nurse cells tended to be adjacent to the cell membrane, or close to the intercellular bridges, rather than next to the nucleus. Whether centriole release from its characteristic juxtanuclear position is in the causal chain of events is not clear. As will be seen below, all aspects of the cell cycle appear to continue in the nurse cells, except for nuclear membrane breakdown, spindle formation, and cytokinesis.

Another approach to the problem of how n is determined has utilized developmental analyses of female sterile mutants. An example in which mitotic control is lost and tumours rather than follicles are generated was described by King *et al.* (1957) and has since been extensively analysed (literature summarized by Johnson and King, 1972). A key feature of the mutant germarium is that stable intercellular bridges sometimes fail to form. Johnson and King proposed that the wild-type gene specifies a substance stabilizing the bridges, and that formation of the proper number and distribution of bridges is an essential component of the mechanism that stops cell division at 16 cells. The case serves to illustrate that a search for mutants that affect the number of divisions is a promising approach. Knaben (1934) found that some individuals of a species of *Macrothylacia* contained 15 nurse cells rather than the usual 7. The individuals had been collected in nature, however, and Hirschler (1942) was unable to repeat the observation. Gill (1963) described five additional female sterile mutants in *Drosophila* in which the number of cells in the sibling cluster was 8, 32, or 48, but a developmental analysis of these cases has not been completed. A recent effort to produce mutations in *Drosophila* with ethylmethanesulfate (Baaken, 1973) resulted in 98 recessive female sterile mutations, which turned out to be at 38 different autosomal loci. Two of these were recognized to affect germarial function, but in such a profound way that neither follicles nor ovarioles were produced.

Where multiples of the normal number of cells are produced, it is of course possible that the defect is in the control not of mitosis, but of the number of sibling clusters invested by the follicular epithelium in one follicle. Mutations affecting the control of n will probably prove to be more easily identified in spermatogenesis where processes comparable to follicle formation and oöcyte-nurse cell differentiation do less to obscure the outcome of the synchronous mitoses.

3.5 OÖCYTE-NURSE CELL DETERMINATION

The fidelity with which only one cell in a sibling cluster ultimately forms an oöcyte is an additional aspect of developmental control in the germarium. There are indications that a particular cell in each sibling cluster is predetermined to be the oöcyte for, as has already been noted, the extrachromosomal DNA body and the fusome in *Dytiscus*, and a high number of intercellular bridges in all other species, are always associated with this cell. There have been several proposals as to how this particular cell becomes committed to form the oöcyte, while all others in the cluster transform into nurse cells. The first was by Giardina (1901), and was based on the behaviour of the extrachromosomal DNA body. As discussed earlier, this body arises in the interphase prior to the first division of the cystoblast and remains throughout the synchronous phase of development as a single structure that is passed to only one spindle pole at each of the four successive mitoses. Since after the final division it was seen to lie in the only cell that would become an oöcyte, the chromatin body was an obvious candidate for the oöcyte determinant. The formation and distribution of the extrachromosomal chromatin body were confirmed in *Dytiscus* by Debaisieux (1909) and Gunthert (1910) and were found by the latter and by Hegner and Russell (1916) to occur in other genera of the same family. As will be seen in a later section of this review, it contains amplified chromosomal DNA, including the cistrons for ribosomal RNA.

Giardina's discovery stimulated a search for similar structures in other groups of insects. While chromatin bodies were soon discovered in a panoistic ovary (Buchner, 1909), comparable phenomena were not seen in other polytrophic systems until a variant on the dytiscid story was found in the crane fly *Tipula* (Bayreuther, 1952). Extrachromosomal DNA bodies are produced in this insect, but they arise one division later than in *Dytiscus*. In conformity with the general rule that members of a sibling cluster behave identically during the mitotic phase of development, the two cells that remain bridged together after the first cystoblast division both produce DNA bodies (Fig. 14). At the 16-cell stage, the oöcyte retains one of these, while the other lies in one of the nurse cells. An additional determinant must therefore be postulated to explain why only one of the two cells with DNA bodies becomes an oöcyte. This behaviour, in combination with the absence of DNA bodies in most other polytrophic systems, suggested to Bauer (1952) and to Bayreuther (1956) that when such structures do occur, they are followers of an as yet unrecognized oöcyte determinant, rather than themselves the primary cause of oöcyte differentiation.

A second proposal arose from the demonstration that the oöcyte is formed in the lepidopteran eight-cell cluster by one of the two cells with three intercellular bridges. A trophic mechanism of oöcyte determination was proposed on the basis of this information (Hirschler, 1942), with the oöcyte developing from what was considered by virtue of its bridges to be one of the two best fed siblings. This postulate, however, assumed a pre-existing polarity of nutrient movement across the bridges, for otherwise the oöcyte with its three bridges could as well be the most heavily taxed donor of nutrients as the wealthiest recipient. It is not clear how polarity with a proper direction could arise between cells that are identical in all regards except bridge number, so that as with Giardina's theory, it is necessary to postulate an additional factor.

The most plausible model thus far proposed is that of Koch, Smith and King (1969) who suggested that oöcyte determination is a consequence of activities in the cortical cytoplasm associated with the four intercellular bridges attached to this cell. Koch and King (1966) found that the nurse cell attached to four bridges forms synaptonemal complexes early in its development in *Drosophila*, and this suggests that the particular bridge attaching it to the oöcyte contains factors important in oöcyte differentiation. The theory is consistent with Knaben's and Hirschler's observations on the number of bridges connected to the oöcyte. It is also consistent with Giardina and Bayreuther's findings that the DNA body goes to the oöcyte, for Giardina (1901), Debaisieux (1909), and Gunthert (1910) all observed that during each mitotic division in *Dytiscus* this structure moves to the spindle pole

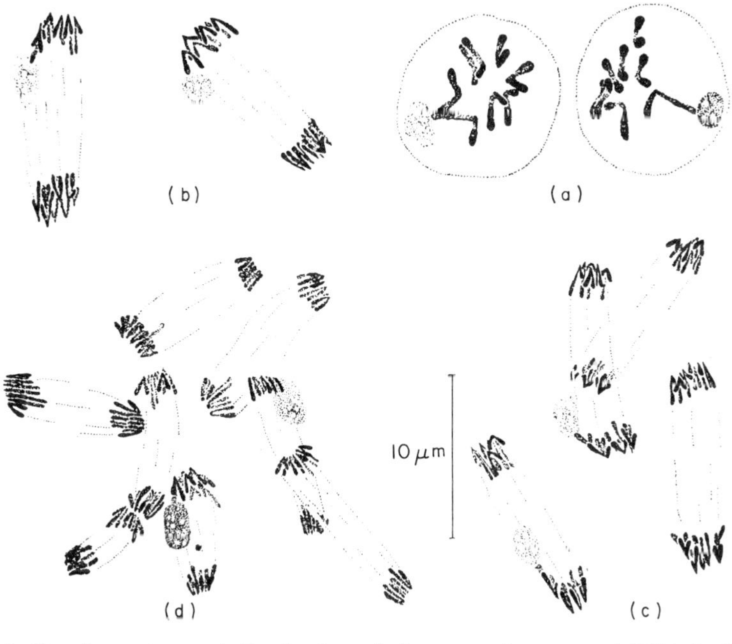

Fig. 14. The formation and distribution of the extra-chromosomal DNA bodies in *Tipula*. (a) Prophase nuclei of a two-cell cluster. The DNA bodies had formed in synchrony during the preceding interphase. The cytoplasms and intercellular bridge connecting the two cells are not shown. (b) Anaphase of the second division, showing that the DNA bodies are not divided, but that each goes to one pole of their respective spindles. (c) and (d) Similar behaviour during the third and fourth anaphases. The 16-cell cluster thus has two cells containing extra-chromosomal DNA bodies. This results from the fact that DNA body formation occurs after the first cystoblast division, rather than before as in *Dytiscus*.

toward the centre of the rosette where the fusome and pre-existing bridges are located. All observations so far are therefore consistent with the concept of an oöcyte determinant that is associated with the first formed intercellular bridge. As will be seen below, many other observations can be explained if it is assumed that one of the earliest tasks of the oöcyte determinant is to polarize the net movement of materials in this bridge.

The possibility that inductive interactions with other cells in the germarium are important in oöcyte determination has also been considered.

TABLE 1

The stage of sibling cluster development at which synchrony is lost and differentiation begins

Stage	Insect	Evidence
Cystoblast		
Mitosis I		
←	*Apis*	$2^n \simeq 48$; not all siblings join in the last set of mitoses
Mitosis *n*		
Premeiotis S-phase		
←	*Drosophila, Habrobracon, Aedes*	*n* is a whole number; synaptonemal complexes formed in some, but not all siblings
Synapsis		
←	*Carabids,* Lepidoptera	synaptonemal complexes formed in all siblings
Germinal vesicle growth		
First meiotic metaphase		

The clearest case is the suggestion by Koch and King (1969) that the decision as to which of the two cystoblasts with four bridges will become the oöcyte depends on which is the first to contact the profollicle cells at the base of the germarium. It is easy to imagine that interaction with follicle cells may be an important early step in oöcyte differentiation, but the behaviour of the DNA body in *Dytiscus* suggests that the oöcyte may be determined from the outset by differential mitosis, and it is difficult to see how the epithelium could be involved in this particular aspect of the process.

As long as it is based exclusively on descriptive and comparative morphology, any theory of oöcyte determination will necessarily remain speculative. Indeed, considerations presented in an ensuing section (Table 1) make the question seem in one respect misleading, for an argument can be made that all cells in the sibling cluster are programmed at the outset to form oöcytes. If this is correct, the "oöcyte determinant" in the sense used here serves the function of preventing one cell in the sibling cluster from being programmed at a later time to differentiate as a nurse cell.

4 Homologies with other insect ovaries

Cystoblasts, or their equivalent, in telotrophic and panoistic ovarioles necessarily develop in ways that differ fundamentally from that described here for polytrophic systems. It can be imagined, for instance, that telotrophic ovarioles would result from the retention of prospective nurse cells in the sibling clusters by the germarium, and by attenuation of intercellular bridges as the oöcytes move basally into the vitellarium. Panoistic ovarioles have been assumed to result from the cystoblasts' differentiating directly into oöcytes without prior mitosis. While these particular speculations have appeared now and then in the literature, the ontogeny and function of the germarium in telotrophic and panoistic ovaries remain inadequately described to permit the elimination of other possibilities.

There are indications that telotrophic and panoistic ovaries may in fact both prove to include more than one variant of the polytrophic developmental pattern. Büning (1972) proposed that the trophic chamber of Hemiptera is sufficiently different from that of the polyphagous Coleoptera to suggest an independent evolutionary origin. There is no reason to assume that panoistic ovaries, which occur in a much wider diversity of insect orders, will show any great uniformity.

In polytrophic systems, as we have seen, a germarial product transforms into a cystoblast, which forms a clone of 2^n cells, producing intercellular bridges by incomplete cytokinesis at each successive mitosis. The entire clone is enclosed in a single follicle, and a predetermined member becomes the oöcyte. The question raised here is: How in fact is this developmental programme varied to produce teletrophic and panoistic ovarioles?

4.1 GERMARIAL FUNCTION IN TELOTROPIC OVARIES

The origins of the apical nurse chamber and its relation to the oöcyte in telotrophic ovaries are not as clear as that between polytrophic nurse cells and oöcytes. The studies summarized below have shown that a population of germ cells comprising the nymphal or larval germarium differentiates during metamorphosis to form both nurse cells and oöcytes, and that these are attached to each other in the adult by attenuated cords of cytoplasm. In view of these findings, it is tempting to speculate that the trophic cords are homologous to the intercellular bridges produced by incomplete cytokinesis in polytrophic systems, but the evidence for such an interpretation is still incomplete. The lack of clarity results from the fact that the cell lineage of the oöcyte and nurse cells cannot be traced as they have been in the polytrophic ovariole, for cellular fusion occurs in the nurse chamber during metamorphosis, producing a syncytium which thwarts any effort to reconstruct sibling relationships from adult morphology.

In last instar nymphs of Heteroptera and larvae of polyphagous Coleoptera, the ovariole consists of a germarium attached at its base directly to the pedicel (Wick and Bonhag, 1955; Schlottman and Bonhag, 1957). In an electron microscope study of the larval germarium of the beetle *Bruchidius*, Büning (1972) discovered that the germ cells are connected to each other by intercellular bridges, but these disappear during metamorphosis when cell fusion converts the nurse cells to a more intimate syncytium. In *Rhodnius* also the cells of the premetamorphic germarium are linked by intercellular bridges (Huebner and Anderson, 1972a), but here there is an additional morphological element. The cells of the germarium produce cytoplasmic processes which converge in a central zone containing a tangled mass of such processes. The configuration reminded the authors of the neuropyle in a ventral ganglion. Many of the intercellular bridges in the germarium are between these processes, but others connect the more peripherally located cell bodies. The possibility that intercellular bridges present during premetamorphic stages of development are precursors of the trophic cords connecting the nurse tissue and oöcytes of the adult was raised in both of these studies, though a direct description of such a transformation has thus far not been possible.

After metamorphosis in Heteroptera the nurse chamber differentiates into an acellular core with a fibrous texture (Fig. 15(a)) surrounded by strands of fusing and autolysing cells (Gross, 1903; Schrader and Leuchtenberger, 1952; Bonhag and Wick, 1955; Anderson and Beams, 1956; Eschenberger and Dunlap, 1966). From the central region, which has been termed the trophic core, fibrous branches arise that penetrate the surrounding nurse tissue. At the base of the nurse chamber the trophic core is continuous with the cytoplasmic cords leading to the oöcytes. Huebner and Anderson (1972a) described the genesis of the trophic core from the neuropyle-like tangle of cell processes in the nymphal germarium. Autolytic breakdown of nurse cells is one component of the process, and this results in a disappearance of cell membranes and intercellular bridges. Another striking feature of the transformation is the production of an exceptional concentration of microtubules (Hamon and Folliot, 1969; Brunt, 1970; Huebner and Anderson, 1970; Macgregor and Stebbings, 1970). These structures traverse the trophic core and its branches between the nurse cells, as well as coursing the length of cytoplasmic cords to the oöcytes. A discussion of their function is included in a later section.

In beetles the adult nurse chamber is a relatively homogeneous syncytium in which a trophic core does not develop. Büning (1972) showed that the syncytium of *Bruchidius* is more complete than had been thought from light microscope studies of other species (e.g. Schlottman and Bonhag, 1957). A meshwork of interstitial cells is embedded in the syncytium, and with low resolution this gives the chamber an appearance of being compartmented into individual nurse cells. The cytoplasmic cords to the oöcytes attach at the basal end of the syncytium (Fig. 15(b)), and are much narrower than those of the Hemiptera, so that they have sometimes been overlooked altogether.

As in other insects, the prefollicular oöcytes lying at the base of the nurse tissue after adult eclosion subsequently move basally, one by one, become invested with a follicular epithelium (Eschenberg and Dunlap, 1966) and finally commence yolk deposition. The configuration has led to a search for oögonia at the base of the nurse chamber where the most immature identifiable oöcytes occur (Fig. 15), but mitotic activity has never been observed in the germ cells of this region (Bonhag and Wick, 1955; Nath *et al.*, 1959; Huebner and Anderson, 1972b). Tritiated thymidine, though readily incorporated by other cells in the adult ovary (Vanderberg, 1963; Zinmeister and Davenport, 1971b) does not appear in young oöcytes even after seven to eight days of incubation (Büning, 1972; Mays, 1972). If oögonia or premeiotic interphase cells are present in the adult, therefore, they require over a week for transformation into recognizable meiotic prophase cells.

Nurse tissue too shows no indication of mitotic activity in polyphagous Coleoptera adults (Schlottmann and Bonhag, 1957), but in many Heteroptera a mitotic zone persists at the apex of the nurse chamber (Schrader and Leuchtenberger, 1952; Bonhag, 1955; Nath *et al.*, 1959; Vanderberg, 1963; Eschenberg and Dunlap, 1966; Brunt, 1971; Huebner and Anderson, 1972a). The morphology of the system suggested to Schrader and Leuchtenberger that cells produced at the apex move basally,

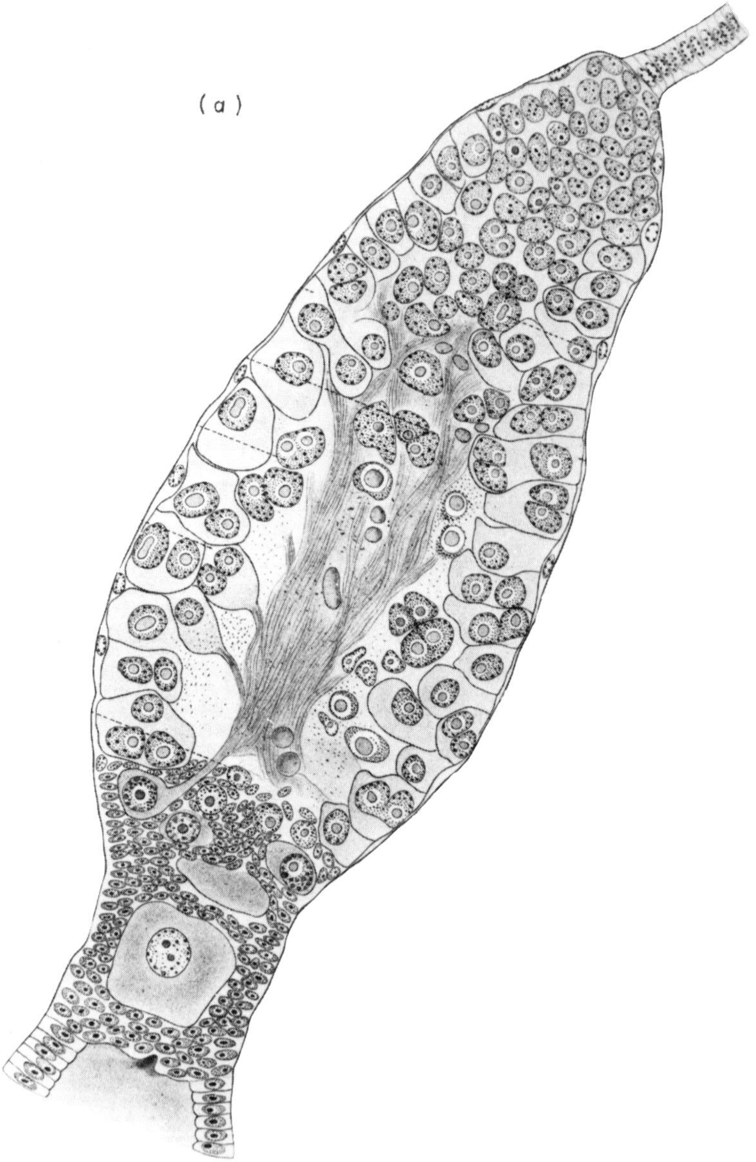

Fig. 15. Nurse chambers of telotrophic ovarioles. (a) The heteropteran chamber. A central, fibrous core containing degenerating cells is attached to oöcytes in the prefollicular and vitellarial regions by strands of cytoplasm. There is apparently a cellular turnover, with mitosis at the apex of the nurse chamber replacing cells that are lost to the trophic cord. (From Gross, 1903.) (b) The nurse chamber of the beetle, *Bruchidius*. In polytrophic Coleoptera the nurse cells form a stable syncytium in which cell turnover and a fibrous trophic core do not occur. (From Büning, 1972.)

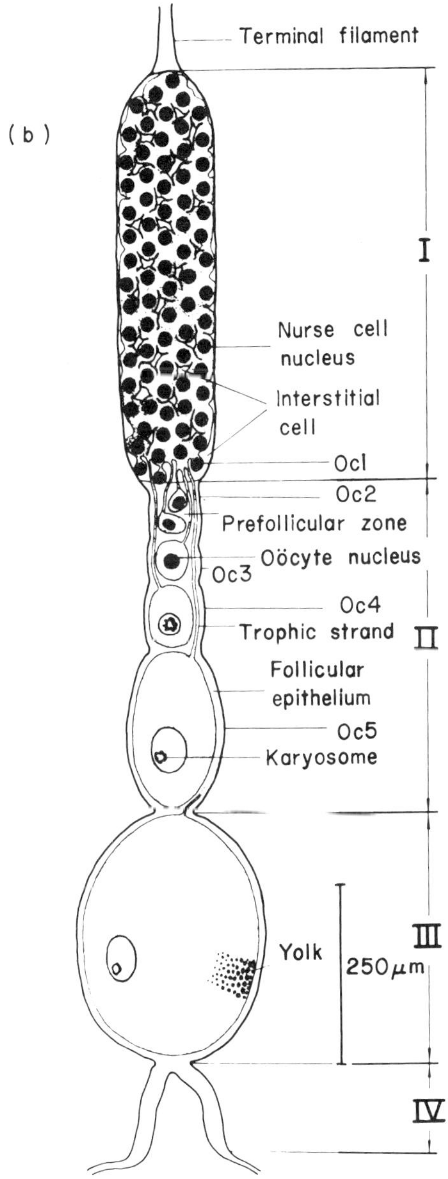
(b)
Terminal filament
I
Nurse cell nucleus
Interstitial cell
Ocl
Oc2
Prefollicular zone
Oöcyte nucleus
Oc3
Oc4
Trophic strand
II
Follicular epithelium
Oc5
Karyosome
III
Yolk
250μm
IV

Fig. 15(b).

fuse with each other, and finally break down, yielding autolytic products to the trophic core. Time course studies of thymidine labelled nurse chambers confirmed that cells generated in the apex tend to move basally (Mays, 1972). While a turnover of cells in the heteropteran nurse chamber is therefore well established, there are no indications of such a cycle in the coleopteran nurse chamber.

Returning finally to the question of how the trophic cords arise, there is one indication that secondary fusion rather than attenuation of pre-metamorphic intercellular bridges may be important in their origin. This is a reference by Mays (1972) to unpublished evidence that the youngest oöcytes in the newly eclosed adult *Pyrrhocoris* are not yet connected to the trophic core. Demonstrating that these oöcytes in fact make such connections at a later time would establish the occurrence of fusion in this species.

The salient parallels between polytrophic sibling clusters and telotrophic ovarioles include the origin of trophic tissue from germ cell derivatives, and the maintenance of cytoplasmic channels for polarized transport. There will be little ground for pressing homology beyond these facts, however, until more extensive information is available on the ontogeny of the nurse chamber and the origin of the oöcytes. In gauging the extent of the departure from polytrophic systems, it will be particularly important to know, for instance, whether the intercellular bridges of the immature germarium allow synchronous mitotic activity, whether there are discrete sibling clusters in the immature germarium and, if so, how they are arranged, whether differentiation between oöcytes and nurse cells occurs within sibling clusters, and whether differentiation between these cells commences with the intercellular bridges still intact. The observations of Büning and of Huebner and Anderson have only recently made these questions apparent, and it is too early to know whether they can be easily answered.

The telotrophic ovary has not yet had its Giardina and *Dytiscus*, but this is not surprising in view of the morphological confusion that results from cellular fusion during its metamorphosis.

4.2 GERMARIA IN PANOISTIC OVARIES

In making homologies with polytrophic systems, it has been traditional since Perez introduced the 2^n rule in 1886 to set $n = 0$ for panoistic germaria. The argument implies that the oögonia in these systems transform directly into oöcytes with no intervening cystoblast divisions. The fine structure of the transformation received scant if any attention, however, and it is reasonable to wonder whether a comparative survey would in fact reveal such a direct relationship in all panoistic species.

Identifying a species as panoistic has generally not been based on a detailed analysis of germarial development, but on the more easily observed fact that the oöcyte in a maturing follicle is not bridged to nurse cells. There are, however, several ways in which such an end could be reached in addition to direct transformation of oögonia into oöcytes. Early disintegration of nurse cells was proposed by Buchner (1909) in *Gryllus*, for instance, and one can also speculate that sibling clusters might separate into individual cells before follicles are formed at the base of the germarium.

One as yet unexplored approach to this question would be to establish whether intercellular bridges occur in panoistic germaria. These structures are basic to the definition of the cystoblast in polytrophic systems and their presence in a panoistic ovary would be inconsistent with the concept that $n = 0$. Vertebrate ovaries, which are panoistic in the sense that nurse cells are not produced, nevertheless contain at pre-follicular stages of development clusters of synchronously developing germ cells connected by intercellular bridges (Franchi and Mandel, 1962; Ruby *et al.*, 1970; Skalko *et al.*, 1972). The transition from sibling clusters to oöcytes isolated in follicles is accompanied by much cell death, and nothing comparable to the insect nurse cell ever develops. Though light microscopy has revealed no comparable phenomenon in panoistic insects, Buchner's suggestion for *Acheta* fits this pattern exactly. It is unfortunate that there are as yet no published surveys of germarial fine structure that have paid explicit attention to the question of whether intercellular bridges are formed, and, if so, what the fate of the siblings may be.

A second approach to germarial function in panoistic ovarioles concerns the origin and mitotic distribution of extrachromosomal DNA bodies. Light microscopy, particularly after Feulgen staining, has shown that such structures occur in many panoistic ovaries (Buchner, 1909; Johnson, 1938; Bayreuther, 1957; Bier *et al.*, 1967; Allen and Cave, 1972). In *Gryllus* its origin in the germarium was described by Buchner (1909), and some aspects of his findings were more recently confirmed by Lima-da-Faria *et al.* (1968). The DNA body first appears during an oögonial interphase and at the subsequent metaphase lies at one side of the chromosomal plate without apparent attachment to the spindle. During anaphase it moves to one pole and is incorporated into a nucleus at telophase. The division product receiving the DNA body differentiates into an oöcyte and in this regard the system behaves identically to the cystocytes of *Dytiscus* and *Tipula*. Neither Buchner nor later workers clarified the developmental fate of the cell not receiving the DNA body. Operating under the influence of Giardina's findings on *Dytiscus*, Buchner regarded the second cell as a potential nurse cell that undergoes premature degeneration, but his observations did not rule out the possibility that it remains an oögonial stem cell. Without clarification of this question, it will be impossible to say whether n

is zero in *Gryllus*, or whether incipient nurse cells are actually formed but fail to differentiate.

Bayreuther (1957) described in the siphonapteran, *Nosopsyllus*, a DNA body that is distributed at mitosis in a manner different from that of *Gryllus*. The structure lies within the equatorial plate during metaphase rather than at one side as in *Gryllus, Tipula* and *Dytiscus*, and actually subdivides during anaphase, each daughter cell inheriting part of it. Finally, there is the case of *Periplaneta* in which Zinmeister and Davenport (1971a) could find no evidence of extrachromosomal DNA body formation.

The contrast between *Gryllus*, *Nosopsyllus* and *Periplaneta* suggests that panoistic ovaries may prove to be a very mixed lot. A comparative survey entailing a search for intercellular bridges, the behaviour of DNA bodies and any other clues to post-oögonial cell lineages and differentiation is still badly needed.

5 Differentiation of nurse cells and oöcyte

As a rule, the chromosomes of animal oöcytes proceed to the pachytene or diplotene stage of meiosis before cytoplasmic growth and yolk deposition are launched. There is then a pause in the meiotic process, while the nucleus, in support of cytoplasmic growth, enters a period of accelerated RNA synthesis that may last for days or weeks, according to the species. Nuclear volume increases nearly apace with the cytoplasm, so that a germinal vesicle, vastly greater in size than an ordinary diploid nucleus, is formed. At the end of egg formation, meiosis resumes, RNA synthesis terminates, the germinal vesicle breaks down, and the chromosomes finally become aligned as tetrads on a metaphase plate.

These general characteristics of oögenesis occur in insects as well as in other animals, with the singular exception that in meroistic ovaries RNA synthesis usually fails to accelerate in the germinal vesicle. The nurse cells usurp this crucial function of the meiotic chromosomes, and the RNA they produce is passed to the oöcyte through the intercellular bridges or trophic cords. In all other regards, including meiotic arrest and germinal vesicle enlargement, oöcytes associated with nurse cells develop in accord with the more generalized pattern.

Nurse cells, as will shortly be seen, also initiate meiosis in many species, but their chromosomes are at an early stage diverted from the oöcyte pathway and commence instead a programme of endomitosis. In contrast to the lock-step development characterizing the establishment of the sibling cluster, oöcyte and nurse cell nuclei begin an independent behaviour, despite the persistence of their intercellular bridges.

By the time the sibling cluster has been enveloped in an epithelium and entered the vitellarium, the oöcyte has already begun its meiotic arrest and the nurse cells have commenced endomitosis. The system is still too small and too secluded to have been available for physiological analysis. As a consequence, the physiological basis of the transition from synchrony to asynchrony and the differentiation of nurse cells from the oöcyte pathway remain unexplored. There is, however, a substantial number of morphological descriptions of the transition in a variety of insects, and a summary of these can do much to put the problem in focus.

5.1 THE END OF SYNCHRONY

The period of synchrony in a polytrophic sibling cluster, having yielded an exact number and arrangement of cells, is followed by meiosis and the differentiation of the nurse cells. There are species differences in the programming of these events, and in particular a substantial variation in the stage of development at which synchrony is finally lost. In Lepidoptera (Grünberg, 1903; Dederer, 1915; Colombo, 1957), and probably in Trichoptera (Marshall, 1907), all cells in the sibling cluster initiate meiosis together. Similar behaviour was recently described in three species of carabid beetles (Ribbert and Weber, 1970) where electron microscopy showed that there are synaptonemal complexes between the homologous chromosomes. In all of these species the homologous chromosomes of all of the siblings undergo synapsis, so that prospective nurse cells are seen to progress at least as far as the zygonema stage of meiotic prophase. This condition is maintained for about two weeks during the last larval instar of *Hyalophora* (Mandelbaum, 1974) and oöcyte-nurse cell differences do not become evident until just before the animal metamorphoses.

A different course of events occurs in some Diptera and Hymenoptera. While synaptonemal complexes occur in prospective nurse cells of *Aedes* (Roth, 1966) and *Drosophila* (Koch and King, 1966), they were found to be restricted in the latter to only two of the 16 siblings. These two cells, which Koch and King termed pro-oöcytes, proved to be the only members of the cluster that are joined to four intercellular bridges. One of the two *Drosophila* pro-oöcytes later abandons its differentiation as an oöcyte and becomes a nurse cell instead. All of the remaining 14 cells move directly from mitosis to the nurse cell pathway without detectably initiating meiosis. Direct differentiation into nurse cells apparently also can occur in the 32-cell complex of *Habrobracon*, but as in *Drosophila* some of the prospective nurse cells also appear to form synaptonemal complexes (Cassidy and King, 1973).

The observations suggest a generalized developmental scheme in which the cystoblast and its progeny are programmed to undergo *n* mitotic divisions and then enter meiosis (Table 1). At some stage of development, varying with the species, many cells are diverted from this programme and redirected to form nurse cells. Phylogenetic variation is manifested in this scheme not only by the value of n, but also by the precise point in the programme at which the diversion of cells from the oögenic pathway occurs.

Alternative schemes could be proposed, including an early, noncommittal programme of development, followed by segregation of the cells along two divergent pathways. In view of the demonstrated tendency for prospective nurse cells to enter meiosis, however, the scheme shown in Table 1 best fits the facts that are presently known.

The model raises the question of whether the loss of synchrony might occur in some species at an even earlier stage than it does in *Drosophila*. A good candidate for this would seem to be *Apis* in which the number of nurse cells has been estimated to be about 48, a clear violation of the 2^n rule (Paulcke, 1900; Engels, 1968). Unless some cystoblast progeny are lost from the sibling cluster in the germarium prior to follicle formation, this result could only be achieved if some cells terminate division before the mitotic programme of the follicle as a whole has been concluded. Paulcke suggested that the prospective oöcyte terminates mitotic activity in the four-cell cluster, and that the remaining three cells, which are thus pro-nurse cells, each divide four more times to produce a total of 48 nurse cells. Engels (1968), however, found that the oöcyte of *Apis* is attached to five intercellular bridges, which means that it must have divided five times rather than twice, as Paulcke's model suggests. Assuming that Paulcke's model is wrong, and that all cells divide synchronously five times, a sibling cluster of 32 cells would result. To generate 48 cells, half of these, excluding the prospective oöcyte, would have to divide again. While a developmental analysis sufficiently detailed to identify the cells undergoing the extra mitosis has yet to appear, the evidence so far points strongly to the onset of asynchrony before the mitotic programme has been completed. *Apis* can therefore best be fit into the model shown in Table 1 by diversion to asynchrony at an even earlier stage than *Drosophila.*

The same sort of diversity apparently occurs in telotrophic ovarioles. Büning (1972) discovered that at the end of the larval period in the beetle *Bruchidius* all germ cells contain synaptonemal complexes, and it can be concluded from this that the prospective nurse cells of this species initiate meiosis before proceeding on their special course of differentiation. By contrast, a comparable analysis of metamorphosis in the *Rhodnius* germarium (Heubner and Anderson, 1972a) did not reveal synaptonemal complexes. These two species thus appear to vary in the stage at which nurse cell differentiation begins, in the way suggested for polytrophic ovarioles in Table 1.

5.2 ASYNCHRONY IN NURSE CELL DEVELOPMENT

The divergence in behaviour of oöcyte and nurse cell nuclei is one symptom of asynchrony, but in addition to this is the evidence that the nurse cells become out of phase with each other in their endomitotic cycles. Early evidence that the endomitotic cycles are asynchronous was reported in cytological examinations of *Drosophila* (Painter and Reindorp, 1939) and *Calliphora* (Bier, 1957). As the nurse cells undergo endomitosis, their chromosomes show a typical cell cycle behaviour. In material fixed for microscopic examination some nurse cell nuclei contain diffuse, interphase chromatin, while others in the same follicle contain prophase chromosomes in various degrees of condensation. What were interpreted as metaphase and anaphase chromosomes were also seen, but since a mitotic spindle does not form and the nuclear envelope does not break down, these remain within one nucleus, falling apart and becoming diffuse again at the next interphase. Although reports of this behaviour were primarily concerned with piecing together the steps in the chromosome cycle, the descriptions indicate that the chromosomes of neighbouring cells within one follicle are often at very different stages in their cycle of replication. A similar situation was shown by King (1970) in his Fig. VI-1, S-5.

Labelling the S-phase of the endomitotic cycle with thymidine-H^3 has also demonstrated a lack of synchrony between nurse cells in *Dytiscus* (Urbani and Russo-Caia, 1964; Urbani, 1970), *Drosophila* (King and Burnett, 1959; Chandley, 1966), *Hygrobia* (Urbani and Pezzoli, 1970), and *Hyalophora* (unpublished observations). A single injection of tritiated thymidine into the haemocoele results in the labelling of some, but not all, nurse cells in the same follicle. The thymidine behaves as a pulse of label that is utilized or destroyed before those cells not initially in the S-phase have an opportunity to incorporate it. In *Hyalophora* a further distinction can be made between dispersed chromatin which may be labelled in one cell, and condensed, heteropycnotic DNA which may be labelled in another. Here as in other systems (Lima-de-Faria, 1959; Guelin, 1968), heterochromatin is apparently replicated at a different time in the cell cycle than euchromatin.

Other evidence for asynchrony is seen in the fact that the nurse cells in several species of Diptera contain amounts of DNA that vary with their position in the follicle. In the later stages of follicle development in *Drosophila* the nurse cells adjacent to the oöcyte achieve a ploidy that may be 2 or 4 times greater than those of the more distally located cells (Jacob and Sirlin, 1959). A similar relationship can be seen in the volume of the nuclei in several species of Diptera (Verhein, 1921; Hertwig, 1935; Brown and King, 1964). Since volume is related to the number of endomitotic replication cycles that have been completed (Painter and Reindorp, 1939; Bier, 1957), and correlates fairly well with DNA content (Jacob and Sirlin, 1959), it is a convenient morphological index to these properties. Brown and King (1964) mapped nuclear volume as a function of position in a 16-cell *Drosophila* cluster, and confirmed that the nuclei of the four cells that are directly bridged to the oöcyte have volumes 2-3 times greater than some of those that are indirectly bridged. Among these four cells, the largest nucleus was in the nurse cell that was attached to no other bridge and the smallest was in the cell with three other bridges, i.e. the former pro-oöcyte. The existence of such a pattern suggests that, while synchronizing mechanisms no longer operate in the sibling cluster, there are nevertheless intercellular influences affecting in a systematic way the rate at which nurse cells progress through their cycles of chromosome replication.

A remarkable exception to the rule of nurse cell asynchrony in the vitellarium was found just before the termination of trophic function in *Drosophila* (Mukenthaler and Mahowald, 1966). Injections of tritiated thymidine resulted in all nurse cell nuclei being labelled at this stage of development. There was at the same time a rise in cytoplasmic label, much of it attributable to mitochondrial DNA synthesis. It was, therefore, proposed that a general stimulus for DNA synthesis (to which the germinal vesicle was presumably refractory) occurs in the differentiated sibling cluster, just before the final atrophy of the nurse cells. The period is very brief, however, probably lasting no more than 10 min.

5.3 THE PHYSIOLOGY OF SYNCHRONY AND ASYNCHRONY

That nuclei residing in a common cytoplasm develop synchronously is an established principle for a wide variety of plant and animal cells. Naturally occurring syncytia, such as plasmodial slime moulds and pre-blastoderm insect eggs, contain hundreds or thousands of nuclei, all of which may undergo mitosis at the same time. Mitotic synchrony in the slime mould was found by Rusch *et al.* (1966) to be a response of the nuclei to a generally distributed cytoplasmic condition, for when the edges of two plasmodia were fused with each other, all nuclei in the region of mixing underwent the next mitosis simultaneously. It could be shown that the nuclei from one plasmodium had accelerated mitosis while those from the other had delayed in order to achieve synchrony. Nuclei distant from the site of fusion also accelerated or delayed mitosis, but there were still differences in their times of division from that of the region of fusion. To explain these results it was postulated that there are cytoplasmic control factors whose concentrations vary with the time after the last division. Such factors can apparently be diluted or concentrated by mixing plasmodia at different stages of the cell cycle, and they have a finite diffusion time across the diameter of the plasmodium. Analogous controls have been demonstrated by cell fusion and nuclear transplantation in many other cells, and Johnson and Rau (1971), in reviewing the subject, suggested that a succession of factors such as the "step enzymes" appearing during the S-phase, and the proteins responsible for chromosome condensation during mitotic prophase may prove to be responsible.

While fusion experiments have not been possible in germarial sibling clusters, the consistency with which mitotic synchrony has been found in these systems is most easily explained by assuming that integrating factors cross the intercellular bridges during the cystoblast cell cycles. Such factors not only control the mitotic phase of cluster formation, but in germaria where the onset of asynchrony is delayed until after synapsis has occurred, they also appear to control the prophase of meiosis. There is even stronger evidence for transmissible meiotic control factors in spermatocytes, for synchrony between bridged siblings is maintained at all stages of the two maturation divisions in a wide variety of organisms (Fawcett, 1961, 1970). Evidence that the intercellular bridges are necessarily the routes of exchange of synchronizing cues is primarily circumstantial, but one remarkable electron micrograph by Dym and Fawcett (1971) provides a unique confirmation. In this figure a bridge connecting two goat spermatogonia is occluded by several layers of flattened septa, and the nuclei of these cells are slightly out of step with each other, one being in metaphase, and the other in early telophase. Intercellular bridges thus appear to result in both mitotic and meiotic synchrony, and it seems fair to presume that the agents exchanged are the same as those accounting for synchrony in other syncytia.

The explanation of synchrony will need to deal as well with the loss of synchrony in polytrophic follicles that have entered the differentiating phase of development. There is no indication of membranous permeability barriers in the bridges at this time. Many electron micrographs have been published of intercellular bridges in follicles that had reached the vitellarium and were therefore well into the asynchronous period of development at the time of fixation (Brown and King, 1964; Steinert and Urbani, 1969; Mahowald and Strassheim, 1970; Kinderman and King, 1973). The fusome has invariably disappeared by this stage of development and has usually been replaced by a cytoplasm containing ribosomes, mitochondria, vesicles of endoplasmic reticulum and other organelles. There are few indications that barriers to diffusion obstruct the bridges. Meyer (1961) figured a plug of homogeneous material in a bridge of *Drosophila hydei*, but this appears to have been an unusual configuration. Estimates of the electrical resistance of the intercellular bridges of vitellogenic *Hyalophora* follicles indicated a resistivity within the range of 100-200 Ω-cm (Woodruff and Telfer, 1974), and this too suggests an absence of barriers to diffusion.

That asynchrony prevails despite these relatively open channels of communication is a question that could be answered in one of three different ways: either synchronizing cues are no longer produced at this stage of follicle formation, or the nuclei have become refractory to them, or else polarized transport restricts their movement through the bridges. While there is at present no experimental basis for distinguishing between these three possibilities, an argument can be made on general grounds that physiological isolation resulting from bridge polarity is the most likely cause of asynchrony. As we have seen, cytoplasmic control of nuclear function is a general principle of cellular physiology, rather than a special property of syncytia, and may in fact result from the diffusibility of enzymes or ligands that play essential roles in the cell cycle (Johnson and Rau, 1971). Such mechanisms would presumably be necessary for the endomitotic cycles of the nurse cells as well, and on these grounds it is unlikely that the sibling clusters fail to produce, or are insensitive to, the synchronizing factors.

Whatever the mechanisms of polarization prove to be, they would be expected to have a profound effect on the transmissibility of such factors. One can speculate, for instance, that ionic synchronizing agents produced by the vitellogenic follicle in *Hyalophora*, would be unable to move freely through the electrical potential gradient that has been shown to be focused in the intercellular bridges (Woodruff and Telfer, 1973). Negatively charged proteins in the oöcyte are unable to reach the nurse cells, and positively charged nurse cell products would presumably be unable to reach the oöcyte, though this has not yet been demonstrated experimentally. As will be shown below, polarity in *Hyalophora* also prevents a free exchange of proteins between nurse cells so that the same mechanism giving rise to the autonomy of the oöcyte would also result in nurse cell asynchrony. Johnson and Rau (1971) summarized the evidence that positively charged substances cause the condensation of prophase chromosomes in tissue culture cells; if such materials were also involved in nurse cell endomitosis, they would not move freely from cell to cell, and asynchrony would therefore be possible.

5.4 ENDOPOLYPLOIDY, DNA AMPLIFICATION, AND UNDER-REPLICATION IN THE NURSE CELL

Comparative studies of chromosome morphology and RNA synthesis in insect ovaries have shown that an important function of the nurse cells is to supplement the synthetic functions of the oöcyte nucleus by producing

RNA for storage in the egg (Zalokar, 1965; Bier *et al.*, 1967). Endomitotic chromosome replications which enhance the capacity of the nurse cells to carry out this function occur in polytrophic ovarioles to an extreme degree (Fig. 16). Feulgen-positive nuclei measuring 100 μm or more in diameter are produced in many of these systems, with DNA accumulating throughout the entire trophic period (Jacob and Sirlin, 1959). There have been several efforts to estimate the degree of endopolyploidy attained, particularly in *Drosophila*. Painter and Reindorp (1939) made an estimate based on nuclear volume, Freed and Schultz (1956) used a microspectrophotometric measure, and Jacob and Sirlin (1959) utilized DNAse-sensitive grain counts in autoradiograms of ovaries from flies raised in the presence of ^{14}C-nucleic acid precursors. In all of these cases the degree of ploidy was estimated to be in the general range of 2^{10}. As has already been discussed, the exact figure varies with the position of the nurse cell in the sibling cluster, and it will no doubt also show phylogenetic variation when more species have been examined.

Quantitative measures of the amount of DNA in a nucleus can be directly related to ploidy only if it is known that all components of the genome are replicated equally. There is evidence that, while much of the genome is in fact replicated, there may also be special instances of both gene amplification and under-replication as well, and for this reason the amount of DNA in a nurse cell nucleus may not be an accurate index to its ploidy.

The evidence for general genomic replication comes from *Diptera*, and is based on the structure of the polytene chromosomes that sometimes occur in the nurse cells. Painter and Reindrop (1939), Bier (1957) and others, have reported that nurse cell chromosomes tend to be polytene during the first few endomitotic cycles, but that later in development the chromatids become more diffusely related to each other (Fig. 16). There is a strain of *Calliphora* which, if raised at a low temperature, retains its nurse cell chromosomes in a polytene condition (Bier, 1959; Ribbert and Bier, 1969). A similar condition occurs in some of the nurse cell nuclei of a female sterile mutant of *Drosophila* (Koch and King, 1964). When reared at 25° C, most sibling clusters in this mutant became tumorous, but at 18° C nurse cells differentiated with polytene chromosomes and a single giant nucleolus. While detailed cytological comparisons with better known chromosomes, such as those in the larval salivary glands, have not been published, it is clear from these studies that at least a large portion of the chromomeres making up the complete genome are replicated during dipteran nurse cell endomitosis.

In the nurse cells of *Ephestia*, a heteropycnotic structure which has been attributed in the lepidoptera to the sex chromosome (Gillot, 1968) was shown both to incorporate thymidine and to increase greatly in number

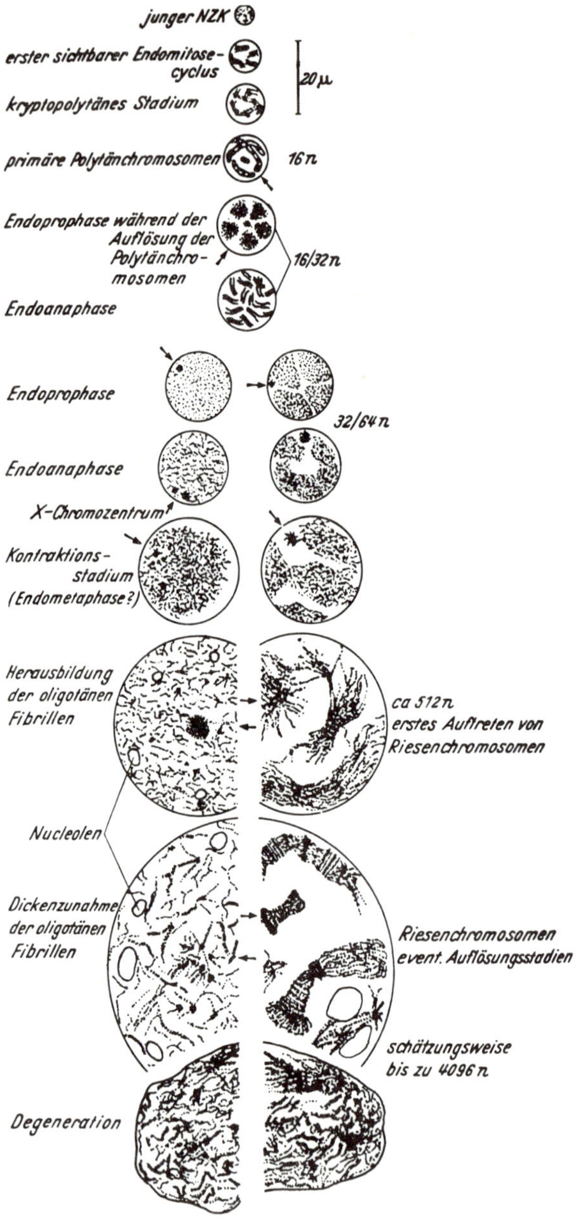

Fig. 16. Development of the endomitotic nurse cell nuclei in *Calliphora*. Normal endopolyploidy is shown in the left-hand column. On the right, the formation of polytene giant chromosomes with a ploidy of up to 4096 n is shown. The latter is a temperature-dependent process, occurring more commonly when the animals are raised in the cold. (After Bier, 1957.)

during endomitotic growth (Guelin, 1968). Unlike the euchromatin, which also increases in amount, these structures do not label with ^{3}H-uridine. These observations, too, are consistent with a general replication of nurse cell DNA, including even components that will not be heavily involved in transcription.

The case for amplification comes from Ribbert and Bier (1969) who studied the nucleoli of *Calliphora* nurse cells that had been rendered polytenic by rearing in the cold. In adult trichogen cells, which also contain polytene chromosomes, there is a single large nucleolus connected to the smallest of the six chromosomes. The nurse cells begin development with a single nucleolus associated with the same chromosome, but at later stages dozens of ring-shaped nucleoli bud off this and float in the nucleoplasm without attachment to chromosomes. Autoradiograms showed that the ring nucleoli incorporate uridine, so it is probable that DNA transcription occurs in them. Extrachromosomal DNA transcription is typical of amplified ribosomal DNA in other systems (Lane, 1967), and it appears likely that a similar phenomenon occurs here.

In *Drosophila*, by contrast, polytenization of the nurse cell chromosomes is accompanied by formation of a single giant nucleolus (Koch and King, 1964). Nucleolar gene amplification in the germinal vesicle of Orthoptera does not always lead to mutiple nucleoli, but may in some species result in a single giant nucleolus (Bier *et al.*, 1967). The absence of multiple nucleoli in *Drosophila* nurse cells, therefore, does not necessarily reflect an absence of over-replication. When the nurse cell chromosomes are dispersed instead of polytenized, multiple nucleoli routinely occur in *Drosophila* (e.g. Koch and King, 1964; Dapples and King, 1970). Under these circumstances it is clear that the nucleolar genes have been replicated during endomitosis, but whether they have been amplified cannot be easily determined.

The possibility of amplification is also suggested by the molecular hybridization studies of Gamborini and Meneghini (1972) in *Rhynchosciara*. The number of rRNA genes per haploid genome was estimated to be 100 for the 4th instar salivary gland and 220 for the adult ovary. This result indicates either under-replication in the salivary glands or amplification in the ovary. On the other hand, Gall, McGregor and Kidston (1969), in an equilibrium density gradient centrifugation study of *Dytiscus* DNA found that a low-density satellite component containing ribosomal RNA cistrons was equal in size in somatic tissue and in maturing follicles whose DNA is primarily from the nurse cells. Phylogenetic variation can therefore be anticipated in the occurrence of nurse cell rDNA amplification.

Finally, under-replication was indicated for part of the genome in the nurse cells of *Hyalophora* (Crippa and Telfer, 1971). Reannealing kinetics of sheared and melted somatic DNA indicated that about 20 per cent of the genome in this species consists of highly redundant base sequences and that this component may not be replicated in the nurse cells. To demonstrate this, DNA was extracted from nurse cells isolated manually from follicles that had been labelled *in vivo* with tritiated thymidine or ^{32}P. This material readily hybridized with the fraction of somatic DNA having a low degree of base sequence redundancy, or with that reannealing with the kinetics of unique base sequences. There was no detectable reannealing, however, with the highly redundant somatic cell DNA, and this suggests that such sequences are not accumulated during the endomitotic phase of nurse cell development. The situation is reminiscent of the *Drosophila* polytene salivary chromosomes in which the highly redundant base sequences associated with the hetero-chromatin fail to replicate (Gall *et al.*, 1971; Botchan *et al.*, 1971).

Comparing the amount of DNA in a nurse cell with the amount in a known haploid or diploid cell will therefore not permit an exact estimate of the degree of ploidy without correction being made for any amplification or under-replication that may have occurred. In this sense, the estimate of ploidy by Bier (1957), based on the number of observable endomitotic cycles in *Calliphora*, is more direct. For present purposes, however, an exact estimate of the ploidy of the nurse cells is not required; the relevant points are that endomitosis continues throughout the period of nurse cell function in the vitellarium, that it results in a vast increase in the amount of DNA available for RNA transcription, and that much of the genome, including rDNA, is replicated during this period.

Endomitosis is neither as clear nor as extreme in the telotrophic nurse chamber, though evidence for its occurrence is nevertheless strong in hemipteran nurse chambers. Schrader and Leuchtenberger (1952) showed by microspectrophotometry that the DNA content of nurse cell nuclei in the cell fusion zone of *Acanthocephala* climbs to levels as high as 17 times greater than that of the diploid trophoblasts giving rise to them. They proposed that nuclear fusion may account for these high values. Recent evidence shows, however, that endomitosis is also involved, for the nuclei of cells that have left the mitotic zone continue to incorporate tritiated thymidine (Vanderberg, 1963; Mays, 1972). The cells of the adult coleopteran nurse chamber also incorporate thymidine (Büning, 1972), and here too DNA synthesis in the absence of mitosis was interpreted as signifying the occurrence of endomitosis.

5.5 GENE AMPLIFICATION AND THE OÖCYTE NUCLEUS

In the oöcyte four copies of a complete genome must be preserved in a condition that will not interfere with the meiotic divisions resuming at the end of the egg formation. It is not surprising, therefore, that this cell does not engage in the endomitotic activity which characterizes the nurse cells.

In several groups of insects the oöcyte does, however, inherit an extra-chromosomal DNA body that has recently proven to be a product of gene amplification. Among polytrophic systems this material was described by Giardina (1901) in *Dytiscus*, by Bayreuther (1952) in *Tipula*, and recently by Gruzova, Zaichikova and Sokolov (1972) in the neuropteran *Chrysopa*. The oöcyte nuclei of panoistic ovaries may also contain such a structure (Buchner, 1909; Johnson, 1938; Bayreuther, 1957; Nilsson, 1966).

There is remarkable uniformity between these species in the behaviour of the DNA body. Lima-de-Faria (1962) demonstrated in *Tipula* that it can be labelled with tritiated thymidine, beginning with the time of its appearance in a cystoblast interphase, and extending well into the early meiotic prophase. Thymidine incorporation was confirmed for both *Dytiscus* (Urbani and Russo-Caia, 1964) and *Acheta* (Lima-de-Faria *et al.*, 1968). The DNA body grows throughout early meiotic prophase until the end of pachytene when it attains in many cases a diameter of over 6 μm. Spectrophotometric measurements of Feulgen stained oöcytes indicated that it contains 59 per cent of the nuclear DNA at pachytene in *Tipula* (Lima-de-Faria, 1962), 25 per cent in *Acheta* (Cave and Allen, 1969), and up to 95 per cent in *Dytiscus* (Kidston, cited by Gall, Macgregor and Kidston, 1969). It therefore represents a substantial increase over the normal diploid DNA content.

During late pachytene and early diplotene, when the oöcyte initiates its most rapid period of growth, there is agreement for all species that the DNA body becomes more diffuse and in some cases disappears, as was already recognized in the earliest papers on the phenomenon (Giardina, 1901; Buchner, 1909). Johnson (1938) observed that it becomes associated with a second kind of basophilic material just prior to its dispersion, and correctly interpreted the structure on these grounds as having become nucleolar. There is wide agreement from subsequent work that at diplotene, the fragmenting DNA body becomes a site of uridine incorporation and RNA accumulation (Urbani and Russo-Caia, 1964; Bier *et al.*, 1967; Lima-de-Faria *et al.*, 1968). The fine structure of the DNA body and its diplotene derivatives were examined by Lima-de-Faria and Moses (1966), Bier *et al.* (1967), Kato (1968), Favard-Sereno (1968), and Allen and Cave (1969), all of whom agree that it becomes a site of nucleolus formation. Hundreds of ring-shaped nucleoli are released to the nucleoplasm in *Gryllus* (*Acheta*) *domesticus* (Kunz, 1969), while in *Blatta* they remain fixed in a single compact structure (Bier *et al.*, 1967). In all cases the Feulgen positive material never again recondenses; it is lost to the cytoplasm during vitellogenesis (Jaworska and Lima-de-Faria, 1973) and when, at the end of vitellogenesis, the nuclear envelope breaks down and the oöcyte chromosomes form a first meiotic metaphase.

The DNA body reflects at least in part a proliferation of genes coding for ribosomal DNA. This conclusion is derived from molecular studies made feasible by the fact that the DNA bodies include a high proportion of the total Feulgen positive material in the germarium. Advantage has been taken of this fact in two studies that sought to identify the nature of the DNA being amplified. Lima-de-Faria *et al.* (1969) achieved this by studying the DNA extractable from whole ovaries of 36-40 day old nymphs of *Acheta domestica*; Gall, Macgregor and Kidston (1969) in a simultaneous study analysed the DNA from the germaria of the ovaries of several dytiscid beetles. In both cases, ultra-centrifugation in CsCl revealed a "satellite DNA" with a buoyant density lower than that of the main component of DNA (Fig. 17). The satellite was also present in DNA extracted from somatic tissue, from testes, or from more mature ovaries in which the nurse cells or follicular epithelium contributed more prominently to the total DNA present. In all of these tissues, however, the satellite was substantially smaller, relative to main-band DNA, than that produced by ovaries rich in DNA bodies. The latter in *Acheta* produced a satellite 18 times larger than that in comparable amounts of DNA from other tissues. In the dytiscid *Colymbetes* the amplification factor was 7 and in *Dytiscus* the figure was about two. If it had been possible to analyse isolated oöcyte nuclei rather than whole tissues, the amplification factor for satellite DNA would presumably have been substantially greater.

In both *Acheta* and the dytiscids labelled ribosomal RNA was found to hybridize with satellite DNA and not with that in the main band. Gall *et al.* showed that the amount of DNA binding ribosomal RNA was proportional to the size of the satellite peak so that ribosomal genes are amplified to the same extent as satellite DNA.

Finally, saturation hybridization analyses revealed that the satellite DNA of *Colymbetes* and *Dytiscus* includes components other than that complementary to ribosomal RNA. Only 3 per cent of the satellite in *Colymbetes* was calculated to hybridize ribosomal RNA. Since ribosomal DNA retains a constant proportion of the satellite during amplification, it is necessary to conclude that parts of the genome other than ribosomal cistrons are involved in DNA body formation in these two dytiscids. The nature and function of the additional amplified DNA are still unknown. More recently the same conclusion was reached by Lima-de-Faria *et al.* (1973) for *Gryllus*. Karyotyping of pachytene nuclei showed that the DNA bodies are associated with several chromomeres in different chromosomes. Since only one of these loci is the nucleolar organizer, it is probable the rDNA is not the only amplified form.

The variety of techniques that have been applied to DNA body analysis and their uniform agreement in pointing to ribosomal gene amplification is

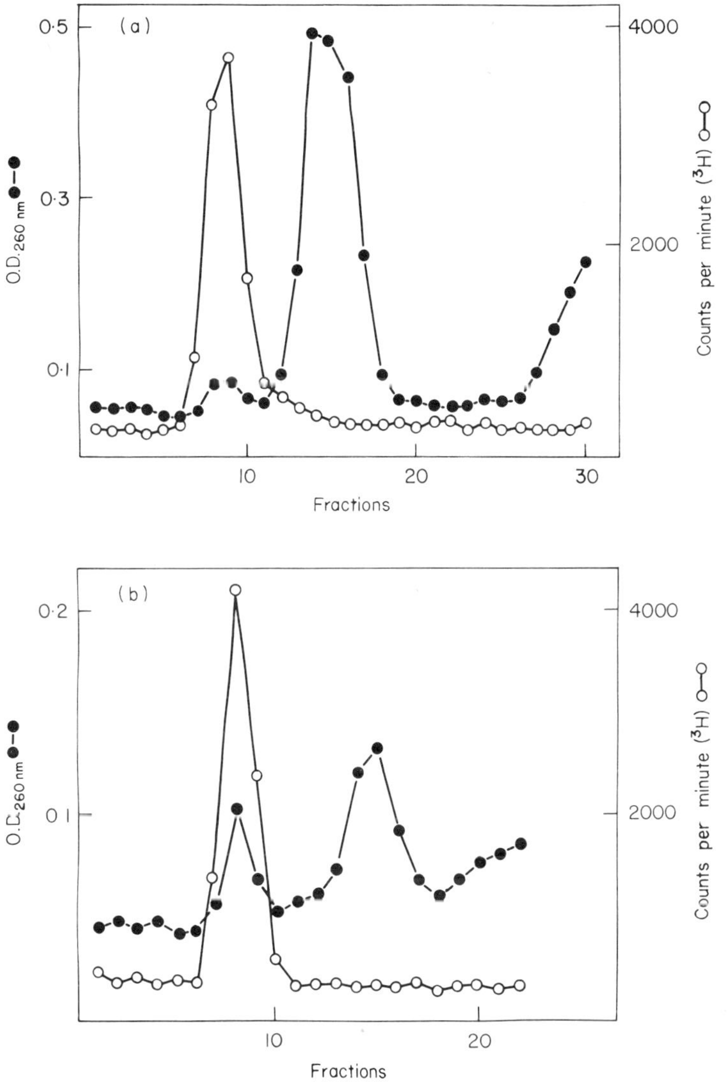

Fig. 17. CsCl density gradient centrifugation of *Colymbetes* DNA extracted from sperm and vas deferens (a) and from adult germaria (b). The solid points are optical density readings, and show that a satellite peak (fractions 8 and 9) is much larger relative to the main band in the germarial DNA. The difference is interpreted as due to the extrachromosomal DNA body in the germarium. The open circles indicate the amount of ^{3}H-labelled ribosomal RNA from *Xenopus* that hybridized with the DNA in each fraction. It appears from this that DNA complementary to rRNA is localized in the satellite. (After Gall, 1969.)

impressive. The function of this process in oögenesis is not so clear, however, except in panoistic ovarioles where the oöcyte is not associated with nurse cells. Amplified nucleoli are a fairly general feature of panoistic ovaries and would be expected to serve a crucial function as the only apparent source of ribosomes for this exceptionally large cell. In polytrophic systems, it has already been noted that DNA bodies generating multiple nucleoli in the oöcyte occur only in a few scattered families. Even among the dytiscids, the DNA body is lacking in some genera (reviewed by Urbani and Russo-Caia, 1964). It is clearly not a requirement for oögenesis in most polytrophic ovaries, and there is no obvious explanation of why it occurs where it does.

6 Synthetic functions of the nurse cell and oöcyte nuclei

Nurse cell and oöcyte differentiation results in nuclei that contain different levels of ploidy and, in some cases, different degrees of amplification. The evidence reviewed below suggests that they may in addition vary in their transcriptional patterns, so that not only the amounts, but also the kinds of RNA produced may differ. RNA's from both sources are apparently deposited in the oöcyte, where they can be presumed to engage in protein synthesis in support of vitellogenesis, and where at least some components are sequestered for later use during embryogenesis. Cytoplasmic growth, including organelle formation, is also notable in the growing sibling cluster, and this too can be presumed to be supported by nuclear RNA synthesis. Whether the cytoplasm of the nurse cell is a special locus for the assembly of particular organelles that are not formed in the oöcyte has not been clarified. The oöcyte by contrast is definitely a site of organelle formation, for all components of the yolk appear to be assembled primarily, if not exclusively, in this cell.

Present concepts of the function of the nurse cells in oögenesis have come primarily from descriptive electron microscopy and from autoradiography of ovarioles labelled by injecting females with tritiated precursors. These methods have been supplemented by a relatively few studies in which the cells making up the follicle were isolated for separate study by molecular techniques. From the discussion that follows it will be apparent that the latter approach has become increasingly necessary for progress in understanding the division of labour between the nurse cell and the oöcyte.

6.1 AUTORADIOGRAPHY OF NURSE CELL RNA SYNTHESIS AND TRANSPORT

Autoradiographic studies of polytrophic ovaries in a large number of species have demonstrated that the nurse cells synthesize RNA with the intensity that would be anticipated from their high degree of ploidy. The earliest observations were made on *Drosophila* (King and Burnett, 1959; Zalokar, 1960), and these were followed by an extensive analysis of blowfly ovaries by Bier (1963a and 1963b). The latter work entailed time-course studies of labelled RNA in follicles from adult females following single injections of tritiated precursors. Within five minutes RNAse sensitive label was visible in the nurse cell nuclei of vitellogenic follicles. Nuclei of the follicular epithelium also incorporated the precursor, but the germinal vesicle of the oöcyte did not. With 10-30 min incubation

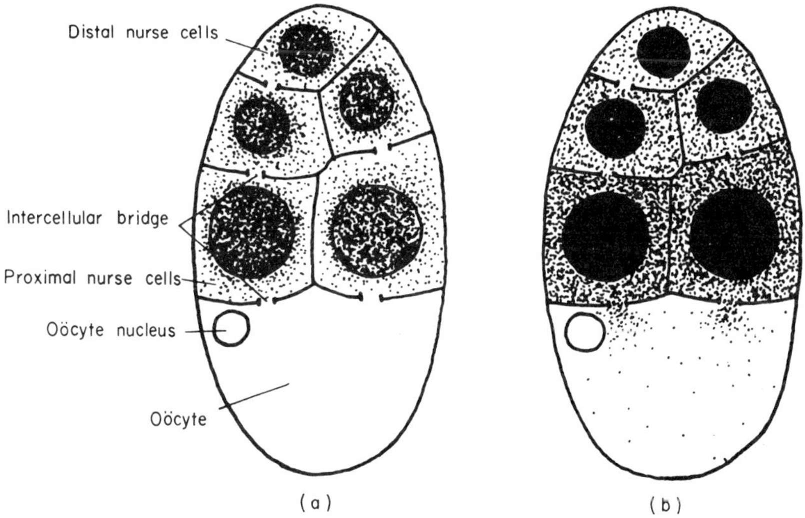

Fig. 18. A summary of autoradiographic results following *in situ* uridine labelling of *Calliphora* follicles. The incubation time was 20-30 min in (a) and 45-90 min in (b). While the nurse cell nuclei were heavily labelled, the oöcyte nucleus had not detectably incorporated the uridine in these early studies. (The karyosphere in the germinal vesicle was later found to be a site of incorporation.) In (b) label has begun to enter the oöcyte, and there is a distal-proximal gradient in the amount of cytoplasmic label in the nurse cells. (After Bier, 1964.)

the nurse cell nuclei were extremely heavily labelled, and silver grains could also be detected for the first time in the surrounding cytoplasm. By 45 min the cytoplasm of the oöcyte directly adjacent to the intercellular bridges also contained label; this fact, combined with the absence of detectable incorporation by the germinal vesicle, established the principle that the nurse cells are the main source of RNA deposited in the oöcyte (Fig. 18).

Many other groups of polytrophic insects exhibit a similar pattern of incorporation. The observations have thus far been made in species of Mecoptera (Ramamurty, 1963), adephagic Coleoptera (Bier, 1965; Bier *et al.*, 1967), Hymenoptera (Bier, 1965; Engels, 1968; Meng, 1970), Lepidoptera (Bier, 1965; Zalokar, 1965; Pollack and Telfer, 1969), Dermaptera (Engels, 1969; Zinmeister, 1973), Collembola (Krzysztofowicz, 1971), Neuroptera (Gruzova *et al.*, 1972), and a variety of other Diptera (Zalokar, 1965; Kunz *et al.*, 1970). There are some exceptions to the general rule of low or undetectable uridine incorporation by the oöcyte nucleus, which are detailed in the next section. Even here, however, the nurse cells are generally the primary source of the RNA amassed in the growing oöcyte.

While the nuclei of the follicular epithelium are also sites of uridine incorporation, evidence of RNA transfer from these cells to the oöcyte is still lacking. A small contribution would be difficult to rule out in a decisive way since its route of entry would be dispersed over the entire oöcyte surface and thus not easily detected by autoradiography. The fine structure of the epithelium suggests, however, that it contributes to oögenesis by synthesizing and secreting proteins. During the growth of the oöcyte the epithelial cells contain well-developed endoplasmic reticulum and Golgi bodies (Koch and King, 1963; Huebner and Anderson, 1972a; Cummings and King, 1970), and the adjacent surfaces of the oöcyte have been shown by autoradiography to incorporate epithelial secretion products into yolk spheres in the oöplasm (L. Anderson and Telfer, 1969; L. Anderson, 1971). Exocytosis and endocytosis are mechanisms for the transport of proteins kept separate from the cytoplasm, and it is therefore not surprising that this activity occurs without detectable transfer of epithelial RNA.

Bier (1963a) described two phases of nurse cell RNA transport to the oöcyte of *Calliphora*. During the initial phase (Fig. 18) the nurse cells are growing in size, and labelling indicates that they retain many of their synthetic products. Some uridine-labelled products cross the intercellular bridges to the oöcyte at this time, and these, he felt, are unstable, since time course studies did not indicate a progressive increase in silver grain density in the oöplasm. At the termination of trophic function the labelled cytoplasm accumulated by the nurse cells flowed *en masse* into the oöcyte and this material, Bier proposed, included much of the stable RNA's deposited in the mature egg. Consistent with this functional pattern are the data on cell volumes in *Drosophila* (Koch and King, 1966) which show a progressive increase in the size of nurse cells until the terminal injection of their cytoplasm.

In *Hyalophora*, by contrast, terminal injection plays a quantitatively secondary role. This was shown by measuring the amount of RNA in perchloric acid extracts of isolated oöcytes and nurse cells (Pollack and Telfer, 1969). In these first experiments on the RNA of isolated cellular components of an insect follicle, the epithelium and nurse cells were loosened by pronase treatment and then manually separated from the oöcyte. The nurse cells, along with the squamous epithelium that covers them, could be dissected free-hand from untreated follicles. In the smallest vitellogenic follicle that could be handled in this way the seven nurse cells together contained 1 μg of RNA and the oöcytes about the same amount. In contrast to what is presumably the case in *Calliphora* and *Drosophila*, the nurse cells maintained a constant 1 μg throughout their subsequent development, while oöcyte RNA gradually increased to a value of just over 2 μg. At the terminal injection of the nurse cell cytoplasm the RNA content of the oöcyte reached an average value of 3 μg, which was maintained throughout the remainder of oögenesis, and until one or two days after the egg had been fertilized, at least five days later.

As a general rule, terminal injection involves only the cytoplasm of the nurse cells. The nuclei remain in place, but are now surrounded by a reduced volume of cytoplasm that has lost much of its basophilia (Bier, 1963; Cummings and King, 1969; Pollack and Telfer, 1969). The intercellular bridges are severed, and the nurse cell residues, along with the epithelium in which they are now embedded, are discarded by the mature egg at ovulation. An exception to the rule of nuclear exclusion occurs in *Carabus*. In this species the nuclei are injected into the oöcyte along with the cytoplasm, and even continue to incorporate uridine as, or shortly before, they are transported (Bier, 1965).

In telotrophic ovaries also uridine labelling revealed at first no evidence of incorporation by the germinal vesicle, while epithelial and nurse chamber nuclei were rapidly and heavily labelled (Bier, 1963b; Vanderberg, 1963). As in polytrophic systems, more recent work has indicated a slow, but significant, rate of silver grain accumulation over the germinal vesicle (Mays, 1972; Büning, 1972; Ullman, 1973). Two of these studies utilized coleopterans (Büning, 1972, for *Bruchidius*, and Ullman, 1973, for *Tenebrio*), in which, as we have seen, the nurse chamber is a stable syncytium. Labelled RNA was transferred from nuclei to the cytoplasm of the nurse chambers within 1-2 h, but did not reach the oöplasm until after at least 4 h of incubation. The intensity of oöcyte labelling then increased progressively, particularly at the anterior end where the trophic cord is attached.

In hemipterans, where cell fusion and pycnosis in the nurse chamber generate an acellular trophic core, the transfer of label from nurse cell nuclei to the oöcyte was less direct. All nuclei in the trophic chambers incorporated uridine except those that had initiated pycnosis (Mays, 1972). Label first appeared in the trophic core after two to three hours of incubation (Zinmeister and Davenport, 1971; Mays, 1972), but whether this required cellular breakdown or was due to RNA synthesis by nuclei whose cytoplasm had already joined the trophic core is not clear. Label was not apparent in the oöplasm until after 12 h of incubation, and could be prevented even then by ligating the ovariole between the nurse chamber and the vitellarium (Zinmeister and Davenport, 1971). As in coleopterans, the oöplasm was often most heavily labelled at the end connected to the cytoplasmic cord. In *Notonecta* Macgregor and Stebbings (1970) found an apico-basal labelling gradient in the trophic core and in the cytoplasmic strands connecting it to the oöcytes between 8 and 24 h after uridine injection. After three weeks, by contrast, the oöplasm was overlain by a higher concentration of silver grains than the trophic core, a result suggesting that back-diffusion of label through the cytoplasmic connecting strands does not readily occur.

Mays (1972) felt that his results indicated two forms of label which traversed the cytoplasmic strands at different rates. The slow form labelled much more heavily than the fast and was thus visible as an intensification of silver grain density in the trophic core and connecting strands many hours after the rapid form had already reached the oöcyte. He estimated rates of migration through the trophic core and cytoplasmic strands of 200 $\mu m\ h^{-1}$ for the fast component and 30 $\mu m\ h^{-1}$ for the slow.

The kinds of RNA provided the oöcyte by the nurse cells require identification by molecular methods, but autoradiography has provided at least one clue that, as will be seen below, is fully consistent with the results of such studies. Uridine labelling was concentrated particularly over the nucleoli in the nurse cells of *Drosophila* (King and Burnett, 1959; Mahowald and Tiefert, 1970), *Hyalophora* (Pollack and Telfer, 1969), and *Calliphora* (Ribbert and Bier, 1969), with other regions of the nuclei labelling to a smaller degree. The result suggests that ribosomes are an important component of the RNA exported by the nurse cells. In contrast to the nucleoli, the heterochromatic DNA that has been attributed to the W-chromosome in *Ephestia* failed to label with uridine (Guelin, 1968).

6.2 GERMINAL VESICLE FUNCTION

In many species of animals the oöcyte chromosomes become associated with unusual amounts of protein and RNAse-sensitive basophilia during the periods of cytoplasmic growth and yolk deposition (Callan, 1963). Particularly where large, yolky eggs are produced, the chromosomes uncoil to such an extent that their DNA cannot be readily detected by Feulgen staining. They are so heavily coated with RNA and protein, however, that they are nevertheless easily visualized by light microscopy. Autoradiography has shown that they incorporate labelled uridine (Gall and Callan, 1962; Lane, 1967), and their fine structure also suggests transcriptional activity (Miller and Beatty, 1969). Their basophilia has thus been interpreted as a stage in the processing of the transcriptional products required by an enormous and rapidly growing cytoplasm.

In a comparative study of germinal vesicle development in fifteen species of insects, Bauer (1933) observed nucleoli and basophilic chromosomes in five panoistic forms (Fig. 19). This finding was confirmed by studies of chromosome morphology in the germinal vesicles of several panoistic species (Kunz, 1967a, 1967b, 1969), and chromosomal uridine incorporation is now well documented in this kind of ovary (Zalokar, 1961, 1965; Bier *et al.*, 1967; Zinmeister and Davenport, 1971). In meroistic oöcytes, by contrast, Bauer found that the chromosomes exhibit little or no basophilia, and that

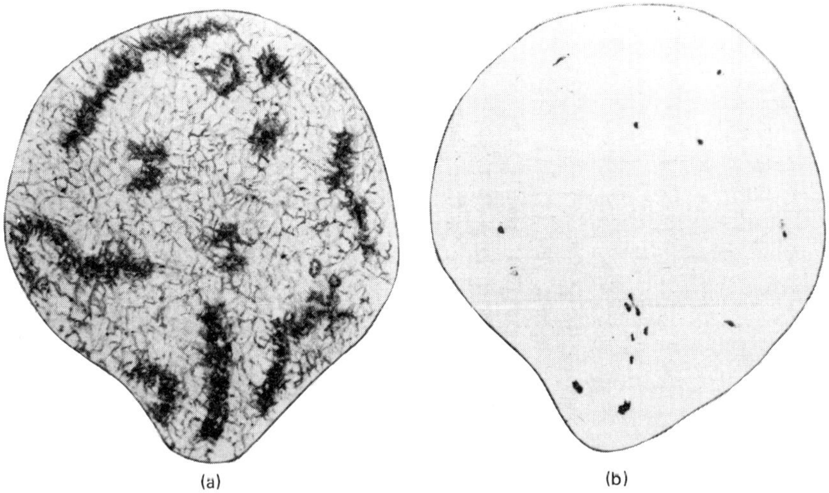

Fig. 19. The germinal vesicle of *Stenobothrus*, an orthopteran, stained for basophilia (a) and for DNA (b). This is a panoistic species and, as in all insects not forming nurse cells, the chromosomes of the germinal vesicle become heavily ladened with RNA, even though their DNA is often too dispersed to be easily seen. (From Bauer, 1933.)

nucleoli are usually absent. The synthetic role usually played by the germinal vesicle is thus assumed in large measure by the nurse cells. The oöcyte chromosomes were found by Bauer to assume one of two very different appearances, according to the species. In the moth *Ephestia* and the wasp *Habrobracon* he reported that the chromosomes become diffuse and finally undetectable by Feulgen staining shortly after the follicle has left the germarium. They remain in this condition until the end of egg formation when they finally recondense to participate in the first meiotic metaphase. A similar behaviour occurs in *Hyalophora*, and here it is accompanied by an absence of detectable uridine incorporation by any component of the germinal vesicle after the onset of vitellogenesis (Pollack and Telfer, 1969).

In species representing the Dermaptera, Mecoptera, Diptera, and Coleoptera, by contrast, Bauer showed that the chromosomes condense after pachytene to form a compact, Feulgen-positive mass which he termed the karyosphere (Fig. 20). The germinal vesicle was essentially empty of

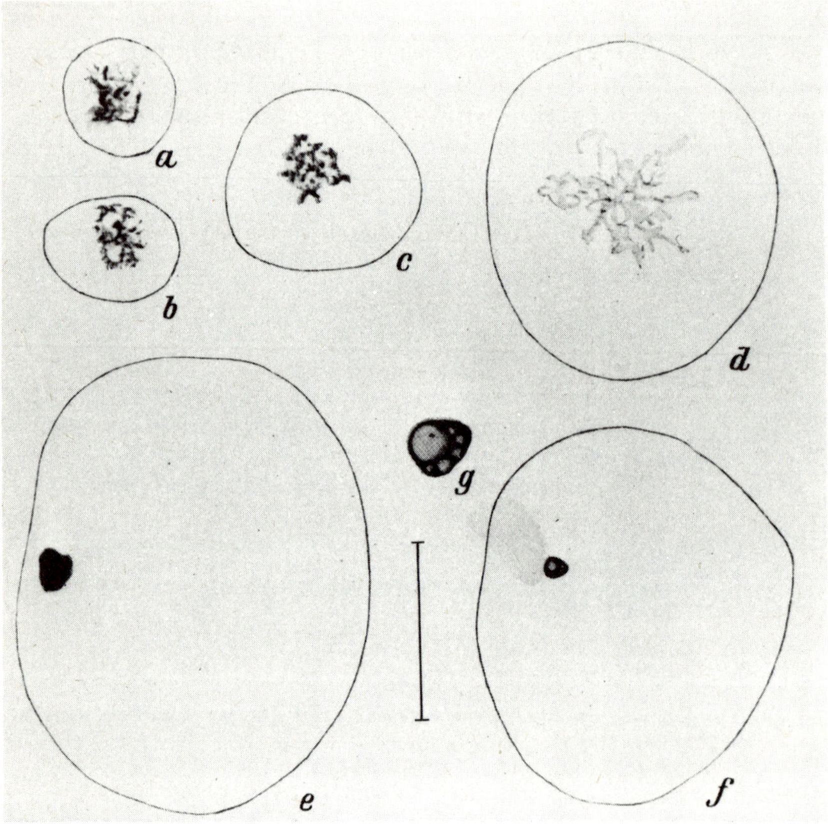

Fig. 20. Germinal vesicle growth and karyosphere formation in *Lucilia*. Feulgen staining shows that the chromosomes gradually condense into a compact body with a diameter of about 5 μm. *f* is drawn to a smaller scale than the other figures; *e*, drawn to the same scale as *f*, would have a diameter shown by the bar. In *f* the germinal vesicle had reached a diameter of 70-80 μm. Karyosphere formation and low basophilia in this case are in contrast to the germinal vesicle shown in Fig. 19 for a panoistic species. (From Bauer, 1933.)

basophilic material during much of egg formation. Karyospheres have recently been described in other Dermaptera (Engels, 1969), Mecoptera (Ramamurty, 1963), Diptera (Bier *et al.*, 1967; Smith and King, 1968; Kunz *et al.*, 1970), Neuroptera (Gruzova *et al.*, 1972), Hymenoptera (Bier,

1965; Engels, 1968), and both telotrophic and polytrophic Coleoptera (Schlottman and Bonhag, 1957; Bier *et al.*, 1967; Büning, 1972). Gruzova *et al.* (1972) showed that in *Chrysopa* the condensed chromosomes become encapsulated by a fibrous material. They noted that this is not a universal phenomenon, however, and suggested that it might prove to occur in those species whose germinal vesicles contain extra-chromosomal DNA. There seems to be no description of a karyosphere in Hemiptera (Eschenberg and Dunlap, 1966; Huebner and Anderson, 1972b), and this group may thus have to be added to *Habrobracon* and the species of moths in which the chromosomes become undetectably diffuse in the germinal vesicle.

The karyosphere has been interpreted as a device for repressing RNA transcription, for uridine incorporation appears to wane as the chromosomes move together (Bier *et al.*, 1967; Gruzova *et al.*, 1972). In autoradiograms of *Drosophila* (Zalokar, 1965), *Musca*, and *Carabus* (Bier *et al.*, 1967), following exposure to tritiated uridine, the karyosphere was overlain by significant concentrations of silver grains only in heavily labelled follicles. This was the only site of uridine labelling in the germinal vesicles of these species. In *Apis* and *Forficula* visible uridine labelling of the karyosphere did not occur (Engels, 1968, 1969), though this does not rule out the possibility of a low level of transcription that was not detectable by the conditions employed. In *Bruchidius* electron microscopy showed the karyosphere to be unusually reticular, and its DNA proved to be sufficiently diffuse to be undetected by Feulgen staining (Büning, 1972). In the same species the germinal vesicle exhibited conspicuous uridine incorporation in autoradiograms, an activity which Büning interpreted as being facilitated by the unusually loose structure of its karyosphere. Mahowald and Tiefert (1970) were able to show that karyosphere labelling in *Drosophila* is particularly prominent at a special stage of oögenesis. At this time it transitorily swells from 3.5 μm to 6 μm diameter, so that here again a loosening of structure is associated with increased RNA synthesis.

Clear evidence that the karyosphere is involved in repression comes from the observations of Kunz *et al.* (1970) on the gall midge *Wachtliella*. This is one of the diptera in which chromosome elimination occurs in somatic cells early in embryonic development; adult somatic cells have only four chromosomes, while the germ cells retain a total of twenty. In the growing oöcyte the four somatic chromosomes become encapsulated in a karyosphere, while the other sixteen remain scattered elsewhere in the nucleoplasm. Autoradiography following labelling with uridine showed that RNA synthesis occurs throughout the germinal vesicle, except in the karyosphere, so that the special germ-line chromosomes are apparently transcribed at a greater rate than the somatic chromosomes. The nucleolar organizer is on a somatic chromosome, and nucleoli as a consequence do not form in the germinal vesicle during vitellogenesis. Oöcyte ribosomes apparently originate in the nurse cells, which also contain the full complement of twenty chromosomes, and in which nucleoli are abundant.

The germinal vesicles of most meroistic species show little evidence of nucleolar function. This is presumably because the karyosphere encloses all chromosomes in the germinal vesicle (Smith and King, 1968; Gruzova *et al.*, 1972) including those carrying the nucleolar organizer. Amplified extra-chromosomal DNA which, as we have seen, includes nucleolar material, escapes the confines of the karyosphere in dytiscids, *Tipula*, and *Chrysopa* and in these species uridine incorporation occurs throughout the nuclear sap surrounding the condensed chromosomes (Urbani and Russo-Caia, 1964; Gruzova *et al.*, 1972). In the many species that do not form extra-chromosomal DNA, whatever low levels of incorporation can be detected in the germinal vesicle are, by contrast, completely confined to the karyosphere.

The absence of detectable incorporation in the germinal vesicle of Lepidoptera, which lack a karyosphere (Zalokar, 1965; Pollack and Telfer, 1969), indicates that here too nucleolar function is probably at a low level. Confirmation of this conclusion comes from molecular studies of Hughes and Berry (1970) who showed that label was absent from oöcyte ribosomes after 4-6 h of incubation in tritiated uridine. Failure to detect ribosomal precursor RNA in isolated oöcytes also supports this conclusion (Iverson *et al.*, 1974).

The extent to which the germinal vesicle chromosomes are specifically repressed with regard to RNA synthesis, as opposed to being simply overshadowed by the polyploidy of the nurse cells, is not yet established. Required for this would be a comparison of the quantity of RNA produced per haploid genome by each cell type during the oöcyte growth period. Several of the autoradiographic observations discussed above, though lacking this quantitative information, nevertheless suggest repression of either all or parts of the genome in the germinal vesicle. Thus, uridine labelling and basophilia are readily seen in each of the many individual nurse cell nucleoli, while in the same follicle not even one nucleolus has been detected in the germinal vesicle (Pollack and Telfer, 1969; Kunz *et al.*, 1970). Finally, where only part of the germinal vesicle DNA becomes enclosed in a karyosphere, direct comparisons are possible which leave little doubt that karyosphere DNA has at best a feeble rate of RNA synthesis relative to that dispersed in the nucleoplasm. It is, therefore, probable that RNA synthesis will in fact prove to occur at lower rates on chromosomes and nucleoli in the germinal vesicle than on their counterparts in the nurse cells.

Additional and as yet poorly understood functions of the germinal vesicle are suggested by several structures that are associated with it. In the oöcytes of many hymenopterans "accessory nuclei" have been described (reviewed by Buchner, 1918) which first appear just before vitellogenesis and may ultimately number in the hundreds. They are believed to arise by nuclear budding and to proliferate as well by pinching in two (Cassidy and King, 1973). They are Feulgen-negative, but contain small amounts of RNA (Hopkins, 1964; Meng, 1970), and in autoradiograms they have not detectably labelled with uridine (Bier, 1965; Meng, 1970). They are covered by a double membrane containing annuli (Hopkins, 1964; P. King and Fordy, 1970), and Cassidy and King (1972) found that they disintegrate in synchrony with the germinal vesicle membrane prior to the first meiotic metaphase in *Habrobracon*. These features too are consistent with their being derivations of the nuclear membrane. Configurations with the same fine structure and histochemistry as the hymenopteran accessory nuclei were recently described in electron microscope thin sections of a collembolan ovary (Palevody, 1972). The description is very brief, however, and it is not yet clear whether their origin and relation to the germinal vesicle are the same as those of Hymenoptera. Accessory nuclei have also been claimed to occur in the moth, *Anagasta* (Cruickshank, 1972), but in this case the envelope is a single membrane that lacks annuli. Comparable structures were cited by Cruickshank as being depicted in a figure of the oöplasm adjacent to the vitelline membrane in *Hyalophora* by Telfer and Smith (1970). As indicated in the text of the latter paper, however, these structures are produced in the oöcyte cortex after the conclusion of yolk formation (Telfer and Anderson, 1968). There is no evidence of their being in any way homologous to the accessory nuclei of Hymenoptera.

A second germinal vesicle product that is substantially more widely distributed is the annulate lamella that is budded from the nuclear membrane in many oöcytes, and often occurs in stacks in the oöplasm (Kessel, 1968; Kessel and Beams, 1969). There have been speculations that these are devices for transmitting and storing RNA in the oöplasm, but evidence of their function is not yet satisfactory.

Finally, in many insects the germinal vesicle contains a proteinaceous sphere that has been variously termed a secondary nucleolus or endobody (Bier *et al.*, 1967; Halkka and Halkka, 1968). It does not stain for DNA and at best only weakly for RNA. It tends to accumulate proteins rich in arginine and lysine. Halkka and Halkka found that it can be seen to label with uridine if precautions are taken to prevent the extraction of low molecular weight RNA during histological fixation. They could not determine whether it is a site of synthesis or of sequestration, however. It occurs in both panoistic and meroistic ovaries, and this makes it especially difficult to relate to chromosomal or nucleolar function. Mahowald and Tiefert (1970) and Kinderman and King (1973), in studies of the germinal vesicle in several species of *Drosophila*, found that the endobody in this genus develops from the nucleolus by shedding the granular component well before vitellogenesis begins. Halkka and Halkka suggested that the secondary nucleus which they characterized in panoistic dragonflies should not necessarily be equated to those observed in other insects.

6.3 THE CLASSES OF RNA PRODUCED

The RNA's stored in the ovulated, unfertilized egg are stable products of the nurse cell nuclei and germinal vesicle, and can at this stage be characterized without contamination by other cells of the ovary. Though there have been surprisingly few attempts to take advantage of this opportunity, information is now available for a few species. The classes detected include ribosomal RNA (28S, 18S, and 5S), transfer RNA (4S), and extraordinary amounts of a small molecular weight fraction, all of which can be labelled in the ovary by uridine. SDS-Acrylamide gel electrophoresis of phenol extracted egg RNA revealed these components in the hemipteran *Dysdercus* (Duspiva *et al.*, 1973), and similar results were obtained in *Hyalophora* (Iverson *et al.*, 1974). Utilizing a procedure developed by Hansen-Delkeskamp (1969) for *Acheta*, Duspiva *et al.* were able to show that ribosomal and 4S RNA are synthesized at an earlier stage of oögenesis than the smaller nucleotides. Newly laid eggs were collected at various times after a female had been injected with labelled uridine. On the third day after the injection label was present only in the small nucleotide fraction of the eggs, while on the fifth to eighth days it was mainly in ribosomal and 4S RNA (Fig. 21). They proposed that the nurse chamber is the origin of the ribosomal and 4S RNA, and showed that high molecular weight ribosomal precursor RNA's are in fact labelled in this zone of the ovary, though they are also produced in the vitellarium, presumably by the epithelial cells. The low molecular weight component, by contrast, is present only in the vitellarium. Duspiva *et al.* therefore suggested that it is produced by the follicular epithelium, though this overlooks a second possibility which has emerged in *Hyalophora*.

In the latter species the low molecular weight component was found to be labelled and accumulated most rapidly by follicles in which the nurse cells had completed their terminal injection of cytoplasm, so that here too the nurse cells cannot be the source of this material (Iverson *et al.*, 1974). When the epithelium was separated from the oöcyte after 1-6 h of *in vivo* incubation with tritiated uridine, however, labelled small nucleotides were found only in the oöcyte. It is therefore likely that they are produced in

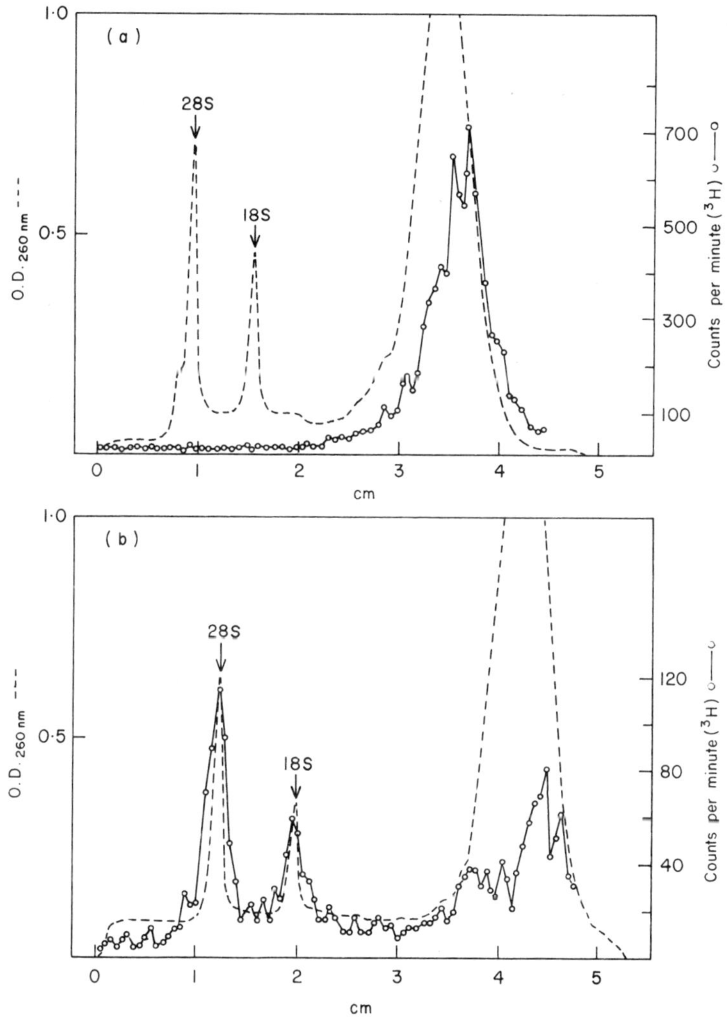

Fig. 21. Electrophoresis in SDS-acrylamide gels of the RNA extracted from newly laid eggs of *Dysdercus.* Scanning for optical density at 260 nm (broken line) revealed three fractions, 28S RNA 18S RNA, and a low molecular weight material that overshadowed all other components. When the eggs had been laid by females 3 days after injection with ^{3}H-uridine, all radioactivity was localized in the low molecular weight-material (a), open circles. Five days after the injection (b), label was recovered from 28S and 18S peaks, and the absence of label in the low molecular weight material made possible the resolution of what are probably the 5S and 4S RNA peaks. The low molecular weight material is thus deposited only late in oögenesis after ribosomes are no longer being received through the cytoplasmic strand from the nurse chamber. (From Duspiva *et al.*, 1973.)

the oöplasm from uridine taken up directly from the blood via the intercellular spaces. There is no evidence at this point suggesting that the epithelial cells produce the material.

The nature of the low molecular weight material is not known in either species. The mature eggs of *Oncopeltus*, however, were shown by Forrest *et al.* (1967) to contain extraordinary amounts of inosine and guanosine, much of which could be shown to serve as RNA precursors during embryogenesis. 1.4 mμmol of inosine and 1.65 mμmol of guanosine were present, compared with only 0.5-0.7 mμmol of RNA-phosphorus per egg. These proportions would give rise to the high optical densities found for the low molecular weight material relative to the ribosomal and 4S RNA peaks found in *Dysdercus* and *Hyalophora*. Whether nucleotides would behave in the same manner as the low molecular weight material in the extraction and analytical procedures employed with these two species has not been reported, however.

The only RNA's thus far detected in mature insect eggs are the 28S, 18S, 5S, and 4S components. It is therefore reasonable to conclude that these are in the main the material whose synthesis by the nurse cells and transport through intercellular bridges has been followed during the last fifteen years by autoradiography. The smaller RNA's are probably too soluble in histological fixatives to have been preserved by the autoradiographic methods most widely employed (Woods and Zubay, 1965), but 28S and 18S ribosomal RNA should have been faithfully preserved. Consistent with the proposal that ribosomes are transported to the oöcyte is the fact that they are almost universally present in intercellular bridges during vitellogenesis in both polytrophic and telotrophic species (e.g. Hamon and Folliot, 1969; Koch and King, 1969; Huebner and Anderson, 1970; Mahowald and Strassheim, 1971; Cassidy and King, 1973; Woodruff and Telfer, 1973). Cytological studies of nucleolar development in the nurse cell nuclei also supports this view (Anderson and Beams, 1956; Dapples and King, 1970; Weber, 1971). And finally, there is support from molecular studies in which labelled monoribosomes were found in oöplasm manually dissected from *Antheraea* follicles that had been labelled 6 h or longer in uridine-^{3}H (Hughes and Berry, 1970). If the nurse cells had been trimmed from the follicles prior to incubation, oöplasmic ribosomes did not become labelled. Since oöplasmic polysomes did not become labelled even in the presence of nurse cells, it was concluded that ribosomes not yet in the act of protein synthesis were the primary form transported.

In *Hyalophora* females that had been injected with uridine 28S, 18S, and 4S RNA were labelled in the nurse cells and epithelium with 30-60 min of *in vivo* incubation, but not in the oöcyte until several hours later (Iversen *et al.*, 1974). Labelled high molecular weight ribosomal precursor RNA was found in the nurse cell-epithelium extracts during the first hour of incubation, but was never detected in the oöcyte. These studies therefore confirmed that ribosomes are not detectably synthesized in the oöcyte, and showed in addition that the same is true of 4S RNA.

Hansen-Delkeskamp (1969) found in the panoistic *Acheta* that eggs laid during the first five days after females had been injected with uridine-^{14}C contained no labelled RNA. Between 6 and 17 days newly laid eggs showed a constant pattern of label in 28S, 18S, and 4S RNA separated by sucrose density gradient centrifugation. The germinal vesicle of this panoistic species thus resembles the nurse cells of meroistic systems in the predominant kinds of stable RNA that are deposited in the oöcyte.

In none of these studies was there evidence for synthesis and transport of RNA in the size range between 18S and 4S, where most messengers might be expected to be located. Assuming that such materials in fact occur, their amount is too small or their heterogeneity is too great to make them evident as discrete peaks by the methods thus far employed. The low level or polysome label in *Antheraea* oöplasm seen by Hughes and Berry (1970) confirms that in quantitative terms the transported products do not include significant amounts of functional messenger RNA. In contrast to this conclusion are the suggestions from autoradiography that the RNA transported to the oöcyte during the nurse cell growth phase in Diptera is largely unstable, and hence primarily messenger RNA (Bier, 1963). The conclusion was drawn from the relatively low silver grain densities seen over vitellogenic oöcytes that had been labelled for long periods of time. However, no attempts to actually measure time-course changes in the amount of label in the oöcyte have been reported. To further complicate the problem, vitellogenic oöcytes grow rapidly at the stages involved, and correction must therefore be made for a progressive dilution of label by the yolk that is formed during the *in vivo* incubation period. The case for unstable RNA transport during the growth phase has therefore not been well established, and the suggestion that autoradiography of intercellular transport has in the past detected primarily ribosomal RNA, as well as whatever fraction of 4S RNA is not extracted by chemical fixatives, appears to be plausible.

In none of the cases where uridine incorporation can be detected in the chromosomes of the oöcyte nucleus of meroistic ovaries have the species of RNA being synthesized been identified. The presence of nucleoli in the germinal vesicle of insects with DNA bodies implies that ribosomal RNA is synthesized, but the demonstration by Gall *et al.* (1969) that ribosomal DNA is only a small part of the amplified portion of the genome indicates that other kinds of RNA may be produced as well. In *Chrysopa* label tended to be retained longer by the germinal vesicle than by the nurse cell nuclei (Gruzova *et al.*, 1972), and a similar impression was expressed for the label in the karyosome of *Drosophila* by Zalokar (1965). Whether this reflects the kinds of RNA being synthesized or a slower transport mechanism has not been determined. An RNA that never reaches the cytoplasm until germinal vesicle breakdown is an added possibility, and it is not clear how, with the limited amounts of material available for experimental material, a distinction can be made.

6.4 OTHER NURSE CELL FUNCTIONS

The nurse cells at vitellogenic stages have a cytoplasm that contains mitochondria, smooth endoplasmic reticulum, and other organelles. Though all of these appear available for transport into the oöcyte, particularly during terminal cytoplasmic injection, the evidence so far is in most cases circumstantial. Aside from the transport of ribosomes and of the centrioles of *Drosophila*, there are no confirmed cases of organelle transport.

There is no comparable experimental evidence that other oöcyte organelles come from the nurse cells rather than arising primarily by replication *in situ*. Muckenthaler and Mahowald (1965) showed that oöplasmic mitochondria can be labelled with tritiated thymidine at a late stage of oögenesis, even after periods of incubation as short as 10 min. This observation supports the concept of at least some *in situ* replication. In *Hyalophora*, mitochondria abound in the intercellular bridges as well as in all cells of the sibling cluster, beginning with the 5th instar larva, nine months in nature before the oöplasm begins to grow rapidly in volume (Fig. 22). It would seem unlikely that the mitochondria move in one direction through the bridges at a significant rate throughout this period, and this serves to emphasize that electron microscopy without collateral information can say little about the direction or rates of movement of organelles found in the intercellular bridges. Such questions clearly cannot be answered without a marking procedure that would allow nurse cell mitochondria to be identified in the oöplasm after transit through the bridges.

Autoradiography of follicles labelled with amino acids has repeatedly failed to show any concentrating of protein synthesis in the nurse cells relative to the oöcyte (Bier, 1963; Melius and Telfer, 1969; Mays, 1972; Büning, 1972). Amino acid incorporation occurs wherever basophilic cytoplasm is distributed in the sibling cluster. In systems such as *Hyalophora* where the oöcyte contains as much or more RNA than the nurse cells (Pollack and Telfer, 1969), this raises the obverse speculation that the proteins required by the nurse cells for ribosome assembly may be to some degree synthesized in the oöcyte (Zinmeister, 1973). As will be seen below, the net positive charge carried by these proteins would allow them to move from the oöcyte to the nurse cells, provided they could escape adsorption to the oöplasm.

There was a tendency in the early literature to attribute protein yolk formation to the nurse cells. Yolk is now recognized, however, as being

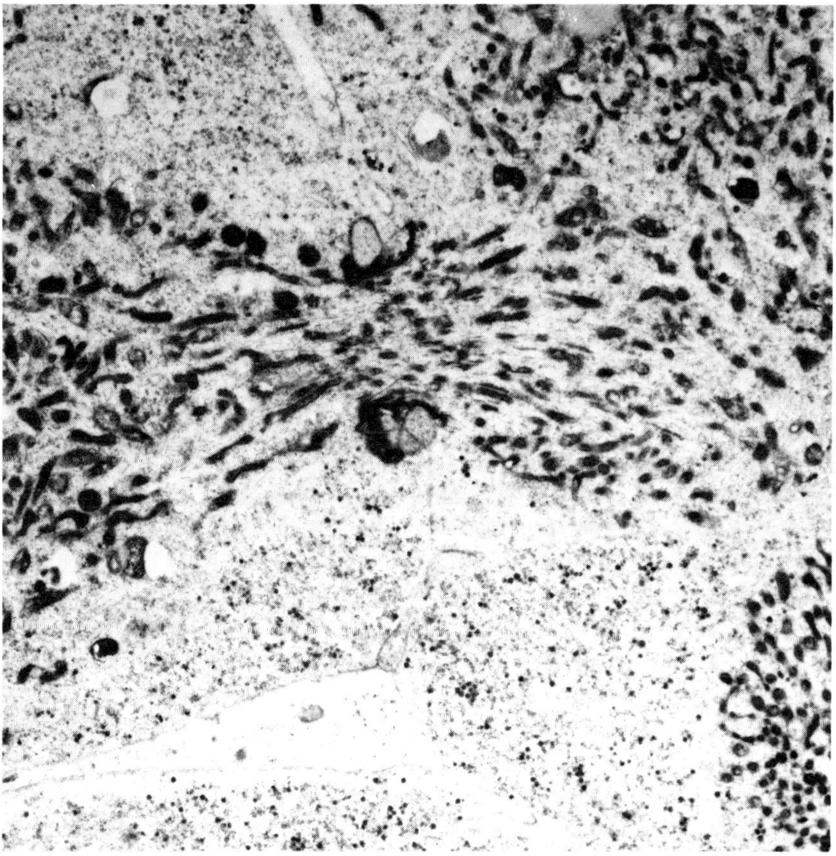

Fig. 22. An intercellular bridge in the vitellarium of a last instar caterpillar of *Hyalophora*. The fusome has disappeared and the bridge as well as the cytoplasm in the adjacent cells have become filled with dense-appearing mitochondria. Though these organelles presumably can cross the bridge, there is no evidence for a polarized movement at this stage of development. x 9200. (Micrograph provided by I. Mandelbaum.)

primarily a mixture of blood proteins and epithelial secretions that are assembled not in the nurse cells, but in the oöcyte cortex (Telfer, 1965; Telfer and Anderson, 1968; L. Anderson, 1971). The only nurse cell contributions that would be required for this function as it is presently understood would be the synthetic machinery for the cell membranes and whatever other components are necessary for the pinocytosis of these materials (Roth and Porter, 1964).

A similar proposal can be made for other types of yolk. Triglyceride-rich lipid droplets were shown by histochemistry to be assembled in the cortex of the oöcyte (e.g. Nath *et al.*, 1958; Bonhag, 1959; Seshachar and Bragga, 1963), and Engels and Drescher (1964) established that most of the glycogen stored in the egg is synthesized in the oöcyte. While glycogen and lipid droplets occur in the nurse cells, evidence for transfer has not appeared, and their quantities are trivial compared to those deposited in the oöcyte. Nath *et al.* (1959) felt that in telotrophic ovaries phospholipid precursor bodies arise in the tropharium and are transmitted to the oöcyte cortex where they become charged with triglycerides. This proposal underscores the possibility that the nurse cells may produce the machinery of yolk formation, even though the yolk spheres, lipid droplets, and glycogen particles, which constitute the yolk, are finally assembled in the oöcyte.

To summarize what is known about the contributions of the nurse cells to oögenesis, it is probably safe to presume that most cytoplasmic constituents in the nurse cells are swept into the oöcyte by the terminal injection process. During the preceding days or weeks of sustained nurse cell function, however, more discriminating kinds of evidence are required to demonstrate intercellular transport. A strong case can be made for the transfer of ribosomes and of 4S RNA to the oöcyte during this period, though in several species with an extra-chromosomal DNA body the germinal vesicle may synthesize a significant proportion of the egg ribosomes. The open nature of the intercellular bridges certainly provides an opportunity for the transfer of organelles such as mitochondria from the nurse cells, but the kinds of experiments necessary to verify this have not as yet been possible, and there is no compelling reason to believe that the oöplasm is not as favourable a site for organelle replication as the nurse cells. In quantitative terms the nurse cells are not a special site for protein synthesis, and they play little if any direct role in the assembly of yolk. Ribosomes, 4S RNA, and centrioles are therefore the only materials known at this time to cross the intercellular bridges prior to the act of terminal cytoplasmic injection.

7 Intercellular transport mechanisms

The polarity of the sibling cluster begins in the germarium with the differentiation of the nurse cells, and culminates in the vitellarium with the terminal injection of their cytoplasm into the oöcyte. During the intervening period of development a net movement of ribosomes, and possibly other cytoplasmic organelles, occurs across the intercellular bridges towards the oöcyte. Intuitively appealing phrases such as "nutrient streaming" have been applied to this process, but other mechanisms are equally possible, and there is as yet little basis for distinguishing among them.

Bulk cytoplasmic streaming is a plausible candidate, particularly in the terminal injection process, for the nurse cells lose most of their volume as well as their basophilia with this event. Unfortunately, terminal injection occurs too rapidly to have been carefully described by the developmental methods so far employed, and no experimental analyses of its mechanism have been published.

Cytoplasmic streaming is also a potential component of bridge transport polarity during the period of sustained nurse cell function. The *Hyalophora* oöcyte attains a volume of about 1 μl just before terminal injection, over half of which is yolk that was formed in the cortical oöplasm, while the remainder is cytoplasm that could potentially have come from the nurse cells. Since the oöcyte grows at an exponential rate, nearly doubling in size every day (Telfer and Rutberg, 1960), over 90 per cent of its volume is accrued during the final five days of nurse cell function. Assuming that the seven nurse cells contribute equally, each can be calculated from these figures to provide the oöcyte with cytoplasm at a rate of 6×10^{-4} μl h^{-1}. With a bridge diameter of 35 μm, this would entail a flow rate of 0.7 mm h^{-1}, which is high enough to suggest that organelles such as ribosomes might be swept into the oöcyte with little chance of back diffusion into the nurse cells. Fluid uptake or extrusion by the oöcyte would affect this rate, however; the oöcyte can in fact be seen to grow primarily by water uptake at a late stage of development when yolk formation no longer obscures the process (Telfer and Anderson, 1968). In the absence of connections to the nurse cells, all incoming materials must enter the oöcyte by permeating its plasma membrane. If this sort of hydration begins during the period of nurse cell function, the flow rate through the bridges would be substantially lower than calculated. Thus, until solvent flow has been directly measured in the bridges, its contribution to transport polarity cannot be evaluated.

A polarized movement of solutes in the absence of solvent flow is also a realistic possibility. In this event both free diffusion and active transport would be conceivable as contributing mechanisms. The result of autoradiography after uridine labelling, for instance, are fully consistent with free diffusion being the primary mode of RNA transport. Since the nurse cell nuclei are the only prominent sites of RNA synthesis in the sibling clusters of most polytrophic species, a diffusion gradient towards the oöcyte would necessarily result. Hydration of the oöcyte would intensify the gradient by diluting the RNA that had already crossed the bridges. Any mechanism that prevented back diffusion, such as adsorption in the oöplasm, would also intensify the gradient. As with solvent flow, however, there is no experimental information allowing an evaluation of the contribution of diffusion gradients to polarized transport.

Active transport, as conventionally conceived of in cellular physiology, could not occur across the intercellular bridges of polytrophic systems because the necessary membrane structures are not present. Nevertheless, a one-way movement of micro-injected proteins that appears to be related to an electrical potential gradient across the intercellular bridges was demonstrated in *Hyalophora* (Woodruff and Telfer, 1973), and this raises the possibility that electrical work performed at remote sites in the nurse cells or the oöcyte may affect the movement of charged particles in the intercellular bridges. Of the three possible modes of polarized movement through the intercellular bridges, therefore, experimental evidence is thus far available only for a mechanism that would promote transport to and retention by the oöcyte of negatively charged molecules and organelles. In this final section on nurse cell function, the information that is known about the physiology of the system is reviewed, along with those aspects of the structure of the sibling cluster that seem relevant to polarized transport.

7.1 INTERCELLULAR PROTEIN TRANSPORT AND ELECTRICAL POLARITY IN THE VITELLOGENIC *HYALOPHORA* FOLLICLE

Fluorescein-labelled rabbit serum globulins microinjected into a nurse cell in vitellogenic *Hyalophora* follicles reached the adjacent oöplasm within 3 min, and had spread throughout the oöplasm by 30 min (Woodruff and Telfer, 1973). Even after 2 h, however, fluorescence remained undetectable in the other seven nurse cells. Similarly, fluorescent globulins injected into the oöcyte spread to the bridge region in less than 30 min, but continued to be excluded from the nurse cells for at least 2 h (Fig. 23). In these experiments, therefore, the labelled protein could move in only one direction through the intercellular bridges. While free diffusion of locally synthesized materials is, as we have seen, a plausible component of polarized RNA transport through the intercellular bridges, the behaviour of fluorescent globulin shows that diffusion must be strongly supplemented by an additional transport mechanism.

In Lepidoptera the centripetal gathering of bridges giving rise to the rosette in the germarium is retained in the vitellarium, so that each nurse cell has a narrow apex where its bridges are concentrated, and an expanded cell body containing the nucleus and most of the cytoplasm (Fig. 24). In all of the 46 cases where the fluorescent serum globulin was injected into a

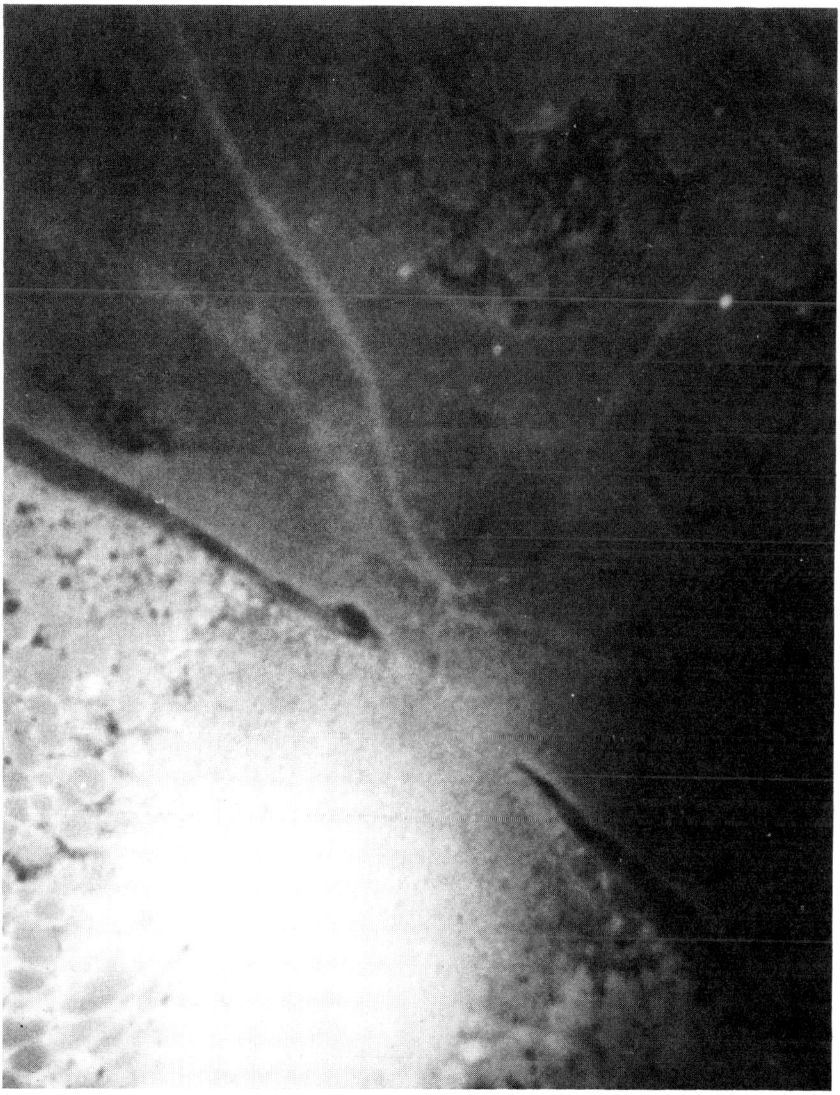

Fig. 23. Fluorescein-labelled rabbit serum globulin injected into an oöcyte (below) diffuses to the region of the intercellular bridges in less than 30 min, but fails to cross the bridges into the nurse cells even after 2 h incubation. This follicle had been incubated 2 h and then fixed in Carnoy's solution prior to embedding and sectioning. The apices of four nurse cells can be seen converging on the bridge region. (Courtesy of R. Woodruff.)

nurse cell, transmission to the oöcyte occurred without detectable back-diffusion into the expanded cell body of the other nurse cells. This result was obtained despite the fact that 4 out of 7 cells in each nurse cell cap are only indirectly bridged to the oöcyte via the apical cytoplasmic extension of one or two siblings (Fig. 10). Though the indirectly bridged cells could not be identified at the time of injection or, in most cases, in the

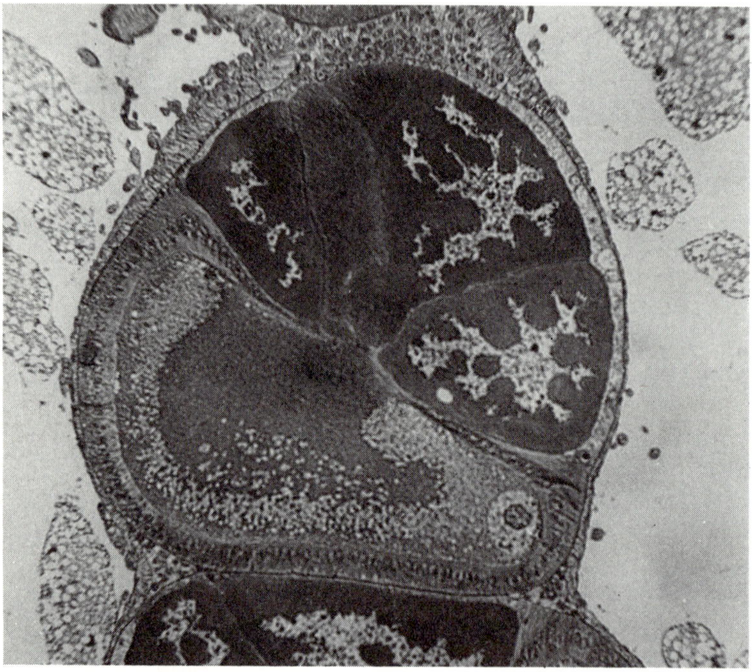

Fig. 24. An early vitellogenic follicle of *Hyalophora.* The nurse cells have extremely irregular, branching nuclei, and appear triangular in sections, with the apex of each directed toward the bridge complex in the centre of the cluster. The oöcyte is at this stage larger than the nurse cells. The follicular epithelium is columnar over the oöcyte and cuboidal or squamous over the nurse cells. The germinal vesicle, on the right, exhibits what is probably an endobody. The section was stained with Ehrlich's hematoxylin and PAS, after Carnoy's fixation.

fluorescent sections, a random selection should have resulted in their being injected in over half of the cases. The result, therefore, suggests that polarized movement must occur not only in the bridges themselves but also in the apical cytoplasmic projections of the nurse cells.

In the 16-cell dytiscid and dipteran sibling clusters the nurse cells have a different arrangement (Fig. 18). Here the rosette configuration and its centralized bridge complex are lost during the growth of the follicles. The new arrangement requires that three of the four nurse cells directly bridged to the oöcyte lie in the route of transport of products from the more distant siblings. Bier (1963a) described a gradient of RNA labelling intensity in the nurse cells of *Calliphora*, with those lying closest to the oöcyte being more heavily labelled than those at the apical end of the cluster (Fig. 18). He proposed that the gradient resulted from the morphological arrangement in which cells directly bridged to the oöcyte must transmit RNA from their more distal siblings, as well as their own synthetic products, to the oöcyte. Way-stations of this sort do not occur in the lepidopteran rosette where the complex of bridges and apical cytoplasms give each nurse cell an equivalent access to the oöcyte.

Correlated with the polarity of protein transport in *Hyalophora* is an electrical potential gradient. Electrophysiological studies demonstrated an average potential difference of 10 mV between the nurse cells and the oöcyte of vitellogenic follicles (Woodruff and Telfer, 1973), and an average resistance of $13 \times 10^3\ \Omega$ (Woodruff and Telfer, 1974). The resting potential of the nurse cells, relative to the blood was about −40 mV, while that of the oöcyte was about −30 mV. It is plausible to assume that the potential gradient between electrodes in the two cells was steepest in the bridge complex at the centre of the rosette, and this was confirmed by the finding that the intracellular potential did not vary as the electrode was moved about within the cytoplasm of the cell bodies. Assuming that the gradient is largely localized within a path length of 100 μm through the complex of bridges and nurse cell apices, a voltage gradient of $1\ \text{V cm}^{-1}$ would result. The direction and the size of the voltage gradient would unquestionably impede the entry of charged proteins such as serum globulin into the nurse cells from the oöcyte. It is therefore reasonable to suggest that electrical polarity may prove to be an important component of the intercellular transport mechanism in this follicle.

As a test of this proposal, experiments were performed in which a ground electrode was inserted into a nurse cell attached to an oöcyte containing fluorescent globulin. A second electrode was inserted into the oöcyte, and a constant current was passed so as to make the nurse cell positive to the oöcyte, reversing its normal polarity. From the known average bridge resistance it could be calculated that the 80 mV current delivered would have rendered the nurse cell approximately 1 mA positive to the oöcyte. In every case 10-30 min of the treatment resulted in a channel of fluorescence through the bridge leading from the oöcyte into the grounded nurse cell (Woodruff and Telfer, 1973). Thus, with a reversed potential gradient that was only one-tenth as steep as the normal gradient, the mechanism barring serum globulin movement into a nurse cell body from the oöcyte had been overcome. Other conditions for fluorescent globulin movement from the oöcyte into the nurse cells were found which were also correlated with reduction or reversal of the potential gradient. These included exposure to 10.3 M dinitrophenol and a spontaneously occurring reversal in animals that had been kept in their pupal dormancy for unnaturally long periods.

If the observed electrical polarity is the primary cause of one-way movement through the bridge complex, positively charged molecules should move more readily from oöcyte to nurse cell than in the reverse direction. Zinmeister (1973) found that histone-like proteins appear to be synthesized throughout the cytoplasm in *Forficula* though their incorporation into ribosomes and chromosomes would be primarily in the nurse cells. Unfortunately, fluorescein-labelled calf thymus histones injected into *Hyalophora* oöcytes adsorbed or precipitated around the tip of the needle so that they were unable to diffuse to the bridge complex, where their behaviour would be a test of the proposal.

Positively charged materials with charge densities lower than that of histone are currently being explored in this laboratory by Dr Woodruff. One of these is fluorescein-labelled myoglobin which contains somewhat smaller proteins with higher isoelectric points than the serum globulins (Paine and Feldherr, 1973). Micro-injections have shown that this preparation contains components which can in fact move from the oöcyte to the nurse cells. Fluorescence also moves from the nurse cells to the oöcytes, but the material is too complex electrophoretically to allow a judgement as to the net electrical charge of the mobile components. A search is continuing for labelled, histologically fixable materials whose mobility in the bridges will allow a final decision on the role of electrical charge. It is already clear, however, that ability to move through the bridges varies with the protein being studied, and the differences are so far consistent with net electrical charge being the crucial parameter. The ability of a component of myoglobin to move from the oöcyte into the nurse cells is in addition evidence against cytoplasmic streaming as the mechanism of intercellular transport in *Hyalophora*, for this mechanism should affect all proteins in an equivalent manner.

The physiology of the polytrophic follicle has not been well enough studied to permit a proposal as to how the intercellular potential difference is generated and maintained. It can be reversibly decreased to 1-2 mV by 10^{-3} M dinitrophenol (Woodruff and Telfer, 1973), so that its maintenance appears to require metabolic energy. Higher concentrations of the inhibitor caused irreversible changes. Two classes of electrical potential generating mechanisms might be envisaged, and a combination of these could be entailed. A Donnan-type equilibrium is one possibility, with a fixed-charge gel in one or both cells rather than a semipermeable membrane across the bridge being required. The energy input would in this case be for the assembly and maintenance of the gel. An alternative model would entail differences between the oöcyte and nurse cell plasma membranes with regard to ion pumping or permeability. An ion pump that ejected cations from the cytoplasm primarily through the nurse cell membranes could generate the observed intercellular potential, for instance. In this case the potential gradient would have to be maintained against the neutralizing flow of ions that would be expected to pass primarily through the intercellular bridges. The energy requirement would thus be for a continuance of ion pumping to replace the effects of diffusion. The distinction between these possibilities has simply not received the required attention as yet.

Information on the physiology of the polytrophic follicle has also been obtained with extracellular recording (Woodruff and Telfer, 1974). Follicles were drawn into tightly fitting capillaries, and the potential difference between the two ends was measured by a system in which the only current flow was through a high impedance oscilloscope and amplifier. Under these conditions an average potential difference of 10 mV was measured across the follicle, with the nurse cell end being negative. That the extracellular potential is generated by the sibling cluster, rather than by the envelope of follicle cells, or by a concerted action of the two, cannot be concluded with certainty. Nevertheless, a relationship to intercellular transport is suggested by the fact that it was generated only by follicles that had not yet undergone their terminal injection of nurse cell cytoplasm. The transfollicular potential was also similar to the intercellular potential difference in its magnitude, its polarity, and its sensitivity to dinitrophenol. It is clearly related to nurse cell function, and it appears to offer a second experimental approach to the electrophysiology of the follicle.

A remarkable feature of the transfollicular potential is its sign. If an electrogenic cation pump situated in the nurse cell membrane were responsible for the potential difference across the intercellular bridges, it would be expected to generate a cation efflux across the nurse cell membrane, extracellular diffusion from the apical to the basal regions of the follicle, an influx of cation through the oöcyte membrane, and diffusion through the bridges to the nurse cells to complete the circuit. In such a system, the extracellular medium would be more electrically positive around the nurse cells than around the oöcyte, and transfollicular potential measured in a capillary tube would necessarily be positive at the nurse cell end. In fact, the opposite is true, and this indicates that the model is either incomplete or incorrect. To proceed further with the question is purely a matter of speculation but, nevertheless, the kinds of phenomena that a physiological analysis of this system may encounter are worth considering. One possible explanation of the sign of the transfollicular potential is that the epithelium supplements the transbridge potential gradient between the oöcyte and the nurse cells. If, for instance, the epithelium over the apical surface of the follicle actively sequestered cations pumped out of the nurse cells and returned them to the oöcyte surface by an intraepithelial route, the extrafollicular medium around the nurse cell end of the follicle could well be negative relative to the oöcyte end. A second possibility is that ion pumps located in the cell membrane do not generate the transbridge potential. A cytoplasmic gel with a net negative charge in the nurse cells, for instance, could also induce a negative charge in the adjacent extrafollicular medium. The existence of these alternatives, while making any prediction about the physiological basis of polarized transport a matter of guesswork, shows that the polytrophic follicle continues to have important possibilities as an experimental object.

7.2 THE STRUCTURAL BASIS OF PHYSIOLOGICAL POLARITY

When the sibling cluster has resided in the vitellarium long enough to initiate yolk deposition, the oöcyte and nurse cells already differ profoundly from each other in their structure. Aside from the chromosomal and nucleolar differences already described, their cytoplasms and cell membranes bear little resemblance to each other. While the nurse cells have smooth contours, the oöcyte surface becomes modified by the production of microvilli and pinocytotic configurations (Roth and Porter, 1964; Stay, 1965; King and Aggarwal, 1965; and many other references). Similar structures are seen in vitellogenic oöcytes of both telotrophic (Beams and Kessel, 1963) and panoistic ovarioles (Favard-Sereno, 1964; Anderson, 1965), and are generally interpreted as related to the process of yolk deposition. From the morphology of pinocytosis and the measured rates of yolk deposition, an oöcyte plasma membrane turnover time of approximately 10 min was calculated (Telfer, 1965). The nurse cell membrane, by contrast, appears to be a morphologically stable structure.

One can anticipate that two cell membranes having such different morphological and physiological characteristics will prove to vary additionally in their permeability and active transport capacities. The surprising fact is not that two such contrasting cells should exhibit differences in electrical potential, but that the difference could be maintained across an open channel of cytoplasm without being neutralized by the diffusion of ions. It has already been noted that the resistivity of the cytoplasm in the bridges is not unusually high during vitellogenesis, and that there is no structural evidence of barriers to diffusion. A 10-mV gradient could thus in principle drive a current of up to 77 μA across the bridge complex between each nurse cell and the oöcyte.

There are two possible answers as to how the potential difference might be maintained. The unusually large size of the cells presumably results in their having sufficient metabolic machinery to pump ions at a rate that would counter the neutralizing effects of a current flow. A second possibility relates to the fact that as a consequence of vitellogenesis, the oöplasm gradually fills with protein yolk spheres, lipid droplets, and glycogen particles, which greatly dilute the basophilic cytoplasm, while nurse cells contain a uniformly basophilic and yolk-free cytoplasm. As has already been suggested, the need for ion pumping could well be reduced if the potential difference were due in part to a Donnan equilibrium resulting from fixed charge differences between a yolky and a yolk-free cytoplasm. A combination of these two factors will presumably account for the ability of the sibling cluster to maintain the average 10-mV potential difference across the bridges.

There is as yet no suggestion from electron microscope studies of a structural basis for active transport mechanisms localized within the bridges of polytrophic systems. In the early literature hematoxylin-staining materials were found to persist in the intercellular bridges throughout the differentiation of the sibling cluster (Giardina, 1901; Gunthert, 1910; Hirschler, 1942), and thus there was a lack of clarity about when and if the fusome disappears. Electron microscopy, by contrast, has generally failed to reveal the fusomal material in the bridges of follicles that have reached the vitellarium. In *Hyalophora*, for instance, Mandelbaum (1974) found that the fusome persists as long as the nuclei remain undifferentiated, but that it disappears as soon as the nurse cells and the oöcyte can be distinguished from each other. At this time microtubules reappear in the bridges, along with a high concentration of mitochondria (Fig. 22). While microtubules are suspected of being involved in transport processes in many biological systems, they are also important in generating shape changes (Porter, 1966), and it is probable that they play this role also in the development of the intercellular bridges.

While they are formed with a diameter of about 1 μm, the intercellular bridges widen progressively during later development, and the dense material lining the outer membrane gradually thickens (Koch and King, 1969; Mahowald, 1971). Their diameter reaches a value of 10 μm in *Drosophila* (Meyer, 1961) of 35 μm in *Hyalophora* (Woodruff and Telfer, 1973). There is thus a 100-fold increase in cross-sectional area in the one case and more than 1000-fold in the other. An extensible ring of striated "leaves", each consisting of 40-70 parallel segments of microtubules, reinforces the perimeter of the bridge in *Habrobracon* during the growth period (Cassidy and King, 1969). Comparable structures were recently found in the bridges of *Drosophila virilis* by Kinderman and King (1973). Both circumferentially and longitudinally oriented microtubules, embedded in places in a dense matrix, line the rim of the intercellular bridges in *Hyalophora* (Fig. 25). Aside from such differentiations of the perimeter, microtubules have not been reported as conspicuous in the cytoplasm occupying the core of the bridges during vitellogenesis (Steinert and Urbani, 1969; Cassidy and King, 1969; Mahowald, 1971; Kinderman and King, 1973). The same is true of *Hyalophora*, though small numbers of these structures are visible in sections of this region (Fig. 26).

In telotrophic hemipterans the trophic core and the cytoplasmic strand connecting it to the oöcytes contain throughout their length longitudinally oriented microtubules (Hamon and Folliot, 1969; Brunt, 1970; Huebner and Anderson, 1970; Macgregor and Stebbings, 1970). Unlike those of polytrophic systems, they appear evenly distributed in cross-sections of the strand, rather than concentrated around the periphery. They almost certainly contribute to the mechanical integrity of the tenuous cytoplasmic strand, and it has been proposed in addition that they play a direct role in polarized transport. They are interspersed by ribosomes and mitochondria and, as we have seen, RNA is known to move through the cord from the nurse cell chamber to the oöcyte. If the currently popular concept turns out to be true that microtubules can effect the directed displacement of cytoplasmic organelles, as they do anaphase chromosomes, the possibility that they transport ribosomes in telotrophic ovaries will have to be taken into account. Huebner and Anderson (1970) showed that the microtubules of the *Rhodnius* ovariole can be destroyed by vinblastine sulphate administered by injection into the animal. The microtubules are replaced by fibrillar granules whose aggregation causes a profound reorganization of the structure of the cytoplasmic strands. Mitochondria and ribosomes are evenly distributed between microtubules in untreated ovaries, while with exposure to vinblastine the cytoplasm forms islands that separate the aggregates of granules. It will be important to combine vinblastine treatment with uridine labelling in order to determine the effects of microtubule disorganization on the transport functions of the cytoplasmic strand.

Unpublished efforts in this laboratory to dissolve the microtubules of the intercellular bridges in *Hyalophora* with vinblastine, colcemid, low temperature, or high pressure have all been ineffective. The microtubules in this system appear to be extremely stable, and it has not been possible to test in this way their possible function in polarized transport. The absence of concentrated microtubular arrays from the cytoplasmic core of the bridge, however, suggests that they are not essential for this function.

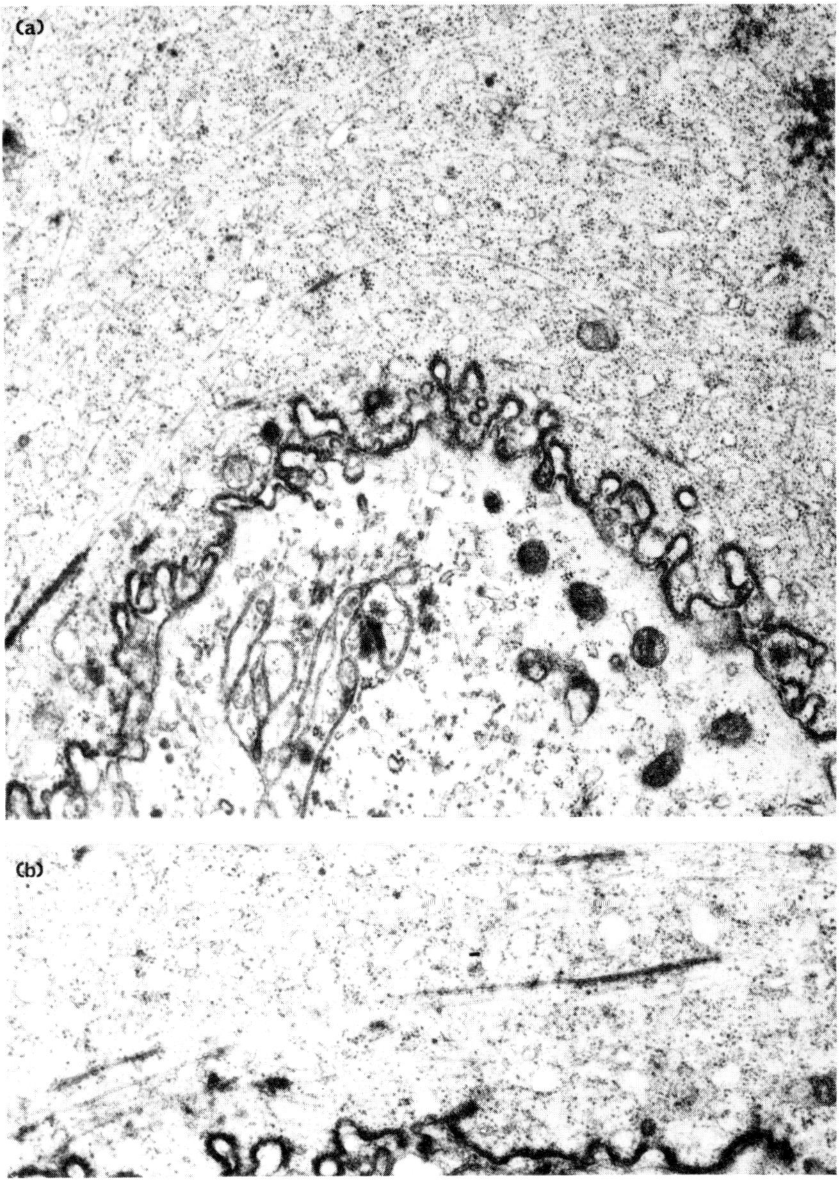

Fig. 25. Electron micrograph of the edges of an intercellular bridge in a vitellogenic *Hyalophora* follicle. Epithelial cells below. The bridge contains, in particular, ribosomes, vesicles and, in this region, microtubules. (a) x 22 800; (b) x 29 500. (Courtesy of I. Mandelbaum.)

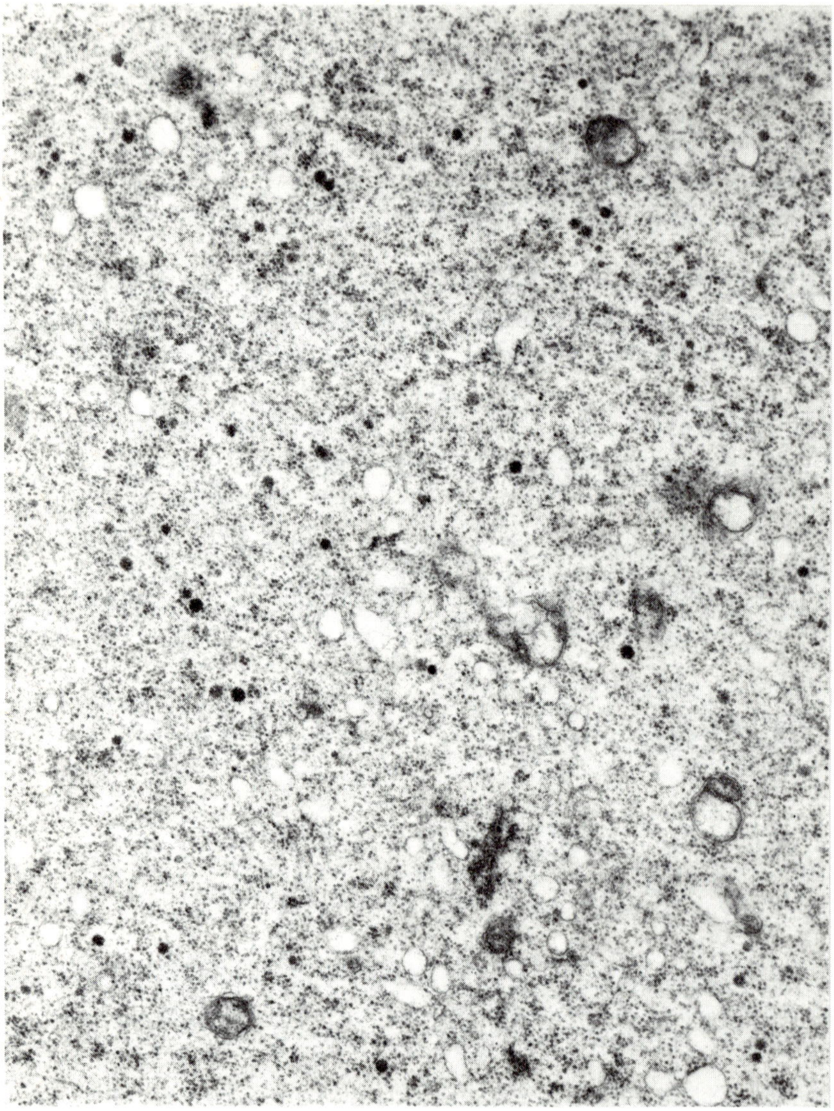

Fig. 26. Electron micrograph of the cytoplasm in the centre of an intercellular bridge in a vitellogenic *Hyalophora* follicle. There are no special features in this zone suggesting barriers to diffusion or sites of polarized transport mechanisms. x 27 500. (Courtesy of I. Mandelbaum.)

Ultrastructural studies of intercellular bridges have thus provided no compelling indication of a mechanism of polarized transport localized within the intercellular bridges in polytrophic systems. At the present time it seems more likely that the physiological basis of polarized transport will be found in the structural and physiological differences between the connected cells rather than in the bridges themselves. In telotrophic ovarioles the structural basis for a transport mechanism involving microtubules in the connecting strand of cytoplasm is present, but experimental evidence that the system actually employs the microtubules in this manner is still lacking.

8 Summary and prospect

In polytrophic ovarioles the oögonia give rise to cystoblasts which are programmed to undergo a defined number of mitotic divisions. Incomplete cytokinesis occurs at each of these divisions, and this results in the formation of a syncytium whose members are connected by stable intercellular bridges. Cell cycle cues are apparently exchanged across the bridges, for the attached members always divide in synchrony with each other. The result is a sibling cluster of 2^n cells, with the value of n being the number of times the syncytium undergoes a set of synchronous divisions. In Dermaptera, n is 1; in Mecoptera, it is 2; in Lepidoptera, 3; in many Diptera and Coleoptera, 4; and in Hymenoptera, it is often 5, though there are many exceptions.

After each set of synchronous divisions, the newly formed bridges in many species are displaced toward the centre of the cluster so that the siblings appear like the petals of a rosette, all attached to a common centre. Prior to the next division the spindle remnant and mid-body which initially occupy each bridge are replaced by a densely staining fibrillar or granular material termed the fusome. Early workers recognized that the fusomes of all bridges merge in the centre of the rosette to form a single structure with branches traversing every bridge and ending in the cytoplasm of every sibling. Electron microscopy has recently confirmed this finding. At the subsequent set of synchronous divisions the mitotic spindles are each oriented with one pole adjacent to the fusome and its associated bridges. This assures that all pre-existing bridges in a cell are retained by only one of its mitotic daughters. A consequence of this behaviour is that in the mature sibling cluster of all species examined two cells are attached to n bridges, two cells to n-1 bridges, four cells to n-2 bridges, and other siblings, if present, to even fewer. The oöcyte invariably develops from one of the two cells with n bridges, while all others become nurse cells.

In telotrophic ovaries also, some germ cell progeny become nurse cells which attach to the oöcytes by bridges of cytoplasm. Reconstruction of cell lineages has not been possible, however, because cell fusion occurs during metamorphosis and siblings cannot be identified in the more synctial nurse chamber that results. Panoistic ovaries, in which no nurse tissue differentiates, might be homologized to polytrophic systems by assuming that the cystoblast differentiates into an oöcyte without first undergoing mitosis and bridge formation. Other models are possible, however, and further work will be required before it is possible to see how cystoblast development is altered in the panoistic ovary.

In many panoistic species an extra-chromosomal DNA body arises by gene amplification before the last oögonial division. A similar structure arises in the cystoblasts of several polytrophic species. The DNA body is inherited by the oöcyte, which highlights the question of how only one cell in the sibling cluster is determined to become the oöcyte. The DNA body was itself once thought to be an oöcyte determinant, but its restricted occurrence in polytrophic species makes this idea of limited use. A more generally occurring structure would be required for this function. At mitosis the DNA body always goes to the daughter cell containing the fusome and bridge complex, so that it necessarily lies in a sibling with n bridges. A region of the cell surface or of the fusome associated with the first intercellular bridge are at present the most plausible candidates.

The differentiation of the oöcyte and nurse cells begins when the n sets of divisions have been completed. In some species all cells in the sibling cluster enter the first meiotic prophase before differentiation commences; in others only the two siblings with n bridges initiate meiosis. In either case, a single cell, the oöcyte, continues on its course into meiosis, while the nurse cells become diverted to a programme of endomitosis. A remarkable feature of this stage of development is that the nurse cells become out of phase with each other in their chromosome replication cycles, even though the intercellular bridges remain intact. The loss of synchrony may prove to result from the onset of polarized transport which, by this scheme, would prevent a free exchange of cell cycle cues across the intercellular bridges.

Nurse cell differentiation is marked by a tremendous increase in nuclear volume and DNA content. Much of the genome is systematically replicated at this time. There is evidence, however, that in some species ribosomal DNA may be amplified, and highly redundant base sequences may be under-replicated. The oöcyte, by contrast, maintains its $4n$, premeiotic complement of chromosomes, though in a few species it too may engage in gene amplification and the production of extrachromosomal DNA. In the latter event nucleolar DNA is included in the amplified fraction.

During vitellogenesis the nurse cells, having greatly increased their DNA content, become the predominant sites of synthesis of oöcyte RNA. Ribosomal and transfer RNA are the only classes thus far identified as originating in the nurse cells, but this does not rule out the possibility that messengers are exported to the oöcyte as well. In the oöcyte nucleus, with its much smaller number of chromosomes, RNA synthesis is much less evident than in the associated nurse cells. In some species the chromosomes become condensed in a karyosphere and there is a suggestion that RNA synthesis is inhibited to a degree related to the compaction of the chromosomes in this structure. When DNA escapes the karyosphere, as in nuclei with extrachromosomal DNA, RNA synthesis is readily apparent. In some polytrophic species a karyosphere is not formed, the chromosomes remaining dispersed, but even here no RNA synthesis is detectable. Oöcyte nuclei in panoistic species, by contrast, have basophilic chromosomes and one or many conspicuous nucleoli. Both these classes of structures incorporate uridine at rates rarely seen in oöcytes associated with nurse cells.

In polytrophic systems, intercellular transport occurs in two distinct phases: a sustained period, measured in days or even weeks, during which a slow deposition of nurse cell RNA occurs in the oöcyte, and a relatively abrupt terminal injection during which much of the residual nurse cell cytoplasm, and in some cases the nuclei, flow across the bridges. During this period of sustained transport there is evidence for RNA movement into the oöcyte, and it is widely presumed that a variety of other organelles are transported as well. Except for ribosomes and a well-documented case of centriole movement, however, there have been no experimental tests of organelle transport prior to terminal injection.

Evidence that transport processes other than free diffusion are involved in the movement of materials across the bridges comes from electrophysiological and micro-injection experiments utilizing *Hyalophora* follicles. These have demonstrated an electrical potential gradient which is steep enough to prevent negatively charged protein molecules from diffusing from the oöcyte to the nurse cells. The same proteins move very readily in the opposite direction, so there is no indication of a diffusion barrier such as a membrane across the bridges.

The fusome disappears with the onset of differentiation, and the bridges widen and become filled with cytoplasm similar in appearance to that in the cells on either side. There is as yet no ultrastructural indication of a transport mechanism in the bridges themselves, and it is therefore presumed that physiological differences between the oöcyte and the nurse cells account for the maintenance of polarity. What the pertinent differences are, when they arise, and how they relate to the early dichotomy in development between the two cell types remains unexplored.

In telotrophic systems the chord of cytoplasm connecting the oöcytes to the syncytial nurse chamber contains a high concentration of longitudinally oriented microtubules. Whether these are purely skeletal elements or are also involved in a transport mechanism has not been resolved.

The control of n, the control of the plane of cytokinesis, the switch from syncytial to asynchronous development despite the persistence of the intercellular bridges, the diversion of the nurse cells from meiosis to endopolyploidy, the protection of the oöcyte from this diversion, a differentiation of the oöcyte and nurse cells with regard to RNA synthesis and organelle formation, and polarized intercellular transport are all clearly expressed and well documented in the polytrophic ovariole. Learning how these processes are achieved and interrelated will require a combination of molecular, cytological and physiological methods that have not traditionally occupied the same laboratory. As a final example of this requirement, it will be necesary to determine the extent to which electrical potential gradients can result from the one-sided distribution of nucleic acid synthesis found in the sibling cluster. The fixation of mobile nucleotides in negatively charged DNA and RNA in the nurse cells, and the absence of this activity in the oöcyte would necessarily have an influence on the potential gradient across the intercellular bridges. The synthesis and transport of proteins, wherever they are localized, would also have such effects. Do these activities in fact contribute significantly, or is the potential gradient generated in the main by mechanisms employing inorganic ions?

The developmental physiology of the oöcyte-nurse cell complex, in raising issues of this sort, deals with characteristics that can surely be regarded as quantitative exaggerations of general cellular properties. Endopolyploidy and large cell volume provide amplification for the experimenter as well as for the oöcyte; the polarity of the complex, with an intercellular bridge providing a constricted equatorial waist, is an extreme version of the apico-basal axiation of many somatic cells. The special insights emerging from this extraordinary system are therefore likely to have a wide applicability. Analysis of the system has developed to the point where new experimental approaches are both feasible and necessary, and an optimistic view is that a less speculative statement on the physiology of the sibling cluster should be possible within a few years.

References

Allen, E. R. and Cave, M. D. (1969). Cytochemical and ultrastructural studies of ribonucleoprotein containing structures in the oöcytes of *Acheta domesticus*. *Z. Zellforsch. mikrosk. Anat.* **101**, 63-71.

Allen, E. R. and Cave, M. D. (1972). Nucleolar organization in oöcytes of gryllid crickets: Subfamilies Gryllinae and Nemobiinae. *J. exp. Zool.* **137**, 433-448.

Anderson, E. (1964). Oöcyte differentiation and vitellogenesis in the roach *Periplaneta americana*. *J. Cell Biol.* **20**, 131-155.

Anderson, E. (1969). Oögenesis in the cockroach, *Periplaneta americana*, with special reference to the specialization of the oolemma and the fate of coated vesicles. *J. Microscopie*, **8**, 721-738.

Anderson, E. and Beams, H. W. (1956). Evidence from electron micrographs for the passage of material through pores of the nuclear membrane. *J. biophys. biochem. Cytol.* 2, Suppl. 439-443.

Anderson, L. M. (1971). Protein synthesis and uptake by isolated *Cecropia* oöcytes. *J. Cell Sci.* **8**, 735-750.

Anderson, L. M. and Telfer, W. H. (1969). A follicle cell contribution to the yolk spheres of moth oöcytes. *Tissue and Cell*, **1**, 633-644.

Anteunis, A., Fautrez-Firlefyn, N. and Fautrez, J. (1966). La structure de ponts intercellulaires "obtures" et "ouvret" entre oögonies et oöcytes dans l'ovaire d'*Artemia salina*. *Archs. Biol., Paris*, 77, 645-654.

Bauer, H. (1933). Die wachsenden Oöcytenkern einiger Insekten in ihrem Verhalten zur Nuklealfärbung. *Z. Zellforsch. mikrosk. Anat.* **18**, 254-298.

Bauer, H. (1952). Die Chromosomen im Soma der Metazoen. *Verh. dt. zool. Ges.* **1952**, 252-268.

Bayreuther, K. (1952). Extra-chromosomale feulgenpositive Körper (Nucleinkörper) in der Oögenese der Tipuliden. *Naturwissenschaften*, 39, 71.

Bayreuther, K. (1956). Die Oögenese der Tipuliden. *Chromosoma*, 7, 508-577.

Bayreuther, K. (1957). Extrachromosomales DNS-haltiges Material in der Oögenese der Flöhe. *Z. Naturf.* **12b**, 458-461.

Beams, H. W. and Kessel, R. G. (1963). Electron microscope studies on developing crayfish oöcytes with special reference to the origin of yolk. *J. Cell Biol.* **18**, 621-649.

Bier, K. (1957). Endomitose und Polytänie in den Nährzellkernen von *Calliphora erythrocephala*. Meigen. *Chromosoma*, **8**, 493-522.

Bier, K. (1959). Quantitative Untersuchungen über die Variabilität der Nahrzellkern-strucktur und ihre Beeinflussung durch die Temperatur. *Chromosoma*, 10, 619-653.

Bier, K. (1963a). Synthese, interzelluläre Transport, und Abbau von Ribonukleinsäure im Ovar der Stubenfliege *Musca domestica*. *J. Cell Biol.* 16, 436-440.

Bier, K. (1963b). Autoradiographische Untersuchungen über die Leistungen des Follikelepithels und der Nährzellen bei der Dotterbildung und Eiwisssynthese im Fliegenovar. *Arch. EntwMech. Org.* **154**, 552-575.

Bier, K. (1964). Gerichter Ribonukleinsäuretransport durch das Cytoplasma. *Naturwissenschaften*, **51**, 418.

Bier, K. (1965). Zur Funktion der Nährzellen im meriostischen Insektenovar unter bersonderer Berücksichtigung der Oögenese adephager Coleopteren. *Zool. Jb. abt. Physiol.* 71, 371-384.

Bier, K. (1967). Oögenese, das Wachstum von Riesenzellen. *Naturwissenschaften*, **54**, 189-195.

Bier, K., Kunz, W. and Ribbert, D. (1967). Struktur und Funktion der Oöcyten-chromosomen und Nukleolen sowie der Extra-DNS während der Oögenese panoistischer und meroistischer Insektern. *Chromosoma*, 23, 214-254.

Bonhag, P. F. (1955). Histochemical studies of the ovarian nurse tissues and oöcytes of the milkweed bug, *Oncopeltus fasciatus* (Dallas). I. Cytology, nucleic acids and carbohydrates. *J. Morph.* **96**, 381-439.

Bonhag, P. F. (1956). The origin and distribution of periodic acid-Schiff-substances in the oöcyte of the earwig, *Anisolabis maritima* (Géné). *J. Morph.* **99**, 433-463.

Bonhag, P. F. (1958). Ovarian structure and vitellogenesis in insects. *A. Rev. Ent.* **3**, 137-160.

Bonhag, P. F. (1959). Histological and histochemical studies on the ovary of the American cockroach, *Periplaneta americana* (L.). *Univ. Calif. Publs Ent.* **16**, 81-124.

Bonhag, P. F. and Wick, J. R. (1953). The functional anatomy of the male and female reproductive systems of the milkweed bug, *Oncopeltus fasciatus* (Dallas). *J. Morph.* **93**, 177-283.

Botchan, M., Kram, R., Schmid, C. W. and Hearst, J. E. (1971). Isolation and chromosomal localization of highly repeated DNA sequences in *Drosophila melanogaster*. *Proc. natn. Acad. Sci. U.S.A.* **68**, 1125-1129.

Brandt, A. (1874). Über die Eiröhren der *Blatta orientalis*. *Meim. Acad. Sci. St. Petersbourg, 7 ser.* **21** (12).

Brown, E. H. and King, R. C. (1964). Studies on the events resulting in the formation of an egg chamber in *Drosophila melanogaster*. *Growth*, **28**, 41-81.

Brunt, A. M. (1970). Extensive System of Microtubules in the Ovarioles of *Dysdercus fasciatus* Signoret (Heteroptera: Pyrrhocoridae). *Nature, Lond.* **228**, 80-81.

Brunt, A. M. (1971). The histology of the first batch of eggs and their associated tissue in the ovariole of *Dysdercus fasciatus* Signoret (Heteroptera: Pyrrhocoridae) as seen with the light microscope. *J. Morph.* **134**, 105-130.

Buchner, P. (1909). Das accessoriche chromosom in Spermatogenese und Oögenese der Orthopteren, zugleich ein Beitrag zur Kenntnis der Reduktion. *Arch. Zellforsch.* **3**, 335-430.

Buchner, P. (1918). Vergleichende Eistudien. I. Die Akzessorischen Kerne der Hymenoptereneis. *Arch. mickrosc. Anat.* **91**, 1-202.

Buck, R. C. and Tisdale, J. M. (1962). The fine structure of the midbody of the rat erythroblast. *J. Cell Biol.* **13**, 109-115.

Büning, J. (1972). Untersuchungen am Ovar von *Bruchidius olotectus* Say (Coleoptera-Polyphaga) Zur Klärung des Oöcytenwachstums in der Prävitellogenese. *Z. Zellforsch. mikrosk Anat.* **128**, 241-282.

Burgos, M. H. and Fawcett, D. W. (1955). Studies on the fine structure of the mammalian testis. I. Differentiation of the spermatids in the cat (*Felis domestica*). *J. biophys. biochem. Cytol.* **1**, 287-300.

Byers, B. and Abramson, D. H. (1968). Cytokinesis in Hela: post-telophase delay and microtubule associated motility. *Protoplasma*, **66**, 413-435.

Callan, H. G. (1963). The nature of lampbrush chromosomes. *Int. Rev. Cytol.* **15**, 1-34.

Cassidy, J. P. and King, R. C. (1969). The dilatable ring canals of the ovarian cystocytes of *Habrobracon juglandis*. *Biol. Bull. mar. biol. Lab., Woods Hole*, **137**, 429-437.

Cassidy, J. D. and King, R. C. (1973). Ovarian development in *Habrobracon juglandis* (Ashmead) (Hymenoptera: Braconidae). I. The origin and differentiation of the oöcyte-nurse cell complex. *Biol. Bull. mar. biol. Lab., Woods Hole*, **143**, 483-505.

Cave, M. D. and Allen, E. R. (1969). Extrachromosomal DNA in early stages of oögenesis in *Acheta domesticus*. *J. Cell Sci.* **4**, 593-609.

Chandley, A. C. (1966). Studies on oögenesis in *Drosophila melanogaster* with ^{3}H-thymidine label. *Expl. Cell Res.* **44**, 201-215.

Claypole, A. M. (1898). The embryology and oögenesis of *Anurida maritima*. *J. Morph.* **14**, 219-300.

Colombo, G. (1957). Ricerche citologiche ed istochimiche sull' oögenesi di *Bombyx mori* L. (Lepidoptera). *Arch. zool. ital.* **42**, 309-347.

Crippa, M. and Telfer, W. H. (1971). Hybridization analysis of DNA replication in nurse and follicle cells in Cecropia moth. *Biol. Bull. mar. biol. Lab., Woods Hole,* **141**, 384.

Cruickshank, W. J. (1972). Ultrastructural modifications in the follicle cells and egg membranes during development of flour moth oöcytes. *J. Insect Physiol.* **18**, 485-498.

Cummings, M. R. and King, R. C. (1969). The cytology of the vitellogenic stages of oögenesis in *Drosophila melanogaster*. I. General staging characteristics. *J. Morph.* **128**, 427-442.

Cummings, M. R. and King, R. C. (1970). Ultrastructural changes in nurse and follicle cells during late stages of oögenesis in *Drosophila melanogaster*. *Z. Zellforsch. mikrosk. Anat,* **110**, 1-8.

Dapples, C. C. and King, R. C. (1970). The development of the nucleolus of the ovarian nurse cell of *Drosophila melanogaster*. *Z. Zellforsch. mikrosk. Anat.* **103**, 34-47.

Debaisieux, P. (1909). Les debuts de l'ovogenese dans le *Dytiscus marginales*. *La Cellule,* **25**, 207-237.

Dederer, P. H. (1915). Oögenesis in *Philosamia cynthia*. *J. Morph.* **26**, 1-41.

Deegener, P. (1928). Geschlechtsorgane. *In* "Handbuch der Entomologie" (Ed. C. Schröder), Vol. 1, pp. 466-533. G. Fischer, Jena.

Duspiva, F., Scheller, K., Weiss, D. and Winter, H. (1973). Ribonucleinsäure-synthese in der telotroph-meroistichen Ovariole von *Dysdercus intermedius* (Heteroptera, Pyrrhoc.). *Arch. Entw. Mech. Org.* **172**, 83-130.

Engels, W. (1968). Extraoocytäre Komponenten des Eiwachstums bei *Apis mellifica*. I. Trophocytäre RNS-Zufuhr. *Insectes Soc.* **15**, 271-288.

Engels, W. (1969). Geschwindigkeit des RNS-Transports in Einährverband der Dermapteren im Vergleich mit anderen Insekten meroistischen Ovartyps. *Zool. Anz. Suppl.* **33**, 30-39.

Engels, W. (1970). Kälte-Wirkungen auf die Glykogenspeicherung in Eizellen von *Musca domestica*. *Arch. Entw. Mech. Org.* **166**, 89-104.

Engels, W. and Drescher, W. (1964). Einbau von H^3-D-Glucose während der Oögenese bie *Apis mellifica* L. *Experientia*, **20**, 445-447.

Eschenberg, K. M. and Dunlap, H. L. (1966). The histology and histochemistry of oögenesis in the water strider, *Gerris remigis* Say. *J. Morph.* **118**, 297-316.

Fawcett, D. W. (1961). Intercellular bridges. *Expl. Cell. Res. Suppl.* **8**, 174-187.

Fawcett, D. W. (1970). Observations on cell differentiation and organelle continuity in spermatogenesis. *In* "The Genetics of the Spermatozoan" (Eds R. A. Beatty and S. Gluecksohn-Waelsch), pp. 37-68. University of Edinburgh.

Fawcett, D. W., Ito, S. and Slautterback, D. (1959). The occurrence of intercellular bridges in groups of cells exhibiting syncytial differentiation. *J. biophys. biochem. Cytol.* 5, 453-460.

Favard-Séréno, C. (1964). Phenomene de pinocytose au cours de la vitellogenese proteique chez le Grillon (Orthoptère). *J. Microscopie,* 3, 323-338.

Favard-Séréno, C. (1968). Evolution des structures nucléolaires au cours de la phase d'accroissement cytoplasmique chez le grillon (Insecta, Orthoptère). *J. Microscopie,* 7, 205-230.

Flemming, W. (1891). Neue Beitrage Zur Kenntniss der Zelle. *Arch. mikrosk. Anat.* 37, 685-751.

Forrest, H. S., Harris, S. E. and Morton, L. J. (1967). Inosine, guanine, uric acid, DNA and RNA in developing eggs of the milkweed bug *Oncopeltus fasciatus* (Dallas). *J. Insect Physiol.* 13, 359-367.

Franchi, L. L. and Mandel, A. M. (1962). The ultrastructure of the oögonia and oöcytes in the foetal and neonatal rats. *Proc. R. Soc. London, Series B,* **157**, 99-114.

Freed, J. J. and Schultz, J. (1956). Effect of the Y chromosome on the DNA content of ovarian nuclei in *Drosophila melanogaster* nuclei. *J. Histochem. Cytochem.* 4, 441-442.

Gall, J. G. and Callan, H. G. (1962). ^{3}H-uridine incorporation into lampbrush chromosomes. *Proc. natn. Acad. Sci. U.S.A.,* **48**, 562-570.

Gall, J. G., Cohen, E. H. and Polan, M. C. (1971). Repetitive DNA sequences in *Drosophila. Chromosoma,* 33, 319-344.

Gall, J. G., Macgregor, H. C. and Kidston, M. E. (1969). Gene amplification in the oöcytes of Dytiscid water beetles. *Chromosoma,* **26**, 169-187.

Gambarini, A. G. and Meneghini, R. (1972). Ribosomal RNA genes in salivary gland and ovary of *Rhynchosciara angelae. J. Cell Biol.* 54, 421-426.

Giardina, A. (1901). Origine dell' oöcite e delle cellule nutrici nei *Dytiscus. Int. Mschr. Anat. Physiol.* **18**, 417-484.

Gill, K. S. (1963). Developmental genetic studies on oögenesis in *Drosophila melanogaster. J. exp. Zool.* **152**, 251-277.

Gillot, S. (1968). Hétérogénéités fontionnelles dans l'ADN de noyaux géants. Étude autoradiographique sur la glande séricigène de *Bombyx mori* L. *Expl. Cell Res.* **50**, 388-402.

Grell, R. and Chandley, A. C. (1965). Evidence bearing on the coincidence of exchange and DNA replication in the oöcyte of *Drosophila melanogaster. Proc. natn. Acad. Sci. U.S.A.* 53, 1340-1346.

Gross, J. (1903). Untersuchungen über die Histologie des Insectenovariums. *Zool. Jb., Abt. Anat. Ontog.* **18**, 71-186.

Grünberg, K. (1903). Keim und Nährzellen in den Hoden und Ovarien der Lepidopteren. *Z. wiss. Zool.* 74, 327-395.

Gruzova, M. N., Zaichikova, Z. P. and Sokolov, I. I. (1972). Functional organization of the nucleus in the oögenesis of *Chrysopa perla* (Insecta, Neuroptera). *Chromosoma,* 37, 353-387.

Guelin, M. (1968). Données preliminaires sur l'existence de vésicules chromatiniennes differenciés dans le noyaux des trophocytes d'*Ephestia kühniella* Z. (Lepidoptère). *C.r. hebd. Séanc. Acad. Sci. Paris,* **266**, 1740-1742.

Gunthert, T. (1910). Die Eibildung der Dytisciden. *Zool. Jb. Abt. Anat. Ontog.* **30**, 301-372.

Halkka, L. and Halkka, O. (1968). RNA and Protein in Nucleolar Structures of Dragonfly Oöcytes. *Science*, **162**, 803-805.

Hamon, C. and Folliot, R. (1969). Ultrastructure des cordons trophiques de l'ovaire de diverse Homopteres Auchenorynches. *C.r. hebd. Séanc. Acad. Sci., Paris,* **268**, Ser. D, 577-580.

Hansen-Delkeskamp, E. (1969). Synthese von RNS und Protein während der Oögenese und frühen Embryogenese von *Acheta domestica. Arch. Entw. Mech. Org.* **162**, 114-120.

Hegner, R. W. (1915). Studies on germ cells. IV. Protoplasmic differentiation in the oöcytes of certain hymenoptera. *J. Morph.* **26**, 495-561.

Hegner, R. and Russell, C. P. (1916). Differential mitoses in the germ cell cycle of *Dineutes nigrior. Proc. natn. Acad. Sci. U.S.A.* **2**, 356-360.

Hertwig, E. (1935). Die Vielwertigkeit der polyploiden Somakerne der Heteropteren durch Chromosomenteilung ohne Kernteilung. *Z. Abstammgslehre*, **70**, 496-501.

Hirschler, J. (1942). Organization und Genese der Ei-Nahrzellen-Verbandes in Ovarien von *Macrothylacia rubi* L.—Lepidoptera. Ein Methodischer Versuch. *Biol. Zbl.* **62**, 555-577.

Hirschler, J. (1945). Gesetzmässigkeiten in den Ei-Nährzellenverbänden. *Zool. Jb. Abt. allg. Zool. Physiol.* **61**, 141-236.

Hirschler, J. (1955). On the cooperation of fusomes in the development of egg-nurse cell complexes in the animal ovary. *La Cellule*, **57**, 67-87.

Hopkins, C. R. (1964). The histochemistry and fine structure of the accessory nuclei in the oöcyte of *Bombus terrestris. Q. J. Microsc. Sci.* **105**, 475-480.

Huebner, E. and Anderson, E. (1970). The effects of vinblastine sulfate on the microtubular organization of the ovary of *Rhodnius prolixus. J. Cell Biol.* **46**, 191-198.

Huebner, E. and Anderson, E. (1972a). A cytological study of the ovary of *Rhodnius prolixus*. I. The ontogeny of the follicular epithelium. *J. Morph.* **136**, 459-494.

Huebner, E. and Anderson, E. (1972b). A cytological study of the ovary of *Rhodnius prolixus*. II. Oöcyte differentiation. *J. Morph.* **137**, 385-416.

Hughes, M. and Berry, S. (1970). The synthesis and secretion of ribosomes by nurse cells of *Antheraea polyphemus. Devl. Biol.* **23**, 651-664.

Huxley, T. (1958). On the agamic reproduction and morphology of aphids. *Trans. Linnean. Soc. London,* **23**, 221-236.

Iverson, R., Telfer, W. H. and Edstrom, J-E. (1974). Electrophoretic analysis of RNA from Cecropia moth ovaries. In preparation.

Jacob, J. and Sirlin, J. L. (1959). Cell function in the ovary of *Drosophila*. I. DNA classes in nurse cell nuclei as determined by autoradiography. *Chromosoma,* **10**, 210-228.

Jaworska, H. and Lima-de-Faria, A. (1973). Amplification of ribosomal DNA in *Acheta*. VII. Transfer of DNA-RNA assemblies from the nucleus to the cytoplasm. *Hereditas,* **74**, 187-204.

Johnson, M. W. (1938). A study of the nucleoli of certain insects and the crayfish. *J. Morph.* **62**, 113-139.

Johnson, J. H. and King, R. C. (1972). Studies of *fes*, a mutation affecting cystocyte cytokinesis, in *Drosophila melanogaster. Biol. Bull. mar. biol. Lab., Woods Hole,* **143**, 525-547.

Johnson, R. T. and Rau, P. M. (1971). Nucleo-cytoplasmic interactions in the achievement of nuclear synchrony in DNA synthesis and mitosis in multi-nucleate cells. *Biol. Rev.* **46**, 97-155.

Kato, K. (1968). Cytochemistry and fine structure of elimination chromatin in *Dytiscidae. Expl. Cell Res.* **52**, 507-522.

Kawamura, K. (1960). Studies on cytokinesis in neuroblasts of the grasshopper, *Chortophaga viridifaciata* (DeGeer). II. The role of the mitotic apparatus in cytokinesis. *Expl. Cell Res.* **21**, 9-18.

Kessel, R. G. (1968). Annulate lamellae. *J. Ultrastruct. Res. Suppl.* **10**, 5-82.

Kessel, R. G. and Beams, H. W. (1969). Annulate lamellae and "yolk nuclei" in oöcytes of the dragonfly *Libellula pulchella. J. Cell Biol.* **42**, 185-201.

Kinderman, N. B. and King, R. C. (1973). Oögenesis in *Drosophila virilis.* I. Interactions between the ring canal rims and the nucleus of the oöcyte. *Biol. Bull. mar. biol. Lab., Woods Hole,* **144**, 331-354.

King, P. E. and Fordy, M. R. (1970). The formation of "Accessory Nuclei" in the developing oöcytes of the parasitoid Hymenopterans *Ophion luteus* (L.) and *Apanteles glomeratus* (L.). *Z. Zellforsch. mikrosk. Anat.* **109**, 158-170.

King, P. E. and Richards, J. G. (1969). Oögenesis in *Nasonia vitripennis* (Walker) (Hymenoptera: Pteromalidae). *Proc. R. ent. Soc. Lond. Ser. A., Gen. Ent.* **44**, 143.

King, R. C. (1970). "Ovarian Development in *Drosophila melanogaster*". Academic Press, New York and London.

King, R. C. and Aggarwal, S. K. (1965). Oögenesis in *Hyalophora cecropia. Growth,* **29**, 17-84.

King, R. C. and Akai, H. (1971). Spermatogenesis in *Bombyx mori.* I. The canal system joining sister spermatocytes. *J. Morph.* **134**, 47-55.

King, R. C. and Burnett, R. G. (1959). Autoradiographic study of uptake of tritiated glycine, thymidine and uridine by fruit fly ovaries. *Science,* **129**, 1674-1675.

King, R. C., Burnett, R. G., and Staley, N. A. (1957). Oögenesis in adult *Drosophila melanogaster.* IV. Hereditary ovarian tumors. *Growth,* **21**, 239-261.

King, R. C. and Koch, E. A. (1963). Studies of ovarian follicle cells of *Drosophila. Q. Jl Microsc. Sci.* **104**, 297-320.

Knaben, N. (1934). Oögenese bei *Tischeria angusticolella* Dup. *Z. Zellforsch. mikrosk. Anat.* **21**, 604-625.

Koch, E. A. and King, R. C. (1964). Studies on the *fes* mutant of *Drosophila melanogaster. Growth,* **28**, 325-369.

Koch, E. A. and King, R. C. (1966). The origin and early differentiation of the egg chamber of *Drosophila melanogaster. J. Morph.* **119**, 283-304.

Koch, E. A. and King, R. C. (1969). Further studies on the ring canal system of the ovarian cystocytes of *Drosophila melanogaster. Z. Zellforsch. mikrosk. Anat.* **102**, 129-152.

Koch, E. A., Smith, P. A. and King, R. C. (1967). The division and differentiation of *Drosophila* cystocytes. *J. Morph.* **124**, 143-166.

Korschelt, E. (1886). Über die Entstehung und Bedeutung der verschiedenen Zellenelemente des Insektenovariums. *Z. wiss. Zool.* 43, 512-720.

Krzysztofowicz, A. (1971). Histochemical and autoradiographic analysis of RNA synthesis in trophic cells of the female gonad in *Tetrodontophora bielanensis* (Waga) (Collembola). *Acta biol. Cracouiensia, Ser. Zool.* 14, 299-305.

Kunz, W. (1967a). Funktionsstrukturen im Oöcytenkern von *Locusta migratoria. Chromosoma,* 20, 332-370.

Kunz, W. (1967b). Lampenbürstenchromosomen und multiple Nucleolen bei Orthopteren. *Chromosoma,* 21, 446-462.

Kunz, W. (1969). Die Entstehung multipler Oöcytennukleolen aus akzessorischen DNS-Körpern bei *Gryllus domesticus. Chromosoma,* 26, 41-75.

Kunz, W., Trepte, H. and Bier, K. (1970). On the function of the germ line chromosomes in the oögenesis of *Wachtliella persicarie* (Cecidomyiidae). *Chromosoma,* 30, 180-192.

Lane, N. J. (1967). Spheroidal and ring nucleoli in amphibian oöcytes. Patterns of uridine incorporation and fine structure features. *J. Cell Biol.* 35, 421-434.

Lee, A. B. (1895). La Regression du Fuseau Caryocinetique. *La Cellule,* 11, 29-51.

Lima-de-Faria, A. (1959). Differential uptake of tritiated thymidine into hetero- and euchromatin in *Melanoplus* and *Secale. J. biophys. biochem. Cytol.* 6, 457-466.

Lima-de-Faria, A. (1962). Metabolic DNA in *Tipula oleracea. Chromosoma,* 13, 47-59.

Lima-de-Faria, A., Birnstiel, M. and Jaworska, H. (1969). Amplification of ribosomal cistrons in the heterochromatin of *Acheta. Genetics,* 61, Suppl, 145-159.

Lima-de-Faria, A., Daskaloff, S. and Enell, A. (1973). Duplication of ribosomal DNA in *Acheta.* I. The number of chromomeres involved in the amplification process. *Hereditas,* 73, 99-118.

Lima-de-Faria, A. and Moses, M. J. (1966). Ultrastructure and cytochemistry of metabolic DNA in *Tipula. J. Cell Biol.* 30, 177-192.

Lima-de-Faria, A., Nilsson, B., Cave, D., Puga, A. and Jaworska, H. (1968). Tritium labelling and cytochemistry of extra DNA in *Acheta. Chromosoma,* 25, 1-20.

Lubbock, J. (1859). On the ova and pseudova of insects. *Phil. Trans. R. Soc. London,* 149, 341-369.

Macgregor, H. C. and Stebbings, H. (1970). A massive system of microtubules associated with cytoplasmic movement in telotrophic ovaries. *J. Cell. Sci.* 6, 431-449.

McGregor, J. H. (1899). The spermatogenesis of amphiuma. *J. Morph.* 15 (suppl.), 57-104.

Mahowald, A. P. (1971). The formation of ring canals by cell furrows in *Drosophila. Z. Zellforsch. mikrosk. Anat.* 118, 162-167.

Mahowald, A. P. (1972). Ultrastructural observations on oögenesis in *Drosophila. J. Morph.* 137, 29-48.

Mahowald, A. P. and Strassheim, J. M. (1970). Intercellular migration of centrioles in the germarium of *Drosophila melanogaster.* An electron microscopic study. *J. Cell Biol.* 45, 306-320.

Mahowald, A. P. and Tiefert, M. (1970). Fine structural changes in the Drosophila oöcyte nucleus during a short period of RNA synthesis. *Arch. Entw. Mech. Org.* 165, 8-25.

Mandelbaum, I. (1974). An electron microscope study of the nurse cells and intercellular bridges of *Hyalophora cecropia.* In preparation.

Marshall, W. S. (1907). The early history of the cellular elements of the ovary of a Phryganid, *Platyphylax designatus* Walk. *Z. Wiss. Zool.* **86**, 214-237.

Mays, U. (1972). Autoradiographische Untersuchungen zum Strofftransport von den Nährzellen zur Oöcyte der Feuerwanze *Pyrrhocoris apterus* L. (Heteroptera). *Z. Zellforsch. mikrosk. Anat.* **123**, 395-410.

Maziarski, S. (1913). Sur la persistance des residus fusariaux pendant les nombreuses generations cellulaires au cours de l'ovogenese de *Vespa vulgaris* L. *Arch. Zellf., Leipzig,* **10**, 507-532.

Melius, M. E., Jr. and Telfer, W. H. (1969). An autoradiographic analysis of yolk deposition in the cortex of the Cecropia moth oöcyte. *J. Morph.* **129**, 1-16.

Meng, C. (1970). Autoradiographische Untersuchungen am oösom in der Oöcyte von *Pimpla turionellae* L. (Hymenoptera). *Arch. Entw. Mech. Org.* **165**, 35-52.

Meyer, G. (1961). Interzellulare Brücken (Fusome) im Hoden und im Ei-Nahrzellverband von *Drosophila melanogaster. Z. Zellforsch. mikrosk. Anat.* **54**, 238-251.

Meyer, H. (1849). Über die Entwicklung des Fettkörpers, der Tracheen, und der Keimbereitenden Geschlechtstheile bei den Lepidopteren. *Z. wiss. Zool.* **1**, 175-197.

Miller, O. L. and Beatty, B. R. (1969). Portrait of a gene. *J. cell comp. Physiol.* **74** (suppl.), 225-232.

Mukenthaler, F. A. and Mahowald, A. P. (1966). DNA synthesis in the oöplasm of *Drosophila melanogaster. J. Cell Biol.* **28**, 199-208.

Nath, V., Gupta, B. and Aggarwal, S. K. (1959). Histochemistry of vitellogenesis in the earwig *Labidura riparia* Pall. and *L. bengalensis* Dohrn. *Res. Bull. Panjab Univ. Sci., New Ser.* **10**, 315-341.

Nath, V., Gupta, B. L. and Lal, B. (1958). Histochemical and morphological studies of lipids in oögenesis. I. *Periplaneta americana. Q. Jl microsc. Sci.* **99**, 315-332.

Nath, V., Gupta, B. and Sareen, M. L. (1959). Histochemistry of vitellogenesis in waterscorpions, *Laccotrephes maculatus* Fahr. and *L. ruber* L. *Res. Bull. Panjab Univ. Sci., New Ser.* **10**, 375-389.

Nilsson, B. (1966). DNA bodies in the germ line of *Acheta domesticus* (Orthoptera). *Hereditas,* **56**, 396-398.

Paine, P. L. and Feldherr, C. M. (1972). Nucleocytoplasmic exchange of macromolecules. *Expl Cell Res.* **74**, 81-98.

Painter, T. S. and Reindorp, E. C. (1939). Endomitosis in the nurse cells of the ovary of *Drosophila melanogaster. Chromosoma,* **1**, 276-283.

Palevody, M. C. (1972). Presence de noyaux accessoires dans l'oöcyte der Collembol. *Folsomia candida* (Insecte, Apterygote). *C.r. hebd. Séanc. Acad. Sci., Paris,* **274**, 3258-3261.

Paulcke, W. (1900). Über die Differenzierung des Zellelemente im Ovarium der Biener-königin. *Zool. Jb., abt. Anat, Ontog.* **14**, 177-202.

Paweletz, N. (1967). Zur Funktion des "Flemmings-körpers" bei der Teilung tierischer Zellen. *Naturwissenschaften,* **54**, 533-535.

Perez, J. (1886). Sur l'histogénèse des elementes contenees dans les gaines ovigeres des insectes. *C.r. hebd. Séanc. Acad. Sci., Paris,* **102**, 181-184.

Phillips, D. M. (1970). Insect sperm: their structure and morphogenesis. *J. Cell Biol.* 44, 243-277.

Platner, G. (1886). Die Karyokinese bie den Lepidopteren als Grundlage für eine Theorie der Zellteilung. *Int. Mschr. Anat. Physiol.* 3, 341-398.

Pollack, S. B. and Telfer, W. H. (1969). RNA in cecropia moth ovaries: sites of synthesis, transport and storage. *J. exp. Zool.* **170**, 1-24.

Porter, K. R. (1966). Cytoplasmic microtubules and their functions. In "Ciba Foundation Symposium on Principles of Biomolecular Organization." (Ed. G. E. W. Wolstenhome), pp. 308-345. J. and A. Churchill, Ltd., London.

Prenant, A. (1888). Observations cytologiques sur les elements seminaux des Gasteropodes pulmones. *La Cellule*, 4, 133-195.

Ramamurty, P. S. (1963). Über die Herkunft der Ribonucleinsäure in den wachsenden Eizellen der Skorpions Fliege *Panorpa communis* (Insecta Mecoptera). *Naturwissenschaften.* **50**, 383-384.

Ramamurty, P. S. (1967). The cytoplasmic inclusions of the trophocytes in the ovary of the scorpion fly, *Panorpa communis* (Mecoptera: Panorpidae). *Proc. R. ent. Soc. Lond., Ser. A., Gen. Ent.* 42, 87-92.

Rasmussen, S. W. (1973). Ultrastructural studies of spermatogenesis in *Drosophila melanogaster* Meigen. *Z. Zellforsch. mikrosk. Anat.* **140**, 125-144.

Ribbert, D. and Bier, K. (1969). Multiple nucleoli and enhanced nucleolar activity in the nurse cells of the insect ovary. *Chromosoma*, 27, 178-197.

Ribbert, D. and Weber, F. (1970). Homologous pairing and synaptinemal complexes in the nurse-cell nuclei of carabid ovaries (Ins. Coleoptera). *Experientia,* **26**, 800-801.

Roth, T. F. (1966). Changes in the synaptonemal complex during meiotic prophase in mosquito oöcytes. *Protoplasma*, **61**, 346-386.

Roth, T. F. and Porter, K. R. (1964). Yolk protein uptake in the oöcyte of the mosquito *Aedes aegypti* (L). *J. Cell Biol.* **20**, 313-332.

Ruby, J. R., Dyer, R. F., Skalko, R. G. and Volpe, E. P. (1970). Intercellular bridges between germ cells in the developing ovary of the tadpole, *Rana pipiens. Anat. Rec.* 167, 1-10.

Rusch, H. P., Sachsenmaier, W., Behrens, K. and Gruter, V. (1966). Synchronization of mitosis by the fusion of the plasmodia of *Physarum polycephalum. J. Cell. Biol.* 31, 204-209.

Schlottman, L. L. and Bonhag, P. F. (1957). Histology of the ovary of the adult mealworm *Tenebrio molitor* (Coleoptera, Tenebrionidae). *Univ. Calif. Publs Ent.* **11**, 351-392.

Schrader, F. and Leuchtenberger, C. (1952). The origin of certain nutritive substances in the eggs of Hemiptera. *Expl Cell Res.* 3, 136-146.

Seshachar, B. R. and Bagga, S. (1963). A cytochemical study of oögenesis in the dragonfly *Pantala flavescens* (Fabricius). *Growth*, 27, 225-246.

Skalko, R. G., Kerrigan, J. M., Ruby, J. R. and Dyer, R. F. (1972). Intercellular bridges between oöcytes in the chicken ovary. *Z. Zellforsch. mikrosk. Anat.* **128**, 31-41.

Smith, P. A. and King, R. C. (1968). Genetic control of synaptonemal complexes in *Drosophila melanogaster. Genetics,* **60**, 335-351.

Stay, B. (1965). Protein uptake in the oöcytes of the Cecropia moth. *J. Cell Biol.* 26, 49-62.

Steinert, G. and Urbani, E. (1969). Communications intercellulaires dans les ovarioles de *Dytiscus marginalis* L. *J. Embryol. exp. Morph.* **22**, 45-54.

Telfer, W. H. (1965). The mechanism and control of yolk formation. *A. Rev. Ent.* **10**, 161-184.

Telfer, W. H. and Anderson, L. M. (1968). Functional transformations accompanying the initiation of a terminal growth phase in the cecropia moth oöcyte. *Devl. Biol.* **17**, 512-535.

Telfer, W. H. and Rutberg, L. D. (1960). The effects of blood protein depletion on the growth of the oöcytes in the Cecropia moth. *Biol. Bull. mar. biol. Lab., Woods Hole,* **118**, 352-366.

Telfer, W. H. and Smith, D. S. (1970). Aspects of egg formation. *In* "Insect Ultrastructure" (Ed. A. C. Neville), pp. 117-134. Royal Entomological Society, London.

Ullmann, S. L. (1973). Oögenesis in *Tenebrio molitor*: Histological and autoradiographical observations on pupal and adult ovaries. *J. Embryol. exp. Morph.* **30**, 179-217.

Urbani, E. (1970). A survey of some aspects of oögenesis in *Dytiscus, Cybister* and *Hygrobia* (Coleoptera). *Acta Embry. Exp.* **3**, 281-297.

Urbani, E. and Pezzoli, V. A. (1970). Differential mitosis and oögenesis in *Hygrobia tarda* Herbst. *Monitore Zool. Ital. (N.S.)* **4**, 233-241.

Urbani, E. and Russo-Caia, S. (1964). Osservazioni citochimiche e auto-radiografiche sul metabolismo degli acidi nucleici nella ovogenesi di *Dytiscus marginalis* L. *Rend. Inst. Sci. Camerino* **5**, 19-50.

Vanderberg, J. P. (1963). Synthesis and transfer of DNA, RNA and protein during vitellogenesis in *Rhodnius prolixus* (Hemiptera). *Biol. Bull. mar. biol. Lab., Woods Hole,* **125**, 556-575.

Verhein, A. (1921). Die Eibildung der Musciden. *Zool. Jb. abt Anat. Ontog.* **42**, 142-212.

Weber, F. (1971). Korrelierte Formveränderung von nukleolus und nukleolusassoziertem Heterochromatin bie der Gattung *Carabus* (Coleoptera). *Chromosoma,* **34**, 261-273.

Weisenberg, R. C. (1972). Changes in the organization of tubulin during meiosis in the eggs of the surf clam, *Spisula solidissima. J. Cell Biol.* **54**, 266-278.

Woltereck, R. (1898). Zur Bildung und Entwicklung des Ostrakoden-Eies Kerngeschichtle und biologische Studien an parthenogenetischen Cypriden. *Z. wiss. Zool.* **64**, 596-623.

Woodruff, R. I. and Telfer, W. H. (1973). Polarized intercellular bridges in ovarian follicles of the Cecropia moth. *J. Cell. Biol.* **58**, 172-188.

Woodruff, R. I. and Telfer, W. H. (1974). Electrical properties of ovarian cells linked by intercellular bridges. *Proc. N.Y. Acad. Sci.* In press.

Woods, P. S. and Zubay, G. (1965). Biochemical and autoradiographic studies of different RNA's: evidence that transfer RNA is chromosomal in origin. *Proc. natn. Acad. Sci. U.S.A.* **54**, 1705-1712.

Zaffagnini, F. (1969). Vitellogenesi e differenziamento ovocitario negli artropodi. *Boll. Zool.* **36**, 263-289.

Zalokar, M. (1960). Sites of ribonucleic acid and protein synthesis in *Drosophila*. *Expl Cell Res.* 19, 184-196.

Zalokar, M. (1965). Études de la formation de l'acide ribonucléique et des proteines chez les insectes. *Revue suisse Zool.* 72, 241-261.

Zinmeister, P. P. (1973). RNA and protein synthesis in the earwig ovary. *J. Insect Physiol.* 19, 1765-1770.

Zinmeister, P. P. and Davenport, R. (1971a). RNA and protein synthesis in the cockroach ovary. *J. Insect Physiol.* 17, 29-34.

Zinmeister, P. P. and Davenport, D. (1971b). An autoradiographic and cytochemical study of cellular interactions during oögenesis in the milkweed bug, *Oncopeltus fasciatus*. *Expl Cell Res.* 67, 273-278.

Major Patterns of Gene Activity During Development in Holometabolous Insects

John A. Thomson

Department of Genetics, University of Melbourne, Parkville, Victoria, Australia[1]

1 Introduction

Testimony to the current intensive work in the field of insect development is provided by the number of excellent recent reviews on aspects of the subject. These include, to cite only a few representative accounts, papers by Anderson (1972a, 1972b), Jura (1972) and Counce (1973) on embryology; Ashburner (1970, 1972) on the cytology of gene function and chromosome

[1] Present address: *Division of Plant Industry, CSIRO, P.O. Box 1600, Canberra City, A.C.T., Australia.*

puffing; Fristrom (1970) and Wright (1970) on the genetic analysis of developmental processes at various levels; Wigglesworth (1970), Wyatt (1972) and Doane (1973) on insect hormones and their role: Fristrom (1972), Gehring and Nöthiger (1973) and Oberlander (1972) on imaginal discs; Chen (1971) and Price (1973) on the biochemistry of development with emphasis on proteins and on nucleic acid metabolism; Ilan and Ilan (1973) on the control of protein synthesis; Lawrence (1970, 1973) and Waddington (1973) on pattern formation in morphogenesis.

Our knowledge of insect development has reached the point where each aspect such as those mentioned can be treated in considerable detail, but in so doing it is easily possible to overlook the general stratagem underlying insect ontogeny. It is, therefore, the purpose of the present article to attempt a rather general overview of the broader features of the patterns of gene activity seen in insect development. The central theme, for reasons originating in favourable genetic and cytogenetic features of the organisms, will be woven around the development of the Diptera, especially of *Drosophila*, *Calliphora* and *Chironomus*, with additional data drawn from studies of Lepidoptera (notably *Bombyx*, *Hyalophora* and *Galleria*), Coleoptera (especially *Tenebrio*) and where possible, Hymenoptera.

In particular, it is the aim of the discussion to clarify the relationship between the holometabolous pattern of development and that of the Hemimetabola. Differences in morphogenesis in these two groups are quantitative rather than qualitative. As Hinton (Hinton and Mackerras, 1970) points out, the issue is really one of a contrast between *tissues* and *organs* showing hemi- or holometabolous modes of development. There are, of course, wide differences in the degree of tissue replacement amongst the endopterygote groups. Hinton sees the evolution of the endopterygotes as simply involving the sharper separation of the larval stages into a series concerned with feeding and a final stage, the pupa, bridging the feeding stages and the adult specialized for reproduction and dispersal. The pupal stage has thus evolved as a modified larval instar with external wings, connecting a form with internal wings with the adult (see Hinton, 1963, for comparison with other theories of the origin of the endopterygote pupa). The differences between the pupa and the preceding larval instar are not necessarily greater than those distinguishing successive larval stages in extreme cases of specialization of two instars for different modes of life, as in larval heteromorphosis (Snodgrass, 1954; Chapman, 1969). When developmental stages are defined as extending from apolysis (retraction of the epidermis from the cuticle of the previous instar) to apolysis, rather than from ecdysis (moulting) to ecdysis, the relation of the pupa (*sensu stricto*) to adult is much clearer. The pupal stage proper is quite short in most endopterygotes. Most of the period inside the pupal cuticle precedes the pupal-adult apolysis, so that the organism is then actually in the adult stage, although *pharate* (covered by the cuticle of the preceding instar: Hinton, 1958). Terminology used in the following discussion reflects this distinction between pupa and pharate adult.

The significance of the evolutionary origin of the pupal stage of the Holometabola in the present context lies in the patterns of gene control which might be anticipated. It is to be expected that the genetic architecture of the holometabolous insects will be based on partial separation, and distinct control of, two "sets" of genes: one determining the larval phenotype and one specifying the adult phenotype (e.g. Wigglesworth, 1961). Such a concept does not, of course, mean that loci are restricted to activity in one phase or the other; undoubtedly many, or even most, are common to both. But is there also a third, "pupal", set (Williams and Kafatos, 1971)? Clearly, from a consideration of the evolutionary origin of complete metamorphosis in the endopterygotes, characters which are strictly pupal should be controlled by genes of the larval set, and any separate control of these is likely to be little more marked than that specific to earlier larval instars. It is in this context that the major significance of Hinton's (1958) emphasis on distinction of the pupal and pharate adult stages becomes apparent.

The larval epidermis, fat body, alimentary system and associated organs, and muscles may all be destroyed and replaced at metamorphosis to a varying extent. Such replacement may be complete as in some Diptera-Cyclorrhapha, or at the other extreme, the adult organ may differ little from that of the larva. Hinton and Mackerras (1970) present a brief comparative summary of the extent of the changes seen at metamorphosis in the main insect groups. These authors point out in dealing with the musculature: ". . . there is no difference between exo- and endopterygotes in the kind or in the degree of metamorphosis of the muscles, but only in the extent of the more drastic changes." Again, in the case of the mid-gut, Hinton and Mackerras conclude: "In most endopterygotes the mid gut is renovated after the larval apolyses precisely as in the exopterygotes . . ." The significant features of these comparisons from the genetic standpoint are, firstly, the absence of a sharp distinction between the kinds of genetic system involved in insects with complete versus incomplete metamorphosis, and secondly, the implication that the degree of overlap in functional separation of larval and adult gene sets may be quite different and variable amongst even fairly closely related groups.

Two strongly contrasting patterns of cell replacement are evident during growth and metamorphosis. Growth may involve either cell multiplication without major increase in cell size, or else increase in cell size without cell division. Typically the first type of growth is continued to provide for cell replacement at metamorphosis, leading to a situation in which cells of the adult organ are derived from cells very closely and directly related to those which formed larval gene products. On the other hand, in the second situation the adult tissue is formed by proliferation of replacement cells of embryonic type set aside early in development as imaginal discs or imaginal histoblasts, and separated by many mitotic cycles from any stem cell programmed to form larval gene products.

The high level of replacement of larval by imaginal tissue and abrupt change in phenotype through pupal stages in the higher endopterygotes provides for maximum contrast between the characters determined by the larval and imaginal gene sets, against a background of activities common to both developmental stages. It is to the pattern of gene activities involved in the development of two contrasting phenotypes during the ontogeny of the holometabolous insects that we now turn. The discussion will show that, on quantitative grounds, the fat body must be assigned the key role in insect metamorphosis, and it is in this organ that key sets of gene activities in larval and imaginal life are manifested.

2 Size and organization of the genome

If the larval and imaginal phenotypes of the Holometabola were determined by separate gene sets of any major size, and especially if a third "pupal" gene set (Williams and Kafatos, 1971) were included in the genome, it would be reasonable to anticipate that the total size of the unique-sequence portion of the genome should be larger in holometabolous than in hemimetabolous insects. Few estimates of genome size in insects are available. Determinations of DNA content give values for the haploid genome of *Drosophila hydei* of 0.2 pg (Mulder *et al.*, 1968), for *Chironomus tentans* of 0.25 pg (Daneholt and Edström, 1967) and for *Dytiscus marginalis* 2.7 pg (Gall *et al.*, 1969). The haploid genome of six orthopterans tabulated by White (1973) ranges from 3.8 to 9.5 pg DNA. In *Ch. tentans*, 4.5 per cent of the DNA of embryos is repetitious, with an average multiplicity of 120 copies (Sachs and Clever, 1972). On the basis of DNA renaturation kinetics, Laird and McCarthy have calculated the genome size and complexity of *D. melanogaster* (see also Wu *et al.*, 1972) and *D. simulans* as 7×10^{10} daltons with 5-10 per cent repetitious sequences of an average multiplicity of 60. The haploid genome of *D. funebris* was estimated at 14×10^{10} daltons. *Sarcophaga bullata*, however, has a much larger minimum genome size: 40×10^{10} daltons, again with 5-10 per cent repetitious sequences of an average multiplicity of 80 copies (Laird and McCarthy, 1969). No comparable determinations of sequence diversity seem to be available for hemimetabolous species. For the few species known, then, the Holometabola have *less* DNA per haploid genome than representatives of the Hemimetabola.

Discussion of the arrangement of the repetitious and unique DNA sequences in the genome is outside the scope of the present article. Recent contributions include those of Crick (1971), Sorsa *et al.* (1973), Bonner and Wu (1973), and of Peacock *et al.* (1973).

Estimates of the number of loci in *Drosophila* vary widely according to the argument adopted. Judd and his co-workers (Judd *et al.*, 1972) have provided strong evidence for a correspondence of the chromomeres of the polytene chromosomes with genetic complementation groups, suggesting about 5000 functional genes over the whole genome. This figure is of the same order of magnitude as estimates based on the number of loci at which lethal mutations can be induced (see O'Brien, 1973). O'Brien, on the other hand, cites observations made by Logan of the annealing of *Drosophila* larval RNA to DNA, as indicating that transcripts of 15-20 per cent of the unique nuclear genome could be present in larval RNA. This would correspond to 30-40 000 gene transcripts of "average" size, and suggests that the number of functional loci in the total genome could be at least one order of magnitude higher than the approximate number of chromomeres. Such an approach might well be extended to the problem of overlap of larval and imaginal gene sets by comparison of the populations of RNA transcripts present at each of the two life stages.

Again, if distinct, or partially distinct, constellations of genetic loci are activated during larval and imaginal development, is there any evidence of clustering of loci affecting the larval phenotype, or perhaps more readily detected, of loci determining imaginal characteristics? Unfortunately remarkably few loci concerned with "larval" characters have been studied. Most larval lethals cannot be included in such an analysis, for they might also be manifested in the adult phenotype if their effect could be tested at that stage alone. Elston and Glassman (1967) have concluded that the apparent clustering of loci affecting various adult characteristics in *D. melanogaster* can be accounted for by the known clustering of all genes within certain regions of the genome. A listing of the known linkage-map positions of enzymes of *Drosophila* (Fox *et al.*, 1971) does not suggest any significant grouping of enzymes likely to be stage-specific in their major roles, but the data are still very limited. The quite uniform effects of segmental aneuploidy over the whole genome in *D. melanogaster* (Lindsley *et al.*, 1973) is a further line of evidence against the existence of functional grouping of genes on any large scale. A considerable number of larval characters have been described in *Bombyx* (Tazima, 1964); these are also clustered into the best-known chromosome segments but spread over much of the genome. The large number (28) of linkage groups would hinder detailed analysis of this point in the silkworm. As the genetics of special larval proteins such as the larval haemolymph and storage proteins, salivary gland secretory products or silk proteins etc., become better known, more detailed analysis of the distribution of "larval" and "imaginal" loci within the genome will become possible.

Grouping of functionally related, imaginal loci may exist in the particular case of the Y-chromosome loci concerned with sperm development in *Drosophila* where lampbrush-like structures have been followed during phases of gene activity (Hess and Meyer, 1963; Hess, 1965a, 1965b; Meyer and Hess, 1965). The similarity of sequence diversity in diploid embryonic cells, in polytene nuclei from the larval salivary gland, and in predominantly diploid cells from pharate adults (Dickson *et al.*, 1971) argues against major differential representation in larval and imaginal tissues of that portion of the genome comprising the structural genes.

3 Patterns of gene activity in replication and transcription

Different patterns of gene activity amongst the various tissues of the holometabolous insects are often suggested by widely differing gene contents. Firstly, specialization of the genetic material in particular somatic tissues may be reflected in changes in ploidy and from various patterns of differential replication of portions of the genome (section 3.1). Secondly, differential activation of those loci represented in each tissue may take place, and can be detected by cytological and biochemical observation in those dipteran tissues with polytene or lampbrush chromosomes (section 3.2) or by examination of specific gene products at the RNA (section 3.3) or protein levels (sections 4, 5). In the last two cases, analysis may be assisted by studies on the time of manifestation in development of the deranging effects seen in mutants and their phenocopies, or on the time of action of metabolic inhibitors.

3.1 GENE CONTENT

3.1.1 *Genetic specialization by modification of the cell cycle*

Growth in certain tissues by cell enlargement without cell division appears to be a general adaptation in holometabolous insects to provide for continuity of cell function without the disruptions necessitated by cytokinesis. An especially significant factor in development of this growth pattern seems to have been the necessity in a number of such tissues for very large-scale protein synthesis during a short feeding period in larval life. The phenomenon is also seen in adult tissues, particularly when highly specialized, terminally differentiated cells are involved. Reduction or abolition of cell division during growth may be especially important in maintaining bulk protein synthesis involving massive ribosomal RNA synthesis as a preliminary (section 3.3) and specialized cytoplasmic organization for secretion or sequestration of the gene products. The following brief account is intended as merely an outline of some of the main cytological features of larval and imaginal tissues.

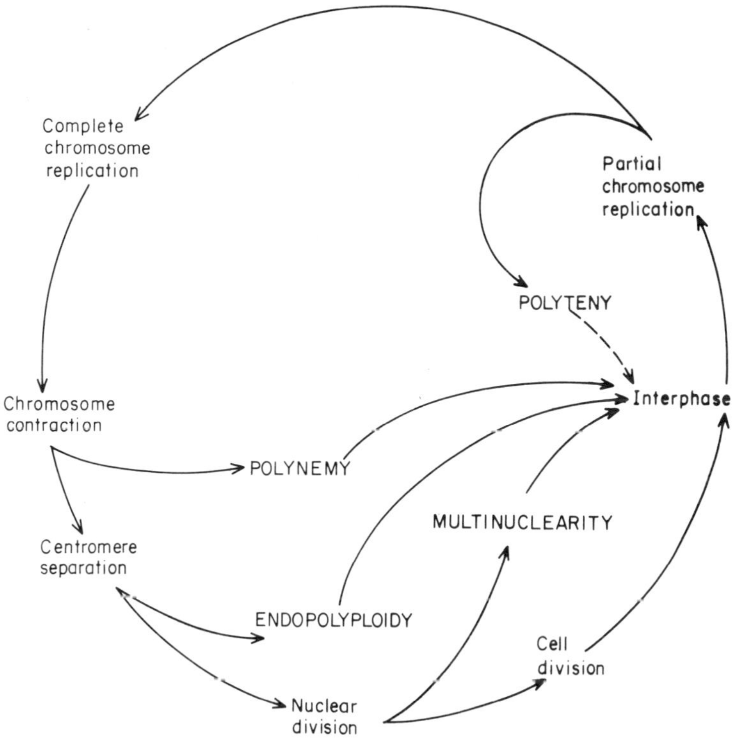

Fig. 1. The relationship of the genetic structure of the nucleus to the cell cycle (after Rudkin, 1972).

Most of the logically obvious short cuts in the cell cycle are seen in insect tissues growing by cell enlargement (Rudkin, 1972) and are summarized in Fig. 1. Genetically speaking, two contrasting situations can be distinguished. Polyteny involves only partial replication of the haploid genome typically through 8 to 10 successive replications without separation of the centromeres (for detailed discussion see Beermann, 1962; Rudkin, 1972). The highly repetitious (rapidly renaturing) polynucleotide sequences are relatively under-represented after these replication cycles. In *D. hydei*, for example, DNA from polytene nuclei of the salivary gland contains only 5 per cent fast-renaturing sequences, in contrast to 20 per cent in tissues of the embryo and pharate adult (Dickson *et al.*, 1971). These authors find from the renaturation kinetics of DNA from embryos and pharate adults that 80 per cent of sequences of 400-600 nucleotides are unique, i.e. represented by a single copy in each genome. DNA from polytene nuclei has this same sequence diversity; all the unique nucleotide sequences of the diploid genome are replicated during polytenization. It is the repetitive DNA sequences located in the chromocentral heterochromatin of polytene nuclei, and in the centromeric regions of the mitotic chromosomes (Gall *et al.*, 1971, and references cited therein) which are under-replicated during polytenization in *Drosophila*. In some Diptera, polyteny may occur in tissues which are either exclusively larval or exclusively imaginal (e.g. footpad or trichogen nuclei) in origin and function, as well as in tissues common to both life stages (e.g. Malpighian tubules).

On the other hand, polynemy, endopolyploidy and the development of multinucleate cells, seem best regarded as involving full replication of the genetic material. Nuclei with polynemic (multiply stranded) chromosomes, such as those of certain ganglionic cells in the larva of *Drosophila*, may remain mitotically active, but show 4C (instar 1) to 8C or even 16C (instar 3) DNA values, with an unaltered heterochromatin to euchromatin ratio (Gay *et al.*, 1970). Endopolyploidy is very widespread in larval tissues of all the endopterygote orders: here the chromosomes are fully replicated and centromeres separate, but the nuclear membrane remains intact. Endopolyploidy is probably the most general specialization of the genetic material in larval secretory and digestive cells, and in the accessory and sheath cells of the gonads in the adult (general: Geitler, 1953; Coleoptera: Romer, 1966). Endopolyploidy may apparently occur together with polynemy (as in the epidermal nuclei of *Apis*: Risler and Romer, 1968) or in the Cecidomyiidae where polyneme (described in the original literature as polytene) chromosomes fibrillate into individual elements to produce an endopolyploid nucleus (see Ashburner, 1970; White, 1973, for detailed discussion). While highly endopolyploid nuclei are generally terminally differentiated, an exception occurs in portions of the larval gut of the mosquito (Berger, 1936, 1937, 1938; review in Clements, 1963). These cells have in mature larvae of *Culex pipiens* 48, 96, or occasionally 192 chromosomes derived by endomitotic division of the diploid complement of 6. These chromosomes pair during interphase soon after pupation, and at the subsequent prophase are seen as the diploid number of bundles of loosely associated threads (the haploid number in *Aedes*, where maternal and paternal homologues must synapse as well). A rapid series of cell divisions takes place with intervening replication phases, producing a large number of small diploid cells from which portions of the imaginal gut are formed. Clearly, the intermediate structures formed by chromosome synapsis before the "somatic reduction" divisions were polyneme rather than polytene. Such modification of the larval genetic material to form that of imaginal cells is known only in *Culex*, *Anopheles*, *Aedes*, and *Orthopodomyia*:

Endopolyploidy and polyteny appear to be related in the phenomenon of differential replication in the Malpighian tubule nuclei of *Dermestes* (Fox, 1970, 1972; see also White, 1973) in which endopolyploidy involves a non-doubling series of DNA values apparently resulting from under-replication of heterochromatic segments of the genome. These cells are, of course, terminally differentiated.

Multinucleate cells are also found in specialized tissues of both larval and imaginal Holometabola. For example, the ventral nephrocytes of *Calliphora* larvae are binucleate, with polytene chromosomes (Thomson and Gunson, 1970). Imaginal oenocytes and fat body cells in the Diptera are often multinucleate (e.g. *Dacus*: Evans, 1967).

3.1.2 *Genetic specialization by differential replication of specific genetic loci*

Three observations led to the examination of the replication patterns of ribosomal (r-)DNA cistrons in insects showing high levels of polyteny in larval tissues. Firstly, if polyteny is itself an adaptation permitting programmed specialization of certain larval cells for large-scale synthesis of a limited range of proteins, the under-replication of rDNA known to occur during the replication cycles leading to polyteny (Hennig and Meer, 1971; Sibatani, 1971; Spear and Gall, 1973) would seem to run counter to the requirement for extensive ribosome synthesis. Secondly, the cytological development of much nucleolar material precedes both the main periods of polytenization and of protein synthesis in these tissues (Thomson, 1973a; section 3.2). Thirdly, as Spear and Gall (1973) point out, although the 250 repeated rRNA cistrons (Ritossa *et al.*, 1966) are located in the nucleolus organizer region of the X and Y chromosomes (Ritossa and Spiegelman, 1965; Pardue *et al.*, 1970) in heterochromatin, the rDNA is genetically active in terms both of rRNA synthesis and in forming a nucleolus.

Saturation values for hybridization of radiolabelled rRNA with DNA from the predominantly diploid nuclei of the larval ganglia, and the polytene nuclei of the salivary glands has established that diploid cells have amounts of rDNA proportional to the number of nucleolus organizers in *D. melanogaster*. In contrast, in polytene nuclei the amount of rDNA is independent of the number of nucleolar organizers (Spear and Gall, 1973). These important and decisive results were based on comparison of the

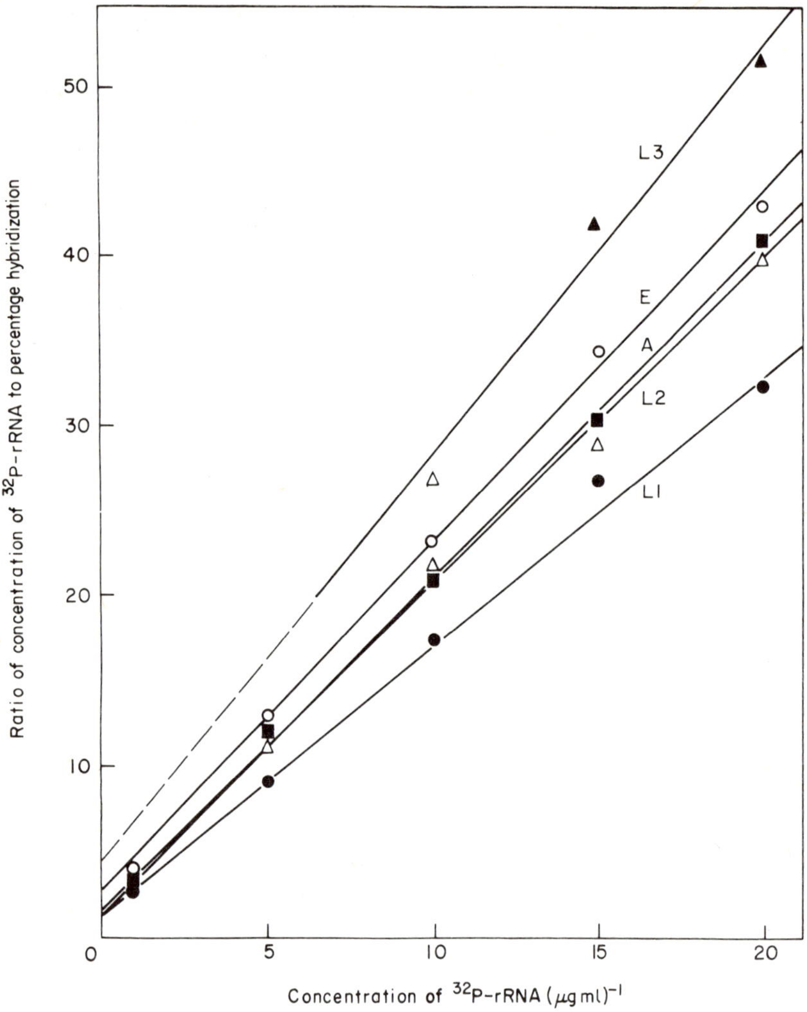

Fig. 2. Hybridization of ^{32}P-rRNA at the indicated concentrations with total DNA from *Calliphora stygia*, shown as the ratio of concentrations of ^{32}P-rRNA to percentage hybridization plotted against concentration of ^{32}P-rRNA (see Thomson, 1973b). DNA from the following developmental stages was used: embryos (E) 12 ± 3 h after oviposition; first-instar larvae (L1) at hatching; second-instar larvae (L2) 2 days after hatching; third-instar larvae (L3) at quiescent stage 12 h before pupariation; adult flies (A) 2-6 h after emergence. Each point represents the mean of 2-4 replicates.

tissues of XO and XX larvae and led Spear and Gall to the conclusion that replication of the rDNA is in polytene tissues under independent control, so permitting a relative increase in the rDNA in XO flies.

A similar conclusion was reached by Thomson (1973b), working with *Calliphora*. Saturation values for hybridization of rRNA from *C. stygia* with DNA from whole insects were compared at six developmental stages (Fig. 2). Total DNA from hatching larvae, before polytenization is evident, contains approximately 20 per cent more rDNA than does total DNA from either mid-embryos or adults. DNA from late third-instar larvae close to pupariation contains about 30 per cent less rDNA than does that of hatching larvae. Larvae of intermediate ages have intermediate proportions of rDNA to total DNA. The tissues of adults are predominantly (but not exclusively, Ashburner, 1970) diploid, whereas many larval tissues are highly polytene. Thus Thomson concluded that rRNA cistrons replicate more rapidly than the rest of the genome in the late embryonic and/or earliest larval stages, but more slowly during the later replication cycles in those larval tissues, destined to become highly polytene. These data support those of Spear and Gall (1973) on *Drosophila*, and suggest that larval development in polytene chromosomes is programmed sequentially: rDNA synthesis precedes rRNA synthesis and polytenization of the rest of the genome largely precedes the main phase of protein synthesis (section 4.3). Extra-chromosomal DNA bodies containing large numbers of cistrons coding for rRNA are seen in the oöcytes of several holometabolous insects, including *Tipula* and *Dytiscus* (review: White, 1973). Amplification of these cistrons apparently provides for the large-scale synthesis of rRNA accumulated in the oöplasm. Once again, although the details are different, a phenomenon occurs in these imaginal cells which is fundamentally similar to that seen in larval cells engaged in particularly massive protein synthesis.

Apparently isolated instances of independent control of DNA replication, presumably of a highly specific and localized kind, are the "DNA puffs" in the salivary glands of certain sciarid flies (Pavan and da Cunha, 1969; Ashburner, 1970, for reviews). In the polytene chromosomes of the imaginal footpads of *Sarcophaga bullata* (Whitten, 1965) and *Hybosciara* (da Cunha *et al.*, 1969), specific bands release extra DNA granules synthesized there, whereas typically in species such as *Rhynchosciara* the extra DNA is retained in the band where it is formed. While in both situations it is reasonable to suppose that the extra DNA is the result of gene amplification, its function and fate are unknown.

Gene amplification does not occur in the case of the extraordinarily active protein synthesis in the silk glands of *Bombyx* (Suzuki *et al.*, 1972).

An interesting series of studies by Lang and his co-workers (summarized by Lang, 1972) document changes in proportion of a special class of "soluble" (s-)DNA in development of the mosquito (*Aedes*). This low molecular weight fraction (500 000 daltons) comprised up to 40 per cent of the total DNA in 3-day larvae, but declined rapidly to about 7 per cent in the late larva. During pharate adult and adult stages sDNA remained about 2-3 per cent of the total DNA (Lang and Meins, 1966). The base composition of sDNA is 39.5 ± 0.92 mol% (G + C), and that of total nuclear DNA 39.1 ± 0.52 mol% in *Aedes*. Concordant results were obtained by differential spectrophotometry, hydrolysis and TLC and buoyant density comparisons (Lang, 1972).

The function of sDNA of *Aedes* is apparently unclear (Lang, 1972) and Lang considers it to be a "replication intermediate". Lang argues that sDNA is not mitochondrial DNA, rDNA, nor a satellite DNA, on the basis of its nuclear location and buoyant density. Definite proof is lacking that sDNA is not, at least in part, extrachromosomal rDNA, for hybridization analyses have not yet been reported. In any event, the sDNA seems to be quantitatively too large a fraction of the total DNA to represent a tissue-specific amplification product and may therefore have a general function.

3.2 CHROMOSOME STRUCTURE AND FUNCTION

The occurrence, biochemistry and interpretation of polytene chromosomes, and especially local "puffing" as evidence of gene activity, have recently been excellently reviewed by Pavan and da Cunha (1969), Ashburner (1970, 1972), Berendes (1972), Panitz (1972), Ribbert (1972) and by others. Localized chromosomal puffing is therefore not examined in the present article, except as part of a general perspective on evidence of the control of gene activity in the Holometabola.

Cytological observations of gene activity have been limited in two different and significant ways by the requirements of this experimental approach (Thomson, 1969). Ability to position a given band or puff on chromosome reference maps is clearly an important feature in cytological studies. This normally requires reasonable, if not complete, linear integrity of the individual chromosome arms; such a condition is found in relatively few tissues of relatively few species (Fig. 3). The size of the chromosomes, reflected in the level of polyteny attained and the degree of longitudinal condensation realized during development, has been a second major criterion. As replication continues throughout larval life in tissues such as the salivary gland of *Drosophila*, most cytological studies have concentrated on the latter half of larval life. Changes in gene activity associated with pupariation and pupation have been especially analysed in relation to control by moulting hormone. It should not be forgotten, however, that cells such as those of the salivary glands in *Drosophila* and *Chironomus* are at this time in the final, and not necessarily most significant, stage of their functional lives. Major phases of gene activity in cyclorrhaphous Diptera are generally completed in tissues such as the fat body and salivary gland by the end of the first two-thirds of larval life (section 4.3.5). Accordingly, attention should be given to analyses of chromosome behaviour during this early phase of development.

The development of polytene chromosomes in fat body and salivary gland nuclei of *Calliphora* has been illustrated by Thomson (1972a). At hatching, the chromatin of nuclei in both tissues surrounds a prominent nucleolus, and includes a strongly condensed heteropycnotic body. The chromatin disperses as the nuclei rapidly enlarge, and about 30 h after hatching, separate irregular and very extended chromatin threads can be

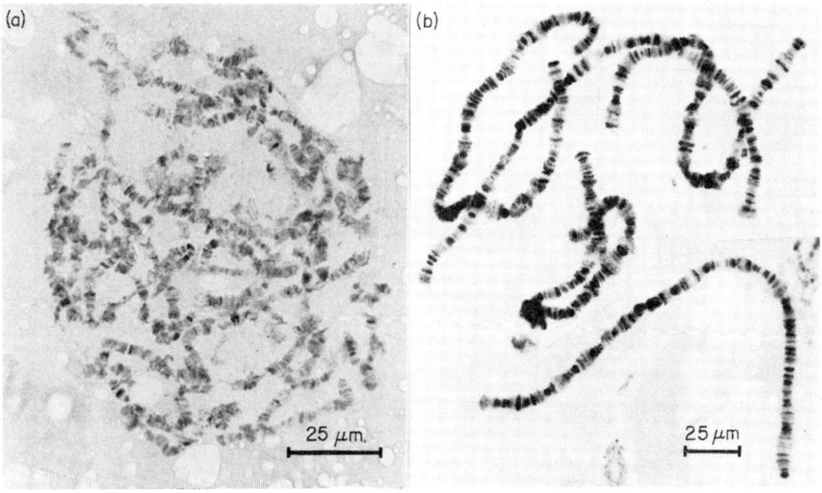

Fig. 3. The polytene chromosome set of nuclei from *C. stygia*: (a) larval fat body cell during the wandering phase of third instar; (b) scutellar trichogen cell of the pharate adult. The trichogen cell complement includes five linearly continuous elements. In fat body nuclei, short-banded segments alternate with diffuse chromosomal regions. Lactic-orcein stain.

distinguished. At approximately 48 h recognizable banding can be detected, the appearance being of side by side approximation of chromatin fibres (see Fig. 2 in Thomson, 1973a), apparently comparable to that seen in polytenization of the ovarian nurse cell chromosomes by Bier (1957, 1959). Homologous chromosomes are paired in some banded sections and separated in others. As growth continues with further replication cycles, the chromosomes appear to contract lengthwise and come progressively to occupy less of the total nuclear volume. The nuclei now (mid-instar 2) consist of three distinct regions: (1) short chromosomal segments of varying length, with the characteristic band-interband elements of the fully developed polytene structure; (2) diffuse chromatin between the banded chromosomal segments; and (3) the condensed heterochromatic body seen more prominently in the prepolytene stage at hatching. The diffuse unbanded regions of chromatin are separated by banded segments of widely varying lengths in fat body (Fig. 3(a)) and salivary gland nuclei. Such regions are absent from developing polytene chromosomes in the trichogen nuclei of pharate adults of *C. stygia*; in these cells the chromosomes appear linearly continuous throughout development (Fig. 3(b)). From mid-instar 2 to the end of feeding (day 6), the rate of protein synthesis is maximal in both the fat body and salivary glands and during this period further replication results in increasing chromosome width. Progressive longitudinal contraction of banded segments occurs so that individual bands become more closely stacked together. Replication ceases in fat body nuclei at the end of feeding in *C. stygia* (as in *C. erythrocephala*, Dahlhelm, 1967). Subsequently the dispersed, unbanded chromosome segments in this tissue become less conspicuous, but do not develop banding. In the salivary gland, protein synthesis slows in post-feeding stages and changes qualitatively. One or two further replication cycles occur, the last shortly before pupariation (as in *Drosophila*: Rodman, 1967).

In late feeding and wandering larvae, and especially in the quiescent stage just before pupariation, typical localized puffing of individual bands or small groups of adjacent bands, are seen in salivary gland chromosomes.

The chromosomes of the larval fat body and salivary gland in *C. stygia* differ in respect of the total number of major bands present, and both differ from those of the trichogen and footpad nuclei of the pharate adult (Thomson, 1969) in which five of the six elements of the haploid chromosome set are present. Comparison of the length of banded sequences in the salivary gland and fat body nuclei with those of trichogen nuclei establishes that the average length of these segments is less than that of whole chromosome arms. One particularly easily recognizable band sequence can be distinguished in all three tissues in *C. stygia* and other chromosome segments are probably common to them.

Segments of the chromosomes which remain dispersed in salivary gland cells are associated with intense uptake of ^{3}H-uridine in the feeding stage. In both the fat body and salivary gland these diffuse regions are closely associated with nucleoli (Fig. 5), into which the chromatin fibres often extend (Thomson, 1973). Continuity of banding patterns intermediate between those of the larval salivary gland and imaginal trichogen nuclei are seen in the pericardial cells (see figure in Thomson and Gunson, 1970).

Chromosomal configurations showing discontinuous banded polytene segments are regular and normal features of many tissues; Ribbert (1972) refers to these as characteristic of the larval polytene chromosomes of the Calyptratae. The unbanded sections of such chromosomes do not appear to arise from previously banded segments, but may rather represent chromosomal regions which did not undergo contraction, condensation and lateral synapsis of component chromatids when these processes affected adjacent regions (Thomson, 1973). According to this interpretation, then, the dispersed sections of polytene chromosomes in tissues such as the larval fat body and salivary gland of *Calliphora* represent a condition entirely analogous to that of the interphase chromatin of non-polytene nuclei, while the tightly banded sequences represent blocks of reversibly inactivated loci. The dispersed chromosome segments in early larval life must include the loci involved in massive transcriptional activity prior to, or during, the main phase of protein synthesis in the first half of larval life. The banded segments may be formally equivalent to condensed and reversibly inactivated sequences such as those represented by karyosomes (usage of Pavan and da Cunha, 1969) such as the inactivated X-chromosome of somatic cells in female mammals, or the heterochromatic bodies of maturing avian erythrocytes.

It is emphasized that nothing in the present argument detracts from significance of isolated puffs. These may be viewed as being restricted sites of gene transcription. Thus a fine control exists, superimposed on a basic coarser pattern of genomic switching, in which whole blocks of genes are inactivated during the main phase of cytodifferentiation. Whether a fully developed polytene chromosome contains dispersed segments is thus seen as a function of the extent and timing of gene activity in that cell. The occurrence and scale of transcription in early cytodifferentiation, and especially the time of switching off of this activity relative to the bundling and contraction of the chromatids into the condensed, banded state seems especially important in determining final chromosome morphology.

In summary, at least four levels of pre-translational control appear to be recognizable in nuclei such as the fat body of *C. stygia*. These involve the replication and transcription of rDNA independently of the rest of the genome; the differential replication of euchromatic relative to heterochromatic sequences; the large-scale condensation and inactivation of chromosome segments, or even of entire chromosomes; and finally, the localized activation of individual loci or small groups of loci, in puffing.

Da Cunha, Pavan and co-workers (1969) have shown that four separate classes of cell exist within the salivary gland of *Bradysia* judged by their histochemically distinct cytoplasmic granules and very different patterns of synthetic activity. In this case, as in others (Clever, 1966), the intensive production of secretory proteins and polysaccharides could not be correlated with changes in puff patterns. Each cell type displayed, however, a chromosomal morphology characteristic of its activity and involving differences in ". . . size, degree of condensation of the chromatic material and the rates and periods of DNA and RNA synthesis" (da Cunha *et al.*, 1969).

Well-documented studies of developmental changes in local chromosome puffing include particularly those dealing with dipteran tissues, especially the larval salivary glands (e.g. *Camptochironomus* (*Chironomus*): Beermann, 1952; Grossbach, 1969; *Acricotopus*: Mechelke, 1953; Panitz, 1972; *Rhynchosciara*: Breuer and Pavan, 1955; *Drosophila hydei*: Berendes, 1965; *D. melanogaster*: Becker, 1959, 1962; Ashburner, 1967, 1969, 1972), imaginal footpads (*Sarcophaga*: Whitten, 1969a, 1969b; Bultmann and Clever, 1969, 1970), imaginal trichogen cells (*Calliphora*: Ribbert, 1967, 1972).

In larval tissues few puff-phenotype correlations have been established (Ashburner, 1970, 1972; Panitz, 1972) and broadly this is to be expected if the main phase of gene-readout for larval salivary function occurs very early in development prior to, or during, the phases of initiation of salivary gland functions in feeding and digestion. The association of markedly changed puff patterns with the onset of pupation in all larval tissues so far investigated reveals major changes in the patterns of gene activity with the initiation of metamorphosis (e.g. Ashburner, 1970, 1972, 1973). It is to be presumed that these changes reflect the initiation in tissues such as the salivary glands, of the cell activity necessary for release of puparial glue in certain *Drosophila* spp. or of cocoon proteins in Nematocera, as well as the mobilization of the enzymes involved in cell death (e.g. salivary glands) or cell reorganization (e.g. Malpighian tubules) which occurs during metamorphosis. There is now accumulating evidence that α- and β-ecdysones may have distinguishable roles in controlling this programme (Clever *et al.*, 1973), as in imaginal cells (Oberlander, 1972). Sequential gene switching may also be achieved by differences amongst loci in the threshold for response to MH as this rises near pupation (Ashburner, 1973).

Potentially the development and function of the polytene nuclei of trichogen and footpad cells in the pharate adult of *Calliphora* and *Sarcophaga* offer the possibility of using puff analysis to follow gene activities during cytodifferentiation involving a series of definite steps occurring in sequence over a number of days (Ribbert, 1972). The sequential pattern of the appearance and regression of chromosome puffs over this period is equally well defined. These are major developmental alterations compared with those through which the salivary gland of *Chironomus*, for example, passes in larval development. A peak of puffing activity in footpad cells of *Sarcophaga* (days 5-6, Whitten, 1969b) coincides with a burst of intensive protein synthesis as indicated by incorporation of ^{14}C-leucine *in vivo*. Qualitatively new protein species are produced in these cells after the peak in puffing activity (Goldberg *et al.*, 1969), which also coincides with deposition of some cuticle layers. The whole programme is completed about 5 days later when the endocuticle is laid down. The major portion of the cuticle protein and chitin is synthesized in the footpad cells between days 9-11 following pupariation (Bultmann and Clever, 1970, based on incorporation of radiolabelled precursors into footpads *in vitro*), and a second peak in total puffing activity. At no time does the scale of synthetic events in such cells approach that seen in the specialized secretory cells of the larva, such as those of fat body or salivary gland.

Further analysis of gene activity in footpad and trichogen nuclei of pharate adults of the blowflies and fleshflies could be extremely valuable. As Ribbert (1972) suggests, the best prospects involve isolation of mutants in which particular aspects of the processes involved are strongly affected. In view of the favourable genetic and cytological features of *Lucilia* (Childress, 1969; Whitten *et al.*, 1975) such studies might well be extended to this organism.

3.3 NUCLEOLAR STRUCTURE AND FUNCTION

Intensive protein synthesis in certain tissues of the larvae of holometabolous insects growing by cell enlargement without division is reflected in a complex developmental sequence affecting the nucleoli. The ontogeny of the nucleolus has been studied in detail in the polytene tissues of *Calliphora* (Thomson and Gunson, 1970; Thomson, 1973a). It is, however, clear from other work that similar events occur in other species including *Drosophila*. Comparisons can also be made with patterns of nucleolar activity in imaginal tissues showing intensive protein synthesis, such as the nurse cells of the ovary of *Calliphora* (Ribbert and Bier, 1969).

The development of nucleoli in the larval fat body and salivary gland of *Calliphora* is shown diagrammatically in Fig. 4. At hatching, in both tissues, nuclei contain a single, more or less central, nucleolus. During instar 1 and the first half of instar 2 (days 1-3) the nucleoli enlarge as active RNA synthesis takes place. Concentration of the ribonucleoprotein (RNP) occurs around the periphery of the expanding nucleolar area and by the end of this period multiple aggregates of nucleolar material are seen in each tissue. As the main period of protein synthesis starts from mid-instar 2 the nucleolar RNP in fat body forms into regular spheroidal masses which become chromosomally associated (Fig. 5). At first these are fairly uniform in size (Fig. 4), but in early instar 3, during peak protein synthesis, five to seven larger masses and many small bodies can be seen (Thomson, 1973). Fragmentation and vacuolation of the nucleolar aggregates continue, until at the commencement of the wandering phase (day 7) of larval life after

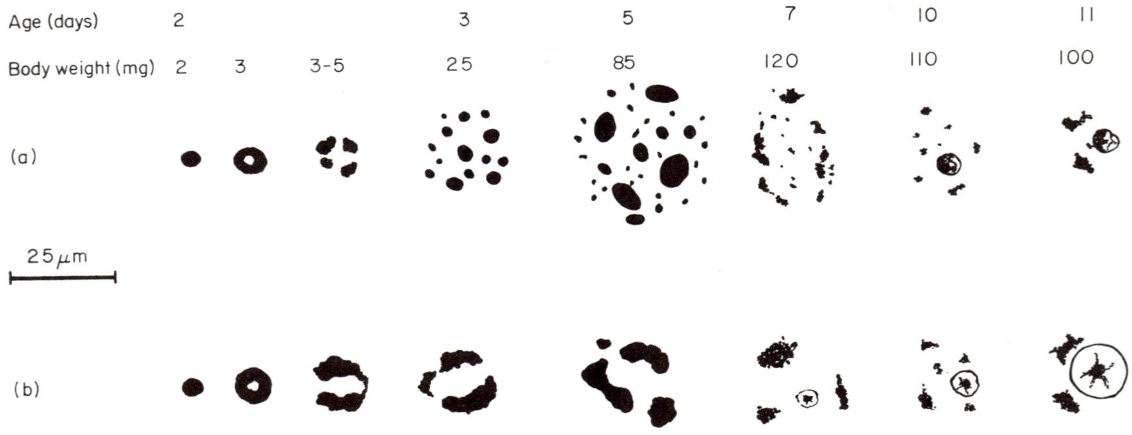

Fig. 4. Diagrams showing stages in the formation of nucleolar RNP bodies in *C. stygia*. Tissue-specific patterns in: (a) larval fat body; (b) larval salivary gland. Type 2 nucleoli (see text) are shown forming in 7-11 day nuclei after the Type 1 material characteristic of the main phase of protein synthesis commences to regress. Based on silver impregnated preparations.

feeding ceases, only small, irregular fibrous remnants of these RNP inclusions remain. During the same period parallel changes in both protein synthesis and nucleolar synthesis, fragmentation and dispersal, occur in salivary gland cells, except that in this tissue the number of separate nucleolar masses is smaller and more varied, being typically 2-5. Fragmentation of the nucleolar aggregates such as that seen in the fat body does not

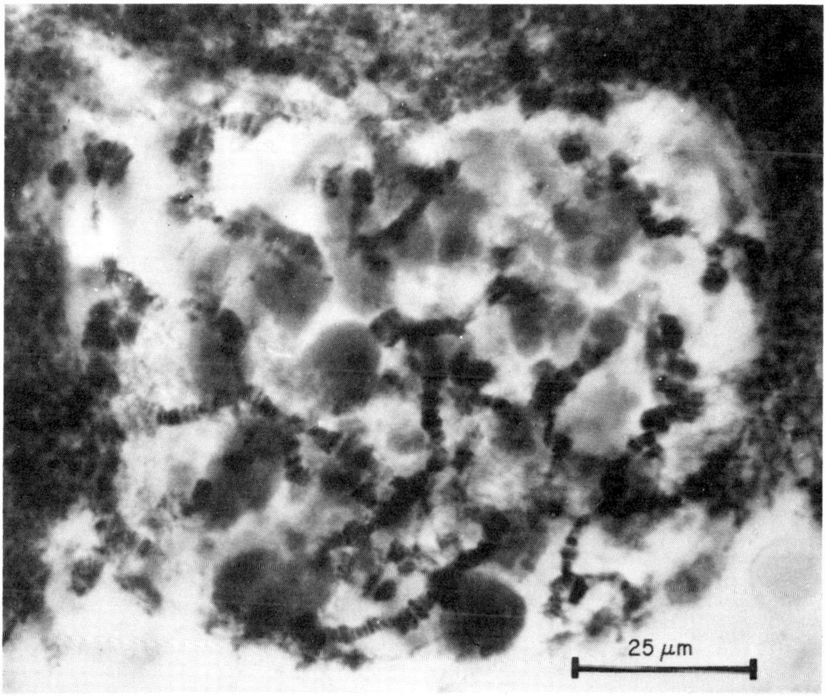

Fig. 5. Multiple Type 1 nucleoli in a larval fat body nucleus of *C. stygia* during the main phase of calliphorin synthesis (day 5) in early third instar, showing association of the nucleolar masses with dispersed regions of the chromosomes. Lactic-orcein stain, undehydrated.

occur. Other organs such as the Malpighian tubules similarly show multiple nuclear inclusions (Type 1, Thomson and Gunson, 1970), but in tissue specific patterns. Although greatly reduced in amount, traces of Type 1 nucleolar material can be seen throughout the wandering stage. The fragments appear to be drawn together and to become more conspicuous during the brief small spurt in protein synthesis detected in salivary gland and fat body (Martin *et al.*, 1969) in the late quiescent larva close to pupariation.

During the wandering stage (days 7-10) in *C. stygia* a second kind of nucleolar structure becomes especially conspicuous in larval salivary-gland cells (Fig. 4(b)) (Type 2 nuclear inclusion body, Thomson and Gunson, 1970). This structure has a regular boundary and a granular core of RNP surrounded by a lightly staining cortex. Two such nucleoli form in some cells, apparently depending on whether the maternal and paternal nucleolar organizer regions are tightly synapsed. Where two of these nucleoli are present, each is about one-half the volume of the corresponding single structure. This Type 2 nucleolus grows rapidly larger and occupies about 10 per cent of the nuclear volume at pupariation. A somewhat similar second nucleolus (Type 3, Thomson and Gunson, 1970) can with difficulty be found in fat body nuclei during the same period. In this tissue a coarsely granular central region is surrounded by very little cortical material. The regular spheroidal, somewhat bladder-like nucleolus (Type 2) is the characteristic structure seen in mature larval salivary gland cells in *Drosophila* and the imaginal trichogen cells of *Calliphora* (Ribbert and Bier, 1969). Such nucleoli are associated with the chromatin only at the nucleolar organizer regions.

The Type 1, chromosomally associated nucleolar inclusions of larval tissues engaged in bulk protein synthesis are highly active in RNA synthesis (Thomson, 1973a). Their development correlates with the appearance of extractable pre-rRNA (Thomson and Schloeffel, cited in Thomson, 1973a). Painter and Biesele (1966) record the occurrence in salivary gland cells of *Drosophila virilis* of multiple nucleoli during instar 2, and the variable number and size, or apparent absence, of nucleoli in instar 3. Electron microscopy revealed in this work active ribosome maturation in the nucleolar bodies of the second instar salivary gland. Highly involuted surfaces covered in ribosome-producing tassels form fringes around the fibrillar internal centres at this stage. Break up and vacuolation of these multiple nucleoli is accompanied by increasing release of the ribosomes to the cytoplasm leaving fibrous core material as small nucleolar vestiges as in *Calliphora*. Other authors have reported nucleolar development in fat body cells to involve lobulation and enlargement, then decrease in size (*Sarcophaga*: Benson, 1965; *Drosophila*: Butterworth *et al.*, 1965), but these descriptions are likely to have been affected by fixation difficulties. "Profuse larger aggregates" of nucleolar material are seen in the developing salivary glands of *Bradysia* and *Rhynchosciara* (Sirlin, 1962). Several other cases of chromosomally associated multiple "micronucleoli" have also been described in sciarids (e.g. Jacob and Sirlin, 1963; Gabrusewycz-Garcia and Kleinfeld, 1966; Gabrusewycz-Garcia, 1972). These "micronucleoli" will bind labelled rRNA (Pardue *et al.*, 1970) and so may well be true nucleoli, the DNA being derived from the nucleolar organizer regions at an early stage of development, and not from the bands to which the "micro-nucleoli" adhere (Rudkin, 1972).

Jacob and Danieli (1970) have demonstrated that the nucleolus of the midge *Smittia* contains DNA segments which replicate autonomously and asynchronously with the rest of the genome. That independent control of ribosomal gene replication occurs in *Drosophila* (Spear and Gall, 1973) and in *Calliphora* (Thomson, 1973b) is strongly suggested by the evidence already discussed above. The working hypothesis may therefore be advanced that two generations of ribosomal-DNA replication are involved in programming of cells specialized for massive protein synthesis, including those of the larval tissues such as fat body and salivary gland as well as imaginal tissues such as the ovarian nurse cells of *Drosophila* (Dapples and King, 1970) and of *Calliphora*. In these latter cells, Ribbert and Bier (1969) describe multiple nucleoli of a type different from the single nucleoli of imaginal trichogen cells. The latter correspond in all respects to Type 2 nucleoli of the present account. Each of the nurse cell nucleoli contains DNA. Together these nucleoli synthesized about 72 per cent of the total nuclear RNA at the stage examined, whereas in trichogen nuclei the single nucleoli were proportionately less active at the time of analysis, synthesizing only 13 per cent of the total nuclear RNA (Ribbert and Bier, 1969). The situation is similar in the nurse cells of *Drosophila* (Dapples and King, 1970). One series of rDNA replications may release from the nucleolar organizer region templates for the bulk synthesis of rRNA in a loose aggregation which separates into tissue specific chromosomally associated masses utilized in a particular and intensive synthetic programme. The second series of rDNA replications may lead to formation of a structure in which continuity with the nucleolar organizer region is maintained (Nash and Plaut, 1965; Barr and Plaut, 1966; Olvera, R., 1969; Rodman, 1969). The nucleolar organizer regions certainly contain rDNA sequences in *Drosophila* salivary glands (Pardue *et al.*, 1970), as in adult tissues (Ritossa and Spiegelman, 1965). The Type 2 nucleolus is in the Diptera far more prominent in the final stages of cell life near pupariation, and before regression of trichogen or footpad nuclei in adult development, than is warranted by its limited content of pre-r- and rRNAs. It may be speculated that a second function exists for such nucleoli, perhaps connected with nucleotide salvage at cytolysis.

Multiple Type 1 nucleoli, which develop before and during intense periods of protein synthesis in certain tissues, are tissue specific in number, distribution and size range. Such characteristic nucleolar patterns cannot be due merely to differences in the total amount of nucleolar material in cells of different function. The maximum volume of the nucleolar RNP in the fat body and salivary glands of feeding larvae of *Calliphora*, for example, is very similar. It seems probable that the number and size of nucleolar masses in such tissue reflect the spacing and chromosomal distribution of the active gene sequences with which the nucleoli associate (Thomson, 1973a). If these active sections of the genome are in close proximity, the aggregates of RNP may not separate into so many discrete bodies, or nucleoli associated with adjacent chromosomal regions may fuse, giving rise to such contrasting nucleolar patterns as those of the fat body and salivary gland in the *Calliphora* larva.

Their relationship to chromosomal regions other than the nucleolus organizer, and their tissue-specific behaviour, suggests that Type 1 nucleoli participate in some aspect or aspects of the processing, stabilization or transport of gene products. Certain ribosomal precursors become chromosomally associated in the salivary gland nuclei of *Chironomus* (Ringborg *et al.*, 1970; Ringborg and Rydlander, 1971) and there are cytological pointers to a role for ribosomal materials in transport of puff products in *Chironomus* (Lezzi, 1967). It seems possible that a series of cellular processes coupling ribosome synthesis to messenger synthesis is being observed here, leading to the question of whether specific ribosome populations programmed for particular synthetic activities are produced at such stages in development. Evidence for ribosomal specificity in eukaryotic tissues is now accumulating from several sources (general: Redman, 1969; Tata, 1973, who includes other references: insects; Boshes, 1970), but the picture is complicated. Differences in protein composition of ribosome populations from larvae and adults in *Drosophila* (Lambertsson *et al.*, 1970) and of moulting hormone treated *versus* untreated tissues of the pharate adult in *Tenebrio* (Patel, 1972) are of unknown relevance, if any, to ribosome function. Also unknown are the nature, origin, and site of complexing with other ribosomal components, of stage-specific initiation factors (Ilan and Ilan, 1971) found in the functional 80S ribosomal complex of *Tenebrio*.

4 Translation of the larval gene set

The gene expression during the larval life of the holometabolous insects is reflected in the synthesis of proteins fulfilling three main roles. The first group comprises the structural proteins and enzymes of the larval tissues themselves. The second consists of the specialized larval secretory proteins involved in digestion, and the silks, puparial glues, and cuticular proteins not carried over into adult development. A third group of larval proteins comprises those bulk proteins of the fat body and haemolymph which seem to be specifically storage proteins, providing reserves for imaginal development in the same way that plant seed proteins, or the yolk proteins of cleidoic eggs, provide for subsequent embryonic development.

The role of protein in imaginal development of the Holometabola is a major one, protein being the main nitrogen reserve (Birt and Christian, 1969). In *Calliphora erythrocephala* at puparium formation, protein nitrogen exceeds 80 per cent of total nitrogen in the animal (Agrell, 1953). The relative roles in metamorphosis of the specialized storage proteins and of the proteins, peptides and amino acids made available by histolysis of larval tissues must vary greatly from species to species. But the scale of synthesis of one particular storage protein, calliphorin from *C. erythrocephala*, is such that it represents more than 60 per cent of the insect's total protein at the onset of metamorphosis (Munn and Greville, 1969). It is in this perspective that so much emphasis is placed on the haemolymph and fat body storage proteins in the following discussion.

Protein nomenclature poses some difficulties in insects. Albumins do not appear to have been unequivocally or regularly identified in insects (Chen, 1971, p. 67; see also Lensky, 1971a). The use of the terms albumin and glubulin for insect proteins seems to serve no useful purpose. These names have been avoided in this discussion. The storage granules of insect fat body often termed "albuminoid granules", and which contain glycoprotein, phospholipid and RNA, are here described as proteinaceous spheres, following Price (1973).

4.1 THE MAJOR PROTEINS AND PEPTIDES OF HAEMOLYMPH

4.1.1 *Proteins*

The plasma protein content of haemolymph from a range of holometabolous insects shows a rapid rise during the mid-larval stages, then falls during pupation and the early pharate adult stages. In the Lepidoptera such as *Pieris* (van der Geest and Borgsteede, 1969; Chippendale and Kilby, 1969) or *Diatraea* (Chippendale, 1970), there is a 6- to 8-fold increase during the last larval instar; similar patterns of increasing protein concentration have been observed in *Bombyx* (Wyatt *et al.*, 1956), *Samia* (Laufer, 1960), *Hyalophora* (Patel, 1971), *Ephestia* (Cölln, 1973) and other species. A peak protein content of 6-8 per cent is common in the Lepidoptera. Amongst the Diptera, a maximum protein concentration is reached in late larval life at a point determined in relation to the end of the active feeding period. Species such as *Calliphora* which have a quite prolonged wandering phase when feeding ceases, show a maximum protein content of about 20 per cent (Phormia: Chen and Levenbook, 1966a; *Calliphora*: Kinnear *et al.*, 1968) and the protein concentration then falls at first slowly, and then rapidly for 24 h before and 24 h after pupariation (Kinnear *et al.*, 1968; Kinnear, 1973). *Drosophila*, in which feeding continues until *relatively* closer to pupariation, shows a fourfold increase in haemolymph protein concentration in the 24 h before pupariation, and a similar pattern is apparently seen in *Culex* (Chen, 1959).

The individual protein species of haemolymph have been extensively studied in a wide range of insects. One to three protein species typically comprise the bulk of the plasma protein when this is separated by methods which do not dissociate the complex molecules found *in vivo*. In larvae of *Bombyx mori*, Oda (1956) observed three main components on ultracentrifugation and by moving boundary electrophoresis; species sedimenting at 17.4S (MW 630 000 daltons) and 2.8S (160 000 daltons) were reasonably stable. The third species with a sedimentation constant of 7S (absent in pupal haemolymph) apparently readily dissociates and might represent an aggregation of the small molecular weight entities. Loughton (1965) found a generally similar situation in *Malacosoma* in sixth instar haemolymph. On gel filtration, pupal haemolymph of *Galleria* was found by Marek (1969) to contain major fractions of MW 470 000 to 500 000 daltons and 12 000-12 900 daltons respectively. Four main components account for the bulk of protein in the haemolymph of *Hyalophora* (Patel, 1971).

A study of the plasma proteins of worker honeybees (*Apis*) showed three main components on cellulose acetate electrophoresis (Lensky, 1971a, 1971b). One major (14S), two lesser (9S, 17S) and one minor peak (13S) showed in ultracentrifuge patterns. Photographs of acrylamide gel separations of the worker bee plasma show a number of very similar components (see Lensky, 1971a, Fig. 9, bands 2-6) which may represent dissociation products of the major component of the plasma. It is therefore possible that this protein in *Apis* may show behaviour comparable to calliphorin in the Diptera.

The quantitatively major plasma proteins of *C. stygia* (Table 1) include two larval proteins (protein A, a lipoprotein of high MW; protein B, apparently the protein II of Munn and Greville, 1969) which are increasingly significant in proportion to the total haemolymph protein from instar 2 to adult life after eclosion. Of the two other major components, calliphorin (protein C) is a larval protein in terms of its origin and synthesis (section 2.4), it constitutes only 5 per cent of the plasma protein at emergence of the adult and disappears from the blood entirely in adults aged for one week. Protein D (Imaginal protein 1; Table 1) appears in the plasma only in the adult fly after emergence, and becomes quantitatively significant in both sexes a few days later (Kinnear, 1973; Kinnear and Thomson, 1975). Although calliphorin is shown as a trimer in Table 1, it seems to occur *in vivo* as a hexamer, for which Munn *et al.* (1971) suggest a noncyclic structure with the 6 subunits arranged as a trigonal prism.

Protein A cannot be predominantly a storage protein. It is synthesized

TABLE 1

The principal plasma proteins of *Calliphora stygia* (based on Kinnear, 1973). NI, no information

	Protein A	Protein B	Protein C	Protein D
Name	(High molecular weight lipoprotein)	Protein II (Munn and Greville, 1969)	Calliphorin (Munn *et al.*, 1967)	(Imaginal 1)
Presence in plasma	Early instar 2 onwards	Early instar 2 onwards	Late instar 2 to early adult	Adult after emergence
Time of synthesis	Instar 2 to adult after emergence	Instar 2 to mid-pharate adult	Instar 2 to mid-instar 3	Emergence of adult onwards
Concentration in plasma				
(i) Mid-instar 3	4%	8%	75%	0
(ii) At adult emergence	41%	24%	5%	0
Conjugated materials				
(i) Lipid	+++	0	+	NI
(ii) Carbohydrate	0	++	+	NI
Molecular weight	400 000	240 000	250 000	NI
Quaternary structure				
(i) Subunits	Dimer (?)	Trimer	Trimer*	NI
(ii) Interchain bonding	Non-covalent, hydrophobic	Electrostatic, non-covalent	Electrostatic, non-covalent	NI
(iii) Ca^{2+}-dependence	—	+	—	NI
Subunit properties				
(i) Molecular weight	240 000	81 000	83 000	70 000
(ii) Heterogeneity				
(a) Electrophoretic	NI	Identical	Diverse	NI
(b) Immunological	NI	Identical	Identical	NI

* Probably a family of heterohexamers *in vivo* (Munn *et al.*, 1971).

continuously through larval, pharate adult and adult stages and accumulates in the plasma. A role in lipid metabolism and transport seems likely, but is unproven.

Protein B is immunologically and electrophoretically distinguishable from calliphorin, from which it is also different in regard to pattern of synthesis and accumulation. Protein B decreases in amount in the post-emergence phase of adult development but does not appear to function as a storage protein for imaginal development in the early pharate stages. Protein B is a homopolymer (Table 1), association of subunits being Ca^{2+} dependent. It is distinct from calliphorin in amino acid composition; the relative content of proline, serine and alanine is higher and that of arginine, methionine and phenylalanine is lower.

Protein C, the major protein of larval and pharate adult *Calliphora* was isolated and named calliphorin by Munn and his co-workers (Munn *et al.*, 1967). Calliphorin has a sedimentation constant of 19.4S and MW of 528 000 daltons. Above pH 6.5, and especially at low ionic strength, calliphorin dissociates into a number of protomeric units which are not identical. The bulk protein is a complex of closely related molecules, probably heterohexamers (Munn *et al.*, 1971). The component subunits are diverse and genetic polymorphism (section 4.4) for these is quite extreme (Kinnear, 1973; Thomson *et al.*, 1975). Although the rules, if any, for association of the monomers to trimers and hexamers are not yet known, even random polymerization would result in a large but certainly finite number of hexameric species.

Calliphorin comprises 75 per cent of the plasma protein of *C. stygia* at the end of the feeding period of larval life; this is a total of 7 mg per animal (live weight 120 mg). The concentration of calliphorin in the plasma then begins to fall. At pupariation the plasma contains 3 mg calliphorin and at emergence of the adult, only 0.03 mg (Kinnear and Thomson, 1975). The level of calliphorin in the whole animal falls during imaginal development, especially early and late in intrapuparial development. The relation of calliphorin to the fat body and to the deposition of proteinaceous spheres is considered in sections 4.2, 4.3.

Calliphorin is a conjugated protein containing 0.5 per cent carbohydrate (Munn *et al.*, 1971) and lipid (Kinnear, 1973). The purified protein contains about 4 m atom calcium per mmol protein, whereas protein B contains 2-3 times as much. X-ray fluorescence spectrography reveals no magnesium, zinc, manganese, iron, nickel, cobalt or copper (Kinnear and Thomson, 1975). In contrast to protein B, calliphorin is not irreversibly dissociated in the presence of calcium chelators. Reducing agents do not affect the nature of the subunits produced on dissociation of calliphorin, and there is no evidence of covalent disulphide links between chains. The effect of pH on aggregation is reversible.

An amino acid analysis of calliphorin from *C. erythrocephala* was published by Munn *et al.* (1971), and values obtained for calliphorin from *C. stygia* are in close agreement (Kinnear, 1973). Calliphorin is unusually rich in aromatic amino acids, tyrosine and phenylalanine amount to 442 and 400 mol mol^{-1} protein respectively. Methionine is also high; Munn and his co-workers give 162 mol mol^{-1} protein. These authors also cite cystine or cysteine as very low (18 mol mol^{-1} of protein) but half-cystine residues were not detected at all in calliphorin from *C. stygia* (Kinnear, 1973). Glutamic and aspartic acids are very high (361 and 426 mol mol^{-1} protein, respectively).

Protein species immunologically cross-reactive with calliphorin have been observed in other Cyclorrhapha (Munn and Greville, 1969). Proteins giving reactions of "identity" with calliphorin were found in extracts of *Sarcophaga* and *Lucilia*. Partially cross-reactive proteins were obtained from *Gastrophilus* and *Drosophila*, and amongst the Brachycera, from *Chrysopilus*, *Rhagio* and *Rhamphomyia*. Of the Nematocera examined, *Simulium* extracts contained a partially cross-reactive component, but *Tipula* and *Chironomus* did not. *Tenebrio*, *Pieris* and *Locusta* proteins were also not immunologically cross-reactive with calliphorin. By negative staining, rather symmetrical particles can be seen by electron microscopy in extracts containing calliphorin (Munn *et al.*, 1967) as well as in the haemolymph of a variety of hemi- and holo-metabolous insects consistent with the general occurrence of a large molecular weight protein of the calliphorin type in each of these (Munn and Greville, 1969). The list includes *Pieris* (estimated as 50 per cent of the larval protein), *Bombyx*, *Antheraea* and *Galleria*, and *Tenebrio* (30 per cent of the larval protein). Purified calliphorin from *Calliphora* appears as right prisms 105 Å wide and 65 Å high, rectangular when viewed from the side, and forming curvilinear equilateral triangles in surface view (Munn *et al.*, 1971).

The subunits of calliphorin from *C. vicina* (*C. erythrocephala*), *C. stygia*, and *Lucilia cuprina* are similar in molecular weight (83 000 ± 5 per cent) but each species has a distinct complement, significantly different in electrophoretic mobility. The differences involve both net charge and conformation (Thomson *et al.*, 1975). Extensive genetic polymorphism has been seen in *C. stygia* and *Lucilia cuprina*. The larval proteins homologous with calliphorin differ quantitatively amongst species in such a way that no boundary can be easily set. The term calliphorin would seem better reserved for the protein from *Calliphora*; the equivalent proteins from other species may then be named in similar fashion. Thus the name lucilin is used here for the major larval plasma protein of *Lucilia*.

In certain chironomid midges the bulk of the larval plasma protein is "haemoglobin" (Manwell, 1966); this is estimated to amount to 40 per cent of the total in *Chironomus thummi* (Braun *et al.*, 1968). These plasma haemoglobins show a high degree of genetic polymorphism (section 4.4) as do calliphorin and lucilin subunits. In *Ch. tentans* 10-12 monomers of MW *c.* 15 900 daltons (Thompson *et al.*, 1968; Tichy, 1970) are found. In plasma of *Ch. pallidivittatus*, the 8 distinct haemoglobins are monomers of similar MW (Tichy, 1970) but in *Ch. plumosus* dimers of 34 000 daltons are observed (Svedberg and Eriksson-Quensel, 1934), while in *Ch. thummi* the haemoglobins occur as 5 monomers and their dimers (Braunitzer and Braun, 1965; Plagens *et al.*, 1972). In *Ch. strenzkei*, Plagens and co-workers have found that monomers comprise about 30, dimers about 40, and tetramers about 30 per cent of the total haemoglobin. The four types of monomer in *Ch. strenzkei* fall into two groups with respect to MW (15 100 and 17 600 daltons). Treated as primarily respiratory proteins, the chironomid haemoglobins pose a considerable evolutionary problem. The occurrence of these purely larval proteins, in association with haem is sporadic amongst the Chironomidae. The difference in resistance to oxygen lack is not well correlated with possession of haemoglobin but several lines of evidence do suggest a role for this protein in oxygen transport at very low oxygen tensions (Gilmour, 1961, p. 126; Wigglesworth, 1965, p. 349 *et seq.* for summaries). The haemoglobins are degraded during imaginal development (Iuga, 1935) although some globins remain, conjugated with a green pigment (Manwell, 1966). It appears likely that the plasma haemoglobin of the chironomids has evolved opportunistically by association of haem with the major plasma protein which, as in other species, serves as a storage protein. Like calliphorin, but unlike other haemoglobins, haemochironomin has a low cysteine content and is high in phenylalanine (see Braunitzer, 1965). The haem prosthetic group of the chironomid molecule is apparently identical with that of vertebrates (Kirrmann, 1930), but the apoprotein is very distinct in chain length and sequence (Braunitzer and Braun, 1965; Buse *et al.*, 1969). The haemoglobins of the chironomids are therefore treated here primarily as storage proteins. It is suggested that these proteins would be more appropriately referred to as haemochironomins, rather than haemoglobins or erythrocruorins. Molecules unconjugated with haem, especially in those species lacking this pigment, can then be designated as chironomins.

Munn and Greville (1969) report that electron microscopy of *Chironomus* reveals no particles similar to calliphorin. This is consistent with the postulated role of the haemochironomins as the main plasma storage protein, for these molecules are small and polymerize little (see above).

Minor plasma proteins, especially those with enzymatic activity have been very widely studied, but these are *quantitatively* insignificant in the protein economy of metamorphosis as a whole. The bulk proteins of haemolymph are often complex heteropolymeric conjugates. There is often one or a family of high molecular weight lipoproteins (e.g. *Calliphora*, Table 1; *Hyalophora*: Thomas and Gilbert, 1968) and one or more families of heterogeneous but closely related conjugated proteins such as calliphorin and its homologues. Subsidiary functions of the bulk plasma proteins may include various enzymatic activities. A real difficulty here is that while the heteropolymeric aggregates found *in vivo* may be enzymatically active, loss of function during purification may make it difficult to determine which components are involved. It is clear that large molecular aggregates in insect plasma and tissue deserve more detailed study. Riechers *et al.* (1969) have shown that several pigment-enzyme complexes occur in the plasma of *Hyalophora*. Such complexes may have esterase, acid phosphatase, leucine amino peptidase, deoxyribonuclease and ribonuclease activity, or even all of these. One complex examined by Riechers and colleagues had a sedimentation constant of 23S. A point of particular interest is that complexes of the same electrophoretic mobility and enzymatic activity were extracted from fat body and mid-gut during imaginal development. These workers therefore raise the possibility that multi-enzyme complexes, like certain other proteins (see below), may be taken up intact from the blood. Calliphorin may also participate in formation of a multi-enzyme complex. In the plasma of *C. stygia*, calliphorin is associated with both deoxy- and ribo-nuclease functions; and at least ribonuclease activity appears to be retained after uptake of the complex into the fat body.

4.1.2 *Peptides*

Plasma peptides also appear to function as reserves during development in many insects. In larvae of *Ephestia* (Chen and Kühn, 1956) and *Phormia* (Levenbook, 1966), for example, the peptide content of the plasma greatly exceeds that of the tissues. Some of the peptides are lipid-amino acid complexes in *Drosophila* (Wren and Mitchell, 1959) and in *Bombyx* (Sissakian, cited by Chen, 1971). In general such peptides are quantitatively more significant in early rather than later larval life (Chen, 1971) and their significance in particular development events is unknown. Exceptions include the dipeptides β-alanyl-L-tyrosine in *Sarcophaga* (Levenbook *et al.*, 1969; Bodnaryk and Levenbook, 1969). γ-L-Glutamyl-L-phenylalanine in *Musca* (Bodnaryk, 1970a), and the phosphate ester, tyrosine-*O*-phosphate (Mitchell and Lunan, 1964). These increase in concentration in the haemolymph until pupariation, and then decline rapidly at pupariation with the sclerotization of the puparium. These compounds are apparently hydrolysed at this time. β-Alanine is a component of the puparium of *Sarcophaga* (Bodnaryk, 1971a) and some other flies (Hackman and Goldberg, 1971), but probably not in peptide linkage (Bodnaryk, 1971b). In *Musca*, the γ-glutamylphenylalanine is hydrolysed to provide phenylalanine for quinone metabolism in sclerotization (Bodnaryk and Skillings, 1971), while in *Drosophila*, tyrosine-*O*-phosphate provides tyrosine for this process (Lunan and Mitchell, 1969). It appears likely that other peptides will be identified and related to particular developmental steps; the examples mentioned are highly group or species specific.

4.2 FAT BODY AND THE STORAGE OF LARVAL PROTEIN

The larval fat body is involved in two ways in the metabolism of plasma proteins. Not only is this organ the predominant site of synthesis of these proteins, but it serves in the late larval, pupal and pharate adult stages in lipid, carbohydrate and protein storage (Kilby, 1963; Price, 1973). This tissue is thus the site of major and highly diverse gene activities.

The fat body is most conspicuous in the holometabolous insects (Buys, 1924), consistent with the greater storage of protein for imaginal reconstruction in these species. Price (1973) mentions that in *Calliphora*, the fat body can comprise about 50 per cent of the wet weight of the late larva, and draws attention to observations by Bishop (1923) on *Apis* which showed that the fat body in that species accounts for about 65 per cent of the fresh weight of the mature larva. The proportion of the body weight of the mature larva made up by the fat body varies quite extremely with nutritional conditions, and is strongly correlated with the size of the imago. Spheroidal bodies containing principally protein, but also phospholipid and RNA, appear in the fat body in late larval development, and may increase through pupal or pharate pupal life. These bodies are described here as proteinaceous spheres, following Price (1973). During or before the pharate pupal or pupal stages, the proteinaceous spheres gradually replace the well-developed endoplasmic reticulum and numerous mitochondria typical of the early larval fat body. Electron microscopic studies documenting this replacement include those on *Philosamia* by Ishizaki (1965) and Walker (1966), on *Calpodes* by Locke and Collins (1968), on *Hyalophora* by Bhakthan and Gilbert (1972), on *Drosophila* by von Gaudecker (1963), on *Sarcophaga* by Benson (1965) and on *Calliphora* by Price (1969).

A number of attempts have been made to distinguish different classes of proteinaceous sphere based on origin or content. Locke and Collins (1966, 1967, 1968) distinguish multivesicular bodies in which protein is sequestered and which may form by coalescence of lysosomes with mitochondria and parts of the endoplasmic reticulum, as early stages in the formation of the large proteinaceous spheres of pupal fat body in *Calpodes*. These build up from the multivesicular bodies by fusion of pinocytotic vesicles containing protein first concentrated in the intercellular spaces. The fat body of this insect shows three distinct phases of activity in the final instar:

successively growth, synthesis for storage and export, and finally uptake and storage in preparation for pupation (compare *Calliphora*, see below). Other protein granules in *Calpodes* fat body appear to develop by packaging of protein in the Golgi complex, by isolation of cellular components by paired membranes or by fusion of granules with each other or microvesicles (Locke and Collins, 1965).

Attempts have also been made to classify the proteinaceous spheres according to their time of appearance. Schmieder (see his Table 1, 1928) collected the observations on time of appearance of "albuminoid spheres" up to that data and found instances in the Lepidoptera, Coleoptera and Hymenoptera, but not Diptera, in which small granules, perhaps of protein, were claimed to occur in the first half of larval life. In more recent studies Nair *et al.* (1967) and Nair and Karnavar (1968) have emphasized the same contrast. *Anthrenus* (Nair and George, 1964), *Oryctes* (Nair *et al.*, 1967), *Trogoderma* (Nair and Karnavar, 1968) and *Drosophila* (Butterworth *et al.*, 1965) are species in which small protein granules have been observed by about the end of the first half of larval life. Larger proteinaceous spheres appear at a later stage in larval development, and especially close to pupation. There is no evidence that the "early" and "late" granules are developmentally connected. In the case of *Drosophila* some dispute has arisen over humoral influences on the formation of proteinaceous spheres (section 6.1; and see Thomasson and Mitchell, 1972) which might be resolved when the relationship between the several types of storage granule are elucidated.

The uptake of plasma proteins into the proteinaceous spheres of the late larval and pupal stage has now been widely demonstrated, particularly in *Pieris* (Chippendale and Kilby, 1969), *Galleria* (Collins and Downe, 1970), *Ephestia* (Cölln, 1973), *Sitotroga* (Chippendale, 1971), *Hyalophora* (Patel, 1971) and *Drosophila* (Thomasson and Mitchell, 1972). The predominant protein taken up and sequestered in *Calliphora* is the plasma protein calliphorin. It is in this genus that patterns of synthesis and uptake of storage protein are best documented, and a detailed description can be given.

The fat body is already fully delimited by the time of hatching in *Calliphora*, and no cell division occurs during the larval instars. It contains about 11 500 cells in *C. stygia* (Thomson, 1973a). These are arranged in three sections almost filling the haemocoele in the mature larva. Paired anterior and dorso-lateral lobes consist of flat sheets one cell layer thick. The cells of these lobes are flattened, irregular pentagons measuring about 100 x 150 μm at pupariation. The posterior fat body is a highly fenestrated lacework of more rounded and columnar cells closely associated with the Malpighian tubules. Price (1969) has shown that proteinaceous spheres are absent from actively feeding larvae (3-4 days old) in *C. erythrocephala*, but rapidly accumulate in the maturing, wandering larvae (see also Pérez, 1910; Dahlhelm, 1967), contemporaneously with a drop in plasma protein from 140 mg ml^{-1} to 90 mg ml^{-1} within 12 h of pupariation. Similar observations of falling plasma protein levels over this period were made by Kinnear *et al.* (1968) on *C. stygia*. In the latter species uptake of protein was demonstrated *in vivo* and *in vitro* using radiolabelled plasma protein (Martin *et al.*, 1971). The rate of uptake is not uniform, but proceeds slowly (0.6 mg d^{-1}) in the earlier wandering stage (days 8-9), and more rapidly (0.9 mg d^{-1}) over the last 24 h before pupariation (days 10-11) and the 24 h following this. The protein content of 100 fat body cells from the posterior region of the organ at the white puparial stage is about 10-16 μg; 24-h pharate pupa, 26-36 μg and in the 48-h pharate adult 24-32 μg (Kinnear, 1973). Thus protein uptake ceases by 24 h after pupariation. (Pupal-adult apolysis is completed in *C. stygia* 28-30 h after pupariation, at 25° C.) By this time the posterior cells of the fat body have about the same protein content as did those of the anterior lobe at pupariation. The predominant protein sequestered in the fat body (bands H15-19 of Martin *et al.*, 1971) is calliphorin, recognized from its subunit structure and electrophoretic behaviour as well as by immunological means (Kinnear, 1973). The subunits of ^{14}C-labelled calliphorin are taken up into fat body cells intact (Martin *et al.*, 1971). The haemolymph of each larva contains 7 mg of this protein in mid-third instar, but only 3 mg at pupariation, when the main store is in the proteinaceous spheres of the fat body. Calliphorin is readily extracted from such spheres isolated from fat body by gradient centrifugation according to the method Thomasson and Mitchell (1972) developed for *Drosophila* (Thomson, unpublished). In *Drosophila*, Thomasson and Mitchell find that protein sequestered in the proteinaceous spheres becomes highly recalcitrant to solubilization in the usual protein extractants, including those containing urea, detergents etc.

Typical proteinaceous spheres form in the late larval fat body of *Chironomus* (Miall and Hammond, 1900), even though the molecular weight of the plasma storage protein (haemochironomin) is unusually low.

In holometabolous insects in which the larval fat body is not entirely replaced at metamorphosis, the proteinaceous spheres are reduced in number during pharate adult development. Tiegs (1922) has described the depletion of the fat body reserves in the chalcid wasp *Nasonia*. Portions of the fat body in head and thoracic regions of the pharate adult are gradually but fully destroyed, their remnants finally being phagocytosed. Abdominal fat body cells lose many of the large and small granules packing the cytoplasm at pupation, and continue to show depletion throughout the remainder of the life of the insect. In the weevil *Calandra*, Murray and Tiegs (1935) found much less evidence of cellular destruction, but the fat body does break up into individual cells and clumps of cells distributed by the haemolymph. Individual cells shrink to about two-thirds their former diameter as reserves are utilized during imaginal development, but the cells do not break down until some weeks after emergence of the adult. Schmieder (1928) describes similar patterns of depletion of fat body cells with respect to proteinaceous granules in two tenthredinid and one ichneumonid wasp. This process has been observed at the fine structural level by electron microscopy in the fat body of *Hyalophora* (Bhakthan and Gilbert, 1972).

Breakdown of the fat body in cyclorrhaphous dipterans, in which the larval fat body is completely replaced after adult emergence, commences in the anterior cells at the time of pupariation. Disintegration of the cell membrane liberates proteinaceous spheres singly and in rafts of cytoplasm, or in partly adherent groups from several cells (compare Teunissen, 1937) into the haemolymph. The process gradually extends posteriorly (*Sarcophaga* and *Phormia*: Fraenkel and Hsiao, 1968; *Drosophila*: Whitten, 1962) as in *Calliphora*. In *C. stygia* even the posterior section of the fat body is represented only by scattered cells or rafts of cells 48 h after pupariation (Kinnear, 1973), but some cells retain their integrity until after emergence of the adult. Proteinaceous spheres liberated at cytolysis are often packed into the interstices of developing imaginal structures such as the flight musculature, where they adhere tightly to cell surfaces. Breakdown of cell membranes in the larval fat body was described by Teunissen (1937) as leading to a syncytial condition in the pharate pupal fat body of *Calliphora*. Similarly the fat body of the late larva in the beetle *Trogoderma* has been described as syncytial (Karnavar and Nair, 1968). Breakdown of the larval fat body into single cells after pupation also has been documented in *Musca* (Wiesmann, 1962) and *Drosophila* (Bodenstein, 1950).

The extent to which larval fat body cells or their remnants persist in the adult, and the extent of their replacement by imaginal cells, is especially variable from group to group. The disappearance of larval fat body components during imaginal development is subject to endocrine modification (section 6.1; and other factors may also be involved, section 6.3) and is slower in autogenous than in anautogenous genetic strains (e.g. *Culex*: Twohy and Rozeboom, 1957; *Lucilia*: Williams, 1972).

4.3 SYNTHESIS OF LARVAL STORAGE PROTEINS

Much circumstantial evidence from a wide range of holometabolous insects implicates the fat body as the source of the bulk haemolymph proteins of the larva. These will include storage proteins, operationally defined as species which are selectively sequestered into proteinaceous spheres in the late larval or pupal fat body, and utilized during imaginal development. A number of immunologically and electrophoretically similar proteins occur in both fat body and plasma in those species so far examined (reviewed by Price, 1973), and the time of peak synthesis of protein in the fat body coincides with steeply rising haemolymph protein concentrations (e.g. *Calliphora*: Martin *et al.*, 1969). Fat body incubated *in vitro* releases to the medium proteins similar by several criteria to those in the plasma, provided the tissue is collected at a time when it is active synthetically *in vivo* (*Bombyx*: Shigematsu, 1960; *Calliphora*: Price and Bosman, 1966; Kinnear *et al.*, 1971; *Drosophila*: Ruegg, 1968). The proteins detected as released under such conditions are of course only the quantitatively most major species. It must be emphasized that the total protein spectrum of the haemolymph is very complex; there are many minor components with a diversity of enzyme activities, and probably origins.

Detailed information on the synthesis of a specific storage protein is at present available only for calliphorin (protein C, Table 1) from *Calliphora* (Munn *et al.*, 1969; Kinnear, 1973; Kinnear and Thomson, 1975) and the homologous protein from *Lucilia*, lucilin (Thomson *et al.*, 1975). In *C. erythrocephala*, two predominant plasma proteins, calliphorin and protein II, were shown by Munn and his colleagues (1969) to be released from the fat body *in vitro*. Protein II is homologous with protein B of *C. stygia*, in which species it is not a storage protein for imaginal development according to the criteria used here, and shows a different pattern of synthesis (Table 1 for summary). Protein A (Table 1) of *C. stygia*, and its homologue in *C. erythrocephala* (Kinnear, 1973), is also synthesized in the larval fat body, but probably in other tissues as well, since its synthesis continues beyond the time of histolysis of the larval fat body. Again this protein is not considered to be a storage species.

Synthesis of calliphorin starts in the second half of instar 2. All subunits are synthesized simultaneously in genetically determined proportions. In inbred strains of *C. erythrocephala* and *C. stygia* the subunit patterns of dissociated calliphorin remain constant throughout larval life. Similarly no developmental change occurs in the relative proportion of subunits of lucilin (Thomson *et al.*, 1975). The rate of synthesis increases until, in the first one-third of instar 3, each of the 11 500 cells of the fat body in *C. stygia* make and release 0.4 μg protein per day (Thomson, 1973) of which about 80 per cent is calliphorin. At the conclusion of the feeding period in *Calliphora* (and in *Lucilia*, Fig. 6), there is a coordinated cessation of synthesis of storage protein subunits, over a 24-h period.

The shutdown of translation of calliphorin is differential relative to other proteins. The synthesis of many other protein species in a range of larval tissues slows or ceases about this time (e.g. salivary gland, Fig. 8; see also

Kinnear *et al.*, 1971). Whatever the control over translation of calliphorin subunits (section 6), the synthesis of the apoprotein of protein A, and of protein B subunits, continues in *C. stygia* after that of calliphorin stops. The synthesis of calliphorin is not resumed if the haemolymph protein level is reduced by perfusion to one-fifth of its normal level (200 mg ml^{-1} at the time calliphorin synthesis ceases) (Thomson, unpublished); this procedure merely results in production of miniature adults. A similar pattern of coordinated cessation of synthesis of the storage protein subunits occurs in *C. erythrocephala* (Kinnear and Thomson, 1975). The fat body then commences its main phase of protein uptake and sequestration, the

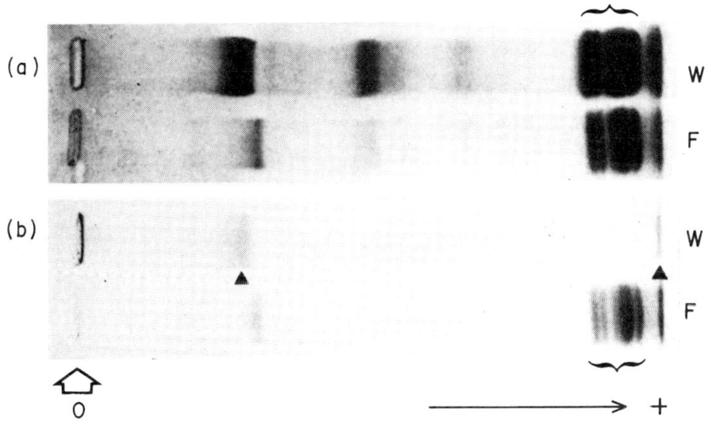

Fig. 6. (a) Electrophoretic separations of plasma protein from *Lucilia cuprina* on 5 per cent polyacrylamide gel in a discontinuous buffer system (0.08 M tris-citrate, pH 8.7 in gel; 0.3 M borate, pH 8.2 in electrode chambers). Haemolymph collected 3 h after injection of ^{14}C-amino acids into third-instar larvae (F, feeding; W, wandering stage; O, origin; + anode). Lucilin subunits are marked by bracket. (b) Autoradiographs of dried gel shown in (a). Labelled amino acids have been incorporated at both feeding and wandering stages into protein species marked by arrows, but into lucilin subunits only during the feeding stage. Exposure, 3 weeks.

nucleolar RNP characteristic of the larval fat body cells regresses, the ribosomal profile alters from polysome-rich to monosome-rich (Sekeri *et al.*, 1968), changes also reflected in the diminution of ribosomes and mitochondria as the proteinaceous spheres commence to form (Price, 1969).

The possibility that tissues other than fat body might contribute significantly to the synthesis of major haemolymph proteins has frequently been raised (e.g. *Samia*, *Hyalophora*: Laufer, 1960; *Diatraea*: Chippendale, 1970a, 1970b). This organ is certainly capable of selective uptake of particular haemolymph proteins at the time of pupation (e.g. *Malacosoma*: Loughton and West, 1965), and might therefore be important as a reservoir for these proteins at metamorphosis. All major blood proteins of *Hyalophora* can be synthesized in the larval fat body present in isolated pupal abdomens lacking the mid-gut (Ruh *et al.*, 1972). As Chippendale (1970b) has pointed out, *in vitro* studies of the mid-gut similar to those already made on fat body in a number of laboratories might help to clarify the role of the mid-gut in both protein synthesis and uptake.

Haemocytes do not contribute significantly to the synthesis of major protein species in holometabolous larvae and pupae (*Bombyx*: Faulkner and Bheemeswar, 1960; *Pieris*: Chippendale and Kilby, 1970; *Diatraea*: Chippendale, 1970b).

4.4 GENETICS OF LARVAL STORAGE PROTEINS

The genetic code specifying proteins and enzymes of ubiquitous eukaryotic cell organelles, the proteins of cell respiration and energetics, the proteins of contractile tissues, and the chromosomal histones are certainly ancient and evolutionarily conservative. By contrast, many of the enzymes of digestion, pigment synthesis and so on may be of comparatively recent origin. Secretory proteins such as silk are perhaps unlikely to have their genetic origins outside the arthropods. Other secretory proteins, like the puparial glue of *Drosophila*, are probably much more recent. The evolution of the storage proteins of the Holometabola is a particularly interesting problem.

Extensive polymorphism of the subunits of calliphorin is seen in Australian field populations of *C. stygia*, and of lucilin subunits in *L. cuprina* (Thomson *et al.*, 1975). Although no population studies have yet been carried out, it is clear that allele frequencies differ markedly from population to population. Genetic analysis is facilitated in these species by the ease with which samples of *c.* 0.15-0.20 μl haemolymph may be withdrawn into a micropipette (serially during development if necessary). Electrophoretic analyses of such samples from single larvae are shown in Fig. 7(a) and (b). Over 90 per cent of larvae bled in this way complete development, so that flies of known larval phenotype may be used for breeding. A series of *Lucilia* strains pure-breeding for patterns consisting of 4 to 7 bands have been established in culture. Eleven principal band positions (designated A to K) have been recognized. The bands differ in dye binding capacity and have been classified as dense (D), medium (M) or light (L) in intensity. The band positions are fairly equally spaced, suggesting unitary charge differences between the subunits occupying several positions, but some inequality of spacing is evident. Separation in the gels employed in this study involves some molecular sieving, so that minor differences in molecular weight, differences in conjugated materials or

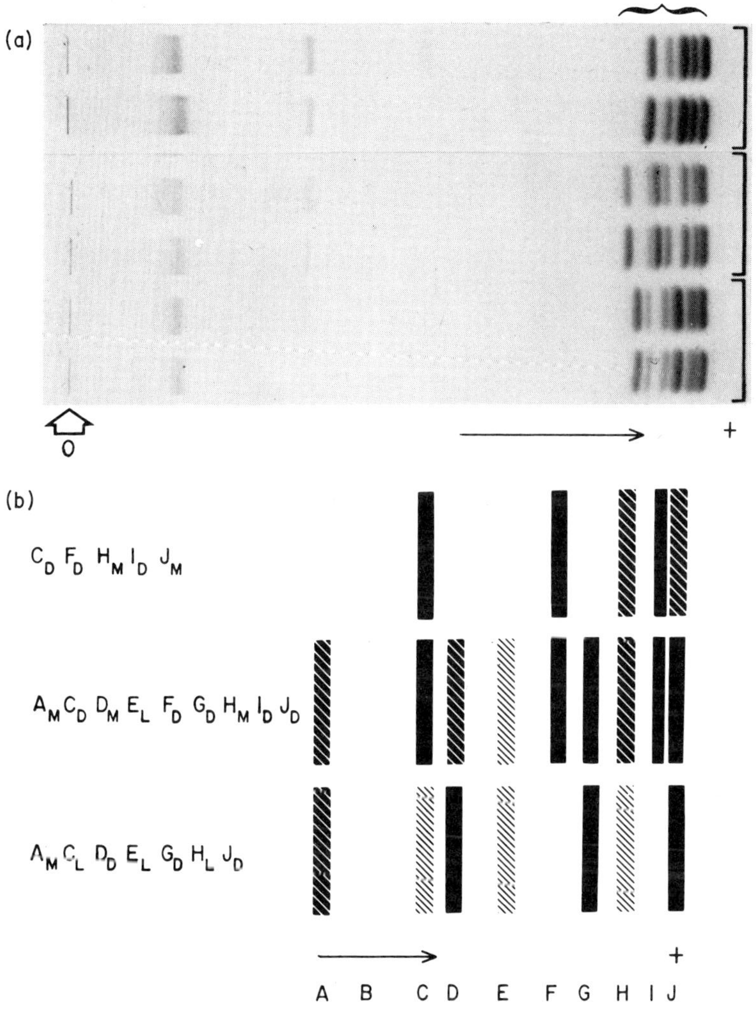

Fig. 7. (a) Electrophoretic separations of larval plasma proteins from *L. cuprina* on 5.5 per cent polyacrylamide gel (buffers as for Fig. 6). The lucilin subunit patterns (bracket) of two individual larvae from each of three representative pure breeding strains illustrate, amongst other differences, contrasting mobilities of the slowest subunits. The latter are determined by various alleles at the *Luc-1* locus. Top: phenotype $C_D E_M F_D H_D J_D$, homozygous for *Luc-1*^{C}. Centre: phenotype $A_M C_D D_D E_L G_D H_L J_D$, homozygous for *Luc-1*^{A}. Bottom: phenotype $B_D C_L E_L F_D H_L J_D$, homozygous for *Luc-1*^{B}. O, origin; +, anode. (b) Diagrammatic representation of lucilin subunits in two pure breeding strains of *L. cuprina* (top and bottom patterns) and in hybrids between them (middle). Phenotypic formulae shown at left. Band patterns are additive in the hybrid. Conditions of separation as for Fig. 7(a).

differences in conformation are indicated. The fastest and slowest components are indistinguishable in molecular weight (83 000 ± 5 per cent daltons) on sodium dodecyl sulphate electrophoresis. Band patterns are consistent within each individual, and amongst the larvae in a pure-breeding strain, from the moult to instar 3 through pharate adult development. Within wide limits, the lucilin subunit pattern is not changed by alterations in handling the samples. Incubation of samples at 35° C for 30 min or repeated freeze-thaw cycles do not affect the band pattern obtained, quantitatively or qualitatively. There is therefore no evidence of differential gene activation or of progressive epigenetic modification during development, and the patterns are not due to partial proteolysis.

The results of crosses between strains are always additive (Fig. 7(b)), giving rise to patterns in which all bands of both parents are represented. In individual heterozygotes, a maximum of nine bands can be resolved in crosses between presently available strains.

The seven loci identified in these preliminary studies (Thomson *et al.*, 1975) have been designated *Luc*(*Lucilin*) 1 to 7. Three alleles have so far been identified at the *Luc-1* locus (*Luc-1*^{A}, *Luc-1*^{B}, *Luc-1*^{C}, where the superscript indicates the relative electrophoretic mobility of the gene product; see Fig. 7(a)). The least variable subunit(s) (with regard to mobility) include that occupying position J, but a faster variant is occasionally seen (position K). The band patterns with few bands (e.g. C E F H J, Fig. 7(a)) can be shown in suitable crosses to have several polypeptides of identical mobility within the heavier bands.

To account for quantitative variation as well as the qualitative subunit pattern, either regulator loci specific to particular structural genes must be postulated, or else additional structural loci whose products overlap those of the cistrons already identified. In the latter case about 12 to 14 structural loci would be required to produce the quantitative and qualitative patterns seen amongst lucilin subunits. The assumption is made here, in the absence of contrary evidence, that the dye-binding capacity of all subunit polypeptides is equivalent. A search for regulator loci affecting *Luc-1* alleles has so far been inconclusive; at present the second hypothesis appears the more attractive, without eliminating the possibility that regulator loci will be identified as the system becomes better known.

Using a multiply marked tester stock constructed by Whitten and his colleagues, *Luc-1*, *Luc-3*, and tentatively the remaining *Luc* loci have been assigned to chromosome 2 on the basis of complete linkage with *black puparium* in males. No crossing over occurs in males of *L. cuprina* (Whitten *et al.*, 1975). The species is ideal for further genetic studies of the *Lucilin* cistrons; single-pair matings are reasonably successful and the formal cytology (Childress, 1969) and genetics of the species quite sophisticated

(Whitten *et al.*, *loc. cit.*). Detailed linkage studies amongst the *Luc* loci are in progress to determine whether more than one gene cluster is involved. Structural comparisons of the *Luc* subunit polypeptides are clearly needed, but origin of the multiple loci through gene duplication certainly appears likely.

The genetics of only one other bulk plasma protein has been examined in detail. This is haemochironomin (haemoglobin) in the Chironomidae. These multiple proteins are not generally polymorphic within field populations, unlike the calliphorins and lucilins. The haemochironomin complement varies, however, between populations of *Ch. tentans* (English, 1969; Tichy, 1970). Hybrids between two differing haemochironomin phenotypes show all bands of both parental species (English, 1969). Unlike calliphorin and lucilin, the haemochironomins show a developmental change in the proportion of the individual molecular species represented (Manwell, 1966; English, 1969; Shrivastava and Loughton, 1970). Multiple genetic loci, which perhaps evolved through gene duplication (Thompson and English, 1966) seem likely to be responsible. A dramatically successful cytogenetic analysis by Tichy was based on detecting recombinant haemochironomin complements in hybrids of *Ch. tentans* and its sibling species *Ch. pallidivittatus*, as well as in hybrids between geographic races of *Ch. tentans* with different protein spectra. The positions of genetic exchange were determined cytologically by examination of the polytene chromosomes of salivary glands from these larvae. *Ch. tentans* showed ten and *Ch. pallidivittatus* eight haemochironomins in Tichy's study, and of these seven were in each case species specific. The cistrons coding for the haemochironomins occupy at least three regions on chromosome 3 of these species (Tichy, 1970).

Another exciting development in the genetic analysis of control of haemochironomin synthesis has been the discovery in *Ch. tentans* of a regulatory locus affecting certain proteins selectively (Thompson and Patel, 1972; Thompson and Horning, 1973). A variant regulatory allele (*Regulator-Jemmerson*, R^{Jem}) was isolated from field-caught heterozygotes. A homozygous stock established from these midges was crossed with *Ch. pallidivittatus* to examine the behaviour of the regulatory locus; individual haemochironomins from either species were recognized by their characteristic tryptic-digest fingerprints. R^{Jem} affects *specifically* just two loci in *Ch. tentans*. These code for two haemochironomins $Hb^{0.50}$ and $Hb^{0.61}$ which are amongst the most similar of the complement of these proteins, and are therefore considered to be due to evolutionarily recent gene duplication (Thompson and Horning, 1973). The regulator locus maps near the structural loci for both haemochironomins in the right arm of chromosome 3. One locus shows increased activity (in terms of the amount of protein, $Hb^{0.61}$, synthesized) when its alleles are arranged in either the *cis* or *trans* configuration with R^{Jem}. The other locus is decreased in activity, but much more markedly so for alleles in the *cis*, which are almost inactivated, compared with those in the *trans* arrangement with R^{Jem}. Thompson and Horning (1973) postulate that the regulatory mutant may represent a lesion in a promoter site for $Hb^{0.50}$. These authors postulate a model of transcriptional regulation similar to that in human β-thalassemia. They suggest that competition may occur between the two closely related structural loci for a common RNA polymerase or for a factor determining specificity of an RNA polymerase. More rapid accumulation of promoter mutations at one of the two loci might then alter their initially equal ability to compete for a particular RNA polymerase, giving rise to a situation where the product ($Hb^{0.50}$) of one locus is more abundant than that of the other ($Hb^{0.61}$). This is the normal situation in the *Ch. tentans* protein spectrum. R^{Jem} might then reflect an additional mutational change in the promoter site for $Hb^{0.50}$ structural gene such that initiation of transcription of the cistron adjacent to R^{Jem} is impaired. No effect would be seen on an allele in the *trans* relationship with its own promoter. Thompson and Horning go on to suggest that if transcription was already maximal at the intact $Hb^{0.50}$ locus, limited only by availability of the initiating nucleoside triphosphate, the presence of R^{Jem} would stimulate transcription at both *cis* and *trans* alleles at the related $Hb^{0.61}$ locus, through greater availability of the common transcriptional factor. Details of the model may require modification to fit further data, but it is difficult to escape the conclusion that here there is evidence for the significance of highly specific controlling factors determining transcription rates.

Braunitzer (1965) has postulated a common evolutionary origin for the vertebrate haemoglobins and the haemoproteins of the chironomid midges. The extensive and complex alterations to gene structure required to reconcile the sequences of the two proteins make this hypothesis unattractive. An independent origin for the plasma storage protein (chironomin) and convergence to a structure capable of specific binding of haem seems more probable, but the genetic systems controlling the multiple insect haemoproteins do show a remarkable overall resemblance to those of the vertebrates. Finally, it seems significant that the only storage proteins investigated genetically in any depth, viz. those of the calliphorids and chironomids, are controlled by several genetic loci which may have arisen by tandem multiplication.

Probably related to the storage proteins are the genetic polymorphisms of plasma bands seen in *Ephestia* by Egelhaaf (1965a, 1965b) and *Bombyx* by Gamo (1968). Gamo identified a locus (*Alb*), situated 6.2 centimorgans from *narrow breast* on chromosome 20 of *Bombyx*, as responsible for a slow-fast polymorphism of one plasma-protein ("albumin") band. Heterozygotes show both slow and fast components. This polymorphism, and those studied by Egelhaaf in *Ephestia*, might well provide starting points for genetic analysis of the lepidopteran storage proteins.

A number of other studies of the inheritance of haemolymph proteins have been based on the protein bands separated from whole plasma by gel electrophoresis. While these variations may involve subunits of complex storage or other proteins, such relationships and significance will not become apparent until the storage protein molecules are first isolated and purified, prior to separation under conditions leading to their disaggregation. The pattern of *synthesis* (not just *presence* in the plasma) of the subunits may serve as one of the best guides to the recognition of specific gene products, and to establishing the functional relationships between them (section 4.3). The significance of reported null alleles must be evaluated with special care, as proteins of altered mobility are easily disguised by other bands in whole-haemolymph patterns. The reports of Duke and Pantelouris (1963), and Duke (1966) cover a number of the plasma proteins of *Drosophila*.

4.5 OTHER LARVAL PROTEINS

We are concerned here not with enumerating the multitude of different structural and enzymic components which have been identified in holometabolous larvae and their individual tissues, but with establishing the broad patterns seen in the development of these insects. The larval protein spectrum has been used by many workers as a basis for comparisons with that of "pupae" or of adults, in the hope of documenting the extent of changeover in protein complement at metamorphosis. This theme, and some examples, are considered in section 5.

Apart from the larval storage proteins and haemolymph lipoproteins which may have transport functions (Gilbert, 1967, for review), the larval proteins may conveniently be considered in two broad categories. One comprises the specialized secretory proteins of digestion (both intra- and extra-corporeal), the silks, cuticle proteins, puparial glues etc. which are often synthesized in quite large quantities. The other group includes the structural proteins and enzymes of the cells and tissues: the proteins of cell organelles and membrane systems, the contractile proteins and the enzymes of the biosynthetic and metabolic machinery of the organism.

Many tissues of the Holometabola show several phases of protein content and synthesis. Such distinct phases are well illustrated in the dipteran salivary gland. In early larval life the cells of the salivary glands of the Diptera, for example, are concerned predominantly with the synthesis of the structural and enzymatic components of the cell and with digestive enzymes, mostly until mid-way through the final instar, and may thereafter take on additional roles related to the onset of metamorphosis. In *Drosophila*, the first phase of gene activation starts at gastrulation when activation of the zygotic genome has been detected (by synthesis of *rosy*$^+$ complementation factor from the paternal genome; Sayles *et al.*, 1973). In the salivary gland of *Drosophila*, changes in fine structure (von Gaudecker, 1972; Lane *et al.*, 1972) mark such a changeover to accumulation of puparial glue proteins. In *Chironomus*, the salivary gland synthesizes only secretory protein (not intracellular materials) after the last larval moult (Wobus *et al.*, 1972), and does so using a stable mRNA (Clever, 1969). Similar changes in protein synthesis are seen in the salivary gland of *Calliphora* (Fig. 8; see Kinnear *et al.*, 1971); in *C. stygia* some qualitative changes in salivary protein synthesis accompany a major decrease in the overall synthetic activity of the glands in the latter half of instar 3. There is a coordinated drop in gross protein synthesis in all major larval tissues of *Calliphora* at this stage (compare Figs 6 and 8).

A number of different tissues are capable of taking up haemolymph proteins. The salivary secretion of *Chironomus* larvae (Doyle and Laufer, 1969) and the salivary cocoon-silk proteins of *Rhynchosciara* (Terra *et al.*, 1973; Bianchi *et al.*, 1973) are in each case of double origin; in part synthesized *de novo* in the gland, and in part derived from sequestered haemolymph protein. The same process may also occur in *Drosophila* salivary glands (Pasteur and Kastritsis, 1971). Willis (1970) and Koeppe and Gilbert (1973) have obtained evidence that haemolymph proteins are taken up by the cells of the larval and pupal epidermis of *Hyalophora*, and *Manduca* respectively. In *Manduca*, Koeppe and Gilbert combined radiotracer, electrophoretic and immunological analyses of the fate of specific haemolymph proteins to show that these are apparently unchanged on deposition in the cuticle. Descriptive investigations on the proteins of larval integument include enzyme analyses on *Drosophila* (Knowles and Fristrom, 1967) and developmental studies on *Galleria* (Srivastava, 1970).

Because of the massive amounts of protein synthesized, the structure of the silk fibroin locus of the silkworms is of special interest. There is no evidence of either specific amplification or of tandem duplication (contrast lucilin and haemochironomin, section 4.4) of the silk fibroin loci in *Bombyx*. Suzuki *et al.* (1972) find by hybridization of purified silk fibroin mRNA with DNA from several contrasting tissues that each haploid genome contains between one and three fibroin genes. The limits of resolution of the technique prevent a more precise estimate. The cells of the posterior silk gland in *Bombyx* are highly endopolyploid, and accumulate large amounts of stable mRNA for silk fibroin. Although very high

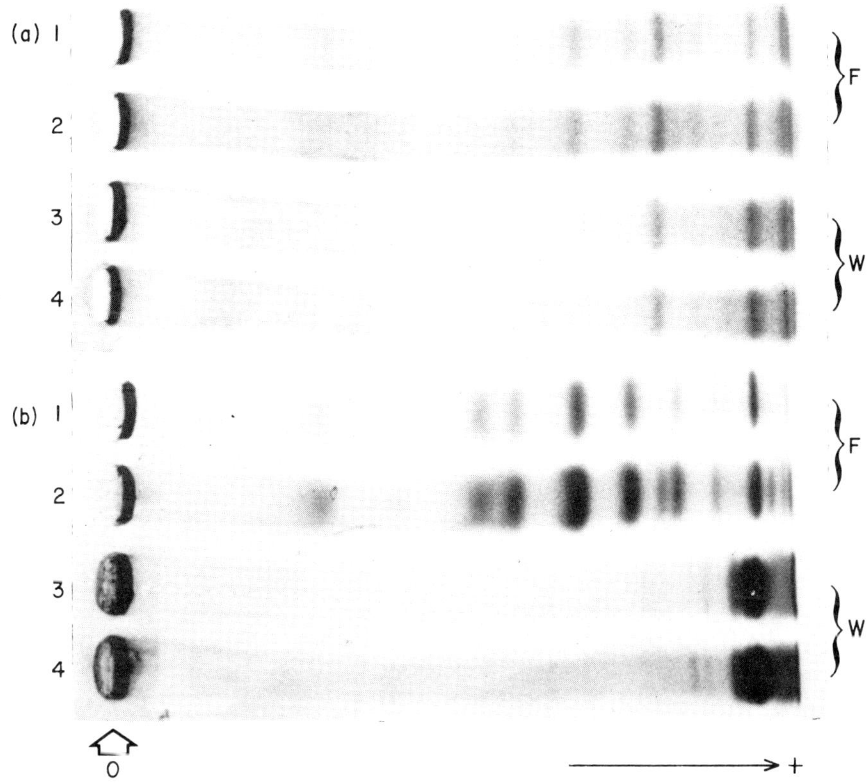

Fig. 8. (a) Electrophoretic separations on 5 per cent polyacrylamide gel (buffers as for Fig. 6) of proteins from larval salivary glands of *C. stygia.* Samples 1 and 2: homogenates prepared 10 and 90 min respectively after injection of ^{14}C-amino acids into feeding third instar larvae. Samples 3 and 4: homogenates prepared 60 and 240 min after such injection (F, feeding; W, wandering stage; O, origin; +, anode). (b) Autoradiographs of dried gel shown in (a). Note changed pattern of incorporation at wandering stage. Exposure, 3 weeks. (Courtesy J. F. Kinnear and M.-D. Martin.)

rates of transcription and translation are required to produce 300 μg fibroin in 3-4 days by each cell (Suzuki *et al.*, 1972), these are within the range of rates recorded in other systems (Kafatos, 1972a, e.g. for transcription of rRNA). Control of fibroin synthesis during development has been considered by Shigematsu and Moriyama (1970).

A new phase of gene activity commences in many larval tissues close to pupariation. This final phase of gene read-out is reflected in marked changes in chromosome puff pattern in dipteran salivary glands (e.g. Ashburner, 1970, 1972, 1973; Berendes, 1965, 1971, 1972, for reviews), and in the temporary small-scale recovery in the level of protein synthesis seen in tissues such as fat body and salivary gland at this time (Martin *et al.*, 1969). Gene activity at this time may reflect changes concerned with the larval-pupal moult, as in the case of rising levels of dopa decarboxylase in the *Calliphora* epidermis (Shaaya and Sekeris, 1965; and perhaps of ribonuclease H: Doenecke *et al.*, 1972, as well as additional RNA polymerases: Doenecke *et al.*, 1973) or with the synthesis or activation of the enzymes of cell autolysis (Clever, 1972; section 6). In *D. hydei* a shift in the spectrum of deoxyribonuclease activities in the salivary cells accompanies changes in puff pattern; the shift seems to be towards a class of activities which may be lysosomal in origin (Boyd and Boyd, 1970). In *Chironomus*, ribonuclease and acid phosphatase increase 9- to 12-fold in specific activity, and a clear change in protease activities takes place at this stage (Laufer and Schin, 1971). The larval secretory protease, pH optimum 5.5, is replaced by a protease of pH optimum 3.5-4.0 (Rodems *et al.*, 1969). Synthesis of the protease zymogen itself is not involved, but the synthesis of new protein is required in mid- but not late pharate pupae, as long-term cycloheximide treatment stops the increase of the protease activity and causes retention of gland cell structure (Clever, 1972; Henrikson and Clever, 1972). This new protein is presumed to be somehow involved in release and activation of previously synthesized lysosomal enzymes. Synthesis of such a protein fraction may also occur prior to cytolysis of intersegmental muscles in the abdomen of adult *Hyalophora* (Lockshin, 1969a, 1969b).

5 Translation of the imaginal gene set

There is general agreement that quantitative differences between larval and adult protein spectra are more conspicuous than qualitative ones (e.g. Chen, 1971, pp. 95-96), although the extent of qualitative change is perhaps correlated with the degree of completeness of metamorphosis (Wyatt, 1968). In general, interpretation of the numerous studies of the changes in the protein spectrum of individuals through metamorphosis appears to have been clouded by an expectation that presence of a protein reflects its synthesis and, even further, activity of the genetic loci coding for that protein species. On such a basis, for instance, it has been concluded that in *Drosophila* ". . . there is no evidence of a massive switch-off of larval genes and of a switching-on of an alternative, adult, set of genes." (Pantelouris and Downer, 1969.) Or again: "Even in the higher Diptera, . . . the data do not imply a switch-over from a larval set of genes to an adult set." (Chen, 1971.) What would be required to substantiate such a conclusion is ultimately evidence of shut down of transcription of the cistrons coding for larval proteins, and of the initiation of synthesis of new mRNAs, culminating in the appearance of "adult" gene products. Fortunately, in the case of many larval proteins, translation stops well before metamorphosis, so that switch-off of particular larval loci (such as those coding for calliphorin) can be monitored by following protein synthesis. Similarly, the switch-on of imaginal loci may be regarded as rigidly demonstrated only where synthesis of a new protein species is demonstrated to depend on immediate mRNA production. This kind of data is available only in a few instances (e.g. imaginal cuticle proteins of *Tenebrio*: Ilan *et al.*, 1966; Ilan and Ilan, 1973). The *de novo* synthesis of a new protein species, however, provides evidence of changed gene activity even if the time of transcription is not known. It seems hardly necessary to emphasize that failure of a larval protein to disappear immediately on pupation affords no information on the activity of the genetic locus coding for it, but a number of authors have failed to recognize this point. Truly "larval" proteins, such as the storage species of fat body and haemolymph, may be utilized gradually throughout imaginal development. Further the time and rate of utilization of such proteins, and hence the protein spectrum of developing insects, may vary between strains, as between autogenous and anautogenous strains of *Lucilia*.

Recognition of changes in the pattern of gene read-out between larva and adult is likely to be obscured in proportion to the number of gene activities common to both larval and imaginal tissues, to the amount of each larval protein carried through metamorphosis, and to the rate of utilization of such larval protein species in the imago. It is not possible to agree with Chen (1971) that ". . . in no case is there a massive alteration of the protein pattern at metamorphosis". At the onset of metamorphosis in *Calliphora*, for example, the larval storage protein calliphorin comprises 60 per cent of the total soluble protein of the animal (Munn and Greville, 1969), whereas in the reproductively mature adult this protein is not represented at all (see section 4.2).

5.1 THE IMAGINAL PROTEINS

Two significant generalizations emerge from a broad range of biochemical and immunological studies of changes in the spectrum of total and of haemolymph proteins during holometabolous ontogeny. Firstly, the proteins can be grouped into larval-like or adult-like kinds, as Butler and Leone (1966) point out in a study of *Tenebrio* proteins. In the interpretation of patterns of gene activity this is a particularly significant point. The evolution of the pupal stage by modification of a larval instar (section 1) is reflected in the *qualitative* similarity of larval proteins, and those of the pharate pupa, e.g. in the body wall of *Drosophila* (Pasteur and Kastritsis, 1971, 1972). The adult integument of *Galleria* contains proteins apparently also present in either or both the larval or early pharate adult stage (Srivastava, 1970). These results are consistent with the existence of overlapping larval and imaginal gene sets, but not with the existence of a separate "pupal" gene set such as that postulated by Williams and Kafatos (1971). Secondly, the transition from one protein set to the other involves both quantitative and qualitative changes (Butler and Leone, 1966, and references cited therein; Chen, 1971, and see above). Often the overall protein pattern of the larva, pupa or pharate pupa and early pharate adult is quite similar, but the adult pattern may diverse strikingly as the quantitative preponderance of the larval fat body and haemolymph storage proteins diminishes with their rapid utilization. These changes are well documented in the Diptera (section 4.1.1) but are also clearly seen in the Lepidoptera as formerly predominant larval protein species disappear during imaginal development (Vinson and Lewis, 1969; Patel, 1971).

A detailed account of the times of appearance and tissues of origin of particular proteins during imaginal development will not be attempted here: some examples are cited by Chen (1971), who also gives a summary of the occurrence of sex-specific peptides in adult insects, and of the synthesis and release of yolk proteins (see Wyatt, 1972; Price, 1973). Other recent additions to knowledge of the occurrence of female specific adult proteins related to yolk formation include studies on *Bombyx* (Kai and Hasegawa, 1971), *Sarcophaga* (Engelmann *et al.*, 1971), *Nasonia* (King *et al.*, 1973) and *Tenebrio* (Laverdure, 1969). Related aspects of endocrine control of gonad development have been fully reviewed by Doane (1973). The yolk proteins, named vitellogenins (Pan *et al.*, 1969) comprise a predominant product of adult fat body in female insects. In *Hyalophora*, vitellogenin is first synthesized in the pharate pupal fat body and by the late pharate pupal stage a considerable amount has been released into the blood (Pan *et al.*, 1969; Pan, 1971). Synthesis then slows until uptake of the protein from the haemolymph into the developing oöcytes takes place during imaginal development. Lipoproteins utilized in the oöcytes are also synthesized in the fat body of *Hyalophora* (Thomas and Gilbert, 1969). Vitellogenin synthesis starts later in those species in which yolk deposition is relatively delayed; it commences after emergence of the adult in the butterfly *Danaus* (Pan and Wyatt, 1971). The variable onset of vitellogenin synthesis relative to reproduction, diapause etc. in *Leptinotarsa* (de Loof, 1969; de Loof and de Wilde, 1970a, 1970b), *Tenebrio* (Pemrick and Butz, 1970a, 1970b) and *Aedes* (Hagedorn *et al.*, 1973), implicates neurosecretory control as well as direct effects of JH.

Where the larval fat body is totally replaced at metamorphosis vitellogenin is a product of the adult fat body. The synthesis of vitellogenin, and its humoral control (section 6) in *Aedes* have been examined in careful detail (Hagedorn and Judson, 1972; Hagedorn *et al.*, 1973; Hagedorn and Fallon, 1973). Other less complete studies also provide evidence that the synthesis of certain imaginal haemolymph proteins awaits development of the adult fat body (*Musca*, "Fraction 4": Bodnaryk and Morrison, 1968). Consistent with this, the larval fat body of many species, especially of anautogenous strains degenerates well before the ovaries mature to the stage of yolk deposition, and the larval fat body contains no proteins corresponding to the major adult-specific haemolymph proteins (*Calliphora*: Kinnear and Thomson, 1975). In *Culex*, Chen (1967) also found no evidence of direct utilization of larval haemolymph proteins in yolk formation, even in an autogenous strain. The larval fat body certainly appears to supply reserve materials for ovarian development (e.g. Adams and Nelson, 1969) in the Diptera, but in the case of the major yolk proteins at least, it does not do so directly. Aspects of the control of replacement of the larval fat body in the adult are discussed in section 6.

A further class of adult proteins to have recently attracted biochemical attention, especially in the Lepidoptera, is represented by the eggshell proteins forming the chorion (Paul *et al.*, 1972a, 1972b).

Little is known at present of the individual protein species in the developing imaginal discs during pharate adult life (Fristrom, 1972). With the advent of mass isolation procedures for these tissues, the prospects for such studies appear good.

Extensive studies have been made by Kafatos and his collaborators of the adult proteins involved in escape of lepidopterans from the cocoon, especially of the cocoonase secreted by the galea of *Hyalophora* (Kafatos, 1970). This protein is synthesized *de novo* in the adult using a long-lived mRNA (Kafatos, 1972a, 1972b). The proteases of the moulting gel of the silkmoth *Antheraea* are secretory products of the early to mid-pharate imago, accumulating in the moulting gel until activated late in the pharate adult stage when the moulting gel becomes fluid, and proteolytic digestion of the pupal endocuticle commences (Katzenellenbogen and Kafatos, 1970, 1971). In *Drosophila*, Berger and Canter (1973) have observed the appearance of a group of esterases in the pharate adult, which disappear abruptly from the insect at eclosion, and which are discarded in the pupal case. Again, a role in eclosion is suggested for these special imaginal proteins.

Much of the bulk of imaginal protein formed in mid- to late pharate adult development, and immediately following eclosion, comprises contractile proteins of the flight musculature, together with the structural and enzymic proteins of its mitochondria (Williams, 1972; Williams and Birt, 1972). The origin, and relationships of these proteins to larval proteins, are considered below.

5.2 THE RELATIONSHIP OF LARVAL AND IMAGINAL PROTEINS

Because larval protein comprises the principal store of nitrogen for imaginal development during the complete metamorphosis of the holometabolous insects (section 4), the way in which larval and adult proteins are related, and the storage and movements of possible intermediates which might be involved, has proved of special interest. Two immediate sources of precursor materials for adult proteins are available. One is provided by those tissues undergoing cytolysis at metamorphosis, including a variable portion of the larval musculature, integument, gut and specialized secretory tissues, which in the mature larva of *Calliphora* account for about 10-20 per cent of the total protein (estimated from the data of Martin *et al.*, 1969). A minor contribution from the fat body should also be included here: some fat body inclusions are formed from degenerating cellular organelles (Locke and Collins, 1968; section 4.2) rather than from sequestered haemolymph proteins. Resorption of moulting fluid at the larval-pupal moult is also a minor source of re-usable proteins in at least certain species (*Apis*: Lensky and Rakover, 1972). Materials of this latter kind derived from breakdown of the pupal cuticle are of course not available until resorption of the moulting fluid just before eclosion (Lensky *et al.*, 1970). The second source of precursors for the adult proteins are the quantitatively major storage proteins of larval haemolymph and fat body.

Reserves from both these sources must become available to developing imaginal tissues as soluble amino acids, peptides or proteins in haemolymph, or as insoluble, granular "packaged" materials in the proteinaceous spheres released by cytolysis of the larval fat body, or as spherules representing engorged granular haemocytes (Whitten, 1964; Crossley, 1965) which have phagocytosed the debris of lysing larval cells. Bodies of these kinds become closely associated with developing imaginal tissues, filling the interstices in flight muscle in the early stages of synthesis of the fibrillar components, but gradually disappear from the thoracic region as adult morphogenesis advances (see Whitten, 1964).

The relation of larval and adult proteins in the Diptera has aroused considerable discussion. The amino acids of isotopically labelled larval haemolymph proteins injected into *Phormia* larvae at the time of pupariation are certainly extensively incorporated into adult proteins (Chen and Levenbook, 1966b, confirmed in *Calliphora* by Kinnear, 1973). ^{14}C-labelled larval haemolymph proteins injected into pupariating larvae are in part metabolized to $^{14}CO_2$ during metamorphosis (Chen and Levenbook, 1966b). Thus a function of the storage proteins is to contribute to the energy requirements of the insect at this stage, presumably via conversion of amino acids to fatty acids and carbohydrate which can be oxidized for ATP generation (D'Costa and Birt, 1966; Crompton and Birt, 1967).

Earlier workers (Agrell, 1964; Chen and Levenbook, 1966b; Dinamarca and Levenbook, 1966; Barritt and Birt, 1971) considered that there was little evidence of extensive *de novo* synthesis of protein during metamorphosis. Thus the question arose whether re-utilization of larval proteins might be possible by reorganization at the peptide level without degradation to free amino acids (reviews by Williams, 1972; Williams and Birt, 1972). Such a process on a large and diverse scale would appear to require the widespread participation of unusual control systems, but precedents do exist for a developmental change in function of a polypeptide by formation of a complex with a specifier sequence (Brew *et al.*, 1968). Alternatively, the change in structure to that of the adult protein might involve limited hydrolysis of larval polypeptides to release new subunits (see Williams and Birt, 1972).

More recently the quantitative importance of the synthesis of imaginal proteins from amino acids has become recognized in the Diptera; a process for which the total amino acid pool of the insect appears to be available, with at least some amino acids turning over very rapidly. Thus, in *Lucilia*, Williams and Birt (1972) find that over the period of imaginal development from pupation to one day after emergence of the adult, about 2.5 mg of a total of 3.5 mg protein in the adult female fly, or more than 65 per cent of the total, must be synthesized *de novo* from amino acids. New synthesis is certainly the major source of imaginal protein.

Direct evidence of the synthesis of certain imaginal proteins during pharate adult development has been obtained for certain haemolymph proteins (*Drosophila*: Boyd and Mitchell, 1966; *Calliphora*: see Table 1, Kinnear and Thomson, 1975), for cytochrome *c* (*Lucilia*: Williams *et al.*, 1972), α-glycerophosphate dehydrogenase (*Lucilia*: Campbell, 1972) and the subunits of the major myofibrillar proteins, myosin, actin and tropomyosin as well as certain sarcoplasmic proteins (*Calliphora*: Kinnear, 1973). Less direct evidence is derived from the observation that the appearance of specific adult proteins can be blocked by injection of cycloheximide in the pharate adult under conditions where precautions against detoxification of the antibiotic is not permitted (Campbell and Birt, 1972). Further, Williams and Birt (1972) document in their review the relative constancy of total and haemolymph-free amino acid pools, and the rapid turnover of at least several amino acids during metamorphosis in the Diptera. The simplest explanation of this situation is the occurrence of rapid protein synthesis coordinated with the replenishment of the amino acid pools by breakdown of larval protein. On the other hand, Kinnear (1973) obtained no evidence of re-utilization of the subunits of calliphorin in the major adult proteins of *Calliphora* in a detailed immunological and structural survey.

Proteolytic enzymes with a low pH optimum have now been implicated in the processes of metamorphosis of *Calliphora* (Agrell, 1951), *Musca* (Chefurka, 1964) and *Lucilia* (Smith and Birt, 1972). Ciavattini *et al.* (1959) and Russo-Caia (1960a, 1960b) emphasize the relation of high proteinase levels to active histogenesis in such insect tissues. Smith and Birt found in *Lucilia* an acid proteinase (pH optimum 4.1) with two peaks of activity during development, the first at pupariation and the second extending over the period 3.5 to 5 days (at 30° C) after pupariation. These peaks correspond strikingly to synthetic maxima for proteins (Williams, 1972). A second proteolytic enzyme (pH optimum 2.8-3.0) is also present during metamorphosis in this fly (Smith and Birt, 1972). Together, these proteinases could provide more than the necessary activity to degrade all the larval proteins destined to be replaced at metamorphosis (Williams and Birt, 1972). The particulate proteinases of metamorphosing *Lucilia* are especially associated with the degradation of larval fat body remnants in the abdomen of the pharate and newly emerged adult. The major larval storage protein available for degradation to provide amino acids for the synthesis of adult proteins is calliphorin or its homologues. Calliphorin from *C. erythrocephala* is reported to contain very small amounts of cysteine, while this amino acid could not be detected in calliphorin from *C. stygia* (section 4.1). Hackman (1956) found only trace amounts of cysteine-cystine in the free amino acid pool of *C. augur* at metamorphosis. Additional cystine is presumably available for imaginal protein synthesis through the usual pathway from methionine, in which calliphorin is quite rich.

Lockshin and Williams (1965) have characterized a proteinase appearing during metamorphosis of the intersegmental muscles of *Hyalophora* as a lysosomal enzyme of cathepsin D type (pH optimum 3.9). Such proteases are presumably normally involved in programmed cell death at metamorphosis, as in the salivary gland of *Chironomus* (Henrikson and Clever, 1972) and probably the mid-gut of *Galleria* (Janda and Krieg, 1969), etc.

The origin of adult proteins other than vitellogenins by synthesis from amino acids does not appear to have been questioned outside the Diptera. Evidence of *de novo* synthesis of adult proteins during metamorphosis is to be found in many studies of imaginal development in the Lepidoptera, including those by Bricteux-Grégoire *et al.* (1957), Telfer and Williams (1960), Stevenson and Wyatt (1962), Chan and Margoliash (1966), Chan and Reibling (1973), and Patel (1971), amongst others. Similar evidence has been obtained for Hymenoptera (*Apis*) by Osanai and Rembold (1970) and for Coleoptera (*Tenebrio*) by Ilan and his colleagues (Ilan *et al.*, 1966; Ilan and Ilan, 1973). The yolk proteins are synthesized *de novo* in the adult (refer section 5.1 for references), but synthesis of these may commence in pharate pupal stages of insects in which the larval fat body is not replaced at metamorphosis.

Where particular proteins occur in both larval and adult tissues, and the cells involved do not share a continuous ontogeny, the question arises whether the information required in the adult stage is derived from the cistrons which are transcribed in larval tissues, or whether entirely separate sets of loci might be utilized. Continuous gradations in the extent of remodelling or replacement of larval by adult tissues seen in the Holometabola, and the apparent relationship to developmental patterns in the Hemimetabola, strongly support the contention that larval and adult phenotypes are determined by variably overlapping, rather than discrete, gene systems. Comparisons of certain larval and adult gene products have also been made, notably by Levenbook and his colleagues, but these revealed no significant difference between larval and imaginal muscle aldolases of *Phormia* as judged by enzymatic properties, pH optima, energy of activation, molecular weight, amino acid composition, tryptic fingerprinting, uv absorption spectra and isoelectric focusing patterns (Levenbook *et al.*, 1973). The larval and adult tropomyosins of this fly are also closely similar, if not identical (Kominz *et al.*, 1962).

The utilization of specialized larval storage peptides such as β-alanyl-L-tyrosine in *Sarcophaga* (section 4.1) also appears to involve cleavage to free amino acids. In this instance the dipeptide is probably cleaved at pupariation to provide free β-alanine and tyrosine for use in sclerotization of the puparium (Bodnaryk, 1970b).

Special opportunities for the direct transfer of proteins from larval to adult cells would appear to arise during reconstruction of such tissues as muscle. Whitten (1964) and Crossley (1965, 1972a, 1972b) have noted two phases in the histolysis of larval muscle in a range of cyclorrhaphous Diptera. The first occurs at pupariation, and the second shortly after eversion of the head. These muscles first vacuolate and are then invaded by granular haemocytes which become engorged with cellular debris. Although breakdown of engorged cellular fragments within the haemocytes may require one or more days for completion, polypeptides of the larval contractile proteins decrease sharply during early imaginal development and only later increase in whole animal extracts as imaginal flight muscle is formed (Kinnear, 1973). Thus the haemocytes do not conserve and transport intact muscle protein subunits. True re-utilization of larval proteins apparently may occur, however, in a group of larval muscles which do not break up completely. In these cells striations disappear and some vacuolation follows, but haemocyte invasion does not take place. The cells are penetrated by myoblast nuclei, the larval nuclei seem to degenerate (Crossley, 1965,

1972a, 1972b), the adult nuclei enlarge and striations reappear suggesting a take-over of the larval contractile machinery by the adult genome.

The quantitative significance of any such process in the formation of imaginal muscle cannot be accurately evaluated for the whole insect on presently available data. In the extreme metamorphosis of *Lucilia, de novo* synthesis of new contractile proteins is certainly the major process in the thorax, as shown by the data of Campbell (1973) and Campbell and Birt (1975). These workers have estimated the actual rate of synthesis of actomyosin from labelled amino acids as averaging 3.4 μg per thorax per h for the 24 h before emergence, and as 5.1 μg actomyosin per thorax per h for the 24 h after emergence. Thus in these two intervals a calculated 82 μg and 122 μg actomyosin would be synthesized. The increase in actomyosin actually measured as taking place in the same periods was 95 μg and 137 μg respectively. Thus more than 85 per cent of the increase in actomyosin over the two days centring on emergence can be accounted for by its synthesis.

In the context of both protein and RNA re-utilization in the Diptera, the occurrence during imaginal development of extracellular ribosomes in the haemocoele of *Calliphora* (Sridhara and Levenbook, 1973) may prove of interest; at present neither the origin (Sridhara and Levenbook, 1974) nor fate of these ribosomes is known with certainty. An additional role of the fat body of pharate pupae in *Bombyx* may be in conservation of nucleic acids for use in imaginal development, these being presumably transported in degraded form from larval tissues such as the silk gland which histolyse early in metamorphosis (Chinzei and Tojo, 1972).

6 Endocrine influences on fat body structure and function

Hormonal control of the synthesis of storage proteins in the pre-metamorphic larval fat body has not yet been demonstrated. In *Calliphora* and *Lucilia* the synthesis of calliphorin and lucilin respectively ceases abruptly (section 4.3), several days before moulting hormone (MH) is detectable in whole-animal extracts. Although this switch-off in translatory activity corresponds to a stage of step-down of synthetic activity in a range of larval tissues, and to some qualitative alteration in the species synthesized (Kinnear *et al.*, 1971), a humoral basis for these changes has not yet been established. Exogenous MH administration at this time in *Calliphora* causes a transitory increase in protein synthesis (Neufeld *et al.*, 1968), but the stimulatory effect is general rather than specific to certain proteins (Thomson *et al.*, 1971). It is tempting to speculate with Price (1973) that juvenile hormone (JH) may be needed to maintain calliphorin synthesis, but neither repeated topical applications nor injection of JH analogues prevent the normal abrupt decline after the end of feeding (Ramakrishnan and Thomson, unpublished results). Again, once calliphorin synthesis has stopped, neither JH administration, nor reduction of the protein concentration in the haemolymph by perfusion, will re-initiate it. Elucidation of the control system involved now appears an urgent and exciting task which might also have practical implications in relation to the development of hormonally based insecticides and their selective use.

MH is undoubtedly involved in selective protein uptake from the haemolymph and in the formation of proteinaceous spheres in the fat body. Detailed analyses of the formation of these bodies in *Calpodes* (section 4.2) led Collins (1969) to the conclusion that MH controls a switch from formation of autophagic vesicles to proteinaceous spheres containing haemolymph storage proteins; later more autophagic vesicles form as autolysis of the major portion of the endoplasmic reticulum takes place. In *Ephestia*, Cölln (1973) finds that ligation of larvae, prior to the critical period for irreversible determination of metamorphosis by MH, prevents reduction in the level of a specific haemolymph protein ("band 2") as well as its appearance in the fat body. Such isolated abdomens respond to injected MH by the selective uptake of band 2 protein (with little effect on the concentration of other components such as band 3). Formation of proteinaceous spheres follows uptake of the plasma protein by the fat body. A similar control of protein uptake may exist in *Plodia* (Pentz and Kling, 1972).

The involvement of MH in the uptake and sequestration of haemolymph storage proteins is much more elusive in the Diptera. In *Drosophila* (von Gaudecker, 1963; Butterworth *et al.*, 1965; Thomasson and Mitchell, 1972) and *Calliphora* (Price, 1969) proteinaceous spheres commence to form in the fat body well before the whole insect contains assayable levels of MH (Shaaya and Karlson, 1965) and before metamorphosis is irrevocably determined. For instance, in *Drosophila*, the proteinaceous spheres appear 17 h before pupariation (Thomasson and Mitchell, 1972) and ligation behind the brain-ring gland complex does not always block sclerotization of the posterior portion of the body until 4 h before puparium formation (Becker, 1962). Low levels of MH may, of course, be present considerably earlier if the hormone formed is turned over rapidly at this stage.

Unknown features of the hormonal environment, and uncertain interactions with lytic factors in adult haemolymph make interpretation of experiments on formation of proteinaceous spheres in larval fat body transplanted into the adult abdomen difficult to interpret. Such data has been used by Butterworth and his colleagues (Butterworth *et al.*, 1965; Butterworth and Bodenstein, 1967) to support the contention that MH is indeed necessary for formation of proteinaceous spheres in *Drosophila*. On the other hand, Thomasson and Mitchell (1972) find that MH stimulates the formation of proteinaceous spheres, and can induce their precocious appearance. Such spheres formed in fat body cultured *in vitro* without MH in some tests, and more regularly if the medium contained calf serum proteins or ovalbumin. Thomasson and Mitchell therefore concluded that MH facilitates formation of proteinaceous spheres in *Drosophila*, but is not a necessary, or the only, trigger. The previous experiments of Butterworth and co-workers may be interpreted along the same lines. The apparent conflict between this aspect of the role of MH in the Diptera and in the Lepidoptera cannot be resolved with existing information. It is tempting to speculate that more than one kind of proteinaceous sphere is involved in the Diptera as in *Calpodes* (see above), and that MH may prove to initiate formation of only certain of these. Certainly the rate of withdrawal of protein from haemolymph increases sharply in *Calliphora* from the time near pupariation (Kinnear *et al.*, 1968; Kinnear, 1973) when MH is known to be present. Prior to this the fat body cells take up considerable plasma protein (Martin *et al.*, 1971). The point at which the protein withdrawn from the haemolymph is actually sequestered into proteinaceous spheres has not been established.

In the adult Colorado potato beetle (*Leptinotarsa*), proteinaceous spheres form in the internal fat body before diapause. Formation of these protein reserves is induced by lack of JH, according to de Loof and Lagasse (1970).

An essential aspect of restructuring the phenotype during metamorphosis involves the death of those larval cells destined to be replaced in adult development. Programmed cell death is neither unique to holometabolous species, nor to larval tissues: it is simply a part of the normal repertoire of morphogenetic processes commonly seen in higher organisms (Saunders, 1966). Imaginal as well as larval tissues are susceptible to cell death at appropriate times (*Calliphora*: Spreij, 1971; *Sarcophaga*: Whitten, 1969c, 1969d; *Drosophila*: Fristrom, 1969, 1972).

The significance of humoral influences on cell death in larval fat body is as difficult to understand as are those on protein storage in this tissue. Aspects of the fate of the larval fat body, especially in *Calliphora* and *Lucilia* have already been discussed in sections 4.2 and 5.2. Cell death in the larval fat body is highly variable in onset in different body regions and amongst different species. Especially in abdominal regions, cell death follows a slow, progressive course during pupal and imaginal development in most holometabolous groups. Disaggregation of the larval fat body into single cells appears to be mediated quite directly by MH, as shown *in vitro* for *Manduca* (Judy and Marks, 1971).

In the Diptera, there is an extreme antero-posterior gradient in the pattern of histolysis, as there is earlier in protein uptake in species such as

Calliphora and *Lucilia*. The existence of this gradient of lysis, together with differences in the timing of histolysis in autogenous and anautogenous strains of the same species (section 4.2), confirm that cell death in the larval fat body is not programmed irrevocably by the hormonal milieu initiating metamorphosis as it is in the adult intersegmental muscles of the saturniids (Lockshin and Williams, 1964, 1965). The environment provided by the haemolymph of adult *Drosophila* is normally highly lytic for mature larval fat body cells, recalling the rapidly accelerated destruction of larval fat body in anautogenous *Lucilia* which Williams (1972) observed from 1-2 days after emergence of the adult. Implanted fat body cells from young third-instar larvae are not competent to respond to the lytic factor(s) (Butterworth, 1973). Transplantation experiments show that the abdominal environment becomes more permissive with increasing age, and is always so in certain mutants such as *apterous*^{4} (Butterworth, 1972; Butterworth and La Tendresse, 1973). That the difference in the rate of lysis of ap^4/ap^4 and wild-type fat body cells is a function of the environment of the cells, not a property of the cells themselves, is also convincingly shown by Butterworth (1972) in his transplantation studies. In wild-type hosts, ap^4/ap^4 cells degenerate as rapidly as wild-type cells. Postlethwait and Weiser (1973) have recently shown that ap^4/ap^4 homozygotes lack, or fail to produce, vitellogenic levels of JH. Low levels of JH would be expected to have a sparing effect on larval fat body (cf. allatectomy, see below). If JH were the only factor controlling regression of the fat body, however, the lytic potential of the abdominal haemolymph for larval fat body cells should increase as the corpus allatum becomes active in stimulating vitellogenin synthesis, whereas Butterworth's (1972) observations indicate amelioration of the lytic tendency with age. Slowed cell death in the larval fat body of ap^4 homozygotes seems to be unrelated to the enhanced cell death in imaginal wing discs which leads to the wingless phenotype of these flies (Fristrom, 1969). Comparative analyses of the haemolymph of newly emerged adults of anautogenous and autogenous strains of such species as *Lucilia* or *Culex* might prove a useful approach in isolation and characterization of lytic factors in the haemolymph.

In a number of dipteran species (references cited by Adams and Nelson, 1969), the disappearance of the larval fat body in adult females can be correlated with ovarian development. Adams and Nelson (1969), analysing this phenomenon in *Musca*, report that allatectomy slows the degradation of the larval fat body in addition to blocking ovarian development. These observations confirm the earlier work of Day (1943) on other flies. The effect is not due to JH, or not to JH acting alone, because ovariectomized *Musca*, in which the corpora allata are continuously active, also show retarded degradation of the larval fat body. Adams and Nelson concluded that the developing ovary may produce a hormone affecting cytolysis of the larval fat body, as well as acting as a nutrient sump for its degradation products (cf. earlier work cited by Adams and Nelson, 1969). It is not known whether any relationship exists between such an ovarian factor and (i) the lytic factors postulated by Butterworth and his colleagues (see above), (ii) the ovarian hormone proposed by Doane (1961) as a regulator of lipid utilization in the fat body of *Drosophila* and (iii) the ovarian oöstatic hormone considered by Adams (1970) and Adams *et al.* (1968) to control the cycles of ovarian development in *Musca* via the corpus allatum.

Complementary to the problem of the control of regression of the larval fat body is that of the development of the adult fat body, which in turn is correlated with yolk synthesis (section 5.1) and the export of yolk proteins to the ovaries via the haemolymph. That JH has a role in the development of the imaginal fat body independent of ovarian maturation is apparent from the increased size of the fat body following administration of the hormone to adult male *Drosophila* (Butterworth and Bodenstein, 1969).

7 Conclusion

The fat body has besides another great function—that of storing reserve materials. . . . It is this capacity of storing food materials that is so important in insect metabolism, and it is largely this that has enabled the insect metamorphosis to be evolved.

Tiegs, 1922

The extreme metamorphoses of *Calliphora* and *Lucilia* which have been emphasized in the present account are built around single protein species, albeit complex heteropolymeric ones: calliphorin and lucilin respectively. Such proteins serve for the transfer of nitrogenous reserves, in the form of amino acids, from the feeding larva stage to the adult. At the commencement of metamorphosis in *Calliphora*, over 60 per cent of the total soluble protein of the insect is composed of calliphorin (Munn and Greville, 1969) synthesized in the larval fat body and stored conjointly in haemolymph and in the fat body proteinaceous spheres. Proteins from histolysis of larval tissues play a subordinate but significant role in metamorphosis. The larval proteins are utilized during metamorphosis both metabolically and as the source of amino acids for the synthesis of adult proteins. The imaginal proteins are predominantly synthesized *de novo* from amino acids, and there is no evidence of special genetic or epigenetic systems involved in large-scale "remodelling" of larval proteins without degradation to the amino acid level (Williams and Birt, 1972). While metamorphosis in these species may involve a particularly complete degree of dependence on a specific larval storage protein, the pattern of development seen here appears common to all Holometabola.

What is known of the haemolymph proteins and the production of proteinaceous spheres in the fat body suggests that the synthesis of one or a few predominant storage proteins in the larval fat body, and subsequent build-up of storage proteins in haemolymph as well as in the fat body, is general throughout the Holometabola. Emphasis has mostly been placed on analytical techniques which tend to separate the subunits of complex heteropolymeric proteins; functional relationships and subunit interactions have therefore often been missed. Further studies of the major plasma proteins present at pupation, and quantitative studies of their economy during metamorphosis, are urgently needed in the Lepidoptera, Coleoptera and Hymenoptera.

The patterns of gene activity in the larval Holometabola reflect synthesis of storage proteins, synthesis of structural and enzymic components of the larval cells, synthesis of specialized larval secretions such as silks, puparial glues etc., and finally the synthesis of the enzymes of cell destruction. Imaginal gene activities are concerned with synthesis of similar structural and enzymic proteins, including those involved in wing structure and flight, and also with the vitellogenic and other proteins of the developing gonads. The characteristic gene systems of the larval stages are those controlling the synthesis of the storage proteins and the specialized larval secretions. The gene sets most distinctive of imaginal development are those concerned with adult epidermal structures, with special mechanisms involved in eclosion (such as cocoonase), with flight, and with the vitellogenic proteins. Many gene activities are qualitatively common to both larva and adult; the contractile and enzymic proteins of muscle, and the structural and enzymic proteins of the cell organelles provide examples of gene readout common to both life stages. There is no distinct "pupal" gene set; characters of this stage can be viewed as late acting "larval" genes. The mechanisms of gene regulation, including those involving specialization of gene content and organization, have a similar basis in larval and adult tissues.

Acknowledgements

Work from my own laboratory described in this article owes much to the technical skills of K. R. Radok and H. S. Revell, whose help is gratefully acknowledged. It is a pleasure to acknowledge my extensive debt to J. F. Kinnear and M.-D. Martin, both in the accumulation of data bearing on our common interest in the development of *Calliphora*, and for helpful discussions. I thank L. M. Birt, A. J. Campbell and K. L. Williams for access to a great deal of information on the metamorphosis of *Lucilia*, some of it before publication, and for their infectious enthusiasm. Studies on the nucleolus included here represent part of a detailed analysis made in collaboration with M. M. Gunson, D. C. Rogers and A. G. Willis. I am similarly indebted to L. M. Birt (Australian National University) and to G. G. Foster and M. J. Whitten (CSIRO, Canberra), who have participated in various phases of the investigations on the larval storage protein of *Lucilia*. I thank J. Martin for suggesting some helpful references, and B. T. O. Lee, for valuable discussions on the genetics of lucilin. I am grateful to P. Hutton for typing the manuscript.

Financial assistance for this work has been received through the Australian Research Grants Committee (D65/15167), and I express my thanks to this organization for continuous support over the duration of my programme on insect development at the University of Melbourne.

References

Adams, T. S. (1970). Ovarian regulation of the corpus allatum in the housefly, *Musca domestica. J. Insect Physiol.* 16, 349-360.

Adams, T. S. and Nelson, D. R. (1969). Effect of corpus allatum and ovaries on amount of pupal and adult fat body in the housefly, *Musca domestica. J. Insect Physiol.* 15, 1729-1747.

Adams, T. S., Hintz, A. M. and Pomonis, J. G. (1968). Oöstatic hormone production in houseflies, *Musca domestica*, with developing ovaries. *J. Insect Physiol.* 14, 983-993.

Agrell, I. (1951). A contribution to the histolysis-histogenesis problem in insect metamorphosis. *Acta physiol. scand.* 23, 179-186.

Agrell, I. (1953). The aerobic and anaerobic utilization of metabolic energy during insect metamorphosis. *Acta physiol. scand.* 28, 306-335.

Agrell, I. (1964). Physiological and biochemical changes during insect development. *In* "The Physiology of Insecta" (Ed. M. Rockstein), Vol. I, pp. 91-148. Academic Press, New York and London.

Anderson, D. T. (1972a). The development of hemimetabolous insects. *In* "Developmental Systems: Insects" (Eds S. J. Counce and C. H. Waddington), Vol. 1, pp. 96-163. Academic Press, London and New York.

Anderson, D. T. (1972b). The development of holometabolous insects. *In* "Developmental Systems: Insects" (Eds S. J. Counce and C. H. Waddington), Vol. 1, pp. 165-242. Academic Press, London and New York.

Ashburner, M. (1967). Pattern of puffing activity in the salivary gland chromosomes of *Drosophila*. I. Autosomal puffing patterns in a laboratory stock of *D. melanogaster*. *Chromosoma (Berlin)*, 398-428.

Ashburner, M. (1969). Pattern of puffing activity in the salivary gland chromosomes of *Drosophila*. II. X chromosome puffing patterns in *D. melanogaster* and *D. simulans*. *Chromosoma (Berlin)*, 27, 47-63.

Ashburner, M. (1970). Function and structure of polytene chromosomes during insect development. *In* "Advances in Insect Physiology" (Eds J. W. L. Beament, J. E. Treherne and V. B. Wigglesworth), Vol. 7, pp. 1-95. Academic Press, London and New York.

Ashburner, M. (1972). Puffing patterns in *Drosophila melanogaster* and related species. *In* "Developmental Studies on Giant Chromosomes" (Ed. W. Beermann), pp. 101-151. Springer-Verlag, Berlin, Heidelberg and New York.

Ashburner, M. (1973). Sequential gene activation by ecdysone in polytene chromosomes of *Drosophila melanogaster*. I. Dependence upon ecdysone concentration. *Devl. Biol.* 35, 47-61.

Barr, H. J. and Plaut, W. (1966). Comparative morphology of nucleolar DNA of *Drosophila*. *J. Cell Biol.* 31, C17-22.

Barritt, L. C. and Birt, L. M. (1971). Development of *Lucilia cuprina*: Correlation of biochemical and morphological events. *J. Insect Physiol.* 17, 1169-1183.

Becker, H. J. (1959). Die Puffs der Speicheldrüsenchromosomen von *Drosophila melanogaster*. I. Beobachtungen zum Verhalten des Puffmusters im Normalstamm und in zwei Mutanten, *giant* und *lethal giant larvae*. *Chromosoma (Berlin)*, **10**, 654-678.

Becker, H. J. (1962). Die Puffs der Speicheldrüsenchromosomen von *Drosophila melanogaster*. II. Die Auslösung der Puffbildung, ihre Specifität und ihre Beziehung zur Funktion der Ringdrüse. *Chromosoma (Berlin)*, 13, 341-384.

Beermann, W. (1952). Chromomerenkonstanz und specifische Modifikationen der Chromosomenstruktur in der Entwicklung und Organdifferenzierung von *Chironomus tentans*. *Chromosoma (Berlin)*, 5, 139-198.

Beermann, W. (1962). "Riesenchromosomen". Protoplasmatologia, Handbuch der Protoplasmaforschung, Bd. IVd, pp. 1-161. Springer-Verlag, Wien.

Benson, K. A. (1965). An investigation of some developmental changes in the larval fat body of *Sarcophaga bullata* prior to metamorphosis. Ph.D. Thesis, University of Virginia. University Microfilms, Ann Arbor.

Berendes, H. D. (1965). Salivary gland function and chromosomal puffing patterns in *Drosophila hydei*. *Chromosoma (Berlin)*, 17, 35-77.

Berendes, H. D. (1971). Gene activation in dipteran polytene chromosomes. *In* "Control Mechanisms of Growth and Differentiation", Society for Experimental Biology, Symposium 25, pp. 145-161. Cambridge University Press, London.

Berendes, H. D. (1972). The control of puffing in *Drosophila hydei*. *In* "Developmental Studies on Giant Chromosomes" (Ed. W. Beermann), pp. 181-207. Springer-Verlag, Berlin, Heidelberg and New York.

Berger, C. A. (1936). Observations on the relation between salivary gland chromosomes and multiple chromosome complexes. *Proc. natn. Acad. Sci. U.S.A.* 22, 186-187.

Berger, C. A. (1937). Additional evidence of repeated chromosome division without mitotic activity. *Am. Nat.* 71, 187-190.

Berger, C. A. (1938). Multiplication and reduction of somatic chromosome groups as a regular developmental process in the mosquito, *Culex pipiens*. *Publ. Carnegie Instn*, 476, 209-232.

Berger, E. and Canter, R. (1973). The esterases of *Drosophila*. I. The anodal esterases and their possible role in eclosion. *Devl. Biol.* 33, 48-55.

Bhakthan, N. M. G. and Gilbert, L. I. (1972). Studies on the cytophysiology of the fat body of the American silkmoth. *Z. Zellforsch. mikrosk. Anat.* 124, 433-444.

Bier, K. (1957). Endomitose und Polytänie in den Nahrzellkernen von *Calliphora erythrocephala* Meigen. *Chromosoma* (*Berlin*), **8**, 493-522.

Bier, K. (1959). Quantitative Untersuchungen über die Variabilität der Nahrzellkernstruktur und ihre Beeinflussung durch die Temperatur. *Chromosoma* (*Berlin*), **10**, 619-653.

Birt, L. M. and Christian, B. (1969). Changes in nitrogenous compounds during the metamorphosis of the blowfly *Lucilia cuprina*. *J. Insect Physiol.* **15**, 711-719.

Bishop, G. H. (1923). Cell metabolism in the insect fat body. II. A functional interpretation of the changes in structure in the fat-body cells of the honey-bee. *J. Morph.* **37**, 533-553.

Bodenstein, D. (1950). The postembryonic development of *Drosophila*. *In* "The Biology of *Drosophila*" (Ed. M. Demerec), pp. 275-367. John Wiley, New York.

Bodnaryk, R. P. (1970a). Biosynthesis of gamma L-glutamyl-L-phenylalanine by the larva of the housefly *Musca domestica*. *J. Insect Physiol.* **16**, 919-929.

Bodnaryk, R. P. (1970b). Effect of dopa decarboxylase inhibition on the metabolism of β-alanyl-L-tyrosine during puparium formation in the fleshfly *Sarcophaga bullata* Parker. *Comp. Biochem. Physiol.* **35**, 221-227.

Bodnaryk, R. P. (1971a). Studies on the incorporation of β-alanine into the puparium of the fly, *Sarcophaga bullata*. *J. Insect. Physiol.* **17**, 1201-1210.

Bodnaryk, R. P. (1971b). N-Terminal β-alanine in the puparium of the fly *Sarcophaga bullata*: Evidence from kinetic studies of its release by partial acid hydrolysis. *Insect Biochem.* **1**, 228-336.

Bodnaryk, R. P. and Levenbook, L. (1969). The role of β-alanyl-L-tyrosine (sarcophagine) puparium formation in the fleshfly *Sarcophaga bullata*. *Comp. Biochem. Physiol.* **30**, 909-921.

Bodnaryk, R. P. and Morrison, P. E. (1968). Immunochemical analysis of the origin of a sex-specific accumulated blood protein in female houseflies. *J. Insect Physiol.* **14**, 1141-1146.

Bodnaryk, R. P. and Skillings, J. R. (1971). γ-Glutamyl transpeptidase catalyses the synthesis of γ-glutamylphenylalanine in the larva of the housefly *Musca domestica*. *Insect Biochem.* **1**, 467-479.

Bonner, J. and Wu, J.-R. (1973). A proposal for the structure of the *Drosophila* genome. *Proc. natn Acad. Sci. U.S.A.* **70**, 535-537.

Boshes, R. A. (1970). *Drosophila* ribosomes. The characterisation of two populations by cell fractionation and isotopic labeling with nucleic acid and protein precursors. *J. Cell Biol.* **46**, 477-490.

Boyd, J. B. and Boyd, H. (1970). Deoxyribonucleases of the organs of *Drosophila hydei* at the onset of metamorphosis. *Biochem. Genetics*, **4**, 447-459.

Boyd, J. B. and Mitchell, H. K. (1966). Turnover of the haemolymph proteins of *Drosophila melanogaster*. *Archs Biochem. Biophys.* **117**, 310-319.

Braun, V., Crichton, R. R. and Braunitzer, G. (1968). Über monomere und dimere Insektenhämoglobine (*Chironomus thummi*). *Hoppe-Seyler's Z. physiol. Chem.* **349**, 45-53.

Braunitzer, G. (1965). (With Braun, V., Hilse, K., Hobom, G., Rudloff, V. and Wettstein, G. von) Constancy and variability of structure in respiratory and viral proteins. *In* "Evolving Genes and Proteins" (Eds V. Bryson and H. J. Vogel), pp. 183-192. Academic Press, New York and London.

Braunitzer, G. and Braun, V. (1965). Untersuchungen an Insekten-Hämoglobinen (*Chironomus thummi*). *Hoppe-Seyler's Z. physiol. Chem.* 340, 88-91.

Breuer, M. E. and Pavan, C. (1955). Behaviour of polytene chromosomes of *Rhynchosciara angelae* at different stages of development. *Chromosoma (Berlin)*, 7, 371-386.

Brew, K., Vanaman, T. C. and Hill, R. L. (1968). The role of α-lactalbumin and the A protein in lactose synthetase: a unique mechanism for the control of a biological reaction. *Proc. natn Acad. Sci. U.S.A.* 59, 491-497.

Bricteux-Grégoire, S., Verly, W. G. and Florkin, M. (1957). Protein synthesis in *Sphinx ligustri* pupae. *Nature*, 179, 678-679.

Bultmann, H. and Clever, U. (1969). Chromosomal control of footpad development in *Sarcophaga bullata*. I. The puffing pattern. *Chromosoma (Berlin)*, 28, 120-135

Bultmann, H. and Clever, U. (1970). Chromosomal control of footpad development in *Sarcophaga bullata*. II. Cuticle formation and tanning. *Devl. Biol.* 22, 601-621.

Buse, G., Braig, S. and Braunitzer, G. (1969). Die Konstitution einer Hämoglobin-Komponente (Erythrocruorin) der Insekten (*Chironomus thummi thummi*, Diptera). *Hoppe-Seyler's Z. physiol. Chem.* 349, 1686-1690.

Butler, J. E. and Leone, C. A. (1966). Antigenic changes during the life cycle of the beetle *Tenebrio molitor*. *Comp. Biochem. Physiol.* 19, 699-711.

Butterworth, F. M. (1972). Adipose tissue of *Drosophila melanogaster*. V. Genetic and experimental studies of an extrinsic influence on the rate of cell death in the larval fat body. *Devl. Biol.* 28, 311-325.

Butterworth, F. M. (1973). Adipose tissue of *Drosophila melanogaster*. VI. Non-susceptibility of the immature larval fat body to the lytic environment of the young adult. *Wilhelm Roux Arch. EntwMech.* 172, 263-270.

Butterworth, F. M. and Bodenstein, D. (1967). Adipose tissue of *Drosophila melanogaster*. II. The effect of the adult internal environment on growth, protein deposition and histolysis in the larval fat body. *J. exp. Zool.* 164, 251-266.

Butterworth, F. M. and Bodenstein, D. (1969). Adipose tissue of *Drosophila melanogaster*. IV. The effect of the corpus allatum and juvenile hormone on the tissue of the adult male. *Gen. comp. Endocr.* 13, 68-74.

Butterworth, F. M. and La Tendresse, B. L. (1973). Quantitative studies of cytochemical and cytological changes during cell death in the larval fat body of *Drosophila melanogaster*. *J. Insect Physiol.* 19, 1487-1499.

Butterworth, F. M., Bodenstein, D. and King, R. C. (1965). Adipose tissue of *Drosophila melanogaster*. I. An experimental study of larval fat body. *J. exp. Zool.* 158, 141-153.

Buys, K. S. (1924). Adipose tissue in insects. *J. Morph.* 38, 485-527.

Campbell, A. J. (1972). Soluble protein synthesis during the adult development of *Lucilia* (Diptera: Calliphoridae). *Abstracts,* 14 *Int. Congr. Entomol.*, pp. 180-181, Canberra.

Campbell, A. J. (1973). "Biosynthesis of Thoracic Protein during the Metamorphosis of *Lucilia cuprina*". Ph.D. Thesis, Australian National University.

Campbell, A. J. and Birt, L. M. (1972). Studies on the appearance of soluble α-glycerophosphate dehydrogenase activity during the development of the sheep blowfly *Lucilia*. *Insect Biochem.* 2, 279-296.

Campbell, A. J. and Birt, L. M. (1975). Thoracic muscle protein biosynthesis in *Lucilia cuprina. Insect Biochem.* In press.

Chan, S. K. and Margoliash, E. (1966). Biosynthesis of cytochrome *c* in developing pupae of *Samia cynthia. J. biol. Chem.* **241**, 2252-2255.

Chan, S. K. and Reibling, A. (1973). Cell-free synthesis of myosin and the regulation of the ribosomal activities in the developing thoracic muscle of the tobacco hornworm. *Insect Biochem.* **3**, 317-324.

Chapman, R. F. (1969). "The Insects, Structure and Function". English Universities Press, London.

Chefurka, W. (1964). Intermediary metabolism of nitrogenous and lipid compounds in insects. *In* "The Physiology of Insecta" (Ed. M. Rockstein), Vol. 2, pp. 669-768. Academic Press, New York and London.

Chen, P. S. (1959). Studies on the protein metabolism of *Culex pipiens* L. III. Comparative analysis of the protein contents in the larval haemolymph of autogenous and anautogenous forms. *J. Insect Physiol.* **3**, 335-344.

Chen, P. S. (1967). Electrophoretic patterns of larval haemolymph-proteins in autogenous and anautogenous forms of *Culex pipiens* L. *Nature*, **215**, 316-317.

Chen, P. S. (1971). Biochemical aspects of insect development. "Monographs in Developmental Biology", Vol. 3. Karger, Basel.

Chen, P. S. and Kühn, A. (1956). Vergleichende Untersuchung der freien Aminosäuren und Peptide während den Raupen- und Puppenentwicklung verschiedenen Genotypen von *Ephestia kühniella. Z. Naturf.* **11b**, 305-314.

Chen, P. S. and Levenbook, L. (1966a). Studies on the haemolymph proteins of the blowfly, *Phormia regina*—I. Changes in ontogenetic patterns. *J. Insect Physiol.* **12**, 1595-1609.

Chen, P. S. and Levenbook. L. (1966b). Studies on the haemolymph proteins of the blowfly, *Phormia regina*—II. Synthesis and breakdown as revealed by isotopic labelling. *J. Insect Physiol.* **12**, 1611-1627.

Childress, D. C. (1969). Polytene chromosomes and linkage group—chromosome correlations in the Australian sheep blowfly *Lucilia cuprina* (Diptera: Calliphoridae). *Chromosoma (Berlin)*, **26**, 208-214.

Chinzei, Y. and Tojo, S. (1972). Nucleic acid changes in the whole body and several organs of the silkworm. *Bombyx mori* during metamorphosis. *J. Insect Physiol.* **18**, 1683-1698.

Chippendale, G. M. (1970a). Metamorphic changes in fat body proteins of the southwestern corn borer, *Diatraea grandiosella. J. Insect Physiol.* **16**, 1057-1068.

Chippendale, G. M. (1970b). Metamorphic changes in haemolymph and midgut proteins of the southwestern corn borer, *Diatraea grandiosella. J. Insect Physiol.* **16**, 1909-1920.

Chippendale, G. M. (1971). Selective protein storage by the fat-body of the Angoumois grain moth, *Sitotroga cerealella. Insect Biochem.* **1**, 122-124.

Chippendale, G. M. and Kilby, B. A. (1969). Relationship between the proteins of the haemolymph and fat body during development of *Pieris brassicae. J. Insect Physiol.* **15**, 905-926.

Chippendale, G. M. and Kilby, B. A. (1970). Protein biosynthesis in larvae of the large white butterfly, *Pieris brassicae. Comp. Biochem. Physiol.* **34**, 241-243.

Ciavattini, M. G., Rimatori, A. and Sileoni, P. (1959). Proteinasi e lipasi nello sviluppo larvale e nella metamorfosi di *Musca domestica* L. *Ric. Sci.* 29, 1874-1879.

Clements, A. (1963). "The Physiology of Mosquitoes". Pergamon Press, Oxford, London, New York and Paris.

Clever, U. (1966). Gene activity patterns and cellular differentiation. *Am. Zool.* 6, 33-41.

Clever, U. (1969). Chromosome activity and cell function in polytenic cells. II. The formation of secretion in the salivary glands of *Chironomus. Expl. Cell. Res.* 55, 317-322.

Clever, U. (1972). The control of cellular growth and death in the development of an insect. *In* "Molecular Genetic Mechanisms in Development and Aging" (Eds M. Rockstein and G. T. Baker III), pp. 33-69. Academic Press, New York and London.

Clever, U., Storbeck, I. and Romball, C. G. (1969). Chromosome activity and cell function in polytenic cells. I. Protein synthesis at various stages of larval development. *Expl. Cell Res.* 55, 306-316.

Clever, U., Clever, I., Storbeck, I. and Young, N. L. (1973). The apparent requirement of two hormones, α- and β-ecdysone, for molting induction in insects. *Devl Biol.* 31, 47-60.

Collins, J. V. (1969). The hormonal control of fat body development in *Calpodes ethlius* (Lepidoptera, Hesperiidae). *J. Insect Physiol.* 15, 341-352.

Collins, J. V. and Downe, A. E. R. (1970). Selective accumulation of haemolymph proteins by the fat body of *Galleria mellonella. J. Insect Physiol.* 16, 1697-1708.

Cölln, K. (1973). Über die Metamorphose der Proteinspektren von Hämolymphe und Fettkörper bei *Ephestia kühniella* Z. *Wilhelm Roux Arch. EntwMech.* 172, 231-257.

Counce, S. J. (1973). The causal analysis of insect embryogenesis. *In* "Developmental Systems: Insects" (Eds S. J. Counce and C. H. Waddington), Vol. 2, pp. 1-156. Academic Press, London and New York.

Crick, F. (1971). General model for the chromosomes of higher organisms. *Nature,* 234, 25-27.

Crompton, M. and Birt, L. M. (1967). Changes in the amounts of carbohydrates, phosphagen, and related compounds during the metamorphosis of the blowfly *Lucilia cuprina. J. Insect Physiol.* 13, 1575-1592.

Crossley, A. C. (1965). Transformations in the abdominal muscles of the blue blowfly *Calliphora erythrocephala* (Meig.). *J. Embryol. exp. Morph.* 14, 89-110.

Crossley, A. C. (1968). The fine structure and mechanism of breakdown of larval intersegmental muscles in the blowfly, *Calliphora erythrocephala. J. Insect Physiol.* 14, 1389-1407.

Crossley, A. C. (1972a). Ultrastructural changes during transition of larval to adult intersegmental muscle at metamorphosis in the blowfly *Calliphora erythrocephala.* I. Redifferentiation and myoblast fusion. *J. Embryol. exp. Morph.* 27, 43-74.

Crossley, A. C. (1972b). Ultrastructural changes during transition of larval to adult intersegmental muscle at metamorphosis in the blowfly *Calliphora erythrocephala.* II. The formation of adult muscle. *J. Embryol. exp. Morph.* 27, 75-101.

da Cunha, A. B., Morgante, J. S., Pavan, C. and Garrido, M. C. (1969). I. Studies on cytology and differentiation in Sciaridae. III. Nuclear and cytoplasmic differentiation in the salivary glands of *Bradysia* sp. *Studies in Genetics V., Univ. Texas Publ.* 6918, 1-11.

da Cunha, A. B., Pavan, C., Morgante, J. S. and Garrido, M. C. (1969). Studies on cytology and differentiation in Sciaridae. II. DNA redundancy in salivary gland cells of *Hybosciara fragilis* (Diptera, Sciaridae). *Genetics Suppl.* 61, 335-349.

Dahlhelm, D. (1967). Die Entwicklung des Fettkörpers bei der Larve von *Calliphora erythrocephala* Meig. *Biol. Zbl.* 86, 273-301.

Daneholt, B. and Edström, J.-E. (1967). The content of deoxyribonucleic acid in individual polytene chromosomes of *Chironomus tentans. Cytogenetics*, 6, 350-356.

Dapples, C. C. and King, R. C. (1970). The development of the nucleolus of the ovarian nurse cell of *Drosophila melanogaster. Z. Zellforsch. mikrosk. Anat.* 103, 34-47.

Day, M. F. (1943). The function of the corpus allatum in muscoid Diptera. *Biol. Bull. mar. biol. Lab., Woods Hole*, 84, 127-140.

D'Costa, M. A. and Birt, L. M. (1966). Changes in the lipid content during the metamorphosis of the blowfly, *Lucilia. J. Insect Physiol.* 12, 1377-1394.

de Bianchi, A. G., Terra, W. R. and Lara, F. J. S. (1973). Formation of salivary secretion in *Rhynchosciara americana.* I. Kinetics of labelled amino acid incorporation. *J. Cell Biol.* 58, 470-476.

de Loof, A. (1969). Hormonal control of the synthesis of an important vitellogenic hemolymph protein in the Colorado beetle, *Leptinotarsa decemlineata* Say (Abstract). *Gen. comp. Endocr.* 13, 518.

de Loof, A. and Lagasse, A. (1970). Juvenile hormone and the ultrastructural properties of the fat body of the adult Colorado beetle, *Leptinotarsa decemlineata* Say. *Z. Zellforsch. mikrosk. Anat.* 106, 439-450.

de Loof, A. and de Wilde, J. (1970a). Hormonal control of synthesis of vitellogenic female protein in the Colorado beetle, *Leptinotarsa decemlineata. J. Insect Physiol.* 16, 1455-1466.

de Loof, A. and de Wilde, J. (1970b). The relation between haemolymph proteins and vitellogenesis in the Colorado beetle, *Leptinotarsa decemlineata. J. Insect Physiol.* 16, 157-169.

Dickson, E., Boyd, J. B. and Laird, C. D. (1971). Sequence diversity of polytene chromosome DNA from *Drosophila hydei. J. molec. Biol.* 61, 615-627.

Dinamarca, M. L. and Levenbook, L. (1966). Oxidation, utilization, and incorporation into protein of alanine and lysine during metamorphosis of the blowfly, *Phormia regina* (Meigen). *Archs Biochem. Biophys.* 117, 110-119.

Doane, W. W. (1961). Developmental physiology of the mutant female sterile (2) adipose of *Drosophila melanogaster.* III. Corpus allatum complex and ovarian transplantation. *J. exp. Zool.* 146, 275-298.

Doane, W. W. (1973). Role of hormones in insect development. *In* "Developmental Systems: Insects" (Eds S. J. Counce and C. H. Waddington), Vol. 2, pp. 291-497. Academic Press, London and New York.

Doenecke, D., Marmaras, V. J. and Sekeris, C. E. (1972). Increased RNase H (Hybridase) activity in the integument of blowfly larvae during development and under the influence of β-ecdysone. *FEBS Letters*, 22, 261-264.

Doenecke, D., Marmaras, V. J. and Sekeris, C. E. (1973). DNA-dependent RNA polymerases from the integument of blowfly larvae during development and after β-ecdysone action. *Insect Biochem.* 3, 283-287.

Doyle, D. and Laufer, H. (1969). Sources of larval salivary gland secretion in the dipteran *Chironomus tentans*. *J. Cell Biol.* **40**, 61-78.

Duke, E. J. (1966). Further studies on the inheritance of lymph protein in *Drosophila melanogaster*. *Genet. Res. (Camb.)* 7, 287-294.

Duke, E. J. and Pantelouris, E. M. (1963). Ontogenesis of lymph proteins in *Drosophila melanogaster*. *Comp. Biochem. Physiol.* **10**, 351-355.

Egelhaaf, A. (1965a). Zur Genetik und Physiologie den Hämolymph-Proteine von *Ephestia kühniella*. *Verh. dt. zool. Ges.* 256-261.

Egelhaaf, A. (1965b). Zur Genetik stadienspezifischer Proteine bei *Ephestia kühniella*. *Z. Vererbungsl.* 97, 150-156.

Elston, R. C. and Glassman, E. (1967). An approach to the problem of whether clustering of functionally related genes occurs in higher organisms. *Genet. Res. (Camb.)* 9, 141-147.

Engelmann, F., Hill, L. and Wilkens, J. L. (1971). Juvenile hormone control of female specific protein synthesis in *Leucophaea maderae*, *Schistocerca vaga* and *Sarcophaga bullata*. *J. Insect Physiol.* 17, 2179-2191.

English, D. (1969). Ontogenetic changes in hemoglobin synthesis of two strains of *Chironomus tentans*. *J. Embryol. exp. Morph.* 22, 465-476.

Evans, J. J. T. (1967). Development and ultrastructure of the fat body cells and oenocytes of the Queensland fruit fly, *Dacus tryoni* (Frogg.). *Z. Zellforsch. mikrosk. Anat.* 81, 49-61.

Faulkner, P. and Bheemeswar, B. (1960). Studies on the biosynthesis of proteins in the silkworm, *Bombyx mori*. *Biochem. J.* 76, 71-78.

Fox, D. J., Abächerli, E. and Ursprung, H. (1971). *Drosophila* enzyme-genetics. A table. *Experientia*, 27, 218-220.

Fox, D. P. (1970). DNA values in somatic tissues of *Dermestes* (Dermestidae: Coleoptera). II. Malpighian tubules of the adult male. *Chromosoma (Berlin)*, 31, 321-330.

Fox, D. P. (1972). DNA content of related species. *Chromosomes Today*, 3, 32-7.

Fraenkel, G. and Hsiao, C. (1968). Morphological and endocrinological aspects of pupal diapause in a fleshfly, *Sarcophaga argyrostoma*. *J. Insect Physiol.* 14, 707-718.

Fristrom, D. (1969). Cellular degeneration in the production of some mutant phenotypes in *Drosophila melanogaster*. *Molec. gen. Genet.* **103**, 363-379.

Fristrom, D. (1972). Chemical modification of cell death in the *Bar* eye of *Drosophila*. *Molec. gen. Genet.* **115**, 10-18.

Fristrom, J. W. (1970). The developmental biology of *Drosophila*. *A. Rev. Ent.* 4, 325-346.

Fristrom, J. W. (1972). The biochemistry of imaginal disk development. *In* "The Biology of Imaginal Disks" (Eds H. Ursprung and R. Nöthiger), pp. 109-154. Springer-Verlag, Berlin, Heidelberg and New York.

Gabrusewycz-Garcia, N. (1972). Further studies of the nucleolar material in salivary gland nuclei of *Sciara coprophila*. *Chromosoma (Berlin)*, 38, 237-254.

Gabrusewycz-Garcia, N. and Kleinfeld, R. G. (1966). A study of the nucleolar material in *Sciara coprophila. J. Cell Biol.* **29**, 347-359.

Gall, J. G., Cohen, E. H. and Polan, M. L. (1971). Repetitive DNA sequences in *Drosophila. Chromosoma (Berlin)*, 33, 319-344.

Gall, J. G., Macgregor, H. C. and Kidston, M. E. (1969). Gene amplification in the oocytes of dytiscid water beetles. *Chromosoma (Berlin)*, 26, 169-187.

Gamo, T. (1968). The inheritance of electrophoretic patterns of blood albumin in the silkworm, *Bombyx mori*, L. *Jap. J. Genet.* 43, 271-277.

Gaudecker, B. von (1963). Über den Formwechsel einigen Zellorganelle bei der Bildung der Reservestoffe im Fettkörper von *Drosophila*–Larven. *Z. Zellforsch. mikrosk. Anat.* 61, 56-95.

Gaudecker, B. von (1972). Der Strukturwandel der larvalen Speicheldrüse von *Drosophila melanogaster.* Ein Beitrag zur Frage nach der Steuernden Wirkung aktiver Gene auf das Cytoplasma. *Z. Zellforsch. mikrosk. Anat.* 127, 50-86.

Gay, H., Das, C. C., Forward, K. and Kaufmann, B. P. (1970). DNA content of mitotically-active condensed chromosomes of *Drosophila melanogaster. Chromosoma (Berlin)*, 32, 213-223.

Gehring, W. J. and Nöthiger, R. (1973). The imaginal discs of *Drosophila. In* "Developmental Systems: Insects" (Eds S. J. Counce and C. H. Waddington), Vol. 2, pp. 211-290. Academic Press, London and New York.

Geitler, L. (1953). "Endomitose und endomitotische Polyploidiserung". Protoplasmatologia. Handbuch der Protoplasmaforschung. VI. c, pp. 1-89. Springer-Verlag, Berlin.

Gilbert, L. I. (1967). Lipid metabolism and function in insects. *In* "Advances in Insect Physiology" (Eds J. W. L. Beament, J. E. Treherne and V. B. Wigglesworth), Vol. 4, pp. 69-211. Academic Press, London and New York.

Gilmour, D. (1961). "The Biochemistry of Insects". Academic Press, New York and London.

Goldberg, E., Whitten, J. M. and Gilbert, L. I. (1969). Changes in soluble proteins during foot-pad development in *Sarcophaga bullata. J. Insect Physiol.* 15, 409-420.

Grossbach, U. (1969). Chromosomen-Aktivität und biochemische Zelldifferenzierung in den Speicheldrüsen von *Camptochironomus. Chromosoma (Berlin)*, 28, 136-187.

Hackman, R. H. (1965). Changes in the free amino acids of the blood of blowfly larvae at metamorphosis. *Aust. J. Biol. Sci.* 9, 400-405.

Hackman, R. H. and Goldberg, M. (1971). Studies on the hardening and darkening of insect cuticles. *J. Insect Physiol.* 17, 335-347.

Hagedorn, H. H. and Fallon, A. M. (1973). Ovarian control of vitellogenin synthesis by the fat body in *Aedes aegypti. Nature*, **244**, 103-105.

Hagedorn, H. H. and Judson, C. L. (1972). Purification and site of synthesis of *Aedes aegypti* yolk proteins. *J. exp. Zool.* **182**, 367-378.

Hagedorn, H. H., Fallon, A. M. and Lauger, H. (1973). Vitellogenin synthesis by the fat body of *Aedes aegypti.* I. Evidence for transcriptional control. *Devl Biol.* 31, 285-294.

Harlow, P. M. (1956). A study of ovarial development and its relation to adult nutrition in the blowfly, *Protophormia terranovae. J. exp. Biol.* 33, 777-797.

Hennig, W. and Meer, B. (1971). Reduced polyteny of ribosomal RNA cistrons in giant chromosomes of *Drosophila hydei. Nature New Biol.* **233**, 70-72.

Henrikson, P. A. and Clever, U. (1972). Protease activity and cell death during metamorphosis in the salivary gland of *Chironomus tentans. J. Insect Physiol.* **18**, 1981-2004.

Hess, O. (1965a). Struktur-differenzierung im Y-Chromosom von *Drosophila hydei* und ihre Beziehungen zu Gen-Aktivitäten. III. Sequenz und Lokalisation der Schleifenbildungsorte. *Chromosoma (Berlin)*, **16**, 222-248.

Hess, O. (1965b). Struktur-Differenzierungen im Y-Chromosom von *Drosophila hydei* und ihre Beziehungen zu Gen-Aktivitäten. I. Mutanten der Funktionsstrukturen. *Ver. dt. zool. Ges., Zool. Anz. Suppl.* **28**, 156-163.

Hess, O. and Meyer, G. F. (1963). Chromosomal differentiations of the lamp-brush type formed by the Y-chromosome in *Drosophila hydei* and *Drosophila neohydei. J. Cell Biol.* **16**, 527-539.

Hinton, H. E. (1958). Concealed phases in the metamorphosis of insects. *Sci. Prog., Lond.* **46**, 260-275.

Hinton, H. E. (1963). The origin and function of the pupal stage. *Proc. R. ent. Soc.*, Lond. (A) **38**, 77-85.

Hinton, H. E. and Mackerras, I. M. (1970). Reproduction and metamorphosis. *In* "The Insects of Australia" (Division of Entomology, CSIRO, ed.), pp. 83-106. Melbourne University Press, Carlton.

Ilan, J. and Ilan, J. (1971). Stage-specific initiation factors for protein synthesis during insect development. *Devl Biol.* **25**, 280-292.

Ilan, J. and Ilan, J. (1973). Protein synthesis and insect morphogenesis. *A. Rev. Ent.* **18**, 167-182.

Ilan, J., Ilan, J. and Quastel, J. H. (1966). Effects of actinomycin D on nucleic acid metabolism and protein biosynthesis during metamorphosis of *Tenebrio molitor* L. *Biochem. J.* **100**, 441-447.

Ishizaki, H. (1965). Electron microscopic study of changes in the subcellular organization during metamorphosis of the fat-body cell of *Philosamia cynthia ricini* (Lepidoptera). *J. Insect Physiol.* **11**, 845-855.

Iuga, V. G. (1935). La phagocytose du tissue trophocytaire pendant la nymphose chez *Chironomus genuinus.* Evacuation des phagocytes chargé de biliverdine. *C.r. Séanc. Soc. Biol.* **118**, 603-605.

Jacob, J. and Danieli, G. A. (1970). Replication of intranucleolar DNA in *Smittia* (Diptera, Chironomidae). *Experientia*, **26**, 1390-1393.

Jacob, J. and Sirlin, J. L. (1963). Electron microscope studies on salivary gland cells. I. The nucleus of *Bradysia mycorum* Frey (Sciaridae), with special reference to the nucleolus. *J. Cell Biol.* **17**, 153-165.

Janda, V. and Krieg, P. (1969). Proteolytische Aktivität des Mitteldarms von *Galleria mellonella* im Zusammenhang mit Wachstum und Metamorphose. *Z. vergl. Physiol.* **64**, 288-300.

Judd, B. H., Shen, M. W. and Kaufman, T. C. (1972). The anatomy and function of a segment of the X chromosome of *Drosophila melanogaster. Genetics, Princeton,* **71**, 139-156.

Judy, K. J. and Marks, E. P. (1971). Effects of ecdysterone *in vitro* on hindgut and hemocytes of *Manduca sexta* (Lepidoptera). *Gen. comp. Endocr.* **17**, 351-359.

Jura, C. (1972). Development of apterygote insects. *In* "Developmental Systems: Insects" (Eds S. J. Counce and C. H. Waddington), Vol. 1, pp. 49-94. Academic Press, London and New York.

Kafatos, F. C. (1970). Cocoonase synthesis: cellular differentiation in developing silk moths. *In* "Problems in Biology: RNA in Development" (Ed. E. W. Hanly), pp. 111-144. University of Utah Press, Salt Lake City.

Kafatos, F. C. (1972a). The cocoonase zymogen cells of silk moths: A model of terminal cell differentiation for specific protein synthesis. *In* "Current Topics in Developmental Biology" (Eds A. A. Moscona and A. Monroy), Vol. 7, pp. 125-191. Academic Press, New York and London.

Kafatos, F. C. (1972b). mRNA stability and cellular differentiation. *In* "Karolinska Symposia on Research Methods in Reproductive Endocrinology. 5th Symposium. Gene Transcription in Reproductive Tissue" (Ed. E. Diczfalusy), pp. 319-345. Karolinska Institutet, Stockholm.

Kai, H. and Hasegawa, H. (1971). Studies on the mode of action of the diapause hormone with special reference to the protein metabolism in the silkworm, *Bombyx mori* L. I. The diapause hormone and the protein soluble in ethanol containing trichloroacetic acid in mature eggs of adult ovaries. *J. seric. Sci.* (*Tokyo*), **40**, 199-208.

Karnavar, G. K. and Nair, K. S. S. (1968). Observations on the cytology and cytochemistry of the fat body of normal and diapausing larvae of *Trogoderma granarium* Everts (Col., Dermestidae). *J. Anim. Morph. Physiol.* **15**, 153-161.

Katzenellenbogen, B. S. and Kafatos, F. C. (1970). Some properties of the silkmoth moulting gel and moulting fluid. *J. Insect Physiol.* **16**, 2241-2256.

Katzenellenbogen, B. S. and Kafatos, F. C. (1971). Inactive proteinases in silkmoth moulting gel. *J. Insect Physiol.* **17**, 823-832.

Kilby, B. A. (1963). The biochemistry of the insect fat body. *In* "Advances in Insect Physiology" (Eds J. W. L. Beament, J. E. Treherne and V. B. Wigglesworth), Vol. 1, pp. 111-174. Academic Press, London and New York.

King, P. E., Rafai, J. and Richards, J. G. (1972). Formation of yolk protein in the eggs of a parasitoid hymenopteran, *Nasonia vitripennis* (Walker) (Pteromalidae: Hym.). *Z. Zellforsch. mikrosk. Anat.* **123**, 330-336.

Kinnear, J. F. (1973). The origin and inter-relationships of larval and imaginal proteins in *Calliphora*: a contribution to the study of gene action in insect metamorphosis. Ph.D. Thesis, University of Melbourne.

Kinnear, J. F. and Thomson, J. A. (1975). Nature, origin and fate of major haemolymph proteins in *Calliphora. Insect Biochem.* In press.

Kinnear, J. F., Martin, M.-D., and Thomson, J. A. (1971). Developmental changes in the late larva of *Calliphora stygia*. III. The occurrence and synthesis of specific tissue proteins. *Aust. J. biol. Sci.* **24**, 275-289.

Kinnear, J. F., Martin, M.-D., Thomson, J. A. and Neufeld, G. J. (1968). Developmental changes in the late larva of *Calliphora stygia*. I. Haemolymph. *Aust. J. biol. Sci.* **21**, 1033-1045.

Kirrmann, A. (1930). Le groupe prosthétique de l'hémoglobine de chironome. *Bull. Soc. Chim. biol.* 12, 1146.

Knowles, B. and Fristrom, J. W. (1967). The electrophoretic behaviour of ten enzyme systems in the larval integument of *Drosophila melanogaster. J. Insect Physiol.* 13, 731-737.

Koeppe, J. K. and Gilbert, L. I. (1973). Immunochemical evidence for the transport of haemolymph protein into the cuticle of *Manduca sexta. J. Insect Physiol.* 19, 615-624.

Kominz, D. R., Maruyama, K., Levenbook, L. and Lewis, M. (1962). Tropomyosin, myosin and actin from the blowfly *Phormia regina. Biochim. biophys. Acta,* 63, 106-116.

Laird, C. D. and McCarthy, B. J. (1969). Molecular characterization of the *Drosophila* genome. *Genetics, Princeton,* 63, 865-882.

Lambertsson, A. G., Rasmuson, S. B. and Bloom, G. D. (1970). The ribosomal proteins of *Drosophila melanogaster.* I. Characterization in polyacrylamide gel of proteins from larval, adult and ammonium chloride-treated ribosomes. *Molec. Gen. Genet.* **108**, 349-357.

Lane, N. J., Carter, Y. R. and Ashburner, M. (1972). Puffs and salivary gland function: The fine structure of the larval and pre-pupal salivary glands of *Drosophila melanogaster. Wilhelm Roux Arch. EntwMech.* **169**, 216-238.

Lang, C. A. (1972). A DNA replication intermediate in mosquito development. *In* "Molecular Genetic Mechanisms in Development and Aging" (Eds M. Rockstein and G. T. Baker III), pp. 71-78. Academic Press, New York and London.

Lang, C. A. and Meins, F. (1966). A soluble deoxyribonucleic acid in the mosquito *Aedes aegypti. Proc. natn. Acad. Sci. U.S.A.* 55, 1525-1531.

Laufer, H. (1960). Blood proteins in insect development. *Ann. N.Y. Acad. Sci.* 89, 490-515.

Laufer, H. and Schin, K. S. (1971). Quantitative studies of hydrolytic enzyme activity in the salivary gland of *Chironomus tentans* (Diptera. Chironomidae) during metamorphosis. *Can. Ent.* 103, 454-457.

Laverdure, A.-M. (1971). Étude électrophorétique des variations du protéinogramme au cours du développement post embryonnaire de *Tenebrio molitor* (Coleoptère). Étude comparée du cas de *Galleria mellonella* (Lépidoptère). *Annls Endocr.* 31, 504-515.

Lawrence, P. A. (1970). Polarity and patterns in the postembryonic development of insects. *In* "Advances in Insect Physiology" (Eds J. W. L. Beament, J. W. Treherne and V. B. Wigglesworth), Vol. 7, pp. 197-266. Academic Press, London and New York.

Lawrence, P. A. (1973). The development of spatial patterns in the integument of insects. *In* "Developmental Systems: Insects" (Eds S. J. Counce and C. H. Waddington), Vol. 2, pp. 157-209. Academic Press, London and New York.

Lensky, Y. (1971a). Haemolymph proteins of the honey bee. I. Separation and characterization of haemolymph proteins of worker larvae. *Comp. Biochem. Physiol.* **38B**, 129-139.

Lensky, Y. (1971b). Haemolymph proteins of the honey bee. II. Differentiation during the development of bee workers. *Comp. Biochem. Physiol.* **39B**, 325-341.

Lensky, Y. and Rakover, Y. (1972). Resorption of moulting fluid proteins during the ecdysis of the honey bee. *Comp. Biochem. Physiol.* **41B**, 521-531.

Levenbook, L. (1966). Haemolymph amino acids and peptides during larval growth of the blowfly *Phormia regina. Comp. Biochem. Physiol.* **18**, 341-351.

Levenbook, L., Bodnaryk, R. P. and Spande, T. F. (1969). β-Alanyl-L-tyrosine. Chemical synthesis, properties and occurrence in larvae of the fleshfly, *Sarcophaga bullata* Parker. *Biochem. J.* **113**, 837-841.

Levenbook, L., Bauer, A. C. and Shigematsu, H. (1973). Fructose diphosphate aldolase during growth and development of the blowfly *Phormia regina* (Meigen.). III. Physicochemical comparison of the larval and adult muscle enzymes. *Archs Biochem. Biophys.* **157**, 625-631.

Lezzi, M. (1967). RNS- und Protein-Synthese in Puffs isolierter Speicheldrüsenchromosomen von *Chironomus. Chromosoma (Berlin)*, **21**, 72-88.

Lindsley, D. L., Sandler, L., Baker, B. S., Carpenter, T. C., Denell, R. E., Hall, J. C., Jacobs, P. A., Miklos, G. L., Davis, B. K., Gethmann, R. C., Hardy, R. W., Hessler, A., Miller, S. M., Nozawa, H., Parry, D. M. and Gould-Somero, M. (1972). Segmental aneuploidy and the genetic gross structure of the *Drosophila* genome. *Genetics, Princeton,* **71**, 157-184.

Locke, M. and Collins, J. V. (1965). The structure and formation of protein granules in the fat body of an insect. *J. Cell Biol.* **26**, 857-884.

Locke, M. and Collins, J. V. (1966). Sequestration of protein by the fat body of an insect. *Nature, Lond.* **210**, 552-553.

Locke, M. and Collins, J. V. (1967). Protein uptake in multivesicular bodies in the molt-intermolt cycle of an insect. *Science, N.Y.* **155**, 467-469.

Locke, M. and Collins, J. V. (1968). Protein uptake into multivesicular bodies and storage granules in the fat body of an insect. *J. Cell Biol.* **36**, 453-483.

Lockshin, R. A. (1969a). Lysosomes in insects. *In* "Lysosomes in Biology and Pathology. 1" (Eds J. T. Dingle and H. B. Fell), pp. 363-391. North-Holland Publishing Company, Amsterdam and London.

Lockshin, R. A. (1969b). Programmed cell death: Activation of lysis by a mechanism involving the synthesis of protein. *J. Insect Physiol.* **15**, 1505-1516.

Lockshin, R. A. and Williams, C. M. (1964). Programmed cell death—II. Endocrine potentiation of the breakdown of the intersegmental muscles of silkmoths. *J. Insect Physiol.* **10**, 643-649.

Lockshin, R. A. and Williams, C. M. (1965). Programmed cell death—I. Cytology of degeneration in the intersegmental muscles of the pernyi silkmoth. *J. Insect Physiol.* **11**, 123-133.

Loughton, B. G. (1965). An investigation of haemolymph protein in Lepidoptera. *J. Insect Physiol.* **11**, 1651-1661.

Loughton, B. G. and West, A. S. (1965). The development and distribution of haemolymph proteins in Lepidoptera. *J. Insect Physiol.* **11**, 919-932.

Lunan, K. D. and Mitchell, H. K. (1969). The metabolism of tyrosine-O-phosphate in *Drosophila. Archs Biochem. Biophys.* **132**, 450-456.

Manwell, C. (1966). Starch gel electrophoresis of the multiple haemoglobins of small and large larval *Chironomus*—a developmental haemoglobin sequence in an invertebrate. *J. Embryol. exp. Morph.* **16**, 259-270.

Marek, M. (1969). Die Trennung der Hämolymphe-Proteine bei Vorpuppen und Puppen der *Galleria mellonella* L. mittels Gelfiltration auf Sephadex G-200, Bestimmung ihrer isoelektrischen Punkte und ihrer Molekulargewichte. *Z. Naturf.* **24b**, 732-740.

Martin, M.-D., Kinnear, J. F. and Thomson, J. A. (1969). Developmental changes in the late larva of *Calliphora stygia*. II. Protein synthesis. *Aust. J. biol. Sci.* **22**, 935-945.

Martin, M.-D., Kinnear, J. F. and Thomson, J. A. (1971). Developmental changes in the late larva of *Calliphora stygia*. IV. Uptake of plasma protein by the fat body. *Aust. J. biol. Sci.* **24**, 291-299.

Mechelke, F. (1953). Reversible Strukturmodifikationen der Speicheldrüsenchromosomen von *Acricotopus lucidus*. *Chromosoma (Berlin)*, **5**, 511-661.

Meyer, G. F. and Hess, O. (1965). Struktur-Differenzierung im Y-Chromosom von *Drosophila hydei* und ihre Bezeihungen zu Gen-Aktivitäten. II. Effekt der RNS-Synthese-Hemmung durch Actinomycin. *Chromosoma (Berlin)*, **16**, 249-270.

Miall, L. C. and Hammond, A. R. (1900). "The Structure and Life-History of the Harlequin Fly (*Chironomus*)". Clarendon Press, Oxford.

Mitchell, H. K. and Lunan, K. D. (1964). Tyrosine-O-phosphate in *Drosophila*. *Archs Biochem. Biophys.* **106**, 219-222.

Mulder, M. P., van Duijn, P. and Gloor, H. J. (1968). The replicative organization of DNA in polytene chromosomes of *D. hydei*. *Genetica*, **39**, 385-428.

Munn, E. A. and Greville, G. D. (1969). The soluble proteins of developing *Calliphora erythrocephala*, particularly calliphorin, and similar proteins in other insects. *J. Insect Physiol.* **15**, 1935 1950.

Munn, E. A., Feinstein, A. and Greville, G. D. (1967). A major protein constituent of pupae of the blowfly *Calliphora erythrocephala* (Diptera). *Biochem. J.* **102**, 5-6 P.

Munn, E. A., Feinstein, A. and Greville, G. D. (1971). The isolation and properties of the protein calliphorin. *Biochem. J.* **124**, 367-374.

Murray, F. V. and Tiegs, O. W. (1935). The metamorphosis of *Calandra oryzae*. *Q. Jl micros. Sci.* **77**, 405-495.

Nair, K. S. S. and George, J. C. (1964). A histological and histochemical study of the larval fat body of *Anthrenus vorax* Waterhouse (Dermestidae, Coleoptera). *J. Insect Physiol.* **10**, 509-517.

Nair, K. S. S. and Karnavar, G. K. (1968). A cytological study of changes in the fat body of *Trogoderma granarium* during metamorphosis, with special reference to the proteinaceous globules. *J. Insect Physiol.* **14**, 1651-1659.

Nair, K. S. S., Karnavar, G. K. and George, J. C. (1967). Studies on the larval fat body of *Oryctes* with particular reference to the cytochemistry of the proteinaceous globules. *J. Anim. Morph. Physiol.* **14**, 28-54.

Nash, D. and Plaut, W. (1965). On the presence of DNA in larval salivary gland nucleoli in *Drosophila melanogaster*. *J. Cell Biol.* **27**, 682-686.

Neufeld, G. J., Thomson, J. A. and Horn, D. H. S. (1968). Short-term effects of crustecdysone (20-hydroxyecdysone) on protein and RNA synthesis in third instar larvae of *Calliphora*. *J. Insect Physiol.* **14**, 789-804.

Oberlander, H. (1972). The hormonal control of development of imaginal discs. *In* "The Biology of Imaginal Discs" (Eds H. Ursprung and R. Nöthiger), pp. 155-172. Springer-Verlag, Berlin, Heidelberg and New York.

O'Brien, S. J. (1973). On estimating functional gene number in eukaryotes. *Nature New Biol.* **242**, 52-54.

Oda, J. (1956). Studies on the protein of insect haemolymph. II. *Nippon Nogei-Kagaku Kaishi,* **30**, 345-347.

Olvera, R. O. (1969). The nucleolar DNA of three species of *Drosophila* in the *hydei* complex. *Genetics Suppl.* **61**, 245-249.

Osanai, M. and Rembold, H. (1970). Biosynthese von Cytochrom *c* I. In-vivo-Einbau von radioaktivem Eisen ($^{59}Fe^{2+}$) und (^{14}C) Lysin in das Cytochrom *c* der Honigbiene. *Hoppe-Seyler's Z. physiol. Chem.* **351**, 643-648.

Painter, T. S. and Biesele, J. J. (1966). Endomitosis and polyribosome formation. *Proc. natn. Acad. Sci. U.S.A.* **56**, 1920-1925.

Pan, M. L. (1971). The synthesis of vitellogenin in the Cecropia silkworm. *J. Insect Physiol.* **17**, 677-689.

Pan, M. L. and Wyatt, G. R. (1971). Juvenile hormone induces vitellogenin synthesis in the Monarch butterfly. *Science, N.Y.* **174**, 503-505.

Pan, M. L., Bell, W. J. and Telfer, W. H. (1969). Vitellogenic blood protein synthesis by insect fat body. *Science, N.Y.* **165**, 393-394.

Panitz, R. (1972). Balbiani ring activities in *Acricotopus lucidus. In* "Developmental Studies on Giant Chromosomes" (Ed. W. Beermann), pp. 209-227. Springer-Verlag, Berlin, Heidelberg and New York.

Pantelouris, E. M. and Downer, R. G. H. (1969). Phenotypic changes of the esterase pattern in insect metamorphosis. *J. Insect Physiol.* **15**, 2357-2362.

Pardue, M. L., Gerbi, S. A., Eckhardt, R. A. and Gall, J. G. (1970). Cytological localization of DNA complementary to ribosomal RNA in polytene chromosomes of Diptera. *Chromosoma (Berlin)*, **29**, 268-290.

Pasteur, N. and Kastritsis, C. D. (1971). Developmental studies in *Drosophila.* I. Acid phosphatases, esterases, and other proteins in organs and whole fly homogenates during development of *D. pseudoobscura. Devl. Biol.* **26**, 525-536.

Pasteur, N. and Kastritsis, C. D. (1972). Developmental studies in *Drosophila.* IV. Quantitative protein changes in organs and whole fly homogenates during development of *D. pseudoobscura. Experientia,* **28**, 215-216.

Patel, N. G. (1971). Protein synthesis during insect development. *Insect Biochem.* **1**, 391-427.

Patel, N. G. (1972). Hormonal regulation of protein synthesis in insects. *In* "Molecular Genetic Mechanisms in Development and Aging" (Eds M. Rockstein and G. T. Baker III), pp. 145-198. Academic Press, New York and London.

Paul, M., Kafatos, F. C. and Regier, J. C. (1972). A comparative study of eggshell proteins in Lepidoptera. *J. supramol. Struct.* **1**, 60-65.

Paul, M., Goldsmith, M. R., Hunsley, J. R. and Kafatos, F. C. (1972). Specific protein synthesis in cellular differentiation. Production of eggshell proteins by silkmoth follicular cells. *J. Cell Biol.* **55**, 653-680.

Pavan, C. and da Cunha, A. B. (1969). Chromosomal activities in *Rhynchosciara* and other Sciaridae. *A. Rev. Genetics,* **3**, 425-450.

Peacock, W. J., Brutlag, D., Goldring, E., Appels, R., Hinton, C. W. and Lindsley, D. L. (1973). The organisation of highly repeated DNA sequences in *Drosophila melanogaster* chromosomes. *Cold Spring Harb. Symp. quant. Biol.* **38**, 405-416.

Pemrick, S. M. and Butz, A. (1970a). Protein synthesis of the fat body of adult *Tenebrio molitor*. *J. Insect Physiol.* 16, 643-651.

Pemrick, S. M. and Butz, A. (1970b). Common electrophoretic properties of the fat body, haemolymph, and oocytes of adult *Tenebrio molitor*. *J. Insect Physiol.* 16, 1443-1453.

Pentz, S. and Kling, H. (1972). Protein pattern in the haemolymph of the pharate pupa of *Plodia interpunctella* (Lepidoptera) in different age groups. *J. Insect Physiol.* 18, 2277-2283.

Pérez, C. H. (1910). Recherches histologiques sur la métamorphose des muscides *Calliphora erythrocephala* Mg. *Archs Zool. exp. gén.* (5) 4, 1-274.

Plagens, U., Fittkau, E. J., Jonasson, P. M. and Braunitzer, G. (1972). Vergleichende Untersuchungen der Hämoglobine verschiedenen Chironomiden. *In* "VI Internationales Symposium über Struktur und Funktion der Erythrocyten", pp. 183-190. Deutschen Akademie der Wissenschaften zu Berlin, Akademie-Verlag, Berlin.

Postlethwait, J. H. and Weiser, K. (1973). Vitellogenesis induced by juvenile hormone in the female sterile mutant *apterous-four* in *Drosophila melanogaster*. *Nature New Biol.* 244, 284-285.

Price, G. M. (1969). Protein synthesis and nucleic acid metabolism in the fat body of the larva of the blowfly, *Calliphora erythrocephala*. *J. Insect Physiol.* 15, 931-944.

Price, G. M. (1973). Protein and nucleic acid metabolism in insect fat body. *Biol. Rev.* 48, 333-375.

Price, G. M. and Bosman, T. (1966). The electrophoretic separation of proteins isolated from the larva of the blowfly, *Calliphora erythrocephala*. *J. Insect Physiol.* 12, 741-745.

Redman, C. M. (1969). Biosynthesis of serum proteins and ferritin by free and attached ribosomes of rat liver. *J. biol. Chem.* 244, 4308-4315.

Ribbert, D. (1967). Die Polytänchromosomen der Borstenbildungszellen von *Calliphora erythrocephala* unter besonderer Berücksichtigung der geschlechtsgebunden Heterozygotie und des Puffmusters während der Metamorphose. *Chromosoma (Berlin)*, 21, 296-344.

Ribbert, D. (1972). Relation of puffing to bristle and footpad differentiation in *Calliphora* and *Sarcophaga*. *In* "Developmental Studies on Giant Chromosomes" (Ed. W. Beermann), pp. 153-179. Springer-Verlag, Berlin, Heidelberg and New York.

Ribbert, D. and Bier, K. (1969). Multiple nucleoli and enhanced nucleolar activity in the nurse cells of the insect ovary. *Chromosoma (Berlin)*, 21, 296-344.

Riechers, L. A., Meyers, F. W. and Berry, S. J. (1969). DNase as a component of "multi-enzyme complexes" in the blood of saturniid moths. *J. Insect Physiol.* 15, 743-753.

Ringborg, U. and Rydlander, L. (1971). Nucleolar-derived ribonucleic acid in chromosomes, nuclear sap and cytoplasm of *Chironomus tentans* salivary gland cells. *J. Cell Biol.* 51, 355-368.

Ringborg, U., Daneholt, B., Edström, J.-E., Egyházi, E. and Rydlander, L. (1970). Evidence for transport of preribosomal RNA from the nucleolus to the chromosomes in *Chironomus tentans* salivary gland cells. *J. molec. Biol.* 51, 679-686.

Risler, H. and Romer, F. (1968). Chromosomenzahlen und DNS-Gehalt der Epidermiszellen während der Entwicklung der Honigbiene. *Verh. dt. zool. Ges.* 153-160.

Ritossa, F. M. and Spiegelman, S. (1965). Localization of DNA complementary to ribosomal RNA in the nucleolus organizer region of *Drosophila melanogaster*. *Proc. natn. Acad. Sci. U.S.A.* 53, 737-745.

Ritossa, F. M., Atwood, K. C. and Spiegelman, S. (1966). A molecular explanation of the bobbed mutants of *Drosophila* as partial deficiencies of "ribosomal" DNA. *Genetics, Princeton,* 54, 819-834.

Rodems, A. E., Hendrikson, P. A. and Clever, U. (1969). Proteolytic enzymes in the salivary gland of *Chironomus tentans*. *Experientia,* 25, 686-687.

Rodman, T. C. (1967). DNA replication in salivary gland nuclei of *Drosophila melanogaster* at successive larval and prepupal stages. *Genetics, Princeton,* 55, 375-386.

Rodman, T. C. (1969). Morphology and replication of intranucleolar DNA in polytene nuclei. *J. Cell Biol.* 42, 575-582.

Romer, F. (1966). Zytophotometrische Untersuchungen des DNS-Gerhalts in verschiedenen Geweben der Larve und Imago von *Oryzaephilus surinamensis* L. (Cucujidae, Coleoptera). *Biol. Zbl.* 85, 409-438.

Rudkin, G. T. (1972). Replication in polytene chromosomes. *In* "Developmental Studies on Giant Chromosomes" (Ed. W. Beermann), pp. 59-85. Springer-Verlag, Berlin, Heidelberg and New York.

Rüegg, M. K. (1968). Untersuchungen zum Proteinstoffwechsel des Wildtyps und der Letalmutante (*ltr*) von *Drosophila melanogaster*. *Z. vergl. Physiol.* 60, 275-307.

Ruh, M. F., Willis, J. H. and Hollowell, M. P. (1972). Blood protein synthesis in pupae of the silkmoth, *Hyalophora cecropia*. *J. Insect Physiol.* 18, 151-160.

Russo-Caia, S. (1960a). Aspetti biochimici della metamorfosi degli Insetti. *Ric. Sci.* 30, 1861-1907.

Russo-Caia, S. (1960b). Aspetti biochemici della metamorfosi degli Insetti. Il metabolismo delle sostanze azotate durante lo sviluppo larvale e la metamorfosi di *Musca domestica* L. *Riv. Biol.* (*Perugia*), 53, 409-426.

Sachs, R. I. and Clever, U. (1972). Unique and repetitive DNA sequences in the genome of *Chironomus tentans*. *Expl Cell Res.* 74, 587-591.

Saunders, J. W. (1966). Death in embryonic systems. *Science, N.Y.* 154, 604-612.

Sayles, C. D., Browder, L. W. and Williamson, J. H. (1973). Expression of xanthine dehydrogenase activity during embryonic development of *Drosophila melanogaster*. *Devl Biol.* 33, 213-217.

Schmieder, R. G. (1928). Observations on the fat body in Hymenoptera. *J. Morph.* 45, 121-185.

Sekeri, K. E., Sekeris, C. E. and Karlson, P. (1968). Protein synthesis in subcellular fractions of the blowfly during different developmental stages. *J. Insect Physiol.* 14, 425-431.

Shaaya, E. and Karlson, P. (1965). Der Ecdysontiter während der Insektentwicklung. II. Die postembryonale Entwicklung der Schmeissfliege, *Calliphora erythrocephala* Meig. *J. Insect Physiol.* 11, 65-69.

Shaaya, E. and Sekeris, C. E. (1965). Ecdysone during insect development. III. Activities of some enzymes of tyrosine metabolism in comparison with ecdysone titer during the development of the blowfly, *Calliphora erythrocephala* Meig. *Gen. comp. Endocr.* 5, 35-39.

Shigematsu, H. (1960). Protein metabolism in the fat body of the silkworm, *Bombyx mori*, L. *Bull. Sericult. Exp. Stn, (Tokyo)*, **16**, 165-170.

Shigematsu, H. and Moriyama, H. (1970). The effect of ecdysterone on fibroin synthesis in the posterior division of the silk gland of the silkworm, *Bombyx mori. J. Insect Physiol.* **16**, 2015-2022.

Shrivastava, H. N. and Loughton, B. G. (1970). The development of haemoglobins in *Chironomus riparius* (Meig.) (Diptera: Chironomidae). *Can. J. Zool.* **48**, 563-568.

Sibatani, A. (1971). Difference in the proportion of DNA specific to ribosomal RNA between adults and larvae of *Drosophila melanogaster. Mol. Gen. Genet.* **114**, 177-180.

Sirlin, J. L. (1962). The nucleolus. *Prog. Biophys.* **12**, 27-66, 319-326.

Smith, E. and Birt, L. M. (1972). Proteolytic activity during the metamorphosis of the blowfly *Lucilia. Insect Biochem.* **2**, 218-225.

Snodgrass, R. E. (1954). Insect metamorphosis. *Smithson. misc. Collns,* **135**, 1-124.

Sorsa, V., Green, M. M. and Beermann, W. (1973). Cytogenetic fine structure and chromosomal localization of the white gene in *Drosophila melanogaster. Nature New Biol.* **245**, 34-37.

Spear, B. B. and Gall, J. G. (1973). Independent control of ribosomal gene replication in polytene chromosomes of *Drosophila melanogaster. Proc. natn. Acad. Sci. U.S.A.* **70**, 1359-1363.

Spreij, T. E. (1971). Cell death during the development of the imaginal disks of *Calliphora erythrocephala. Neth. J. Zool.* **21**, 221-264.

Sridhara, S. and Levenbook, L. (1973). Extracellular ribosomes during metamorphosis in the blowfly *Calliphora erythrocephala. Biochem. biophys. Res. Commun.* **53**, 1253-1259.

Sridhara, S. and Levenbook, L. (1974). The contribution of the fat body to RNA and ribosomal changes during development of the blowfly *Calliphora erythrocephala* (Meig.). *Devl Biol.* **38**, 64-72.

Srivastava, R. P. (1970). Electrophoretic behaviour of cuticular proteins of different developmental stages of *Galleria mellonella. J. Insect Physiol.* **16**, 2345-2351.

Stevenson, E. and Wyatt, G. R. (1962). The metabolism of silkmoth tissues. I. Incorporation of leucine into protein. *Archs Biochem. Biophys.* **99**, 65-71.

Suzuki, Y., Gage, L. P. and Brown, D. D. (1972). The genes for silk fibroin in *Bombyx mori. J. molec. Biol.* **70**, 637-649.

Svedberg, T. and Eriksson-Quensel, I.-B. (1934). The molecular weight of erythrocruorin. II. *J. Am. chem. Soc.* **56**, 1700-1706.

Tata, J. R. (1973). Ribosome-membrane interaction and protein synthesis. *In* "Karolinska Symposia on Research Methods in Reproductive Endocrinology. 6th Symposium. Protein Synthesis in Reproductive Tissue", pp. 192-224. Karolinska Institutet, Stockholm.

Tazima, Y. (1964). "The Genetics of the Silkworm". Logos-Academic Press, London.

Telfer, W. H. and Williams, C. M. (1960). The effects of diapause, development and injury on the incorporation of radioactive glycine into the blood proteins of the Cecropia silkworm. *J. Insect Physiol.* **5**, 61-72.

Terra, W. R., de Bianchi, A. G., Gambarini, A. G. and Lara, F. J. S. (1973). Haemolymph amino acids and related compounds during cocoon production by the larvae of the fly *Rhynchosciara americana*. *J. Insect Physiol.* 17, 2097-2106.

Teunissen, R. J. H. (1937). Strukturelle Veränderungen im Gewebe der Stoffwechselzellen des "Fettkörpers" von *Calliphora* während des Umbaues der Puppe. *Cytologia*, Tokyo, Fujii Jubilee Volume, 836-844.

Thomas, K. K. (1972). Studies on the synthesis of lipoproteins during larval-pupal development of *Hyalophora cecropia*. *Insect Biochem.* 2, 107-118.

Thomas, K. K. and Gilbert, L. I. (1968). Isolation and characterization of the hemolymph lipoproteins of the American silkmoth, *Hyalophora cecropia*. *Archs Biochem. Biophys.* 127, 512-521.

Thomas, K. K. and Gilbert, L. I. (1969). The hemolymph lipoproteins of the silkmoth *Hyalophora gloveri*: studies on lipid composition, origin and function. *Physiol. Chem. Physics*, 1, 293-311.

Thomasson, W. A. and Mitchell, H. K. (1972). Hormonal control of protein granule accumulation in fat bodies of *Drosophila melanogaster* larvae. *J. Insect Physiol.* 18, 1885-1899.

Thompson, P. E. and English, D. S. (1966). Multiplicity of hemoglobins in the genus *Chironomus* (*Tendipes*), *Science, N.Y.* 152, 75-76.

Thompson, P. and Horning, M. J. (1973). Regulatory interactions involving two haemoglobin loci of *Chironomus*. *Biochem. Genetics*, 8, 309-319.

Thompson, P. E. and Patel, G. (1972). Compensatory regulation of two closely related hemoglobin loci in *Chironomus tentans*. *Genetics, Princeton*, 70, 275-290.

Thompson, P., Blecker, W. and English, D. S. (1968). Molecular size and subunit structure of the hemoglobins of *Chironomus tentans* *J. biol. Chem.* 243, 4463-4467.

Thomson, J. A. (1969). The interpretation of puff patterns in polytene chromosomes. *Curr. mod. Biol.* 2, 333-338.

Thomson, J. A. (1973a). Patterns of gene activity in larval tissues of the blowfly, *Calliphora*. *In* "The Biochemistry of Gene Expression in Higher Organisms" (Eds J. K. Pollak and J. W. Lee), pp. 320-332. Australia and New Zealand Book Company, Sydney.

Thomson, J. A. (1973b). Differential replication of ribosomal DNA during larval development in *Calliphora* (Diptera). *Devl Biol.* 35, 362-365.

Thomson, J. A. and Gunson, M. M. (1970). Developmental changes in the major inclusion bodies of polytene nuclei from larval tissues of the blowfly, *Calliphora stygia*. *Chromosoma* (*Berlin*), 30, 193-201.

Thomson, J. A., Kinnear, J. F., Martin, M.-D. and Horn, D. H. S. (1971). Effects of crustecdysone (20-hydroxyecdysone) on synthesis, release and uptake of proteins by the larval fat body of *Calliphora*. *Life Sci.* 10, 203-211.

Thomson, J. A., Radok, K. R., Shaw, D. C., Whitten, M. J., Foster, G. G. and Birt, L. M. (1975). Genetics of lucilin, a storage protein from the sheep blowfly, *Lucilia cuprina* (Calliphoridae). In preparation.

Tichy, H. (1970). Biochemische und cytogenetische Untersuchungen zur Natur des Hämoglobin-Polymorphismus bei *Chironomus tentans* und *Chironomus pallidivittatus*. *Chromosoma* (*Berlin*), 29, 131-188.

Tiegs, O. W. (1922). Researches on the insect metamorphosis. I. On the structure and post-embryonic development of a chalcid wasp, *Nasonia*. II. On the physiology and interpretation of the insect metamorphosis. *Trans. R. Soc. S. Aust.* **46**, 319-527.

Twohy, D. W. and Rozeboom, L. E. (1957). A comparison of food reserves in autogenous and anautogenous *Culex pipiens* populations. *Am. J. Hyg.* **65**, 316-324.

Van der Geest, L. P. S. and Borgsteede, F. H. M. (1969). Protein changes in the haemolymph of *Pieris brassicae* during the last larval instars and the beginning of the pupal stage. *J. Insect Physiol.* **15**, 1687-1693.

Vinson, S. B. and Lewis, W. J. (1969). Electrophoretic study of protein changes between several developmental stages of three species of *Heliothis* (Lepidoptera: Noctuidae). *Comp. Biochem. Physiol.* **28**, 215-220.

Waddington, C. H. (1973). The morphogenesis of patterns in *Drosophila*. *In* "Developmental Systems: Insects" (Eds S. J. Counce and C. H. Waddington), Vol. 2, pp. 499-535. Academic Press, London and New York.

Walker, P. R. (1966). An electron microscope study of the fat body of the moth *Philosamia*, during growth and metamorphosis. *J. Insect Physiol.* **12**, 1009-1018.

White, M. J. D. (1973). "Animal Cytology and Evolution", 3rd ed. Cambridge University Press, London.

Whitten, J. M. (1962). Breakdown and formation of connective tissue in the pupal stage of an insect. *Q. Jl microsc. Sci.* **103**, 359-367.

Whitten, J. M. (1964). Haemocytes and the metamorphosing tissues in *Sarcophaga bullata, Drosophila melanogaster,* and other cyclorrhaphous Diptera. *J. Insect Physiol.* **10**, 447-469.

Whitten, J. M. (1965). Differential deoxyribonucleic acid replication in the giant foot-pad cells of *Sarcophaga bullata*. *Nature, Lond.* **209**, 1019-1021.

Whitten, J. M. (1969a). Coordinated development in the fly foot: sequential cuticle secretion. *J. Morph.* **127**, 73-104.

Whitten, J. M. (1969b). Coordinated development in the foot pad of the fly *Sarcophaga bullata* during metamorphosis: changing puffing patterns of the giant cell chromosomes. *Chromosoma (Berlin)*, **26**, 215-244.

Whitten, J. M. (1969c). Cell death during early morphogenesis: Parallels between insect limb and vertebrate limb development. *Science, N.Y.* **163**, 1456-1457.

Whitten, J. M. (1969d). Haemocyte activity in relation to epidermal cell growth, cuticle secretion, and cell death in a metamorphosing cyclorrhaphan pupa. *J. Insect Physiol.* **15**, 763-778.

Whitten, M. J., Foster, G. G., Arnold, J. T. and Konowalow, C. (1975). The Australian sheep blow fly *Lucilia cuprina*. *In* "Survey of Genetics" (Ed. R. C. King), Vol. 3. In Press. Plenum, New York.

Wiesmann, R. (1962). Untersuchungen über den larvalen und imaginalen Fettkörper den Imago von *Musca domestica* L. *Mitt. Schweiz. ent. Ges.* **35**, 185-210.

Wigglesworth, V. B. (1961). Insect polymorphism—a tentative synthesis. *In* "Insect Polymorphism" (Ed. J. S. Kennedy), Symposium No. 1, pp. 103-113. Royal Entomological Society, London.

Wigglesworth, V. B. (1965). "Principles of Insect Physiology", 6th ed. Methuen London.

Wigglesworth, V. B. (1970). "Insect Hormones". Oliver and Boyd, Edinburgh.

Williams, C. M. and Kafatos, F. C. (1971). Theoretical aspects of the action of juvenile hormone. *Mitt. Schweiz. ent. Ges.* **44**, 151-162.

Williams, K. L. (1972). Protein synthesis during the metamorphosis of *Lucilia cuprina.* Ph.D. Thesis, Australian National University.

Williams, K. L. and Birt, L. M. (1972). A study of the quantitative significance of protein synthesis during the metamorphosis of the sheep blowfly *Lucilia. Insect Biochem.* 2, 305-320.

Williams, K. L., Smith, E., Shaw, D. C. and Birt, L. M. (1972). Studies of the levels and synthesis of Cytochrome *c* during adult development of the blowfly *Lucilia cuprina. J. biol. Chem.* **247**, 6024-6030.

Willis, J. H. (1970). Juvenile hormone and the cuticular proteins of the cecropia silkworm. *Am. Zool.* **10**, 320.

Wobus, U., Popp, S., Serfling, E. and Panitz, R. (1972). Protein synthesis in the *Chironomus thummi* salivary gland. *Molec. gen. Genet.* **116**, 309-321.

Wren, J. J. and Mitchell, H. K. (1959). Extraction methods and an investigation of *Drosophila* lipids. *J. biol. Chem.* **234**, 2823-2828.

Wright, T. R. F. (1970). The genetics of embryogenesis in *Drosophila. Adv. Genet.* **15**, 261-395.

Wu, J.-R., Hurn, J. and Bonner, J. (1972). Size and distribution of the repetitive segments of the *Drosophila* genome. *J. molec. Biol.* **64**, 211-219.

Wyatt, G. R. (1968). Biochemistry of insect metamorphosis. *In* "Metamorphosis: A Problem in Developmental Biology" (Eds. W. Etkin and L. I. Gilbert), pp. 143-184. North-Holland Publishing Company, Amsterdam.

Wyatt, G. R. (1972). Insect hormones. *In* "Biochemical Actions of Hormones" (Ed. G. Litwack), Vol. 2, pp. 385-490. Academic Press, New York and London.

Wyatt, G. R., Lougheed, T. C. and Wyatt, S. S. (1956). The chemistry of insect haemolymph. Organic components of the haemolymph of the silkworm, *Bombyx mori*, and two other species. *J. gen. Physiol.* **39**, 853-868.

Subject Index

G

P

T

U

V

W

Cumulative List of Authors

Numbers in bold face indicate the volume number of the series

Cumulative List of Chapter Titles

Numbers in bold face indicate the volume number of the series